Rhizosphere Biology

Series Editor

Anil Kumar Sharma, Biological Sciences, CBSH, G.B. Pant University of Agriculture & Technology, Pantnagar, Uttarakhand, India

The Series **Rhizosphere Biology**, emphasizes on the different aspects of Rhizosphere. Major increase in agricultural productivity, to meet growing food demands of human population is imperative, to survive in the future. Along with methods of crop improvement, an understanding of the rhizosphere biology, and the ways to manipulate it, could be an innovative strategy to deal with this demand of increasing productivity. This Series would provide comprehensive information for researchers, and encompass all aspects in field of rhizosphere biology. It would comprise of topics ranging from the classical studies to the most advanced application being done in the field. Rhizoshpere is a dynamic environment, and a series of processes take place to create a congenial environment for plant to grow and survive. There are factors which might hamper the growth of plants, resulting in productivity loss, but, the mechanisms are not very clear. Understanding the rhizosphere is needed, in order to create opportunities for researchers to come up with robust strategies to exploit the rhizosphere for sustainable agriculture.

There are titles already available in the market in the broad area of rhizosphere biology, but there is a major lack of information as to the functions and future applications of this field. These titles have not given all the up-to-date information required by the today's researchers and therefore, this Series aims to fill out those gaps.

More information about this series at http://www.springer.com/series/15861

Suresh Kumar Dubey • Satish Kumar Verma
Editors

Plant, Soil and Microbes in Tropical Ecosystems

Springer

Editors
Suresh Kumar Dubey
Department of Botany
Banaras Hindu University
Varanasi, Uttar Pradesh, India

Satish Kumar Verma
Department of Botany
Banaras Hindu University
Varanasi, Uttar Pradesh, India

ISSN 2523-8442 ISSN 2523-8450 (electronic)
Rhizosphere Biology
ISBN 978-981-16-3363-8 ISBN 978-981-16-3364-5 (eBook)
https://doi.org/10.1007/978-981-16-3364-5

This Springer imprint is published by the registered company Springer Nature Singapore Pte Ltd.
The registered company address is: 152 Beach Road, #21-01/04 Gateway East, Singapore 189721, Singapore

Foreword

The concept of rhizosphere was enunciated by Hiltner in 1904. Rhizosphere biology per se takes into its ambit organisms that belong to large groups of bacteria, fungi, Protozoa, nematodes, and insects. Thus it is interactions in this complex milieu that are to be considered critically in order to arrive at any meaningful conclusions. This inter alia requires that rhizosphere is analysed not only at a single plant level but also at field scale. Available information derived from laboratory and field investigations suggests complexity not only at species level but also at population and community scale. Until recently this information was derived for especially bacteria and fungi through phenotypic studies based on artificial media and the data so generated was therefore difficult to duplicate for a single soil site. This situation has now completely changed with the availability of tools of metagenomics and gene sequencing, which permit exploration of microbes that were prone to artificial cultivation. As a result it is possible now to investigate rhizosphere into its variable components not only into species spectrum but also at temporal and spatial scales. In addition, the availability of other advanced analytical tools including microscopes allows deeper investigations into the realm of in situ localization of microbial communities and diffusivity of nutrients from nutrients that play an important role in interactions that are pivotal to appreciate the close linkages between below ground and above ground components of the plant ecosystem. It is the music generated in the rhizosphere that culminates in improved shoot growth with the resultant better plant productivity. Notwithstanding these attributes, rhizospheric microbial populations are an admixture of neutral, antagonistic, mutualistic, and pathogenic components. Any disturbance by way of use of herbicides and pesticides can disturb the delicate equilibrium that may take long to come back to normalcy. Among these inhabitants, mutualises belonging to the group mycorrhizae extend their hyphae as additional root hairs and thus exert a “mycorrhizospheric” effect. On the other hand nodules in case of leguminous crops are a harbinger of non-nitrogen fixing yet growth promoting bacteria with further influence of the rhizosphere of plants. The inter-linkages of phosphorous and nitrogen economy come to play a decisive influence through these two groups of mutualists with resultant beneficial influence on crop productivity. In tropical ecosystems the role of nutrient turnover can make or mar the survival and fitness of a plant community under such circumstances. Considering limitations and desirability of sustained production systems to meet the food demands of tomorrow, bringing together published information on the subject is always a welcome step. In the present volume, the editors, Prof. S.K. Dubey and Dr. S. K. Verma, have done a useful exercise to fill this gap. I hope it will further the cause of rhizosphere biological assessment with tomorrow's needs and aspirations.

Formerly Head & Dean, Department of Microbiology, CBSH G.B. Pant University of Agriculture & Technology Pantnagar, Uttarakhand, India

B. N. Johri

Preface

Microbial communities have a profound effect on the performance of plants by influencing their growth and development. Plant-associated microbes including mainly archaea, bacteria, and fungi are beneficial as well as harmful to them. Large proportions of these microbes live in the close vicinity of the roots (rhizosphere) and are called rhizospheric microorganism. Actually, plants selectively recruit soil microbes towards the rhizospheric region to get benefits in terms of nutrient acquisition, growth stimulation, and protection from biotic and abiotic stresses. The composition of these rhizospheric microbial communities and the level of interactions with plants largely depend on the type of soils, plant genotype, and abiotic factors. Some of the microbes including endophytes, rhizobia, and mycorhizae enter into the root tissues and develop permanent internal mutualistic association. Rhizospheric microorganisms and mycorhizae have been well studied for their beneficial role in plant development and protection. They improve host plant fitness by mobilising inorganic nutrients such as nitrogen, phosphorus, potassium, and iron, fixing atmospheric nitrogen, producing growth hormones, and suppressing plant pathogens. Also they induce host plant gene expression for better adaptation. Excessive and injudicious use of chemical fertilisers and pesticides already has impacted our ecosystems, posing a great threat to human and animal health. Also these chemicals disturb the composition and interactions of natural microbiota. Research and knowledge of rhizospheric biology would be very critical in manipulating rhizospheric microbes for making strategy for sustainable cultivation of crops and medicinal plants. Nevertheless, a majority of the microbes is non-cultivable and hence new technological approaches such as functional metagenomics and high-throughput sequencing need to be designed to understand better the functional role of non-cultivable microbes. Most of the studies related to plant–microbe interactions pertain to temperate regions despite the fact that tropical ecosystems are richer in diversity and are complex in terms of plant–microbe interactions. It is very crucial to develop a better understanding of how the soil types and abiotic factors influence the plant–soil–microbe interactions in tropics. In view of the aforementioned issues, this volume in the "Rhizosphere Biology" series is titled *Plant, Soil and Microbes in Tropical Ecosystems* to cover the research topics related to tropical ecosystem rhizosphere–microbial interactions.

This volume consists of 17 chapters in which we covered areas ranging from application and development of modern techniques and tools to study rhizospheric biology to basic science and research development, understanding diversity and functional roles, and application of rhizospheric microbiota in developing new alternative technology, i.e., biofertilisers and biocontrol agents for sustainable agriculture.

In this book, attempts have been made to highlight multi-dimensional plant–microbe interactions in tropical agroecosystems. This book provides a glimpse of the basic and advanced perspective towards rhizospheric biology designed for a wide group of readers.

We have great pleasure in bringing this book to a global audience. We express our sincere gratitude to all the contributors for significant contributions towards the completion of this volume. We are also grateful to reviewers for improving the quality of the chapters included in this book. We also thank Ms. Madurima Kahali, Ms. Vaishnavi Venkatesh, and the entire production team at Springer for their cooperation and support in bringing out this book.

Varanasi, Uttar Pradesh, India

Suresh Kumar Dubey
Satish Kumar Verma

Contents

About the Editors

Suresh Kumar Dubey is a professor in the Molecular Ecology Laboratory, Department of Botany, Banaras Hindu University, India. His interests include structure and function of the microbial community involved in controlling methane emission, screening, utilization of microbes for their potential role in bioremediation, and studies on the prevalence of some pathogenic microbes in clinical and food samples and their response to drugs. He has won multiple awards for his work, including Biotech Research Society of India Young Scientist Medal 2004, DST-BOYSCAST Govt. of India fellowship 2006, INSA visiting fellowship 2012 and 2017, and JSPS Invitation fellowship 2012, among others.

Satish Kumar Verma is an Assistant Professor in the Department of Botany, Banaras Hindu University, Varanasi, India. He received his Ph.D. in Botany (Plant Microbes) from Banaras Hindu University, Varanasi, India. Dr. Verma also served as Assistant Professor in the Department of Botany, Visva-Bharati, Santiniketan, India, for about three years. Dr. Verma has also visited the Department of Plant Biology, Rutgers University, New Jersey, USA, for a year. Dr. Verma has published 22 research papers in scientific reputed journals. He has also published 6 book chapters and edited a book on *Seed Endophytes*. His areas of research interest include the functions of plant microbiomes, diversity and ecology, and the roles of microbes in the modulation of plant development and protection of hosts from biotic and abiotic stresses.

Chapter 1
Plant–Rhizobacteria Interactions to Induce Biotic and Abiotic Stress Tolerance in Plants

Raghvendra Saxena, Manish Kumar, and Rajesh Singh Tomar

Abstract Climate change and extreme environmental conditions are recognized as the most challenging threats to agricultural systems, leading to significant limitations in crop production and yield worldwide. It is a big concern to increase or maintain crop productivity under changing climate conditions to cater for increasing food demand. Among abiotic stresses, salinity, drought and extreme heat are the most common stresses. Abiotic stresses contribute to reducing crop plant production by 50% or more. Like the effects of abiotic stress, constant exposure to biotic stresses—which include pathogen infections and pest and insect attacks—contribute to a major drop in crop productivity and wastage of crops. There is also constant pressure from extreme weather conditions due to climate change and the incidence of biotic stresses. There is a great need to develop biotic and abiotic stress resilience in crops to mitigate the adverse effects of stresses. Such resilience can be achieved through development and adoption of eco-friendly approaches in agricultural systems for crop sustainability and food security. The focus on plant–microbe interactions has attracted more attention in recent years for inducing plant resistance and defence against abiotic and biotic stresses. Plant growth promoting rhizobacteria facilitate abiotic stress resilience in plants by several strategies through activation of plant growth regulators (which include ethylene, auxin (indole-3-acetic acid)), activity of enzymes such as 1-aminocyclopropane-1-carboxylate (ACC)–deaminase and production of bacterial products such as exopolysaccharide. Diverse plant–microbe interactions in the rhizosphere also help to regulate plant defence pathways under adverse conditions through induction of systematic resistance or systemic acquired resistance. Moreover, other strategies such as microbial antagonism through production of several compounds such as antibiotics, siderophores, bacteriocins and secondary metabolites further boost disease resistance in plants.

Understanding of the great importance of plant growth–promoting rhizobacteria in agricultural systems and their involvement in induction of plant defence

---

R. Saxena (✉) · M. Kumar · R. S. Tomar
Amity Institute of Biotechnology, Amity University Madhya Pradesh, Maharajpura, Dang, Gwalior, Madhya Pradesh, India
e-mail: rsaxena@gwa.amity.edu

S. K. Dubey, S. K. Verma (eds.), *Plant, Soil and Microbes in Tropical Ecosystems*, Rhizosphere Biology, https://doi.org/10.1007/978-981-16-3364-5_1

mechanisms through various strategies to increase crop resilience against adverse conditions offers a potential tool to maintain sustainability in agricultural systems.

This chapter focuses on the role of plant–microbe interactions and application of plant growth–promoting rhizobacteria to attain comprehensive protection of crops in adverse conditions to address crop sustainability and food security.

Keywords Biotic stress · Abiotic stress · PGPR · Induced systematic resistance (ISR) · Systemic acquired resistance (SAR)

1.1 Introduction

Abiotic and biotic stresses affect plants and animals. Being sessile organisms, plants are greatly affected by these stresses and environmental changes because they cannot escape from these adverse situations and must instead tolerate them. Therefore, perturbations of external environmental conditions that negatively affect plants' physiological and metabolic activities lead to limitations in growth and development. Moreover, such stresses induce several adaptive responses in plants at the cellular and molecular levels to mitigate adverse effects of plant pathogens and environmental stresses (Verma et al. 2013). Extreme environmental conditions and pathogen attacks are important causes of negative effects on crop productivity worldwide (Grover et al. 2011).

With growth in the human population and inflating food demand, food security and production have become major challenges in the current agricultural scenario worldwide. It is estimated that 70% more food crop production will be required to fulfil the food demands of 2.3 billion additional people by 2050 globally (FAO 2009).

Plants are frequently exposed to adverse environmental conditions and consequently experience poor growth and productivity. These environmental stresses are broadly categorized into two groups: biotic stresses and abiotic stresses. Abiotic stresses— salinity, desiccation, high temperatures, floods, cold, heavy metal contamination etc.—put major constraints on crop growth and productivity worldwide. Among abiotic stresses, drought, salinity and extreme temperatures are major stresses that cause huge losses of crop productivity globally because of their adverse impacts on growth, development, yield and seed quality of crop plants. In a wide variety of crops, abiotic stresses result in yield losses ranging from 10% to 50% or more, depending on the crop (Gull et al. 2019). Drought, salinity and extreme temperatures are among the most important abiotic stresses. It was previously estimated that approximately 1.8 billion people would face acute freshwater scarcity in the first quarter of the twenty-first century, while the rest of the population would face water crises to a considerable extent (Nezhadahmadi et al. 2013). Abiotic stresses, especially drought and salinity, are known to cause major reductions in crop yields and economic losses to farmers. Increasing climate change and recurrence of abiotic stresses are major threats to food security and sustainability of crop production systems.

Plant responses to abiotic and biotic stresses are intricate phenomena, governed by multiple complex traits. Therefore, it is important to understand plants' responses and their underlying mechanisms under adverse conditions in order to enhance plant resistance, which is the major concern in the current agricultural scenario (Saxena et al. 2019a, b; Raza et al. 2019).

Abiotic stresses are mainly governed by perturbations in nonliving components of the environment, whereas biotic stresses are those imposed on plants by a wide variety of other organisms, including viruses, fungi, insects, pests, nematodes, arachnids, weeds etc. These organisms' attacks on crop plants cause adverse impacts on the plants by depriving them of nutrients or by changing their physiological and metabolic activities, resulting in poor growth and less development. Moreover, under extreme and severe conditions, they may kill the plants. Biotic and abiotic stresses also severely affect crop productivity and cause major crop losses. Plants do not possess an immune system; therefore, they have evolved various defence strategies governed by their genetic composition to prevent deleterious effects of pathogen attacks (Gull et al. 2019; Verma et al. 2016). Plant–microbe interactions play important roles in strengthening plant defences against abiotic and biotic stresses. Interactions with nonpathogenic bacteria are important in providing effective tolerance or bioprotection against biotic stresses in plants when they are inoculated; similarly, interactions with root-colonizing bacteria enhance abiotic stress tolerance in plants. There is a need to address the issues of abiotic and biotic stresses associated with crop loss by identifying strategies and technological approaches that can promote crop resilience under adverse conditions and help mitigate the adverse effects of those stresses. Further, such approaches should be environmentally friendly and should not require large expenditure. They should be based on promoting adaptations in plant capacity under stressful conditions (Kang et al. 2009).

Microorganisms constitute the most vital component of the earth's living system, since microorganisms are the natural inhabitants of the soil and thus a vital living component of the rhizosphere. Plant–microbe interactions constitute the most delicate system in the agricultural system that contributes directly or indirectly to agricultural crop production. Moreover, microbes contribute to seed germination and growth as natural inhabitants in various symbiotic associations (Chakraborty et al. 2015). Different types of plant–microbe interactions constitute an important component of the ecosystem, and such plant–microbe interactions regulate plant defence mechanisms for better survival under extreme conditions (Kumar et al. 2019; Meena et al. 2017).

Soil microorganisms surviving in different environmental niches exhibit diverse adaptive metabolic attributes that can help to mitigate the adverse impacts of the extreme environments in which they live. Microbes living in extreme conditions show immense potential to adapt under stressful conditions; therefore, exploitation of plant–microbe interactions should be the most promising approach in the agricultural sector to increase and maintain food productivity in order to sustain food security (Kumar et al. 2018). Moreover, utilization of beneficial plant–microbe interactions is the most eco-friendly approach to achieve these goals. Application of plant growth–promoting rhizobacteria (PGPRs) as bioinoculants could offer a great potential strategy to minimize deleterious effects of abiotic threats on crops, which cause significant declines in plant growth and yields (Enebe and Babalola 2018). PGPRs could play an important role in management of salinity and drought stresses in plants, as reports have indicated that such beneficial soil microorganisms have a propensity to colonize the root–soil area (rhizosphere) and the endo-rhizosphere of plants to enhance abiotic stress resistance in plants.

There are several strategies through which microbes promote plant growth, such as increases in 1-aminocyclopropane-1-carboxylate (ACC) deaminase; regulation of ethylene levels; and production of the auxin indole-3-acetic acid (IAA), cytokinin, exopolysaccharide (EPS), volatile compounds etc. Further, there are significant increases in osmolyte accumulation and antioxidant enzyme activity, modulation of stress response gene expression levels and changes in root morphology to improve drought tolerance in plants (Khan et al. 2019). Reports have indicated that ACC deaminase–producing PGPRs not only are involved in improving plant growth but also can induce sufficient protection against abiotic stresses (such as drought, salinity, flooding and inorganic and organic contaminants) and biotic stresses (bacterial and fungal pathogens) in plants (Glick 2014). Moreover, it has been reported that production of IAA by a wide variety of soil microorganisms contributes significantly to plant root system development, thereby helping to reduce drought stress (Sharma et al. 2015). Furthermore, to maintain osmotic balance and homeostasis, PGPRs secrete plant growth regulators and enzymes such as IAA and ACC deaminase, among others, which act as signalling molecules in stress conditions, leading to induction of stress response pathways in plants to improve their stress tolerance (Gayathri and Donald 2018).

Recently, Barra et al. (2016) pointed out the importance of rhizocompetent stress-tolerant bacterial strains with variable activity of ACC deaminase and production of IAA for reducing the effects of salinity stress in plants. This indicates that understanding of plant–microbe interactions and their roles in improving stress tolerance under adverse conditions can be a potential tool in agriculture for sustainable production in adverse conditions through optimization of plant–microbe interactions. PGPRs are economically and environmentally beneficial for plant growth promotion. PGPRs alter physico-biochemical and molecular mechanisms in plants, helping them to withstand adverse environmental conditions. Plant–microbe interactions perform a wide range of functions and confer mutual benefits on the plants and microbes. The plants provide the microbes with reduced carbon and other metabolites for growth; in return, the microbes offer certain advantages to the plants. PGPRs have great importance in agricultural systems because they play important roles in enhancing plant growth and yield through effective nutrient acquisition and assimilation. Moreover, PGPRs improve soil texture and secrete important extracellular signalling compounds, hormones, secondary metabolites etc., which further boost plant growth and tolerance of stress. It has been reported that PGPRs are involved in positively modulating plant responses to both biotic and abiotic stresses.

Therefore, they act as biostimulants that can increase crop resilience against adverse conditions, hence offering a potential tool to be utilized to maintain agricultural sustainability by reducing dependency on agrochemicals.

This chapter discusses the effects of PGPRs in the resilience of plants against biotic and abiotic stresses. It also suggests development of suitable bioinoculants for application to different crops, along with other approaches to provide protection from abiotic stresses and tolerance of biotic stresses.

1.2 Rhizobacteria as Beneficial Agents

Microorganisms are an integral component of the biotic system on earth. As an integral part of the biotic component of the rhizosphere, they establish fine interactions with plants, which play vital roles in agricultural systems. As an important natural partner in the rhizosphere, microbes are capable of establishing diverse symbiotic associations with plants. The rhizosphere is the zone surrounding the root system of the plant, which is enriched with a wide variety of nutrients and exudates composed of amino acids, sugars, carbohydrates etc. These support the growth of microbes; therefore, the rhizosphere has a higher density of microorganisms than those of soils in other places. The diverse bacteria that occupy the natural rhizospheric habitat are referred to as rhizobacteria (Schroth and Hancock 1982).

Depending on their interactions with plants and their impacts on plant growth–promoting attributes, rhizobacteria can be categorized into harmful, beneficial and neutral groups (Dobbelaere et al. 2003). Among the diverse groups of free-living bacteria present in the rhizosphere, those groups of rhizobacteria that exhibit plant growth–promoting characteristics are known as plant growth–promoting rhizobacteria (Kloepper et al. 1989). Those that colonize the rhizosphere, live on root surfaces (also known as the rhizoplane) or live inside the roots exhibit growth-promoting potential. It is estimated that only 1–2% of bacteria exhibit plant growth–promoting features, have beneficial effects on plant growth and strengthen plant tolerance against environmental stresses and biotic threats (Antoun and Kloepper 2001).

Among the different genera of bacteria that have been studied, *Bacillus* and *Pseudomonas* spp. have been identified as the most predominant PGPR genera (Podile and Kishore 2007). PGPRs can help plants to resist stresses and maintain plant growth and normal physiological functions. Although there is an abundance of beneficial soil bacteria in the rhizosphere, they have still not been adequately studied and characterized, because there is a dearth of relevant information. To date, this has limited their application as bioinoculant tools in the agricultural sector to mitigate environmental and biotic stresses (Ojuederie et al. 2019). Rhizobacteria of the genera *Pseudomonas* and *Bacillus* are considered the most effective ones in terms of their ability to trigger plant resistance against stresses through induction of systemic resistance and antagonistic effects on pathogens (Table 1.1) (Kloepper et al. 2004; Van Wees et al. 2008; Beneduzi et al. 2012). Exploitation of the roles of

Table 1.1 Plant growth–promoting bacteria (PGPRs) associated with mediation of systemic resistance against pathogens in different crop plants

PGPR strains	Crops	Diseases	Pathogens	References
Pseudomonas fluorescens GRP3	Rice	Sheath blight	*Rhizoctonia solani*	Pathak et al. (2004)
Pseudomonas fluorescens	Pearl millet (*Pennisetum glaucum*)	Downy mildew	*Sclerospora graminicola*	Raj et al. (2003)
Bacillus spp.	Rice	Bacterial leaf blight	*Xanthomonas oryzae*	Udayashankar et al. (2011)
Pseudomonas sp.	Potato, lettuce	Rhizoctonia diseases	*Rhizoctonia solani*	Schreiter et al. (2018)
Bacillus pumilus, Paenibacillus costume, Mycobacterium immunogenum	Tomato	Root-knot disease	Nematode (*Meloidogyne incognita*)	Cetintas et al. (2018)
Pseudomonas putida strain NH-50	Sugar cane	Red rot	*Glomerella tucumanensis* (Speg.) Arx & E. Müll.	Hassan et al. (2011)

PGPRs as important components in plant–rhizobacteria systems, conferring beneficial effects on agricultural systems, has proved to be an effective strategy in agricultural sustainability and mitigation of biotic and abiotic stresses arising from climate change and other anthropogenic activities. Various types of microbes—*Bacillus* (Kasim et al. 2016), Micrococcaceae HW-2 (Hong et al. 2016), *Pseudomonas*, *Microbacterium*, *Curtobacterium* (Cardinale et al. 2015), *Bradyrhizobium* (Masciarelli et al. 2014), *Pantoea* (Damam et al. 2014), *Variovorax*, *Paenibacillus* (Yolcu et al. 2011) and many others—have shown plant growth–promoting attributes and potential for stress mitigation. Different studies have revealed that soil microorganisms possess the ability to mitigate adverse impacts of abiotic stresses (drought, salinity, extreme temperatures, heavy metal contamination etc.) on plants. Some of these confer tolerance of salinity and drought (*Azospirillum* sp., *Pseudomonas syringae*, *Pseudomonas fluorescens* and *Bacillus* spp.) and nutrient deficiency (*Bacillus polymyxa* and *Pseudomonas alcaligenes*) (Table 1.2) (Chakraborty et al. 2015).

1.3 Plant–Rhizobacteria Interactions and Abiotic Stress Tolerance

Studies have indicated that PGPRs are involved directly or indirectly in increasing crop resilience against various abiotic stresses. In one study, priming of chickpea genotypes with a PGPR consortium culture (*Bacillus subtilis*, *Bacillus thuringiensis* and *Bacillus megaterium*) revealed improved tolerance under drought stress. This

Table 1.2 Plant growth–promoting bacteria (PGPRs) associated with abiotic stress tolerance in different crop plants

Types of stress	PGPR strains	Mechanisms	Crops	References
Drought	*Achromobacter piechaudii* ARV8	ACC deaminase activity	Tomato	Mayak et al. (2004a)
Drought	*Pseudomonas* spp.	ACC deaminase activity	Pea (*Pisum sativum* L.)	Arshad et al. (2008)
Drought	*Bacillus* spp.	Siderophore production, IAA, phosphate solubilization	*Sorghum bicolor*	Grover et al. (2014)
Drought	*Ochrobactrum pseudogrignonense* RJ12, *Pseudomonas* sp. RJ15, *Bacillus subtilis* RJ46	ACC deaminase activity	*Vigna mungo* L., pea (*Pisum sativum* L.)	Saikia et al. (2018)
Drought, salinity	*Burkholderia cepacia*	ACC deaminase activity, exopolysaccharide	*Capsicum annuum*	Maxton et al. (2018)
Drought	*Variovorax paradoxus*, *Pseudomonas* spp., *Achromobacter* spp., *Ochrobactrum anthropi*	ACC deaminase activity	Wheat (*Triticum aestivum* L.)	Chandra et al. (2019)
Drought	*Pseudomonas putida*, *Bacillus amyloliquefaciens*	ACC deaminase activity	Chickpea (*Cicer arietinum* L.)	Kumar et al. (2016)
Salinity	*Bacillus* spp.	IAA, ACC deaminase activities	Rice	Mishra et al. (2017)
Salinity	*Enterobacter* spp.	ACC deaminase activity	Rice	Sarkar et al. (2018)
Salinity	*Mesorhizobium* spp.	ACC deaminase activity	Chickpea (*Cicer arietinum* L.)	Chaudhary and Sindhu (2017)
Salinity	*Streptomyces* spp.	Auxin activity	Wheat (*Triticum aestivum*)	Sadeghi et al. (2012)
Salinity	*Klebsiella* sp. MBE02	Auxin activity	*Arachis hypogea*	Sharma et al. (2016)

ACC 1-aminocyclopropane-1-carboxylate, *IAA* indole-3-acetic acid

improved tolerance correlated with increased relative water content (RWC) and enhanced accumulation of various osmolytes (succinate, leucine, disaccharide, saccharic acid and glyceric acid), along with other metabolites, in chickpea genotypes. PGPRs have the ability to induce plant tolerance under abiotic stress by regulation of various physiological and metabolic pathways (Khan et al. 2019).

Several types of bacteria—such as *Azospirillum*, *Klebsiella*, *Burkholderia*, *Bacillus* and *Pseudomonas*—have been identified as PGPRs in maize cropping systems. The term 'induced systemic tolerance' (IST) refers to increasing tolerance in plants through modulation of physical and chemical processes triggered by microorganisms when the plants are exposed to a stressful situation. One study revealed that PGPRs have immense ability to increase tolerance of salinity stress by approximately 50% in maize and wheat; therefore, application of PGPRs leads to significantly enhanced crop resilience under salinity stress and improved crop productivity in wheat (Orhan 2016). With the frequent incidence of abiotic stress, there is always a major concern to identify and develop strategies that can be used to mitigate the deleterious impacts of abiotic stress on crop growth and yields. Various research activities—involving genetic engineering, plant breeding, resource management practices etc.—are under way to develop stress-tolerant plant varieties, but many of these technologies are time consuming and costly. However, the results of several studies have now supported the potential role of microorganisms in helping plants deal with drought and salinity stress through improved tolerance (Vurukonda et al. 2016).

Plant growth–promoting bacteria (PGPBs), which are bioeffector microbes, can offer several benefits to the agricultural sector with appropriate application. PGPBs can induce plant growth and ameliorate plant resilience against biotic and abiotic stresses (Ventorino et al. 2016). Therefore, exploration of the plant growth–promoting activities of several bacterial strains isolated from different extreme environments may provide important information to broaden the range of applications of PGPRs as a potential tool in agricultural sustainability.

There are various reports available on beneficial soil microorganisms showing PGPR attributes. They note that soil microorganisms in areas where the conditions are extreme show better adaptations to survive under those situations. Such microbes could therefore be of great help if used in agriculture to increase tolerance and crop productivity. Moreover, it is now accepted that beneficial soil microorganisms possess important attributes that can increase crop tolerance and improve plant growth and productivity under abiotic and biotic stresses in several ways such as mobilization of nutrients, improvement of soil texture and health, secretion of plant growth regulators, disease suppression etc. (Verma et al. 2016). PGPRs isolated from places with less rainfall are better able to survive and extend protection to plants by increasing their tolerance of desiccation. Mayak et al. (2004a) noted that PGPRs isolated from areas with low rainfall are more effective in this regard than other similar bacteria isolated from sites with sufficient availability of water. For instance, the bacterial strain *Achromobacter piechaudii* ARV8, isolated from rhizospheric soil in a dry region, exhibited ACC deaminase activity that induced significant drought tolerance in tomato. Other researchers have also demonstrated protective effects of ACC deaminase production by PGPRs on different plants against loss of biomass from drought stress (Belimov et al. 2009; Shakir et al. 2012; Penrose and Glick 2003). The same mechanism is equally effective against salinity stress, which otherwise causes plants to suffer more inhibition of growth and development (Mayak et al. 2004b).

PGPRs produce a variety of primary or low molecular weight secondary metabolites—proline, glycine betaine, sugars, polyamines, amides and other enzymes, EPS etc.—that help plants to enhance their abiotic stress tolerance under adverse conditions (Jha et al. 2011; Kasotia et al. 2016; Kurz et al. 2010; Singh and Jha 2016). Production of various secondary metabolites by salinity-tolerant rhizobacteria has shown the potential capability to induce salinity stress tolerance in plants by improving their physiological conditions. Application of such rhizobacteria therefore has the potential for mitigation of salinity stress to improve crop productivity (Mishra et al. 2018).

PGPRs that express ACC deaminase activity decrease plant ethylene levels, as this enzyme breaks down the ethylene precursor ACC to α-ketobutyrate and ammonium, leading to decreased ethylene concentrations in stressed plants and improved plant tolerance of stress. Notably, ACC deaminase–producing rhizobacteria confer induced tolerance in plants against a wide range of different biotic and abiotic stresses through effective plant–microbe interactions (Glick et al. 2007). Among various different crop management practices used in the agricultural sector, application of PGPRs via different methods (such as seed priming or application to the soil) is important to achieve the desired effects in protecting plants against stress. The underlying mechanism of PGPR involvement in reduction of plant ethylene levels is metabolization of the ethylene precursor at the root–soil interface under stress conditions, thereby improving crop yields (Belimov et al. 2009). The stress-induced increase in plant ethylene levels varies depending on the genotype and the magnitude of the stress. Therefore, it is suggested that opportunities for better management and application of PGPRs in agricultural systems should be explored to improve water use and carbon gains in field crops.

A recent study on drought stress tolerance in two important crops—mung bean (*Vigna mungo* L.) and pea (*Pisum sativum* L.)—found that a consortium of rhizobacteria strains (*Ochrobactrum pseudogrignonense* RJ12, *Pseudomonas* sp. RJ15 and *Bacillus subtilis* RJ46) had the ability to produce ACC deaminase. The results indicated improved tolerance in these crops, due to ACC deaminase activity leading to decreased ACC accumulation and regulation of ethylene levels (Saikia et al. 2018). Grover et al. (2014) conducted a study on sorghum and revealed that inoculation with different strains of *Bacillus* spp. imparted improved tolerance of moisture stress conditions, improving seedling growth and physiological attributes. This improved tolerance was attributed to phosphate solubilization and production of IAA and siderophores. Further, improved drought and salinity stress tolerance were observed in *Capsicum annuum* when it was inoculated with *Burkholderia cepacia*. It was reported that ACC deaminase activity of PGPRs promoted growth and development in conditions of drought and salinity stress (Maxton et al. 2018). Chandra et al. (2019) studied the impact of PGPRs on wheat (*Triticum aestivum* L.) under drought stress. Inoculation of the wheat with *Variovorax paradoxus* RAA3, *Pseudomonas* spp., *Achromobacter* spp. and *Ochrobactrum anthropi* improved seedling growth, which correlated with increased activity of ACC deaminase, siderophore production and phosphate solubilization properties of PGPRs under drought stress (Chandra et al. 2019). Mishra et al. (2017)

conducted a study on rice inoculated with different rhizobacteria (*Bacillus* spp.) collected from various agroclimatic zones under salinity stress. The results indicated that production of ACC deaminase and IAA by these rhizobacteria improved seedling growth under salinity stress.

Abiotic stresses—mainly drought, salinity and extreme temperatures—affect plant growth and limit crop productivity significantly. Plants have an inherent ability to cope with adverse conditions but only to a limited extent. Several genetic engineering tools and breeding methods are available for crop improvement to develop tolerance of abiotic and biotic stresses in plants. The role of soil microorganisms cannot be ignored. Our present understanding of beneficial soil microorganisms in the rhizosphere and their immense potential for improving plant tolerance of both biotic and abiotic stresses offers an alternate eco-friendly approach to develop crop resilience under stress.

Plant–rhizobacterium interactions involve modulation of various physiological, biochemical and molecular pathways under stressful conditions to boost tolerance. We still do not fully understand the exact mechanisms through which PGPRs impart their beneficial effects on plants and modulate different signalling networks to improve tolerance under abiotic stress. It has been suggested that plant–rhizobacteria interactions facilitate increase nutrient uptake, maintain plant water relations and enhance photosynthesis and source–sink relationships to boost plant growth and yields. PGPRs modulate several physiological, cellular, biochemical and molecular processes to improve plant tolerance under abiotic stress (Gayathri and Donald 2018). Diverse groups of microbes have been identified as having the ability to catabolize plant exudates, leading to protection of the plants from drought and salinity stress. PGPRs produce a wide variety of substances—ACC deaminase (Saleem et al. 2015), siderophores (Stajkovic-Srbinovic et al. 2014), plant growth regulators, salicylic acid (Ekinci et al. 2014), the phytohormone IAA (Gujral et al. 2013), phosphate-solubilizing enzymes (Kumari and Khanna 2016) and microbiocidal and biostatic enzymes (Moustaine et al. 2017)—which boost important biochemical and physiological processes involved in plant defence against stresses.

Plant–rhizobacteria interaction increase plant defence by modulating several cellular processes, improving photosynthesis, nutrient uptake and source–sink relationships and thereby improving plant growth. PGPRs exhibit the ability to modulate several factors—such as phytohormones status, protein function, gene expression and metabolite synthesis in plants—improving their defence responses. Enhanced antioxidant activity, accumulation of osmolytes, salt compartmentalization etc. reduce osmotic stress and the effects of ion toxicity in response to salinity stress and drought stress. Moreover, extracellular signalling molecules trigger stress-responsive pathways in plants to help them cope better with adverse conditions (Gayathri and Donald 2018).

1.4 Plant–Rhizobacteria Interaction and Biotic Stress Tolerance

Phytopathogens are the principal causes of biotic stress in crops, leading to substantial decreases in crop yields and crop losses. PGPRs can help plants to resist phytopathogens and biotic stresses by adopting appropriate strategies against such threats, including antagonism and triggering of systemic resistance. The presence of PGPRs in the soil has a profound effect on the soil characteristics. They secrete several different groups of compounds, thereby increasing the quality of the soil for better cultivation (Gouda et al. 2017). It is also important to note that appropriate application of PGPRs in crops also depends greatly on their compatibility with the soil type and with other indigenous microbes in the soil (Singh et al. 2016).

PGPRs possess several plant growth–promoting attributes and secrete groups of compounds that confer plant tolerance of both abiotic and biotic stresses. Different species of PGPRs (such as *Bacillus*) that are present in agricultural fields can promote plant growth and development either by increasing the availability of nutrients or by triggering plant defences against plant pathogens, infections, insect attacks etc. (Kumar et al. 2012; Egamberdieva and Lugtenberg 2014). A study conducted in tomato revealed that methyl jasmonate (MeJA) and the ethylene precursor ACC can boost resistance against *Pseudomonas syringae* pv. *tomato* (Pieterse et al. 1998, 2000). In another study on increased resistance against bacterial canker disease, which is caused by *Clavibacter michiganensis* subsp. *michiganensis* (*Cmm*), it was suggested that treatment of tomato (*Solanum lycopercican* L.) plants with *Pseudomonas* sp. 23S triggered induced systemic resistance (ISR) in the plants and reduced the severity and progression of the disease. It was further suggested that it was salicylic acid that mediated induced systemic resistance in the plants (Takishita et al. 2018). Application of salicylic acid resulted in better tolerance of *Rhizoctonia solani* in cowpea by enhancing phenylalanine ammonia lyase (PAL) activity (Chandra et al. 2007).

Use of PGPRs as biocontrol agents offers an eco-friendly option for control of plant diseases. Presently, several PGPR species of different genera are used as biocontrol agents—*Agrobacterium*, *Azotobacter*, *Azospirillum*, *Bacillus*, *Delftia*, *Burkholderia*, *Rhizobium*, *Paenibacillus*, *Pantoea*, *Pseudomonas* and *Serratia*—to combat plant pathogens and prevent disease progression (Glick 2012). Application of PGPR strains belonging to the important genera *Bacillus* and *Pseudomonas* as biocontrol agents in cannabis plants achieved improvements in yield and growth under stress and provided better tolerance against powdery mildew, which is the most common pathogen affecting cannabis yields (Lyu et al. 2019). The competence of *Pseudomonas* sp. RU47 as a biocontrol agent in the rhizospheres of two important crops—potato (*Solanum tuberosum* L.) and lettuce (*Lactuca sativa* L.)—was studied by Schreiter et al. (2018), who found that its application as a bioinoculant was an effective strategy to control the effects of disease caused by the plant pathogen *Rhizoctonia solani*.

In recent years, biocontrol of plant-parasitic nematodes through antagonism by PGPR application has attracted considerable attention, and studies have been conducted to assess the potential of PGPRs as biocontrol agents to protect plants from disease-causing phytonematodes (Sidhu 2018). Application of PGPRs (*Bacillus pumilus*, *Paenibacillus costume* and *Mycobacterium immunogenum) was found to be an effective biocontrol strategy against the nematode Meloidogyne incognita*, which causes root rot disease in tomato (Cetintas et al. 2018). Similarly, biocontrol effects of different rhizobacterial strains (*R. leguminosarum* and *P. fluorescens*) were observed in different legume crop rhizospheres, leading to decreased pathogenesis due to root-knot nematodes (*Meloidogyne javanica*) and improved seedling growth (Tabatabaei and Saeedizadeh 2017). Application of PGPRs in rice resulted in effective suppression of the phytopathogen *Xanthomonas oryzae* pv. *oryzae* (which is responsible for bacterial blight disease in rice) and also achieved effective resistance to blister blight disease (caused by the phytopathogen *Exobasidium vexans* Massee) in tea (Suryadi et al. 2019). Inoculation with the PGPR *Pseudomonas putida* strain NH-50, which has the ability to produce pyoluteorin, was found to significantly reduce red rot disease in sugar cane by inhibiting growth of *Glomerella tucumanensis* (Speg.) Arx & E. Müll. (Hassan et al. 2011).

1.4.1 Mechanisms of Rhizobacteria-Mediated Phytopathogen Tolerance in Plants

PGPRs are highly diverse, which can also help induce plant resistance against several types of biotic stress caused by pathogen attacks. Several studies have revealed that PGPRs induce biotic stress tolerance in plants either through local antagonism to soilborne pathogens or through induction of systemic resistance against several pathogens. Nonpathogenic rhizobacteria can interact with plants and stimulate substantial increases in plant capabilities for defence against pathogens and plant diseases. The reduction in disease is associated with decreased pathogen growth and reduced colonization of plant tissue, reflecting the ability of the plants to resist the pathogens. This is the mechanism of induced systemic resistance in plants (Van Loon et al. 1998).

It has been reported that PGPRs act as biocontrol agents by producing various compounds—antibiotics, siderophores etc.—that can control pathogen progression and sustain plant growth. Rhizobacterium-mediated induced systemic resistance in plants and pathogen-induced systemic acquired resistance (SAR) induced by bacteria in plants together induce greater resistance to plant pathogens and disease (Van Loon et al. 1998). Studies have revealed that signalling molecules such as salicylic acid, secreted by rhizobacteria, trigger pathogen resistance in plants through salicylic acid–mediated systemic acquired resistance in the plants, which is induced by pathogen attacks and is followed by activation of pathogenesis-related (PR) proteins. Moreover, secretion of other signalling molecules—such as jasmonic acid, ethylene and lipopolysaccharides—leads to triggering of induced systemic resistance in plants. Microbial antagonism is one of the mechanisms through which rhizobacteria reduce the impact of pathogens in plants and improve plant tolerance of biotic stress (Beneduzi et al. 2012; Spoel and Dong 2012; Van Wees et al. 2008). Siderophores, bacteriocins and antibiotics are some of the important compounds produced and released by PGPRs, and they are very effective in reducing disease and limiting progression of pathogens in plants through antagonistic activity (Maksimov et al. 2011). Some of the important antagonistic activities that are likely to be dominant in the rhizosphere include synthesis and secretion of hydrolytic enzymes—such as chitinases, glucanases, proteases and lipases—that restrict the activities of fungal pathogens (Maksimov et al. 2011). Regulation of ACC deaminase activity, control of ethylene levels in plant under biotic stress (Kamilova et al. 2005), siderophore production (Van Loon 2007) and competition for suitable space on root surfaces for colonization and nutrient acquisition are some of the strategies exhibited by PGPRs that help induce plant tolerance of pathogen infections.

1.5 Conclusion

The current reality in the agricultural sector is that climate change and frequent occurrences of biotic and abiotic stresses lead to significant limitations in crop productivity. This has prompted research into development of methods to induce the intrinsic defences of plants against such stresses in order to maintain agricultural sustainability. To date, the concept of plant–microbe interactions and the roles of PGPRs have been underexplored, but there is huge potential for exploitation of plant–microbe interactions as potential tools in abiotic stress tolerance and as biocontrol agents for defence against biotic stresses. Commercial development of single rhizobacterial strains or combinations of different rhizobacterial strains as effective biocontrol agents could be exploited for cost-effective, low-input, eco-friendly and sustainable plant management to reduce dependence on agrochemicals in agricultural systems. Moreover, application of PGPRs offers a long-term eco-friendly option to develop both intrinsic and extrinsic abilities of plants to resist biotic and abiotic stressful conditions and to sustain crop growth and yields.

References

Antoun H, Kloepper JW (2001) Plant growth promoting rhizobacteria. In: Brenner S, Miller JH (eds) Encyclopedia of genetics. Academic, New York, NY, pp 1477–1480

Arshad M, Shaharoona B, Mahmood T (2008) Inoculation with *Pseudomonas* spp. containing ACC-deaminase partially eliminates the effects of drought stress on growth, yield, and ripening of pea (*Pisum sativum* L.). Pedosphere 18:611–620. https://doi.org/10.1016/S1002-0160(08)60055-7

Barra PJ, Inostroza NG, Acuña JJ, Mora ML, Crowley DE, Jorquera MA (2016) Formulation of bacterial consortia from avocado (*Persea americana* Mill.) and their effect on growth, biomass and superoxide dismutase activity of wheat seedlings under salt stress. Appl Soil Ecol 102:80–91

Belimov AA, Dodd IC, Hontzeas N, Theobald JC, Safronova VI, Davies WJ (2009) Rhizosphere bacteria containing 1-aminocyclopropane-1-carboxylate deaminase increase yield of plants grown in drying soil via both local and systemic hormone signaling. New Phytol 181:413–423. https://doi.org/10.1111/j.1469-8137.2008.02657.x

Beneduzi A, Ambrosini A, Passaglia LMP (2012) Plant growth–promoting rhizobacteria (PGPR): their potential as antagonists and biocontrol agents. Genet Mol Biol 35(4 Suppl):1044–1051. https://doi.org/10.1590/s1415-47572012000600020

Cardinale M, Ratering S, Suarez C, Montoya AMZ, Geissler-Plaum R, Schnell S (2015) Paradox of plant growth promotion potential of rhizobacteria and their actual promotion effect on growth of barley (*Hordeum vulgare* L.) under salt stress. Microbiol Res 181:22–32

Cetintas R, Kusek M, Fateh SA (2018) Effect of some plant growth–promoting rhizobacteria strains on root-knot nematode, *Meloidogyne incognita*, on tomatoes. Egypt J Biol Pest Contr 28:7

Chakraborty U, Chakraborty B, Dey P, Chakraborty AP (2015) Role of microorganisms in alleviation of abiotic stresses for sustainable agriculture. In: Chakraborty U, Chakraborty BN (eds) Abiotic stresses in crop plants. CAB International, Wallingford, pp 232–253

Chandra A, Saxena R, Dubey A, Saxena P (2007) Changes in phenylalanine ammonia lyase activity and isozyme patterns of polyphenol oxidase and peroxidase by salicylic acid leading to enhance resistance in cow pea against *Rhizoctonia solani*. Physiol Plant 29:361–367

Chandra D, Srivastava R, Gupta VVSR, Franco CMM, Sharma AK (2019) Evaluation of ACC-deaminase-producing rhizobacteria to alleviate water-stress impacts in wheat (*Triticum aestivum* L.) plants. Can J Microbiol 65:387–403. https://doi.org/10.1139/cjm-2018-0636

Chaudhary D, Sindhu SS (2017) Amelioration of salt stress in chickpea (*Cicer arietinum* L.) by coinculation of ACC deaminase–containing rhizospheric bacteria with *Mesorhizobium* strains. Legum Res 40(1):80–86

Damam M, Gaddam B, Kausar R (2014) Effect of plant growth promoting rhizobacteria (PGPR) on *Coleus forskohlii*. Int J Curr Microbiol App Sci 3(9):266–274

Dobbelaere S, Vanderleyden J, Okon Y (2003) Plant growth–promoting effects of diazotrophs in the rhizosphere. Crit Rev Plant Sci 22:107–149. https://doi.org/10.1080/713610853

Egamberdieva D, Lugtenberg B (2014) Use of plant growth–promoting rhizobacteria to alleviate salinity stress in plants. In: Miransari M (ed) Use of microbes for the alleviation of soil stresses, vol 1. Springer, New York, NY, pp 73–96. https://doi.org/10.1007/978-1-4614-9466-9

Ekinci M, Turan M, Yildirim E, Güneş A, Kotan R, Dursun A (2014) Effect of plant growth promoting rhizobacteria on growth, nutrient, organic acid, amino acid and hormone content of cauliflower (Brassica oleracea L. var. botrytis) transplants. Acta Sci Pol Hortorum Cultus 13 (6):71–85

Enebe MC, Babalola OO (2018) The influence of plant growth–promoting rhizobacteria in plant tolerance to abiotic stress: a survival strategy. Appl Microbiol Biotechnol 102(18):7821–7835. https://doi.org/10.1007/s00253-018-9214-z

FAO (2009) High level expert forum—how to feed the world in 2050. Economic and social development, food and agricultural organization of the United Nations. Rome, Italy

Gayathri I, Donald S (2018) Plant growth promoting rhizobacteria in amelioration of salinity stress: a systems biology perspective. Front Plant Sci 9:1473

Glick BR (2012) Plant growth–promoting bacteria: mechanisms and applications. Scientifica 96:3401

Glick BR (2014) Bacteria with ACC deaminase can promote plant growth and help to feed the world. Microbiol Res 169:30–39

Glick BR, Todorovic B, Czarny J, Cheng Z, Duan J, McConkey B (2007) Promotion of plant growth by bacterial ACC deaminase. Crit Rev Plant Sci 26:227–242

Gouda S, Kerry RG, Das G, Paramithiotis S, Shin HS, Patra JK (2017) Revitalization of plant growth promoting rhizobacteria for sustainable development in agriculture. Microbiol Res 206:131–140

Grover M, Ali SKZ, Sandhya V, Venkateswarlu B (2011) Role of microorganisms in the adaptation of agricultural crops to abiotic stresses. World J Microbiol Biotechnol 27:1231–1240

Grover M, Madhubal R, Ali SZ, Yadav SK, Venkateswarlu B (2014) Influence of *Bacillus* spp. strains on seedling growth and physiological parameters of sorghum under moisture stress conditions. J Basic Microbiol 54:951–961

Gujral MS, Agrawal P, Khetmalas MB, Pandey R (2013) Colonization and plant growth promotion of sorghum seedling by endorhizospheric *Serratia* sp. Acta Biol Indica 2(1):343–352

Gull A, Lone AA, Wani NUI (2019) Biotic and abiotic stresses in plants. In: De Oliveira A (ed) Abiotic and biotic stress in plants. Intech Open, London. https://doi.org/10.5772/intechopen.85832

Hassan MN, Afghan S, Hafeez FY (2011) Biological control of red rot in sugarcane by native pyoluteorin-producing *Pseudomonas putida* strain NH-50 under field conditions and its potential modes of action. Pest Manag Sci 67:1147–1154. https://doi.org/10.1002/ps.2165

Hong SH, Ham SY, Kim JS, Kim I-S, Lee EY (2016) Application of sodium polyacrylate and plant growth–promoting bacterium, *Micrococcaceae* HW-2, on the growth of plants cultivated in the rooftop. Int Biodeterior Biodegrad 133:297–303

Jha Y, Subramanian RB, Patel S (2011) Combination of endophytic and rhizospheric plant growth promoting rhizobacteria in *Oryza sativa* shows higher accumulation of osmoprotectant against saline stress. Acta Physiol Plant 33:797–802

Kamilova F, Validov S, Azarova T, Mulders I, Lugtenberg B (2005) Enrichment for enhanced competitive plant root tip colonizers selects for a new class of biocontrol bacteria. Environ Microbiol 7:1809–1817

Kang Y, Khan S, Ma X (2009) Climate change impacts on crop yield, crop water productivity and food security. Prog Nat Sci 19:1665–1674

Kasim WA, Gaafar RM, Abou-Ali RM, Omar MN, Hewait HM (2016) Effect of biofilm forming plant growth promoting rhizobacteria on salinity tolerance in barley. Ann Agric Sci 61 (2):217–227

Kasotia A, Varma A, Tuteja N, Choudhary DK (2016) Amelioration of soybean plant from saline-induced condition by exopolysaccharide producing *Pseudomonas*–mediated expression of high affinity K^+ transporter (*HKT1*) gene. Curr Sci 111(12):25

Khan N, Bano A, Babar MA (2019) Metabolic and physiological changes induced by plant growth regulators and plant growth promoting rhizobacteria and their impact on drought tolerance in *Cicer arietinum* L. PLoS One 14(3):e0213040

Kloepper JW, Lifshitz R, Zablotowicz RM (1989) Free-living bacterial inocula for enhancing crop productivity. Trends Biotechnol 7:39–43

Kloepper JW, Ryu C-M, Zhang SA (2004) Induced systemic resistance and promotion of plant growth by *Bacillus* spp. Phytopathology 94:1259–1266

Kumar P, Dubey RC, Maheshwari DK (2012) *Bacillus* strains isolated from rhizosphere showed plant growth promoting and antagonistic activity against phytopathogens. Microbiol Res 167 (8):493–499

Kumar M, Mishra S, Dixit V, Kumar M, Agarwal L, Chauhan PS, Nautiyal CS (2016) Synergistic effect of *Pseudomonas putida* and *Bacillus amyloliquefaciens* ameliorates drought stress in chickpea (*Cicer arietinum* L.). Plant Signal Behav 11:e1071004

Kumar A, Singh V, Tripathi V, Singh P, Singh A (2018) Plant growth–promoting rhizobacteria (PGPR): perspective in agriculture under biotic and abiotic stress. In: Prasad R, Gill SS, Tuteja N (eds) New and future developments in microbial biotechnology and bioengineering: crop improvement through microbial biotechnology. Elsevier, Amsterdam, pp 333–342. https://doi.org/10.1016/B978-0-444-63987-5.00016-5

Kumar M, Kour D, Yadav AN et al (2019) Biodiversity of methylotrophic microbial communities and their potential role in mitigation of abiotic stresses in plants. Biologia 74(3):287–308

Kumari P, Khanna V (2016) Biodiversity of *Pseudomonas* and *Bacillus* possessing both bioantagonistic and plant growth promoting traits in chickpea rhizosphere. Int J Sci Nat 7 (1):153–158

Kurz M, Burch AY, Seip B, Lindow SE, Gross H (2010) Genome-driven investigation of compatible solute biosynthesis pathways of *Pseudomonas syringae* pv. *syringae* and their contribution to water stress tolerance. Appl Environ Microbiol 76(16):5452–5462

Lyu D, Backer R, Robinson WG, Smith DL (2019) Plant growth–promoting rhizobacteria for cannabis production: yield, cannabinoid profile and disease resistance. Front Microbiol 10:1761

Maksimov IV, Abizgil'dina RR, Pusenkova LI (2011) Plant growth promoting rhizobacteria as alternative to chemical crop protectors from pathogens. Appl Biochem Microbiol 47:333–345

Masciarelli OL, Lanes A, Luna V (2014) A new PGPR co-inoculated with *Bradyrhizobium japonicum* enhances soybean nodulation. Microbiol Res 169:609–615

Maxton A, Singh P, Masih SA (2018) ACC deaminase–producing bacteria mediated drought and salt tolerance in *Capsicum annuum*. J Plant Nutr 41:574–583

Mayak S, Tirosh T, Glick BR (2004a) Plant growth–promoting bacteria that confer resistance to water stress in tomato and pepper. Plant Sci 66:525–530. https://doi.org/10.1016/j.plantsci.2003.10.025

Mayak S, Tirosh T, Glick BR (2004b) Plant growth–promoting bacteria that confer resistance in tomato to salt stress. Plant Physiol Biochem 2:565–572

Meena KK, Sorty AM, Bitla UM, Choudhary K, Gupta P, Pareek A, Singh DP, Prabha R, Sahu PRPK, Gupta VK, Singh HB, Krishnani KK, Minhas PS (2017) Abiotic stress responses and microbe-mediated mitigation in plants: the omics strategies. Front Plant Sci 8(172):1–25. https://doi.org/10.3389/fpls.2017.00172

Mishra S, Dixit VK, Khan MH, Mishra SK, Dviwedi G, Yadav S, Lehri A, Chauhan PS (2017) Exploitation of agro-climatic environment for selection of 1-aminocyclopropane-1-carboxylic acid (ACC) deaminase producing salt tolerant indigenous plant growth promoting rhizobacteria. Microbiol Res 205:25–34. https://doi.org/10.1016/j.micres.2017.08.007

Mishra J, Fatima T, Arora NK (2018) Role of secondary metabolites from plant growth–promoting rhizobacteria in combating salinity stress. In: Egamberdieva D, Ahmad P (eds) Plant microbiome: stress response. Springer, Singapore, pp 127–163. https://doi.org/10.1007/978-981-10-5514-0_6

Moustaine M, Elkahkahi R, Benbouazza A, Benkirane R, Achbani EH (2017) Effect of plant growth promoting rhizobacterial (PGPR) inoculation on growth in tomato (*Solanum lycopersicum* L.) and characterization for direct PGP abilities in Morocco. Int J Environ Agric Biotechnol 2(2):590–596

Nezhadahmadi A, Prodhan ZH, Faruq G (2013) Drought tolerance in wheat. Sci World J 1:610721. https://doi.org/10.1155/2013/610721

Ojuederie OB, Olanrewaju OS, Babalola OO (2019) Plant growth promoting rhizobacterial mitigation of drought stress in crop plants: implications for sustainable agriculture. Agronomy 9:712

Orhan F (2016) Alleviation of salt stress by halotolerant and halophilic plant growth–promoting bacteria in wheat (*Triticum aestivum*). Braz J Microbiol 47(3):621–627

Pathak A, Sharma A, Johri BN (2004) Pseudomonas strain GRP3 induces systemic resistance to sheath blight in rice. Int Rice Res Notes 29:35–36

Penrose DM, Glick BR (2003) Methods for isolating and characterizing ACC deaminase–containing plant growth–promoting rhizobacteria. Physiol Plant 118:10–15. https://doi.org/10.1034/j.1399-3054.2003.00086.x

Pieterse CMJ, VanWees SCM, VanPelt JA, Knoester M, Laan R, Gerrits H, Weisbeek PJ, Van Loon LC (1998) A novel signaling pathway controlling induced systemic resistance in *Arabidopsis*. Plant Cell 10:1571–1580. https://doi.org/10.1105/tpc.10.9.1571

Pieterse CMJ, Van Pelt JA, Ton J, Parchmann S, Mueller MJ, Buchala AJ, Métraux J-P, Van Loon LC (2000) Rhizobacteria-mediated induced systemic resistance (ISR) in *Arabidopsis* requires sensitivity to jasmonate and ethylene but is not accompanied by an increase in their production. Physiol Mol Plant Pathol 57:123–134. https://doi.org/10.1006/pmpp.2000.0291

Podile AR, Kishore GK (2007) Plant growth–promoting rhizobacteria. In: Gnanamanickam SS (ed) Plant-associated bacteria. Springer, Dordrecht, pp 195–230. https://doi.org/10.1007/978-1-4020-4538-7_6

Raj SN, Chaluvaraju G, Amruthesh K, Shetty HS, Reddy M, Kloepper JW (2003) Induction of growth promotion and resistance against downy mildew on pearl millet (*Pennisetum glaucum*) by rhizobacteria. Plant Dis 87:380–384

Raza A, Razzaq A, Mehmood SS, Zou X, Zhang X, Lv Y, Xu J (2019) Impact of climate change on crops adaptation and strategies to tackle its outcome: a review. Plants 8:34

Sadeghi A, Karimi E, Dahazi PA, Javid MG, Dalvand Y, Askari H (2012) Plant growth promoting activity of an auxin and siderophore producing isolate of *Streptomyces* under saline soil condition. World J Microbiol Biotechnol 28:1503–1509

Saikia J, Sarma RK, Dhandia R, Yadav A, Bharali R, Gupta VK, Saikia R (2018) Alleviation of drought stress in pulse crops with ACC deaminase producing rhizobacteria isolated from acidic soil of northeast India. Sci Rep 8:3560

Saleem AR, Bangash N, Mahmood T, Khalid A, Centritto M, Siddique MT (2015) Rhizobacteria capable of producing ACC deaminase promote growth of velvet bean (*Mucuna pruriens*) under water stress condition. Int J Agric Biol 17:663–667

Sarkar A, Ghosh PK, Pramanik K, Mitra S, Soren T, Pandey S, Mondal MH, Maiti TK (2018) A halotolerant *Enterobacter* sp. displaying ACC deaminase activity promotes rice seedling growth under salt stress. Res Microbiol 169:20–32

Saxena R, Kumar M, Tomar RS (2019a) Plant responses and resilience toward drought and salinity stress. Plant Arch 19(Suppl 2):50–58

Saxena R, Kumar M, Jyoti A, Tomar RS (2019b) Untapped potential of salicylic acid, jasmonic acid and PGPRs to develop abiotic stress resilience in crop plants. Curr Trends Biotechnol Pharm 13 (4):376–390

Schreiter S, Babin D, Smalla K, Grosch R (2018) Rhizosphere competence and biocontrol effect of *Pseudomonas* sp. RU47 independent from plant species and soil type at the field scale. Front Microbiol 9:97

Schroth MN, Hancock JG (1982) Disease-suppressive soil and root-colonizing bacteria. Science 216:1376–1381

Shakir MA, Asghari B, Arshad M (2012) Rhizosphere bacteria containing ACC deaminase conferred drought tolerance in wheat grown under semi-arid climate. Soil Environ 31:108–112

Sharma V, Kamal B, Srivastava N, Negi Y, Dobriyal AK, Jadon VS (2015) Enhancement of in vitro growth of *Swertia chirayita* Roxb. ex Fleming co-cultured with plant growth promoting rhizobacteria. Plant Cell Tissue Organ Cult 121:215–225

Sharma S, Kulkarni J, Jha B (2016) Halotolerant rhizobacteria promote growth and enhance salinity tolerance in peanut. Front Microbiol 7:1600

Sidhu HS (2018) Potential of plant growth–promoting rhizobacteria in the management of nematodes: a review. J Entomol Zool Stud 6(3):1536–1545

Singh RP, Jha PN (2016) The multifarious PGPR *Serratia marcescens* CDP-13 augments induced systemic resistance and enhanced salinity tolerance of wheat (*Triticum aestivum* L.). PLoS One 11(6):e0155026

Singh JS, Abhilash PC, Gupta VK (2016) Agriculturally important microbes in sustainable food production. Trends Biotechnol 34:773–775

Spoel SH, Dong X (2012) How do plants achieve immunity? Defence without specialized immune cells. Nat Rev Immunol 12:89–100

Stajkovic-Srbinovic O, Delic D, Kuzumanovic D, Protic N, Rasulic N, Knezevic-Vukcevic J (2014) Growth and nutrient uptake in oat and barley plants as affected by rhizobacteria. Rom Biotechnol Lett 19(3):9429–9436

Suryadi Y, Susilowati DN, Fauziah F (2019) Management of plant diseases by PGPR-mediated induced resistance with special reference to tea and rice crops. In: Sayyed RZ (ed) Plant growth promoting rhizobacteria for sustainable stress management. Volume 2: Rhizobacteria in biotic stress management. Springer, Singapore, pp 65–110

Tabatabaei FS, Saeedizadeh A (2017) Rhizobacteria cooperative effect against *Meloidogyne javanica* in rhizosphere of legume seedlings. Hellen Plant Prot J 10:25–34

Takishita Y, Charron JB, Smith DL (2018) Biocontrol rhizobacterium *Pseudomonas* sp. 23S induces systemic resistance in tomato (*Solanum lycopersicum* L.) against bacterial canker *Clavibacter michiganensis* subsp. *michiganensis*. Front Microbiol 9:2119

Udayashankar A, Nayaka SC, Reddy M, Srinivas C (2011) Plant growth–promoting rhizobacteria mediate induced systemic resistance in rice against bacterial leaf blight caused by *Xanthomonas oryzae* pv. *oryzae*. Biol Control 59:114–122

Van Loon LC (2007) Plant responses to plant growth–promoting rhizobacteria. Eur J Plant Pathol 119:243–254

Van Loon LC, Bakker PAHM, Pieterse CMJ (1998) Systemic resistance induced by rhizosphere bacteria. Annu Rev Phytopathol 36:453–483

Van Wees SCM, Van der Ent S, Pieterse CMJ (2008) Plant immune responses triggered by beneficial microbes. Curr Opin Plant Biol 11:443–448. https://doi.org/10.1016/j.pbi.2008.05.005

Ventorino V, Parillo R, Testa A, Viscardi S, Espresso F, Pepe O (2016) Chestnut green waste composting for sustainable forest management: microbiota dynamics and impact on plant disease control. J Environ Manag 166:168–177

Verma S, Nizam S, Verma PK (2013) Biotic and abiotic stress signalling in plants. In: Sarwat M, Ahmad A, Abdin MZ (eds) Stress signaling in plants: genomics and proteomics perspective, vol 1. Springer, New York, NY, pp 25–49

Verma P, Saxena R, Tomar RS (2016) Rhizobacteria: a promising tool for drought tolerance in crop plants. Int J Pharm Bio Sci:116–125

Vurukonda S, Vardharajula S, Shrivastava M, Ali SKZ (2016) Enhancement of drought stress tolerance in crops by plant growth promoting rhizobacteria. Microbiol Res 184:13–24

Yolcu H, Turan M, Lithourgidis A, Çakmakçi R, Ali KOÇ (2011) Effects of plant growth–promoting rhizobacteria and manure on yield and quality characteristics of Italian rye grass under semi-arid conditions. Aust J Crop Sci 5(13):1730–1736

Chapter 2
Rhizospheric and Endophytic Microorganisms and Their Role in Alleviation of Salinity Stress in Plants

Pramod Kumar Sahu, Nisha Kumari, Amrita Gupta, and Nazia Manzar

Abstract The acreage under salinity is increasing, and the stress thus generated in the plants causes severe damage to the quality and quantity of the produce. Salinity stress also increases the susceptibility of the plants against pests and diseases. Since the requirement to feed billions of mouths is ever-increasing, shortening of the arable lands is not desirable. Thus, measures to improve plant's tolerance and performance under saline soils could be of great practical significance in crop production. The plant breeding approach for developing salinity-tolerant line has limitations for developing alternatives of the commercial cultivars. The microbial agents hold greater promise and suitability to be used as stress alleviator for wide varieties of crops and their commercial cultivars. This chapter summarizes the interaction of rhizosphere and endophytic microorganisms with plants and their role in improving salt tolerance. Various aspects of plant tolerance are discussed in this chapter that are proven to be enhanced by microbial agents such as nutrient uptake, ion homeostasis, reduction in reactive oxygen species by various antioxidants, membrane integrity, ACC deaminase production, and maintaining the osmotic balance of the plant cells. Exploring and characterizing such potential microbes could be useful tool in developing smart package and practices for increasing agricultural production.

Keywords Salinity · Plant-microbe interaction · Rhizosphere · Endophytes · Stress alleviation

P. K. Sahu (✉) · N. Manzar
ICAR-National Bureau of Agriculturally Important Microorganisms, Maunath Bhanjan, UP, India
e-mail: Pramod.Sahu@icar.gov.in

N. Kumari
Hemvati Nandan Bahuguna Garhwal University, Srinagar Garhwal, Uttarakhand, India

A. Gupta
Amity Institute of Biotechnology, Amity University Uttar Pradesh, Lucknow, UP, India

S. K. Dubey, S. K. Verma (eds.), *Plant, Soil and Microbes in Tropical Ecosystems*, Rhizosphere Biology, https://doi.org/10.1007/978-981-16-3364-5_2

2.1 Rhizosphere as a Site of Plant-Microbial Interaction

The rhizosphere term is referred to as the part of the soil under the vicinity of the plant roots where a large portion of microbial activity takes place. The diversity of microorganisms in the rhizosphere plays a vital role in favoring plant growth and health (Brahmaprakash and Sahu 2012; Sahu et al. 2016a, 2019; Brahmaprakash et al. 2017). The root exudates released from root containing organic nutrients favor the growth of these microbes. The rhizosphere comprises of three parts, endorhizosphere, rhizoplane, and ectorhizosphere. The part of the soil away from the rhizosphere zone makes the bulk soil or non-rhizosphere. Microbial density in the rhizosphere is higher than that in the bulk soil, as a result of which the microbial growth rate and activities are also high in the rhizosphere due to the availability of organic nutrient content of root exudates (Rossmann et al. 2020). The naturally occurring microflora of the rhizosphere has both beneficial and pathogenic microbes with a considerable impact on plant growth and development (Brahmaprakash et al. 2017).

The microbes inhabiting the rhizosphere have crucial ecological importance. They help in nutrient solubilization, mobilization, and fortification, secrete certain plant growth inducers, and improve overall plant health (Sahu et al. 2016a). All the ecological interactions between plant and microbes involving mutualism, commensalism, competition, and parasitism take place in this microhabitat (Brahmaprakash et al. 2017). This rhizosphere microflora also produces antagonistic molecules to suppress the growth of pathogenic microbes (Malviya et al. 2020). In the era of sustainable crop production, the microbial interaction with the plant is of vital importance (Brahmaprakash and Sahu 2012). It avails the plant by making nutrients available that are present in the surrounding, solubilizing the unavailable form of nutrients like phosphorus, zinc, etc. and by converting some toxic compound to less toxic form. Reports have shown that application of microbial consortium could improve and enhance the land production potential (Sahu et al. 2016b).

The type of microbial community to be prominent in the rhizospheric soil is highly influenced by the root exudates of the plant. Rhizosphere harbors bacteria, fungi, algae, nematodes, protozoa, mites, and insects. Since a great amount of microbial activity takes place in rhizospheric soil, both species diversity and species richness are important (Brahmaprakash et al. 2017). Bacteria make the most abundant microorganism class in the rhizosphere affecting plant physiological functions (Rossmann et al. 2020). Their population estimation is in the range of 10^6–10^8 cells per cubic centimeter of the soil. These bacteria are able to produce certain enzymes that help them in performing a set of functions that are of ecological importance, involving ammonification, protein degradation, denitrification, and cellulose degradation. The most prevalent bacterial genera are *Arthrobacter*, *Bacillus*, *Pseudomonas*, *Sarcina*, *Clostridium*, *Achromobacter*, *Enterobacter*, etc. Fungi are eukaryotic organisms that are heterotrophic in nature. The rhizosphere is inhabited by both pathogenic and beneficial fungi. The prevalent genera of fungi present in the soil are

Penicillium, *Verticillium*, *Trichoderma*, *Aspergillus*, *Fusarium*, *Mucor*, *Rhizopus*, *Sclerotium*, etc. (Brahmaprakash et al. 2017).

2.2 Endophytes in Plant-Microbe Interaction

The term "endophyte" was given by de Bary in the year 1866, describing the organisms that colonizing the plants internally. Initially, endophyte's existence was identified by the phenomenon of anti-herbivory (Bastías et al. 2018). It was observed that the grasses with high endophytic population were not attacked by insects. The term "endophyte" has been defined and redefined with subtle changes to the original meaning, over the years. Later endophytes were defined as follows: "microbes which are present in the plants and do not cause any harm to its host plant." Endophytic associations can either be (1) commensal (there is no effect on the host plant by the endophyte), (2) parasitic (the endophyte benefits, but there is no benefit to the host plant), or (3) mutualistic (both the host plant and the endophyte benefit each other).

2.2.1 Fungal Endophytes

Fungal endophytes are reported to be associated with production of several medicinally important secondary metabolites and novel compounds. For example, production of anti-cancerous drugs such as Taxol, camptothecin, etc. was reported to be produced by the fungal endophytes associated with the bark of its host plants (Puri et al. 2005). The history of endophytes was started with "fescue mystery" which was associated with the anti-grazing property of fescue grass imparted by the fungal endophytes present in it. The presence of fungal endophyte *Acremonium coenophialum* was conferring anti-herbivory to the fescue grass infected with it. Endophytic diversity varies in different parts of the plant (Vega et al. 2008; Fürnkranz et al. 2012). Some endophytic fungi have been found to show symbiotic interaction with higher vascular plants (Arnold and Lewis 2005).

2.2.2 Bacterial Endophytes

Bacteria living inside the plant without any ill effect to the plant are the bacterial endophytes. The diversity of endophytic bacterial species exists in different plant parts and different plant species. A large diversity of bacterial endophyte species can be colonized in plant at one time. Broadly the bacterial diversity is classified into two major categories, first "gram-positive" and second "gram-negative," and includes genus such as *Agrobacterium*, *Arthrobacter*, *Azotobacter*, *Bacillus*, *Burkholderia*,

Enterobacter, *Frankia*, and *Pseudomonas*, which are the most common (Rosenblueth and Martínez-Romero 2006). The isolation can be done from the part of surface-sterilized internal plant tissue. It gains entry to the plant through various open passages including root hairs, stomatal opening, foliar damage, and germinating radical (Reinhold-Hurek et al. 2007). Endophytes play a significant role in plant growth promotion under saline stress. It is an advantage for the endophytic bacteria to live in vicinity of host plant as it gets protected from extreme environment (Sturz et al. 2000).

2.3 Salinity Stress in Plants

Changing climate conditions have posed several problems in agriculture (Nair et al. 2017). Salinity is currently emerging as a severe problem limiting agricultural production. The soil having electrical conductivity of $\geq$4 dS/m (40 mM NaCl concentration) is considered a saline soil. Highly saline soil can adversely affect the plant biomass as well as total yield. In some cases, if the plant is exposed to salinity for a longer duration, the plant may die. Salinity in the form of sodium, calcium, and magnesium salts of chloride, bicarbonate, and sulfate can interfere with the mobilization of other essential nutrients from the soil to the plants. Plants are subjected to different stress conditions throughout their life cycle. Based on physiology of the plants, tolerance mechanisms and susceptibility to the salt stress vary. Information regarding plant's response to salinity stress could pave the way for designing a management strategy to improve plant performance under salinity stress. High salinity hampers plants at all the aspects of plant growth such as seed germination, early growth, plant architecture, biomass accumulation, and quality and quantity of the produce (Van Zelm et al. 2020).

The physiological impact of high salt concentration in the soil causes a sharp decline in soil water potential which in turn restricts water entry into the plant. It interferes with nutrient mobilization from the soil despite the availability. Excess Na^+ ions in the rhizosphere cause nutrient imbalance and compete with essential ions for the uptake. These excessive Na^+ ions in the cell lead to altered biochemical processes and cause ion toxicity in the plant cells. Apart from ion toxicity, it also causes osmotic stress to the plants. As a result of all these, the total plant yield reduces. Plants have mechanisms to overcome the salinity stress by which they could reduce the effects of toxicity generated by excessive salt. Van Zelm et al. (2020) have divided plant response towards salinity stress into three major aspects:

1. *Early response to salt stress*, which includes perception of sodium ions and signaling in the plants by calcium spiking and alerting plant cells by generating signal molecules such as reactive oxygen species, phospholipids, and protein kinase signaling. This, at early stage, helps plant in maintaining balance in Na^+/K^+ transport in the cells.

2. *Multiphase growth response*, in which plant adjusts itself by osmotic adjustment in the cells to support normal growth.
3. *Spatial salt stress responses*—in this aspect, spatial exclusion of ions takes place via ion transporters, abscisic acid signaling, etc.

2.4 Mechanisms of Rhizospheric and Endophytic Bacteria for Salinity Stress Alleviation

Given plant's response toward salinity stress, microbes help in improving these stress-mitigating strategies (Meena et al. 2017). Beneficial microorganisms reduce the effects of salinity in plants by solubilization of nutrients; synthesis and secretion of plant growth regulators such as IAA and organic acids; nitrogen fixation; production of protective enzymes such as ACC (1-aminocyclopropane-1-carboxylic acid) deaminase, glucanase, and chitinase; induction of systemic tolerance; and production of compatible solutes (proline, glycine betaine, and mannitol) (Singh et al. 2020a, b).

Many reports suggested that rhizospheric and endophytic bacteria of plants play a key role in alleviating salt stress tolerance in plants (Yang et al. 2009; Meena et al. 2017; Prabha et al. 2018). They protect plants against salinity conditions through various mechanisms and are proven to promote plant growth and crop yield when plant is under salt stress condition (Kearl et al. 2019). Nautiyal et al. (2013) gave the observation that *B. amyloliquefaciens* NBRI-SN13 confers salt tolerance to rice plant by altering the gene expression of at least 14 genes when plant was in salt stress. It is also stated that the strain NBRI-SN13 triggers osmoprotectant mechanism that results in induction of salt tolerance in rice plant. A change in microbial community in rice rhizosphere under salt stress was reported. An actinomycete, *Streptomyces* sp. strain PGPA39 was reported to increase salt tolerance with 180 mM NaCl concentration in tomato plant. The strain was found to increase the plant biomass and chlorophyll content in a notable amount. A study was conducted to analyze salinity effect on *Fusarium* wilt severity in tomato plant.

Daami-Remadi et al. (2009) tested the effect at different salt concentrations (2–10 g/L), and it was observed that six salt treatments showed no effect on controlling *Fusarium* mycelial growth, but at a high salt concentration (8 g/L), increased in sporulation was recorded. An increase in the salinity level from 2 to 8 g/L in tomato plants was found to enhance *Fusarium* wilt severity and notable leaf damage index. There was a significant increase of 55% and 66% in leaf damage index recorded when plant was under high salt stress 8 g/L and 10 g/L, respectively. A significant loss of fruit fresh weight was recorded at 40% and 78% under high salt stress 8 g/L and 10 g/L, respectively. From the above, we can conclude that salt stress affects plant's overall growth and productivity. The following are the key mechanisms by which rhizospheric and endophytic microbes alleviate salinity stress in plants.

2.4.1 ACC Deaminase Production

Elevated salinity levels induce production of ethylene to a higher level, which in turn retards plant growth (Mahajan and Tuteja 2005). ACC deaminase produced by PGPRs plays a positive role in abiotic stress signaling. ACC deaminase cleaves ethylene precursor, i.e., ACC to ammonia and α-ketobutyrate. Consequently, ethylene level lowers down in the plant under stress condition

2.4.2 Maintaining Ion Homeostasis and Detoxification

Maintaining Na^+/K^+ balance in the plant cells is very crucial for growth under salt stress. Plants require a balanced cellular ionic environment for normal functioning of bioprocesses inside the cytoplasm as well as in the nucleus. Exposure to prolonged abiotic stresses leads to disturbance in the ion balance of the cell. In order to cope with the external factors, plants need to maintain ion homeostasis by adjusting the expression pattern of distinct transporters and ion channels in the cell. Interestingly, certain PGPRs are identified which are involved in maintaining ion homeostasis of the plant cell by direct or indirect mechanisms. HKT1 are the transporters that differentially maintain cellular K^+ and Na^+ levels. Under salt stress, *B. subtilis* GB 03 emits VOC which in turn represses the expression of HKT1 in the root while promoting expression in the shoot. The potassium transporter HKT1 enhances the accumulation of K^+ and exclusion of Na^+ in the cells. Upregulation of HKT1 in shoot tissue helps in recirculating Na^+ ions in the plant and plays a dual role in ion homeostasis (reviewed in Qin et al. 2016). Consequently, Na^+ levels are optimally maintained in the plant. Biofilm-producing bacteria also reduce Na^+ uptake as biofilm covers the rhizosphere and acts as a barrier for Na^+ uptake under high salinity (Kasim et al. 2016).

2.4.3 Detoxification of Reactive Oxygen Species (ROS)

PGPRs belonging to genus *Bacillus* are widely used for ROS detoxification in the plants due to the availability of superoxide dismutase (SOD), one of the key detoxifying enzymes. Higher levels of reactive oxygen species are produced due to salinity which further has negative effects on plant physiology by oxidation (Chawla et al. 2013). Oxidative damage caused by these ROS molecules can be overcome by either enzymatic or nonenzymatic ROS detoxification machinery in the cell. Various reports confirm the activation of enzymes such as catalase (CAT), superoxide dismutase (SOD), peroxidase (PO), and ascorbate peroxidase (APX) activity in stressed plants upon PGPR inoculation. Nonenzymatic antioxidants include phenolics and flavonoids which reduce ROS.

Higher Na^+ content increases production of reactive oxygen species (ROS) in the cells which causes lipid peroxidation and thus enhances the permeability of the plant cell membranes, resulting in leakage of the ions (Mahajan and Tuteja 2005). The membrane stability could be enhanced by curbing these ROS. Microbes produce and induce the production of antioxidants to degrade the reactive oxygen species.

2.4.4 Improvement in Nutrient Acquisition

Rhizospheric and endophytic microbes are useful for utilizing nutrients that are present in unavailable form in the soil, e.g., iron and phosphorus. Rhizospheric microbes benefit plants by promoting their growth, inhibiting phytopathogens, and strengthening plant's tolerance to abiotic stresses (Brahmaprakash et al. 2017). High salinity reduces nutrient uptake and thus retards plant growth. Microorganisms/ PGPR supporting higher nutrient uptake can thus be useful in alleviating salinity-induced growth reduction. There are several reports of improving nutrient uptake by application of microbes under salinity stress, especially of N, P, K, Fe, and Zn.

Nitrogen is a very important mineral nutrient for growth of the plants. Even though it is available most abundantly on the earth as dinitrogen (N_2) form, it is a limiting nutrient to plants because the plant can only utilize nitrogen in its ionic form, primarily nitrate ions. Microbes convert this atmospheric nitrogen to available form by nitrogen fixation (Biswas and Gresshoff 2014). These microbes exist in plant in different interaction levels both symbiotic and asymbiotic and help with nitrogen nutrition. A symbiotic association is when a relationship is made between two interacting organisms and an organism is benefited. Symbiosis between legumes and *Rhizobium* is an example of mutualism interaction in which both the organisms benefit each other. Bacteria such as *Azotobacter*, *Derxia*, and *Beijerinckia* are able to fix atmospheric nitrogen asymbiotically in the rhizosphere of nonleguminous crops. *Azotobacter* inoculation has shown to enhance plant growth and health by increasing foliage, roots, chlorophyll, and carotenoid content (Maheswari and Kalaiyarasi 2015). It is well-known producer of plant hormones such as indole acetic acid and gibberellic acid (Barat et al. 2016). Many bacteria such as *Azospirillum lipoferum* are able to fix the atmospheric nitrogen to plant asymbiotically. They fix atmospheric nitrogen in nonleguminous plants (Sahu et al. 2017). *Azospirillum* colonizes most of the agriculturally important crops, therefore making it suitable inoculant in agriculture (Gupta and Sahu 2017a, b).

Phosphorus (P) is another most crucial nutrient element involved in multiple vital roles including energy metabolism and structural genetic material. Despite being present in higher amount, the available form of phosphorus is limited as it is very active in soil solution and tends to get fixed in the soil minerals (Rashid et al. 2004). There is a diverse class of microbes inhabiting the soil that have the potential to dissolve this fixed form of P by various mechanisms making it available in the forms which plants can take up (Gupta and Sahu 2017c). These microbes are known as phosphate-solubilizing microorganisms (PSMs) such as *Bacillus*, *Acinetobacter*,

Pseudomonas, *Rhizobium*, *Glomus*, *Gigaspora*, *Penicillium*, and *Aspergillus* (Krishnaveni 2010).

Iron is required by all life forms including the plants, and it is also one of the most limiting trace nutrient elements (Crowley 2006). Microbes produce Fe-chelating compounds called siderophores, which chelate and supply Fe even at very low availability (Buyer et al. 1994). Fungal endophyte *Phomopsis liquidambari* was reported by Su et al. (2019) to improve iron and molybdenum nutrition in peanut. They also found that Fe and Mo acquisition and transformation genes (AhFRO1, AhIRT1, and AhMOT1) were upregulated. Reports suggested improvement in iron nutrition of crop plants by rhizospheric and endophytic microorganisms under salinity stress.

2.4.5 Maintaining Osmotic Balance

High salinity poses toxic effects by generating osmotic imbalance due to high Na^+ concentration in the surrounding. Since maintaining osmotic balance in the cell is essential life parameter, plant maintains this balance by production of osmoprotectants or compatible solutes or osmolytes. Microbes augment cell osmotic balance by stimulating production of compatible solutes, namely, polyamines, proline, betaine, sugar and amino acid derivatives, quaternary ammonium compounds, etc. (Jha et al. 2011; Singh et al. 2015; López-Gómez et al. 2019; Nadeem et al. 2020).

2.5 Approaches for Stress Mitigation Using Rhizospheric and Endophytic Bacteria

Microbes contribute to mitigation of salt stress in plants following the abovementioned mechanisms (Table 2.1), and inoculating such rhizospheric and endophytic microbes could therefore be a choice in sustaining crop productivity in saline areas. Abdelshafy Mohamad et al. (2020) has studied 117 endophytic bacteria from herb *Thymus vulgaris* and further identified them by 16S rRNA gene sequences. On the basis of PGPR traits, three endophytic bacilli were selected which further improved growth of tomato plants under various NaCl concentration (50–200 mM) by modulating antioxidant enzyme activity (superoxide dismutase, catalase, and peroxidase). Selected strains were found to have antagonistic activity against *F. oxysporum* along with reducing harmful effects of salinity making them significant for use as biofertilizer and biocontrol agent.

Zhang et al. (2019) studied *Trichoderma harzianum* in *Cucumis sativus* plants in mitigating salt stress by inducing antioxidant enzymes including peroxidase (POD), polyphenol oxidase (PPO), phenylalanine ammonia lyase (PAL), catalase (CAT),

Table 2.1 Potential rhizospheric and endophytic microbes identified to be used in salt alleviation

Sn.	Microbes	Plant host	Mechanism	References
1	*Bacillus* sp.	Tomato (*Solanum lycopersicum*)	Plant growth-promoting activities in vitro, including auxin synthesis, diazotrophy, phosphate solubilization, siderophore production, and production of lytic enzymes (i.e., chitinase, cellulase, protease, and lipase)	Abdelshafy Mohamad et al. (2020)
2	*Serratia marcescens*	Wheat (*Triticum aestivum*)	Membrane integrity by minimizing oxidative damages	Singh and Jha (2016)
3	*Trichoderma harzianum*	*Cucumis sativus*	Maintaining osmotic balance and metabolic homeostasis	Zhang et al. (2019)
4	*P. pseudoalcaligenes* and *Bacillus pumilus*	Rice (*Oryza sativa* L.)	Osmoprotectants	Jha et al. (2011)
5	*Bacillus amyloliquefaciens*	Rice (*Oryza sativa* L.)	Membrane integrity, accumulation of osmolytes, photosynthetic activity, and gene expression	Chauhan et al. (2019)
6	*Bacillus cereus*	*Vigna radiata*, *Cicer arietinum*, and *Oryza sativa*	Increased activities of enzymes such as superoxide dismutase, peroxidase, ascorbate peroxidase, catalase defense enzymes such as chitinase, β-1,3-glucanase, and phenylalanine ammonia lyase	Chakraborty et al. (2011)
7	*Rhizobium radiobacter*	Maize (*Zea mays*)	High K^+/Na^+ ratio and total chlorophyll and soluble sugars and peroxidase activity increase	Moussa et al. (2012)
8	*Bacillus polymyxa* and *Azospirillum brasilense*	Maize (*Zea mays*)	PGPRs activity	Abo-Kora (2016)
9	*Pseudomonas* sp.	Maize (*Zea mays*)	Higher proline and POD activity	Fazal and Bano (2016)
10	*Azospirillum brasilense*	Wheat (*Triticum durum* var. waha)	Proline and total sugar accumulation reduced	Nabti et al. (2010)
11	*Bacillus*, *Oceanobacillus*, and *Halomonas* genera	Wheat (*Triticum turgidum* subsp. *durum*)	Nitrogen fixation, ACC deaminase activity, auxin production, inorganic phosphate	Albdaiwi et al. (2019)

(continued)

Table 2.1 (continued)

Sn.	Microbes	Plant host	Mechanism	References
			solubilization, and siderophore production	
12	*Pseudomonas extremorientalis* and *P. chlororaphis*	Common bean (*Phaseolus vulgaris*)	IAA production	Egamberdieva (2011)
13	*Bacillus subtilis* and *Pseudomonas fluorescens*	Radish plants (*Raphanus sativus*)	Increase in photosynthetic pigments, proline, total free amino acids, phytohormones contents (IAA and GA3), and the contents of N, P, K^+, Ca^{2+}, and Mg^{2+}	Mohamed and Gomaa (2012)
14	*Azospirillum lipoferum* and *Piriformospora indica*	Sesame (*Sesamum indicum*)	Enhancement of relative water content (up to 20%), maximum photochemical quantum yield of PSII (Fv/Fm) (up to 25%), antioxidant enzyme activity, nutrient absorption, proline (36–65%) and secondary metabolite contents, DPPH radical scavenging activity (18–22%), and linoleic acid	Khademian et al. (2019)
15	*Halobacillus* sp. and *Bacillus halodenitrificans*	Wheat (*Triticum aestivum*)	Multiple plant growth-promoting traits such as indole-3-acetic acid (IAA) production and siderophore production, ACC deaminase activity, and P solubilization	Ramadoss et al. (2013)
16	*Azotobacter chroococcum*	Maize (*Zea mays*)	Nitrogen-fixing plant growth-promoting bacteria	Rojas-Tapias et al. (2012)
17	Arbuscular mycorrhizal fungi (*Glomus etunicatum*) and *Methylobacterium oryzae*	Maize (*Zea mays*)	Phytohormone production and nutrient uptake to improve plant growth	Lee et al. (2015)
18	*Rhizobium* and *Pseudomonas* sp.	Maize (*Zea mays*)	Decreases in electrolyte leakage and in osmotic potential, an increase in osmoregulant (proline) production, maintenance of relative water content of leaves, and selective uptake of K ions	Bano and Fatima (2009)

(continued)

Table 2.1 (continued)

Sn.	Microbes	Plant host	Mechanism	References
19	*Azospirillum brasilense* and *Rhizobium tropici*	Maize (*Zea mays*)	Antioxidant enzymes that detoxify reactive oxygen species (ROS)—ascorbate peroxidase (APX), catalase (CAT), and superoxide dismutase (SOD), mainly in leaves. Proline contents in leaves and roots and malondialdehyde (MDA) in leaves—plant-stress-marker molecules—were significantly reduced	Fukami et al. (2018)
20	*Bacillus aryabhattai* and *B. mesonae*	Tomato (*Solanum lycopersicum*)	Production of higher levels of proline, abscisic acid (ABA), and antioxidant enzyme activities	Yoo et al. (2019)
21	*Bacillus licheniformis*	Tomato (*Solanum lycopersicum*)	High ACC deaminase activity	Chookietwattana and Maneewan (2012)
22	*Sphingomonas* sp.	Tomato (*Solanum lycopersicum*)	Gibberellic acid (GA4), catalase (CAT), superoxide dismutase, and reduced glutathione were significantly regulated	Halo et al. (2015)
23	*Bacillus amyloliquefaciens*	Rice (*Oryza sativa*)	Production of abscisic acid	Shahzad et al. (2017)
24	*Curtobacterium* sp.	Soybean plants	Indole-3-acetic acid (IAA), abscisic acid (ABA), siderophore, and 1-aminocyclopropane-1-carboxylic acid (ACC) deaminase production	Khan et al. (2019)
25	*Bacillus fortis*	*Capsicum annum*	Microbe-inoculated plants exhibited reduced level of ethylene, lipid peroxidation, and reactive oxygen species (ROS)	Yasin et al. (2018)
26	*Bacillus subtilis*, *Bacillus atrophaeus*, *Bacillus sphaericus*,	Strawberry plants (*Fragaria ananassa*)	Lowered electrolyte leakage of plants under saline conditions	Karlidag et al. (2013)

(continued)

Table 2.1 (continued)

Sn.	Microbes	Plant host	Mechanism	References
	Staphylococcus kloosii, and *Kocuria erythromyxa*			
27	*Klebsiella* sp.	Wheat (*Triticum aestivum* var. C309)	Increase in proline, total soluble sugar, and total protein content of treated plants	Singh and Jha (2017)
28	*Herbaspirillum* sp.	*Brassica rapa* L. ssp. *pekinensis* (Chinese cabbage)	Increased K^+/Na^+ ratio in roots generating balance in the ratio of ion homeostasis	Lee et al. (2016)
29	*Bacillus pumilus* and *Exiguobacterium* sp.	Tomato (*Solanum lycopersicum*)	Increased (threefold) lipid peroxidation, while glutathione, catalase, and peroxidase activities were significantly reduced	Ali et al. (2017)
30	*Sphingomonas* sp.	*Solanum pimpinellifolium*	Relatively high levels of salicylic acid (SA) and low levels of JA	Khan et al. (2017)
31	*Kocuria rhizophila* and *Cronobacter sakazakii*	Wheat (*Triticum aestivum*)	Antioxidant enzymes and the production of 1-aminocyclopropane-1-carboxylic acid (ACC) deaminase	Afridi et al. (2019)

superoxide dismutase (SOD), ascorbate peroxidase, and glutathione reductase (GR). It has also shown increased levels of proline, soluble sugars, soluble protein, ascorbic acid, and chlorophyll as well as improved root activity. *T. harzianum* improved the ratio of glutathione (GSH)/oxidized glutathione (GSSG) and AsA/dehydroascorbate (DHA) and upregulated the expression of gene which is involved in the AsA-GSH cycle and also increased the K^+ content and ethylene level.

Khademian et al. (2019) studied the co-inoculation effects of *Azospirillum lipoferum* and *Piriformospora indica* in sesame (*Sesamum indicum* L.) plants. The inoculation improved tolerance to salinity stress by increasing relative water content, photochemical quantum yield of PSII (Fv/Fm), antioxidant enzyme activity, nutrient absorption, proline content, secondary metabolite content, DPPH radical scavenging activity, and linoleic acid, while decreasing malondialdehyde, electrolyte leakage, Na^+, and oleic/linoleic acid ratio. In this study, co-inoculation was found to be effective in sesame plant.

Fukami et al. (2018) studied the effects of *Azospirillum brasilense* and *Rhizobium* tropici in single and co-inoculation in maize under salinity stress resulting in alleviation of reactive oxygen species (ROS) toxicity by increased antioxidant activity and proline content. There occur upregulation of different antioxidant genes *APX*1, *CAT*1, *SOD*2, and *SOD*4 and downregulation of pathogenesis-related genes *PR*1, *prp*2, and *prp*4 and heat shock protein *hsp*70. The study showed that the single inoculation of *Azospirillum* and co-inoculation of microbes help in salt tolerance in maize plant.

Yoo et al. (2019) found that *Bacillus aryabhattai* and *B. mesonae* act as plant growth-promoting bacteria (PGPB) which help tomato plant against salinity stress. In this study, microbial inoculants were compared for individual as well as co-inoculation effects in plant. Results indicated that there is an increase in carotenoid content by *Bacillus aryabhattai* inoculation as compared to uninoculated plants. In the case of *B. mesonae* inoculation, an increase in proline, abscisic acid, and antioxidant enzyme activities was recorded. Both the bacteria decreased the electrolyte leakage and increased Ca^{2+} content. If the plant is treated with *B. mesonae*, upregulation of 9-cis-epoxycarotenoid dioxygenase 1 (*NCED*1) and abscisic acid-responsive element-binding protein 1 (*AREB*1) genes was reported, whereas *Bacillus aryabhattai* downregulated *AREB*1 gene in tomato. This study indicated that the salinity alleviation is both ABA-independent and ABA-dependent in tomato plants.

Shahzad et al. (2017) found that plant growth-promoting endophytic bacteria (PGPEB) *Bacillus amyloliquefaciens* produces abscisic acid which helps in salinity tolerance of rice plant. It shows that treated plant produces different concentrations of abscisic acid in comparison to untreated plants. Different essential amino acids (glutamic acid, aspartic acid, phenylalanine, proline, and cysteine) were significantly upregulated, and the stress-sensitive endogenous ABA levels were significantly reduced, whereas the levels of endogenous salicylic acid were significantly higher. This showed that production of phytohormone by endophytic bacteria helps plant in tolerance against salinity stress.

Khan et al. (2019) found that endophytic bacterial strain SAK1, isolated from *Artemisia princeps*, produces different phytohormones, antioxidant enzymes, and ACC deaminase enzyme which help plants in plant growth and tolerance against salinity stress in soybean plants. Further, it was found to decrease the production of jasmonic acid which invades reactive oxygen species.

Yasin et al. (2018) found that the plant growth-promoting halotolerant rhizobacteria *Bacillus fortis* help in mitigating salinity stress in capsicum plants. The multi-trait activity of bacteria helps in increasing the physical parameter of plant under salt stress. It also helped plants in increasing proline content and in upregulation of salt stress-related genes along with reducing ethylene, lipid peroxidation, and reactive oxygen species levels. Results of the above activity show that the microbes helped plants by physiological and biochemical process in alleviating salinity stress.

Singh and Jha (2017) studied the mechanism of halotolerant PGPR *Klebsiella* sp. in mitigating salinity stress in wheat. The strain found to help plants in maintaining membrane integrity to grow plant under salt stress. Treatment with the potent bacteria helped plants to cope up with salinity stress by increasing various antioxidant enzymes, proline, total soluble sugar, and total protein content apart from decreasing the concentration of salinity-induced malondialdehyde (MDA) content.

Lee et al. (2016) reported that rhizosphere bacteria *Herbaspirillum* sp. help in the alleviation of salt stress in *Brassica rapa* ssp. *pekinensis* (Chinese cabbage). It produces plant beneficial factors, such as auxin, siderophore, and 1-aminocyclopropane-1-carboxylic acid deaminase. It also increased K^+/Na^+ ratio in plant which helped plants for increasing its biomass. Colonization of bacteria in plant was also confirmed by green fluorescent protein (*gfp*)-tagging approach.

Ali et al. (2017), among various endophytic bacterial diversities, studied two endophytic bacterial strains *Bacillus pumilus* and *Exiguobacterium* sp. for imparting salt tolerance in tomato plant. These strains helped in increasing biomass, photosynthetic rate, and pigment accumulation compared to control plants. Bacterial endophytes helped tomato plants in mitigating the salinity stress by increased levels of glutathione, catalase, peroxidase, lipid peroxidation, and methionine production.

Khan et al. (2017) found that plant growth-promoting endophytic bacteria (*Sphingomonas* sp. LK11) along with exogenous jasmonic acid (JA) application improved growth performance during salinity stress, mainly due to the expression of glutathione-related genes in *Sphingomonas* sp. genome. The interplay effects of JA and PGPEB were assessed in wild-type *Solanum pimpinellifolium* and non-isogenic mutant (Got-3). Synergism effect shows great improvement in physical parameter of plant and also responded well to salinity stress by significantly regulating glutathione contents in wild-type and Got-3 plants. Combined effects of exogenous jasmonic acid and PGPEB help to overcome the salinity effects to *Solanum pimpinellifolium*.

Afridi et al. (2019) showed that plant growth-promoting endophytes (PGPEs) help wheat plants in tolerating salinity stress. Two varieties of wheat Pasban 90 and Khirman were subjected to two levels of salt stress (80 and 160 mM NaCl) in the presence of two potent endophytic bacteria *Kocuria rhizophila* and *Cronobacter sakazakii* having ACC deaminase activity. It was found to reduce the ethylene production which helped plants in tolerating salinity stress. It also has different growth-promoting and antioxidant activities in the plants while decreasing the Na^+ content in comparison to the untreated plants. Further, from the measured morphological and biochemical parameters, it was found that wheat variety Pasban 90 shows more tolerance toward saline stress in comparison to the other variety and the bacterial strain *Cronobacter sakazakii* performed better, compared to the other strain.

2.6 Conclusion and Future Prospects

Microbes are co-evolved with plants during evolution and have developed complex interactions for the survival of each other. Exploring agriculturally important interaction and utilization of them for improving crop production could be a sustainable approach for improving the quality and quantity of crop plants. Faulty conventional farming practice leads to enhance salinity of the soil, making it less suitable for economic crop production. Since plants have their own mechanism for stress tolerance, microbes improving those aspects could be of greater significance for the farming community. Rhizospheric and endophytic microorganisms make the closest microbial communities that affect plant growth and performance; thus mining microbial gold for different stress-alleviating prospective could raise hopes of sustainable farming.

Acknowledgment The authors gratefully acknowledge the support rendered by Indian Council of Agricultural Research-National Bureau of Agriculturally Important Microorganisms (ICAR-NBAIM), under the project "Consortium of endophytic and rhizospheric bacteria for alleviation of salinity stress in tomato."

References

Abdelshafy Mohamad OA, Ma JB, Liu YH, Zhang D, Hua S, Bhute S, Hedlund BP, Li WJ, Li L (2020) Beneficial endophytic bacterial populations associated with medicinal plant *Thymus vulgaris* alleviate salt stress and confer resistance to *Fusarium oxysporum*. Front Plant Sci 11:47

Abo-Kora HA (2016) Endophytic colonization of maize (*Zea mays*) root plants by PGPRs under salinity stress. Nat Sci 14(7):34–51

Afridi MS, Mahmood T, Salam A, Mukhtar T, Mehmood S, Ali J, Khatoon Z, Bibi M, Javed MT, Sultan T, Chaudhary HJ (2019) Induction of tolerance to salinity in wheat genotypes by plant growth promoting endophytes: involvement of ACC deaminase and antioxidant enzymes. Plant Physiol Biochem 139:569–577

Albdaiwi RN, Khaymi-Horani H, Ayad JY, Alananbeh KM, Kholoud M, Al-Sayaydeh R (2019) Isolation and characterization of halotolerant plant growth promoting rhizobacteria from durum wheat (*Triticum turgidum* subsp. *durum*) cultivated in saline areas of the dead sea region. Front Microbiol 10:1639

Ali A, Shahzad R, Khan AL, Halo BA, Al-Yahyai R, Al-Harrasi A, Al-Rawahi A, Lee IJ (2017) Endophytic bacterial diversity of *Avicennia marina* helps to confer resistance against salinity stress in *Solanum lycopersicum*. J Plant Interact 12(1):312–322

Arnold AE, Lewis LC (2005) Ecology and evolution of fungal endophytes, and their roles against insects. Insect-Fungal Associations: Ecology Evolution, Oxford University Press, New York, NY, pp 74–96

Bano A, Fatima M (2009) Salt tolerance in *Zea mays* (L) following inoculation with *Rhizobium* and *Pseudomonas*. Biol Fertil Soils 45(4):405–413

Barat SP, Gupta A, Singh D, Srivastav A (2016) Production of liquid biofertilizer by using *Azotobacter* species and their effect on plant growth. Int J Curr Microbiol App Sci 5(7):654–659

Bastías DA, Alejandra Martínez Ghersa M, Newman JA, Card SD, Mace WJ, Gundel PE (2018) The plant hormone salicylic acid interacts with the mechanism of antiherbivory conferred by fungal endophytes in grasses. Plant Cell Environ 41(2):395–405

Biswas B, Gresshoff PM (2014) The role of symbiotic nitrogen fixation in sustainable production of biofuels. Int J Mol Sci 15(5):7380–7397

Brahmaprakash GP, Sahu PK (2012) Biofertilizers for sustainability. J Indian Inst Sci 92(1):37–62

Brahmaprakash GP, Sahu PK, Lavanya G, Nair SS, Gangaraddi VK, Gupta A (2017) Microbial functions of the rhizosphere. Plant-microbe interactions in agro-ecological perspectives. Springer, Singapore, pp 177–210. https://doi.org/10.1007/978-981-10-5813-4_10

Buyer JS, Kratzke MG, Sikora LJ (1994) Microbial siderophores and rhizosphere ecology. Biochemistry of metal micronutrients in the rhizosphere. Lewis Publishers, Boca Raton, FL, pp 67–80

Chakraborty U, Roy S, Chakraborty AP, Dey P, Chakraborty B (2011) Plant growth promotion and amelioration of salinity stress in crop plants by a salt-tolerant bacterium. Recent Res Sci Technol 3(11):61–70

Chauhan PS, Lata C, Tiwari S, Chauhan AS, Mishra SK, Agrawal L, Chakrabarty D, Nautiyal CS (2019) Transcriptional alterations reveal *Bacillus amyloliquefaciens*-rice cooperation under salt stress. Sci Rep 9(1):1–13

Chawla S, Jain S, Jain V (2013) Salinity induced oxidative stress and antioxidant system in salt-tolerant and salt-sensitive cultivars of rice (*Oryza sativa* L.). J. plant Biochem. Biotech 22 (1):27–34

Chookietwattana K, Maneewan K (2012) Selection of efficient salt-tolerant bacteria containing ACC deaminase for promotion of tomato growth under salinity stress. Soil Environ 31(1):30–36

Crowley DE (2006) Microbial siderophores in the plant rhizosphere. Iron nutrition in plants and rhizospheric microorganisms. Springer, Dordrecht. https://doi.org/10.1007/1-4020-4743-6_8

Daami-Remadi M, Souissi A, Oun HB, Mansour M, Nasraoui B (2009) Salinity effects on *Fusarium* wilt severity and tomato growth. Dyn Soil Dyn Plant 3(1):61–69

Egamberdieva D (2011) Survival of *Pseudomonas extremorientalis* TSAU20 and *P. chlororaphis* TSAU13 in the rhizosphere of common bean (*Phaseolus vulgaris*) under saline conditions. Plant Soil Environ 57(3):122–127

Fazal A, Bano A (2016) Role of plant growth-promoting rhizobacteria (PGPR), biochar, and chemical fertilizer under salinity stress. Commun Soil Sci Plant Anal 47(17):1985–1993

Fukami J, de la Osa C, Ollero FJ, Megías M, Hungria M (2018) Co-inoculation of maize with *Azospirillum brasilense* and *Rhizobium* tropici as a strategy to mitigate salinity stress. Funct Plant Biol 45(3):328–339

Fürnkranz M, Lukesch B, Müller H, Huss H, Grube M, Berg G (2012) Microbial diversity inside pumpkins: microhabitat-specific communities display a high antagonistic potential against phytopathogens. Microb Ecol 63(2):418–428

Gupta A, Sahu PK (2017a) Coinoculation of *Azospirillum* with PGPR for plant growth promotion. Int J Res Appl Sci Engg Tech 5(8):383–385

Gupta A, Sahu PK (2017b) *Azospirillum*: bioformulations, product quality and survivability. Int J Res Appl Sci Engg Tech 5(8):467–470

Gupta A, Sahu PK (2017c) Phosphorus nutrition of plants: a microbial perspective. MOJ Ecol Environ Sci 2(6):00043

Halo BA, Khan AL, Waqas M, Al-Harrasi A, Hussain J, Ali L, Adnan M, Lee IJ (2015) Endophytic bacteria (*Sphingomonas* sp. LK11) and gibberellin can improve *Solanum lycopersicum* growth and oxidative stress under salinity. J Plant Interact 10(1):117–125

Jha Y, Subramanian RB, Patel S (2011) Combination of endophytic and rhizospheric plant growth promoting rhizobacteria in *Oryza sativa* shows higher accumulation of osmoprotectant against saline stress. Acta Physiol Plant 33(3):797–802

Karlidag H, Yildirim E, Turan M, Pehluvan M, Donmez F (2013) Plant growth-promoting rhizobacteria mitigate deleterious effects of salt stress on strawberry plants (Fragaria × ananassa). Hortic Sci 48(5):563–567

Kasim WA, Gaafar RM, Abou-Ali RM, Omar MN, Hewait HM (2016) Effect of biofilm forming plant growth promoting rhizobacteria on salinity tolerance in barley. Ann Agric Sci 61 (2):217–227

Kearl J, McNary C, Lowman JS, Mei C, Aanderud ZT, Smith ST, West J, Colton E, Hamson M, Nielsen BL (2019) Salt-tolerant halophyte rhizosphere bacteria stimulate growth of alfalfa in salty soil. Front Microbiol 10:1849

Khademian R, Asghari B, Sedaghati B, Yaghoubian Y (2019) Plant beneficial rhizospheric microorganisms (PBRMs) mitigate deleterious effects of salinity in sesame (*Sesamum indicum* L.): physio-biochemical properties, fatty acids composition and secondary metabolites content. Ind Crop Prod 136:129–139

Khan AL, Waqas M, Asaf S, Kamran M, Shahzad R, Bilal S, Khan MA, Kang SM, Kim YH, Yun BW, Al-Rawahi A (2017) Plant growth-promoting endophyte *Sphingomonas* sp. LK11 alleviates salinity stress in *Solanum pimpinellifolium*. Environ Exp Bot 133:58–69

Khan MA, Asaf S, Khan AL, Ullah I, Ali S, Kang SM, Lee IJ (2019) Alleviation of salt stress response in soybean plants with the endophytic bacterial isolate *Curtobacterium* sp. SAK1. Ann Microbiol 69(8):797–808

Krishnaveni MS (2010) Studies on phosphate solubilizing bacteria (PSB) in rhizosphere and non-rhizosphere soils in different varieties of foxtail millet (*Setaria italica*). Int J Agric Food Sci Technol 1(1):23–39

Lee Y, Krishnamoorthy R, Selvakumar G, Kim K, Sa T (2015) Alleviation of salt stress in maize plant by co-inoculation of arbuscular mycorrhizal fungi and *Methylobacterium oryzae* CBMB20. J Korean Soc Appl Biol Chem 58(4):533–540

Lee GW, Lee KJ, Chae JC (2016) *Herbaspirillum* sp. strain GW103 alleviates salt stress in *Brassica rapa* L. ssp. *pekinensis*. Protoplasma 253(3):655–661

López-Gómez M, Hidalgo-Castellanos J, Marín-Peña AJ, Herrera-Cervera JA (2019) Relationship between polyamines and osmoprotectants in the response to salinity of the legume–rhizobia symbiosis. In: Osmoprotectant-mediated abiotic stress tolerance in plants. Springer, Cham, pp 269–285

Mahajan S, Tuteja N (2005) Cold, salinity and drought stresses: an overview. Arch Biochem Biophys 444(2):139–158

Maheswari NU, Kalaiyarasi M (2015) Comparative study of liquid biofertilizer and carrier based biofertilizer on green leafy vegetables. Int J Pharm Sci Rev Res 33(1):229–232

Malviya D, Sahu PK, Singh UB, Paul S, Gupta A, Gupta AR, Singh S, Kumar M, Paul D, Rai JP, Singh HV, Brahmaprakash GP (2020) Lesson from ecotoxicity: revisiting the microbial lipopeptides for the management of emerging diseases for crop protection. Int J Environ Res Public Health 17(4):1434

Meena KK, Sorty AM, Bitla UM, Choudhary K, Gupta P, Pareek A, Singh DP, Ratnaprabha, Sahu PK, Gupta VK, Singh HB, Krishanani KK, Minhas PS (2017) Abiotic stress responses and microbe-mediated mitigation in plants: the omics strategies. Front Plant Sci 8(172):1–25

Mohamed HI, Gomaa EZ (2012) Effect of plant growth promoting *Bacillus subtilis* and *Pseudomonas fluorescens* on growth and pigment composition of radish plants (*Raphanus sativus*) under NaCl stress. Photosynthetica 50(2):263–272

Moussa LA, Mohy EA, El Banna Ib M (2012) Zea mays cultivar behavior as affected by Rhizobium radiobacter inoculation in salt-stressed environments. Am J Sci 8(7):743–750

Nabti E, Sahnoune M, Ghoul M, Fischer D, Hofmann A, Rothballer M, Schmid M, Hartmann A (2010) Restoration of growth of durum wheat (*Triticum durum* var. waha) under saline conditions due to inoculation with the rhizosphere bacterium *Azospirillum brasilense* NH and extracts of the marine alga *Ulva lactuca*. J Plant Growth Regul 29(1):6–22

Nadeem M, Ali M, Kubra G, Fareed A, Hasan H, Khursheed A, Gul A, Amir R, Fatima N, Khan SU (2020) Role of osmoprotectants in salinity tolerance in wheat. In: Climate change and food security with emphasis on wheat. Academic Press, New York, NY, pp 93–106

Nair SS, Sahu PK, Brahmaprakash GP (2017) Microbial inoculants for agriculture under changing climate. Mysore J Agric Sci 51(1):27–44

Nautiyal CS, Srivastava S, Chauhan PS, Seem K, Mishra A, Sopory SK (2013) Plant growth-promoting bacteria *Bacillus amyloliquefaciens* NBRISN13 modulates gene expression profile of leaf and rhizosphere community in rice during salt stress. Plant Physiol Biochem 66:1–9

Prabha R, Singh DP, Verma MK, Sahu P, Kumar P (2018) Bacterial diversity in rhizosphere of *Paspalum scrobiculatum* L.(kodo millet) is revealed with shotgun metagenome sequencing and data analysis. Data Brief 20:1653–1657

Puri SC, Verma V, Amna T, Qazi GN, Spiteller M (2005) An endophytic fungus from *Nothapodytes foetida* that produces Camptothecin. J Nat Prod 68(12):1717–1719

Qin Y, Druzhinina IS, Pan X, Yuan Z (2016) Microbially mediated plant salt tolerance and microbiome-based solutions for saline agriculture. Biotech Adv 34(7):1245–1259

Ramadoss D, Lakkineni VK, Bose P, Ali S, Annapurna K (2013) Mitigation of salt stress in wheat seedlings by halotolerant bacteria isolated from saline habitats. Springer Plus 2(1):6

Rashid M, Khalil S, Ayub N, Alam S, Latif F (2004) Organic acids production and phosphate solubilization by phosphate solubilizing microorganisms (PSM) under in vitro conditions. Pak J Biol Sci 7(2):187–196

Reinhold-Hurek B, Krause A, Leyser B, Miché L, Hurek T (2007) The rice apoplast as a habitat for endophytic N 2-fixing bacteria. In: The apoplast of higher plants: compartment of storage, transport and reactions. Springer, Dordrecht, pp 427–443

Rojas-Tapias D, Moreno-Galván A, Pardo-Díaz S, Obando M, Rivera D, Bonilla R (2012) Effect of inoculation with plant growth-promoting bacteria (PGPB) on amelioration of saline stress in maize (*Zea mays*). Appl Soil Ecol 61:264–272

Rosenblueth M, Martínez-Romero E (2006) Bacterial endophytes and their interactions with hosts. Mol Plant-Microbe Interact 19(8):827–837

Rossmann M, Pérez-Jaramillo JE, Kavamura VN, Chiaramonte JB, Dumack K, Fiore-Donno AM, Mendes LW, Ferreira MM, Bonkowski M, Raaijmakers JM, Mauchline TH (2020) Multitrophic interactions in the rhizosphere microbiome of wheat: from bacteria and fungi to protists. FEMS Microbiol Ecol 96(4):032

Sahu PK, Sharma L, Gupta L, Renu (2016a) Rhizospheric and endophytic beneficial microorganisms: treasure for biological control of plant pathogens. In: Santra S, Mallick A (eds) Recent biotechnological applications in India. ENVIS centre on Environ Biotech, Univ of Kalyani, Kalyani, pp 50–63

Sahu PK, Lavanya G, Gupta A, Brahmaprakash GP (2016b) Fluid bed dried microbial consortium for enhanced plant growth: a step towards next generation bioformulation. Vegetos 29(4):6–10

Sahu PK, Gupta A, Sharma L, Bakade R (2017) Mechanisms of *Azospirillum* in plant growth promotion. Sch J Agric Vet Sci 4(9):338–343

Sahu PK, Singh DP, Prabha R, Meena KK, Abhilash PC (2019) Connecting microbial capabilities with the soil and plant health: options for agricultural sustainability. Ecol Indic 105:601–612

Shahzad R, Khan AL, Bilal S, Waqas M, Kang SM, Lee IJ (2017) Inoculation of abscisic acid-producing endophytic bacteria enhances salinity stress tolerance in *Oryza sativa*. Environ Exp Bot 136:68–77

Singh RP, Jha PN (2016) The multifarious PGPR *Serratia marcescens* CDP-13 augments induced systemic resistance and enhanced salinity tolerance of wheat (*Triticum aestivum* L.). PLoS One 11(6):e0155026

Singh RP, Jha PN (2017) Analysis of fatty acid composition of PGPR *Klebsiella* sp. SBP-8 and its role in ameliorating salt stress in wheat. Symbiosis 73(3):213–222

Singh M, Kumar J, Singh S, Singh VP, Prasad SM (2015) Roles of osmoprotectants in improving salinity and drought tolerance in plants: a review. Rev Environ Sci Biotechnol 14(3):407–426

Singh DP, Singh V, Shukla R, Sahu P, Prabha R, Gupta A, Sarma BK, Gupta VK (2020a) Stage-dependent concomitant microbial fortification improves soil nutrient status, plant growth, antioxidative defense system and gene expression in rice. Microbiol Res 239:126538

Singh S, Singh UB, Trivedi M, Sahu PK, Paul S, Paul D, Saxena AK (2020b) Seed biopriming with salt-tolerant endophytic *Pseudomonas geniculata*-modulated biochemical responses provide ecological fitness in maize (*Zea mays* L.) grown in saline sodic soil. Int J Environ Res Public Health 17(1):253

Sturz AV, Christie BR, Nowak J (2000) Bacterial endophytes: potential role in developing sustainable systems of crop production. CRC Crit Rev Plant Sci 19(1):1–30

Su CL, Zhang FM, Sun K, Zhang W, Dai CC (2019) Fungal Endophyte *Phomopsis liquidambari* improves iron and molybdenum nutrition uptake of peanut in consecutive monoculture soil. J Soil Sci Plant Nutr 19(1):71–80

Van Zelm E, Zhang Y, Testerink C (2020) Salt tolerance mechanisms of plants. Annu Rev Plant Biol 71:403

Vega FE, Posada F, Aime MC, Pava-Ripoll M, Infante F, Rehner SA (2008) Entomopathogenic fungal endophytes. Biol Control 46(1):72–82

Yang J, Kloepper JW, Ryu CM (2009) Rhizosphere bacteria help plants tolerate abiotic stress. Trends Plant Sci 14(1):1–4

Yasin NA, Akram W, Khan WU, Ahmad SR, Ahmad A, Ali A (2018) Halotolerant plant-growth promoting rhizobacteria modulate gene expression and osmolyte production to improve salinity tolerance and growth in *Capsicum annum* L. Environ Sci Pollut Res 25(23):23236–23250

Yoo SJ, Weon HY, Song J, Sang MK (2019) Induced tolerance to salinity stress by halotolerant bacteria *Bacillus aryabhattai* H19–1 and *B. mesonae* H20–5 in tomato plants. J Microbiol Biotechnol 29(7):1124–1136

Zhang F, Wang Y, Liu C, Chen F, Ge H, Tian F, Yang T, Ma K, Zhang Y (2019) *Trichoderma harzianum* mitigates salt stress in cucumber via multiple responses. Ecotoxicol Environ Saf 170:436–445

Chapter 3
Rhizospheric Diversity of Cyanobacteria and Their Significance in Tropical Ecosystem

Samadhan Yuvraj Bagul, Ritu Vishwakarma, Shaloo Verma, Hillol Chakdar, and G. S. Bandeppa

Abstract Cyanobacteria are gram-negative diverse group of unicellular to filamentous photoautotrophs. They are found ubiquitously in nature. Ability of the cyanobacteria for nitrogen fixation and phosphorous solubilization makes them promising biofertilizers, and their plant growth-promoting potential in the rhizosphere makes them a suitable candidate for sustainable agriculture. Cyanobacteria in the rhizosphere and their importance in tropical ecosystems have been highlighted in this chapter.

Keywords Cyanobacteria · Diversity · Biofertilizer · Plant growth promotion · Tropical

3.1 Introduction

Cyanobacteria are gram-negative photosynthetic microorganisms involved in global oxygen supply and primary production of biomass in aquatic ecosystem. Cyanobacteria have also been reported for nitrogen fixation (N_2) and carbon dioxide (CO_2) sequestration, thus contributing toward the carbon and nitrogen economy of different ecological habitats (Singh et al. 2016; Wyatt and Silvey 1969). Cyanobacteria have a wider adaptability and are found in diverse ecological niches. Bagul et al. (2018) reported different types of heterocystous and non-heterocystous cyanobacteria from diverse ecological niches of India, including hot water spring of Odisha, cold regions of Leh and Uttarakhand, marine water from Odisha, and arsenic-contaminated field of Ballia, Uttar Pradesh. Cyanobacteria are also capable of tolerating biotic and abiotic stress such as salt, heavy metals, and drought and cold

S. Y. Bagul (✉) · R. Vishwakarma · S. Verma · H. Chakdar
National Bureau of Agriculturally Important Microorganisms, Kushmaur, Uttar Pradesh, India

G. S. Bandeppa
Indian Institute of Rice Research, Hyderabad, India

S. K. Dubey, S. K. Verma (eds.), *Plant, Soil and Microbes in Tropical Ecosystems*, Rhizosphere Biology, https://doi.org/10.1007/978-981-16-3364-5_3

conditions. Comprising about 150 genera with more than 2000 species, they exhibit remarkable diversity in their morphology, ranging from simple unicellular and colonial to complex filamentous forms with or without branching (Van den Hoek et al. 1995). They flourish in nitrogen-deficient environment. Cyanobacteria can grow in purely inorganic medium using light as energy and CO_2 and N_2 as sole carbon and nitrogen sources, respectively (Wyatt and Silvey 1969). Over several years, cyanobacteria have been utilized as biofertilizer in rice crop for nitrogen fixation along with plant growth-promoting activities. Inoculation of cyanobacteria into the rice field has been practiced in many tropical countries which helps in reducing the cost of expensive chemical fertilizers. Cyanobacterial extracts have been reported to have a significant response to food crops like wheat, maize, rice, tomato, and cucumber (Priya et al. 2015; Bidyarani et al. 2016; Gayathri et al. 2017). Cyanobacteria have the ability to produce extracellular substances and modulate pH, temperature, and redox activity, besides playing a role in the volatilization of ammonia and methane generation; therefore, application of cyanobacteria in the rice field directly or indirectly has been utilized in the management and productivity of rice ecosystem (Prasanna et al. 2002). A group of heterocyst-forming cyanobacteria such as *Anabaena*, *Calothrix*, *Hapalosiphon*, and *Nostoc* have also been reported to enhance soil microbial parameters, seed germination, and yield of rice crop (Obana et al. 2007; Prasanna et al. 2013; Hussain and Hasnain 2012; Mazhar et al. 2013; Karthikeyan et al. 2007).

3.2 Rhizospheric Diversity of Cyanobacteria in Tropical Ecosystem

The rhizosphere is the region of large number of microbial population and diversity. Higher metabolic activity in the rhizosphere region related to the successful production of crops and increase soil fertility. However, abundance of cyanobacteria and its diversity are meagerly explored in the rhizosphere of crop plants. The general belief that cyanobacteria are obligate phototrophs has been perhaps the major reason for the dearth of information on these organisms in this niche (Karthikeyan et al. 2009). Soil enzymes play a vital role which regulate elements' transformation in the soil and increase the fertility of the soil. The process of elemental transformation is the result of microbial activity as microorganisms have a role in nutrient cycling; thus, both soil fertility and microbial activity are generally closely related, leading to differences in yields and changes in various soil parameters (Nain et al. 2010). The rhizospheric diversity has been reported by many research groups from different tropical countries including India, Sri Lanka, Iraq, Saudi Arabia, Singapore, etc. The cyanobacteria that are reported to dominate in the rhizosphere are the population of non-heterocystous cyanobacteria. Adhikary and Baruah (2015) studied comparative diversity and composition of nitrogen-fixing cyanobacteria in three different land use systems of upper Assam, i.e., rice field, reserve forest, and coal field. Nitrogen-fixing cyanobacteria belonging to nine genera were isolated which included six heterocystous forms, viz., *Anabaena*, *Nostoc*, *Scytonema*, *Calothrix*, *Rivularia*, *Westiellopsis*, and three non-heterocystous forms, viz., *Lyngbya*, *Phormidium*, and *Oscillatoria*. The dominance of *Nostoc* and *Anabaena* in the reserve forests and rice fields, whereas both were missing in the coal-contaminated sites. *Oscillatoria* was the dominant genus, and the species belonging to this genus were abundant in coal field areas. Jena and Adhikary (2007) reported 56 taxa from eastern and northeastern states of the country belonging to 21 genera: *Chlorococcum (1)*, *Treubaria (1)*, *Pediastrum* (9), *Hydrodictyon (1)*, *Botryococcus (1)*, *Coenochloris (1)*, *Radiococcus (1)*, *Coenocystis (1)*, *Oocystis (1)*, *Glaucocystis (1)*, *Chlorella (1)*, *Kirchneria* (2), *Kirchneriella (1)*, *Ankistrodesmus (10)*, *Coelastrum* (3), *Actinastrum* (2), *Tetrastrum (1)*, *Crucigenia (1)*, *Crucigeniella (1)*, *Desmodesmus* (6), and *Scenedesmus* (9). All these species were recorded first time from this region, and out of these, 16 species were reported first from India (Singh et al. 2018). Paddy ecosystem harbors nitrogen-fixing cyanobacterial species mainly dominated by *Nostoc*, *Anabaena*, *Tolypothrix*, *Aulosira*, *Cylindrospermum*, *Scytonema*, *Westiellopsis*, and several other genera commonly flourishing in Indian paddy (Nayak et al. 2004; Prasanna and Nayak 2007; Saadatnia and Riahi 2009). Bora et al. (2016) have isolated six strains of two closely related genera—*Nostoc* and *Cylindrospermum*—*Nostoc carneum*, *Nostoc hatei*, *Nostoc muscorum*, *Cylindrospermum muscicola* (strain A), *Cylindrospermum muscicola* (strain B), and *Cylindrospermum indicum* from terraced paddy field and jhum land of biodiversity hotspot zone of Assam, Northeast India. Debnath and Bhadury (2016) have isolated five abundant cyanobacteria from the rice fields of arsenic-affected Bengal Delta Plains (BDP) of South Asia and maintained in vitro. The characterized isolates resembled *Leptolyngbya* sp. (isolate LBK), *Nostoc* sp. (isolates NOC and NOK), and *Westiellopsis* sp. (isolates WEC and WEK) based on polyphasic taxonomy. Haider and Haifaa (2018) have identified 96 species, including four heterocystous species represented by *Anabaena*, *Calothrix*, *Cylindrospermum*, and *Nostoc*. However, the non-heterocystous species were represented by 13 species: *Aphanocapsa*, *Aphanothece*, *Arthrospira*, *Chroococcus*, *Gloeocapsa*, *Lyngbya*, *Merismopedia*, *Microcystis*, *Microcoleus*, *Oscillatoria*, *Phormidium*, *Schizothrix*, and *Spirulina*. Soil samples were collected from six different agricultural sites in Al Diwaniyah City, Iraq. The dominant species of cyanobacteria was *Oscillatoria*, followed by *Phormidium*, *Chroococcus*, *Gloeocapsa*, and *Lyngbya*. Several arid zones like Shantiniketan (West Bengal, India) and the Thar Desert along with Achrol, Jaisalmer, Manwar, and Pokhran (Rajasthan, India) have been studied for cyanobacterial diversity in India. In Shantiniketan, a novel cluster of *Scytonema* and *Tolypothrix* cyanobacteria has been found which possessed abundant scytonemin in a sheath of cells for protection from high solar irradiance (Kumar and Adhikary 2015). In the Thar Desert, the dominance of *Phormidium*, *Oscillatoria*, and *Lyngbya* followed by *Nostoc*, *Scytonema*, and *Calothrix* has been reported (Bhatnagar et al. 2008). Several novel strains of Oscillatoriales have been also reported from the Thar Desert by Dadheech et al. (2012). Silambarasan et al. (2012) in his study have isolated marine cyanobacteria from rhizosphere soil samples of the three mangroves, viz.,

Parangipettai, Ariyankuppam, and Mudasal Odai mangroves southeast coast of India. Jing et al. (2015) studied the diversity of the diazotroph communities in the rhizosphere sediment of five tropical mangrove sites with different levels of pollution along the north and south coastline of Singapore by pyrosequencing of the nifH gene and found that *Scytonema* sp. and *Pseudanabaena* sp., which belong to the Nostocales (heterocyst forming) order of cyanobacteria. Moreover, filamentous non-heterocystous cyanobacteria *Microcoleus* were detected at all five sampling sites. Amarawansa et al. (2018) found 13 different cyanobacteria genera from paddy soil crust in the intermediate and dry zones of Sri Lanka based on their morphological characteristics. Among them, six cyanobacteria genera were unicellular (*Chroococcus*, *Aphanocapsa*, *Aphanothece*, *Synechococcus*, *Johannesbaptistia*, *Microcystis*), and seven genera were filamentous types (*Lyngbya*, *Oscillatoria*, *Leptolyngbya*, *Pseudanabaena*, *Anabaena*, *Spirulina*, *Nostoc*).

3.3 Significance of Cyanobacteria in Tropical Ecosystem

3.3.1 Phytohormone Production

Cyanobacteria have been reported to produce phytohormones such as IAA, IBA, gibberellins, cytokinin, abscisic acid, and jasmonic acid (Manickavelu et al. 2006). Table 3.1 shows different cyanobacterial strains reported for phytohormone production. A non-heterocyst *Chroococcidiopsis* sp. MMG-5 has been studied and showed significant amount of IAA (25 μg/mL) production, and when co-treated with wheat, mung beans, and pea crop, it showed significant increase in shoot and root length (Ahmed et al. 2010a). Ahmed et al. (2010b) reported *Arthrospira platensis* MMG-9 with 194.3 μg/mL IAA production and with enhanced root and shoot parameters. *Anabaena vaginicola* has been reported to produce IBA and IAA, 2146.9 ng/g and 9.93 ng/g fresh weight, respectively. The effects of these strains have been evaluated and are found beneficial on several vegetable and herbaceous crops (Hashtroud et al. 2013). Prasanna et al. (2013) investigated the effect of *Anabaena* sp. (RPAN59/8) amended with compost and found enhanced growth parameters as well as enhanced quality of tomato fruit. Co-inoculation of plant growth-promoting rhizobacteria along with cyanobacteria has also been reported to increase the plant growth and grain yield significantly (Nain et al. 2010). Karthikeyan et al. (2007) investigated the potential of cyanobacteria on wheat along with different dose of chemical fertilizers; interestingly. all the treatments showed enhanced plant growth and yield parameters. A study with *Anabaena* and *Trichoderma viride* biofilm showed 12–25% increase in yield of soybean as well as enhanced microbial activity. Cyanobacterial association with *Gunnera* has shown production of arabinogalactan proteins that might have played important role in plant growth and development (Bergman et al. 1996). Kumar and Kaur (2014) studied the germination behavior of wheat seeds with cyanobacterial filtrate and found that germination, vigor index, and number of seedlings were higher as compared to untreated. The abovementioned studies

Table 3.1 Different cyanobacterial strains exhibiting phytohormone production

Cyanobacteria	Habitat	PGPR trait	Tryptophan concentration (μg/ml)	Production	Beneficiary crop	Reference
Phormidium sp. MI405019	Mangrove	IAA	50	11.71 μg/mg Chl a	Tobacco	Boopathi et al. (2013)
Anabaena Ck1	Rice endophyte	IAA	–	199.95 (ng/mL)	*B. oleracea* var. capitata	Hussain and Hasnain (2012)
Chroococcidiopsis Ck4	Rice endophyte	Cytokinin	–	9.20 (ng/mL)	*B. oleracea* var. *capitata*	Hussain and Hasnain (2012)
Fischerella muscicola NDUPC001	Rice field	IAA	500	286.82 μg/mL	Rice	Mishra et al. (2019)
Nostoc Pc	–	IAA	–	23 (pmol/mg Chl a)		Sergeeva et al. (2002)
Anabaena sp. CW1	–	IAA	–	11.43 μg/mL	–	Prasanna et al. (2010)
Anabaena vaginicola	Paddy field	IBA	–	1.275 (μg g^{-1} DW)		Shariatmadari et al. (2013)
Nostoc calcicola	Paddy field	IBA	–	2.958 (μg g^{-1} DW)		Shariatmadari et al. (2013)
Aulosira fertilissima	–	IAA	100	7.1 μg/mL		Kumar and Kaur (2014)

indicate that cyanobacteria could be a potential component in integrated nutrient management. The world is looking for organic farming; certainly cyanobacteria are one of the important catalysts that could play a vital role.

3.3.2 *Cyanobacteria as a Biofertilizer*

Sustainable agriculture is the present trend in agriculture that has gained an attention by reducing the use of chemical pesticides and inorganic fertilizers and increasing the use of biofertilizers as an alternative to improve crop yield (Nain et al. 2010). Cyanobacteria have the ability to fix atmospheric nitrogen. They are important component of rice ecosystem and known to fix 20–25 kg N/ha/season (Prasanna and Kaushik 2006). Cyanobacteria are categorized into heterocystous and non-heterocystous forms. Plate 3.1 depicts confocal and light microscopic images of heterocystous and non-heterocystous cyanobacteria. Heterocyst is a specialized structure and nitrogen fixation site. Filamentous cells form heterocyst when inorganic nitrogen source is deprived of cultivation medium (Fig. 3.1). These cells lack photosystem II and maintain microaerobic environment which is required for nitrogenase enzyme responsible for nitrogen fixation. PS I provides ATP for nitrogen fixation in this process which is an energy intensive process. Non-heterocyst-forming cyanobacteria fix atmospheric nitrogen by temporal (CO_2 fixation during day time and N_2 fixation at night) and spatial separation. However, a new study reveals the constitutive nitrogenase activity in the presence of light and oxygen by *Cyanothece* sp. ATCC 3051142 (Young et al. 2019). These findings could pave the way for auto mode of nitrogen fixation in plants in the future. Cyanobacteria mimic the photosynthesis of plants; however, its metabolism has been regarded as bacterial. Studies have shown that the artificial inoculation of cyanobacteria to marine mangroves has significant effect on germination and nitrogen fixation. The researchers indicated the use of cyanobacteria in affected area to establish mangroves (Toledo

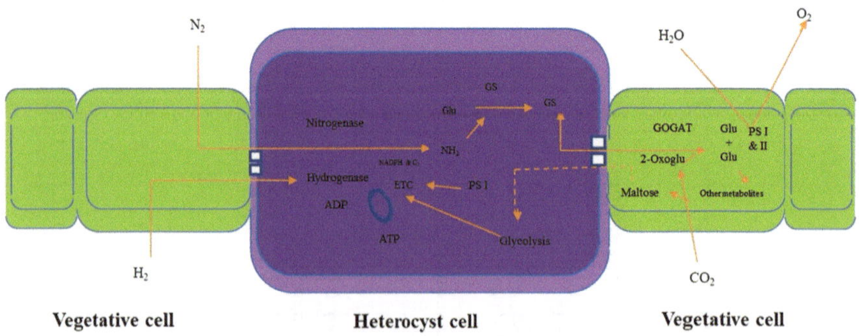

Plate 3.1 Confocal images of cyanobacteria (first row left to right, *Hapalosiphon* sp., *Nostoc* sp., *Calothrix* sp., *Leptolyngbya* sp.). Light microscopic images of cyanobacteria (second row left to right, *Tolypothrix* sp., *Hapalosiphon* sp., *Anabaena* sp., *Leptolyngbya* sp.)

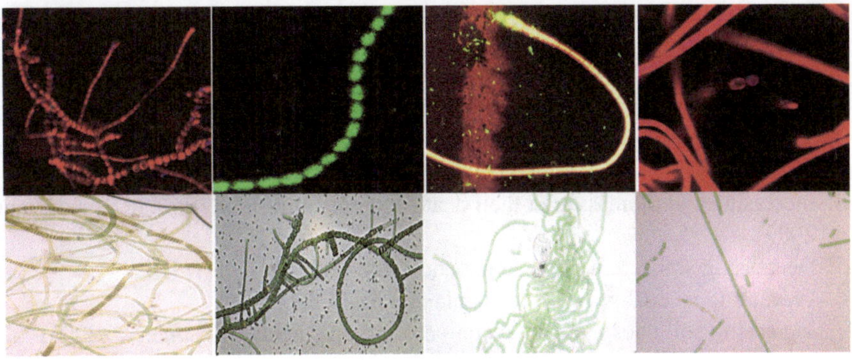

Fig. 3.1 Nitrogen fixation factory (heterocyst cell in cyanobacteria)

et al. 1995). Rice seedling treated with *Anabaena* sp. showed increased root and shoot length as compared to control plant as a result of nitrogen fixation by cyanobacteria (Saadatnia and Riahi 2009).

Phosphorous is another important nutrient required for plant growth after nitrogen. However, solubilization of mineral P is affected by different factors, and plant could take it easily. Reports suggest that cyanobacteria could solubilize inorganic phosphorous such as tricalcium phosphate, $FePO_4$, $AlPO_4$, hydroxyapatite ($Ca_5(PO_4)_3{}^{-}OH$), and rock phosphate. Yandigeri et al. (2011) investigated cyanobacteria for Mussoorie rock phosphate and tricalcium phosphate solubilization and found that *Westiellopsis prolifica* and *Anabaena variabilis* were able to solubilize it. Roychoudhury and Kaushik (1989) also reported P solubilization by cyanobacteria. Chinnusamy et al. (2006) have tested the combination of cyanobacteria, VAM fungi, *Azospirillum*, and PSB and found significant improvement of growth and yield in the plant. Nutritional status and fertility of the soil were also enhanced with these treatments. Stihl et al. (2001) reported alkaline phosphatase activity in *Trichodesmium* sp. Natesan and Shanmugasundaram (1989) studied the *Anabaena* ARM 310 for phosphate solubilization and found that this cyanobacterium was able to solubilize tricalcium phosphate. *Aulosira fertilissima* was able solubilize tricalcium phosphate (4.02 μg/mL) and showed improved seed vigor and growth of the wheat plant (Kumar and Kaur 2014). Cyanobacteria have the dual advantages of nitrogen fixation and P solubilization which make them suitable biofertilizer for agricultural use.

3.3.3 Advantages of BGA Biofertilizers

Cyanobacteria play an important role in the maintenance and build-up of soil fertility, consequently increasing rice growth and yield as a natural biofertilizer (Song et al. 2005). The other roles of cyanobacteria include:

1. Increasing soil pores with having filamentous structure and production of adhesive substances
2. Excretion of growth-promoting substances such as hormones (auxin, gibberellin), vitamins, and amino acids (Roger and Reynaud 1982; Rodríguez et al. 2006)
3. Increasing water-holding capacity through their jelly structure (Roger and Reynaud 1982)
4. Increasing soil biomass after their death and decomposition (Saadatnia and Riahi 2009)
5. Decreasing soil salinity (Saadatnia and Riahi 2009)
6. Preventing weeds' growth (Saadatnia and Riahi 2009)
7. Increasing soil phosphate by excretion of organic acids (Wilson et al. 2006)

3.4 Plant Protection Against Diseases and Pest

Cyanobacteria could produce induced systemic resistance by producing diverse range of biologically active molecules in the rhizosphere which elicit the plant growth under different stresses (Prasanna et al. 2009a, b, 2010). Cyanobacteria could protect plant by providing mechanical and physical strength of the cell wall. Physiochemical reactions are altered by producing defense-related chemicals against the phytopathogens. Major defense enzymes involved in plant growth are chitinase, phenylalanine ammonia lyase (PAL), polyphenol oxidase (PPOs), phenolics, and phytoalexins (Kloepper et al. 1992). Prasanna et al. (2013) showed enhanced production of defense enzymes 189% and 239% of PAL and PPO, respectively, resulting in increased plant growth parameters and bioprotection against *Fusarium* wilt of tomato. Radhakrishnan et al. (2009) studied the effect of cultural filtrate of *Calothrix elenkinii* on fungicidal and algicidal activity which showed promising result. The strain was also evaluated for plant growth promotion, and the dual advantages of plant growth promotion and biocontrol potential make the strain more suitable candidate for agricultural use. Biondi et al. (2004) have reported insecticidal as well as nematicidal activity of *Nostoc* ATCC53789 on *Helicoverpa armigera* and *Caenorhabditis elegans*. Prasanna et al. (2008) showed biocidal activity against phytopathogenic fungi with *Anabaena* strain. Cyanobacterial extracts have been reported to reduce the infection of *Botrytis cinerea* in strawberries and *Erysiphe polygoni* in turnips and tomato seedlings, besides reducing the growth of saprophytic organisms and soilborne fungal pathogens (Kulik 1996; Prasanna et al. 2013). Application of *Calothrix elenkinii* and augmentation with copper nanoparticles (CuNPs) exhibited 76% disease control efficacy in pathogen-challenged plants such as tomato as compared to control. Similarly, augmentation enhanced the chitosanase activity by 10% and 7%, compared to CuNPs and *Calothrix elenkinii* alone. Higher dehydrogenase activity and increased root and shoot length have been also recorded in the rhizosphere soil of diseased plants as compared to healthy plants. Total PLFA content in the soil also increased significantly by 1.4–3.3-fold, compared to the control (Mahawar et al. 2019).

References

Adhikary A, Baruah PP (2015) Comparative diversity and composition of nitrogen-fixing cyanobacteria in three different land use systems of upper Assam. Ethiop J Environ Stud Manag 8:727–737

Ahmed M, Stal LJ, Hasnain S (2010a) Association of non-heterocystous cyanobacteria with crop plants. Plant Soil. https://doi.org/10.1007/s11104-010-0488-x

Ahmed M, Stal LJ, Hasnain S (2010b) Production of indole-3-acetic acid by the cyanobacterium *Arthrospira platensis* strain MMG-9. J Microbiol Biotechnol 20:1259. https://doi.org/10.4014/jmb.1004.04033

Amarawansa RPUI, Balasooriya BLWK, Dandeniya WS, Suganthan B, Dasanyaka T (2018) Identification of cyanobacteria inhabiting paddy fields in intermediate zone and dry zone of Sri Lanka. Trop Agric Res. https://doi.org/10.4038/tar.v29i4.8259

Bagul S, Tripathi S, Chakdar H, Karthikeyan N, Pandiyan K, Singh A, Kumar M (2018) Exploration and characterization of cyanobacteria from different ecological niches of india for phycobilins production. Int J Curr Sci Appl Microbiol 7(12):2822–2834. https://doi.org/10.20546/ijcmas.2018.712.321

Bergman B, Matveyev A, Rasmussen U (1996) Chemical signalling in cyanobacterial-plant symbioses. Trends Plant Sci 1:191. https://doi.org/10.1016/1360-1385(96)10021-2

Bhatnagar A, Makandar MB, Garg MK, Bhatnagar M (2008) Community structure and diversity of cyanobacteria and green algae in the soils of Thar Desert (India). J Arid Environ 72:73. https://doi.org/10.1016/j.jaridenv.2007.05.007

Bidyarani N, Prasanna R, Babu S, Hossain F, Saxena AK (2016) Enhancement of plant growth and yields in Chickpea (*Cicer arietinum* L.) through novel cyanobacterial and biofilmed inoculants. Microbiol Res 188–189:97. https://doi.org/10.1016/j.micres.2016.04.005

Biondi N, Piccardi R, Margheri MC, Rodolfi L, Smith GD, Tredici MR (2004) Evaluation of *Nostoc* strain ATCC 53789 as a potential source of natural pesticides. Appl Environ Microbiol 70 (6):3313–3320

Boopathi T, Balamurugan V, Gopinath S, Sundararaman M (2013) Characterization of IAA production by the mangrove Cyanobacterium *Phormidium* sp. MI405019 and its influence on tobacco seed germination and organogenesis. J Plant Growth Regul 32:758. https://doi.org/10.1007/s00344-013-9342-8

Bora A, Gogoi HK, Veer V (2016) Algal Wealth of Northeast India. In: Bioprospecting of indigenous bioresources of North-East India. Springer, New York, NY, pp 215–228

Chinnusamy M, Kaushik BD, Prasanna R (2006) Growth, nutritional, and yield parameters of wetland rice as influenced by microbial consortia under controlled conditions. J Plant Nutr 29:857. https://doi.org/10.1080/01904160600651803

Dadheech PK, Abed RMM, Mahmoud H, Krishna Mohan M, Krienitz L (2012) Polyphasic characterization of cyanobacteria isolated from desert crusts, and the description of *Desertifilum tharense* gen. et sp. nov. (Oscillatoriales). Phycologia 51:260. https://doi.org/10.2216/09-51.1

Debnath M, Bhadury P (2016) Adaptive responses and arsenic transformation potential of diazotrophic Cyanobacteria isolated from rice fields of arsenic affected Bengal Delta Plain. J Appl Phycol 28:2777. https://doi.org/10.1007/s10811-016-0820-9

Gayathri K, Easwarakumar K, Elias S (2017) Probabilistic ontology based activity recognition in smart homes using Markov Logic Network. Knowl-Based Syst 121:173–184. https://doi.org/10.1016/j.knosys.2017.01.025

Haider AA, Haifaa MJ (2018) Effect of environmental factors on cyanobacteria richness in some agricultural soils. Geomicrobiol J 36:1. https://doi.org/10.1080/01490451.2018.1517196

Hashtroud MS, Ghassempour A, Riahi H, Shariatmadari Z, Khanjir M (2013) Endogenous auxins in plant growth-promoting Cyanobacteria-*Anabaena vaginicola* and *Nostoc calcicola*. J Appl Phycol 25:379. https://doi.org/10.1007/s10811-012-9872-7

Hussain A, Hasnain S (2012) Comparative assessment of the efficacy of bacterial and cyanobacterial phytohormones in plant tissue culture. W J Microbiol Biotechnol 28:1459. https://doi.org/10.1007/s11274-011-0947-4

Jena M, Adhikary S (2007) Chlorococcales (Chlorophyceae) of Eastern and North-eastern States of India. Algae 22(3):167–183. https://doi.org/10.4490/algae.2007.22.3.167

Jing H, Xia X, Liu H, Zhou Z, Wu C, Nagarajan S (2015) Anthropogenic impact on diazotrophic diversity in the mangrove rhizosphere revealed by nifH pyrosequencing. Front Microbiol 6:1172. https://doi.org/10.3389/fmicb.2015.01172

Karthikeyan N, Prasanna R, Nain L, Kaushik BD (2007) Evaluating the potential of plant growth promoting cyanobacteria as inoculants for wheat. Eur J Soil Biol 43:23. https://doi.org/10.1016/j.ejsobi.2006.11.001

Karthikeyan N, Prasanna R, Sood A, Jaiswal P, Nayak S, Kaushik BD (2009) Physiological characterization and electron microscopic investigation of cyanobacteria associated with wheat rhizosphere. Folia Microbiol 54(1):43–51. https://doi.org/10.1007/s12223-009-0007-8

Kloepper JW, Schippers B, Bakker PA (1992) Proposed elimination of the term endorhizosphere. Phytopathology 82:726–727

Kulik MM (1996) Actinomycetes, cyanobacteria and algae. In: Rhizoctonia species: taxonomy, molecular biology, ecology, pathology and disease control. Springer, Dordrecht

Kumar D, Adhikary SP (2015) Diversity, molecular phylogeny, and metabolic activity of cyanobacteria in biological soil crusts from Santiniketan (India). J Appl Phycol 27:339. https://doi.org/10.1007/s10811-014-0328-0

Kumar A, Kaur R (2014) Impact of cyanobacterial filtrate on seed germination behaviour of wheat. Int J Basic Appl Microbiol 1(1):11–15

Mahawar H, Prasanna R, Gogoi R (2019) Elucidating the disease alleviating potential of cyanobacteria, copper nanoparticles and their interactions in *Fusarium solani* challenged tomato plants. Plant Physiol Rep 24:533. https://doi.org/10.1007/s40502-019-00490-8

Manickavelu A, Nadarajan N, Ganesh SK, Ramalingam R, Raguraman S, Gnanamalar RP (2006) Organogenesis induction in rice callus by cyanobacterial extracellular product. Afr J Biotechnol 5:437. https://doi.org/10.5897/AJB05.363

Mazhar S, Cohen J, Hasnain S (2013) Auxin producing non-heterocystous Cyanobacteria and their impact on the growth and endogenous auxin homeostasis of wheat. J Basic Microbiol 53 (12):996–1003. https://doi.org/10.1002/jobm.201100563

Mishra SK, Singh J, Pandey AR, Dwivedi N (2019) Indole-3-acetic acid production by the cyanobacterium Fisherella muscicola NDUPC001. Curr Sci 116

Nain L, Rana A, Joshi M, Jadhav SD, Kumar D, Shivay YS, Paul S, Prasanna R (2010) Evaluation of synergistic effects of bacterial and cyanobacterial strains as biofertilizers for wheat. Plant Soil 331:217. https://doi.org/10.1007/s11104-009-0247-z

Natesan R, Shanmugasundaram S (1989) Extracellular phosphate solubilization by the cyanobacterium *Anabaena* ARM310. J Biosci 14:203. https://doi.org/10.1007/BF02716680

Nayak S, Prasanna R, Pabby A, Dominic TK, Singh PK (2004) Effect of BGA-Azolla biofertilizers on nitrogen fixation and chlorophyll accumulation at different depths in soil cores. Biol Fertil Soils 40:67–72

Obana S, Miyamoto K, Morita S, Ohmori M, Inubushi K (2007) Effect of *Nostoc* sp. on soil characteristics, plant growth and nutrient uptake. J Appl Phycol 19(6):641–646. https://doi.org/10.1007/s10811-007-9193-4

Prasanna R, Kaushik BD (2006) Cyanobacteria in soil health and sustainable agriculture. Health Environ 3:91–105

Prasanna R, Nayak S (2007) Influence of diverse rice soil ecologies on cyanobacterial diversity and abundance. Wetl Ecol Manag 15:127. https://doi.org/10.1007/s11273-006-9018-2

Prasanna R, Kumar V, Kumar S, Yadav AK, Tripathi U, Singh AK, Jain MC, Gupta P, Singh PK, Sethunathan N (2002) Methane production in rice soils is inhibited by cyanobacteria. Microbiol Res 157:1–6. https://doi.org/10.1078/0944-5013-00124

Prasanna R, Nain L, Tripathi R, Gupta V, Chaudhary V, Middha S, Joshi M, Ancha R, Kaushik BD (2008) Evaluation of fungicidal activity of extracellular filtrates of cyanobacteria - possible role of hydrolytic enzymes. J Basic Microbiol 48:186. https://doi.org/10.1002/jobm.200700199

Prasanna R, Jaiswal P, Nayak S, Sood A, Kaushik BD (2009a) Cyanobacterial diversity in the rhizosphere of rice and its ecological significance. Industrial J Microbiol 49:89–97

Prasanna R, Nain L, Ancha R, Shrikrishna J, Joshi M, Kaushik BD (2009b) Rhizosphere dynamics of inoculated cyanobacteria and their growth-promoting role in rice crop. Egypt J Biol 11:26–36

Prasanna R, Gupta V, Natarajan C, Chaudhary V (2010) Allele mining for chitosanases and microcystin-like compounds in *Anabaena* strains. W J Microbiol Biotechnol 26:717–724

Prasanna P, Chaudhary V, Gupta V, Babu S, Kumar A, Singh R, Singh V, Shivay S, Lata N (2013) Cyanobacteria mediated plant growth promotion and bioprotection against *Fusarium* wilt in tomato. Eur J Plant Pathol 136:337–353

Priya H, Prasanna R, Ramakrishnan B, Bindya Rani N, Babu S, Thapa S, Renuka S (2015) Influence of cyanobacterial inoculation on the culturable microbiome and growth of rice. Microbiol Res 171:78. https://doi.org/10.1016/j.micres.2014.12.011

Radhakrishnan B, Prasanna R, Jaiswal P, Nayak S, Dureja P (2009) Modulation of biocidal activity of Calothrix sp. and Anabaena sp. by environmental factors. Biologia (Bratisl). https://doi.org/10.2478/s11756 009 0169 5

Rodríguez AA, Stella AM, Storni MM, Zulpa G, Zaccaro MC (2006) Effects of cyanobacterial extracellular products and gibberellic acid on salinity tolerance in *Oryza sativa* L. Saline Syst 2 (1):7

Roger PA, Reynaud PA (1982) Free — living blue — green algae in tropical soils. In: Dommergues YR, Diem HG (eds) Microbiology of tropical soils and plant productivity. Developments in plant and soil sciences, vol 5. Springer, Dordrecht, pp 147–168. https://doi.org/10.1007/978-94-009-7529-3_5

Roychoudhury P, Kaushik BD (1989) Solubilization of Mussoorie rock phosphate by cyanobacteria. Curr Sci 58:569–557

Saadatnia H, Riahi H (2009) Cyanobacteria from paddy fields in Iran as a biofertilizer in rice plants. Plant Soil Environ 55:207. https://doi.org/10.17221/384-pse

Sergeeva E, Liaimer A, Bergman B (2002) Evidence for production of the phytohormone indole-3-acetic acid by cyanobacteria. Planta 215(2):229–238

Shariatmadari Z, Riahi H, Seyed Hashtroudi M, Ghassempour AR, Aghashariatmadary Z (2013) Plant growth promoting cyanobacteria and their distribution in terrestrial habitats of Iran. Soil Sci Plant Nutr 59:535. https://doi.org/10.1080/00380768.2013.782253

Silambarasan G, Ramanathan T, Kathiresan K (2012) Diversity of marine cyanobacteria from three mangrove environment in Tamil Nadu Coast, South East Coast of India. Curr Res J Biol Sci 4 (3):235–238

Singh JS, Kumar A, Rai AN, Singh DP (2016) Cyanobacteria: a precious bio-resource in agriculture, ecosystem, and environmental sustainability. Front Microbiol 7:529. https://doi.org/10.3389/fmicb.2016.00529

Singh AK, Singh PP, Tripathi V, Verma H, Singh SK, Srivastava AK, Kumar A (2018) Distribution of cyanobacteria and their interactions with pesticides in paddy field: a comprehensive review. J Environ Manag 224:361. https://doi.org/10.1016/j.jenvman.2018.07.039

Song T, Martensson L, Eriksson T, Zheng W, Rasmussen U (2005) Biodiversity and seasonal variation of the cyanobacterial assemblage in a rice paddy field in Fujian, China. FEMS Microbiol Ecol 54(1):131–140

Stihl A, Sommer U, Post AF (2001) Alkaline phosphatase activities among populations of the colony-forming diazotrophic cyanobacterium *Trichodesmium* spp. (Cyanobacteria) in the red sea. J Phycol 37:310. https://doi.org/10.1046/j.1529-8817.2001.037002310.x

Toledo G, Bashan Y, Soeldner A (1995) In vitro colonization and increase in nitrogen fixation of seedling roots of black mangrove inoculated by a filamentous cyanobacteria. Can J Microbiol 41:1012. https://doi.org/10.1139/m95-140

Van Den Hoek C, Mann DG, Jahns HM (1995) Algae: an introduction to phycology. Cambridge University Press, Cambridge, p 623

Wilson AE, Sarnelle O, Tillmanns AR (2006) Effects of cyanobacterial toxicity and morphology on the population growth of freshwater zooplankton: meta-analyses of laboratory experiments. Limnol Oceanogr 51(4):1915–1924

Wyatt JT, Silvey JK (1969) Nitrogen fixation by *Gloeocapsa*. Science 165:908–909

Yandigeri M, Yadav A, Srinivasan R, Kashyap S, Pabbi S (2011) Studies on mineral phosphate solubilization by cyanobacteria *Westiellopsis* and *Anabaena*. Microbiology 80(4):558–565. https://doi.org/10.1134/s0026261711040229

Young J, Gu L, Hildreth M, Zhou R (2019) Unicellular cyanobacteria exhibit light-driven, oxygen-tolerant, constitutive nitrogenase activity under continuous illumination. bioRxiv 619353

Chapter 4
Cyanobacteria in Rhizosphere: Dynamics, Diversity, and Symbiosis

Rupanshee Srivastava, Tripti Kanda, Shivam Yadav, Rajeev Mishra, and Neelam Atri

Abstract Cyanobacteria also known as blue-green algae are primitive prokaryotic organisms capable of thriving in extremes of the environment, capable of oxygenic photosynthesis, and possessing the ability to fix atmospheric nitrogen into a biologically usable form. They have been area of research interest for a long time due to their potential of being exploited in multifarous sphere. Since long they have been recognised as excellent substitute for chemical fertilizer due to their ability to fix nitrogen, therefore, widely acknowledged as biofertilizers. They have been also found to be useful in bioremediation processes like other microbes and thus have been useful in the reclamation of usar land and in remediating various other pollutants either by bioaccumulating or degrading them. Apart from this, they have been useful in pest management as a biocontrol agent. Cyanobacteria alongwith methanotrophs are helpful in reducing the level of emission of greenhouse gases. Either by being free living or in association with other organisms in symbiosis, they are playing key role in the amelioration of various environmental concerns. This chapter is written to describe the diversity of cyanobacteria in the rhizosphere and utilization of cyanobacteria potential in various fields, thus making it an efficient, cost-effective, eco-friendly, and sustainable alternative for a better environment in the future.

Keywords Cyanobacteria · Rhizospheric diversity · Symbiotic system · Biofertilizer · Biocontrol

---

R. Srivastava · T. Kanda · N. Atri (✉)
Department of Botany, MMV, Banaras Hindu University, Varanasi, India

S. Yadav
Department of Botany, Banaras Hindu University, Varanasi, India

R. Mishra
Department of Bioinformatics, MMV, Banaras Hindu University, Varanasi, India

S. K. Dubey, S. K. Verma (eds.), *Plant, Soil and Microbes in Tropical Ecosystems*, Rhizosphere Biology, https://doi.org/10.1007/978-981-16-3364-5_4

4.1 Introduction

Cyanobacteria are a widely distributed group of prokaryotes which are capable of performing oxygenic photosynthesis. Many cyanobacteria have the ability to fix atmospheric nitrogen, and they also aid in converting insoluble phosphorus into a soluble form (Irisarri et al. 2001; Kaushik 1998; Roger et al. 1993; Singh 1961). Recently, it has been found that they can be potential sources of pigments, bioactive metabolites, therapeutic drugs, and nutritional supplements. However, the important research outputs of these organisms are their significant role in improving soil fertility and plant productivity. Cyanobacteria are the natural inhabitants of the agroecosystem and are dominant phototrophic organisms in water (fresh and marine water, hot springs) and the soil. These organisms are cosmopolitan as they are capable of surviving on a minimum requirement of carbon dioxide, water, and light (Castenholz 2001; Woese 1987). They are found to be present in a wide variety of soils like normal, saline, alkaline, acidic soils, etc. These organisms possess various characters contributing to the productivity of the agricultural crops, the fertility of the soils, and maintaining the balance of the ecosystems. They are a vital constituent of paddy fields, harboring morphologically dissimilar wide-ranging flora and representing specific diversity influenced by crop stage and season (Singh 1961).

Since long, the fertility of the soil and sustainable green energy have been known to be contributed by the microbes (Koller et al. 2012). Recent investigations have demonstrated large-scale cultivation of cyanobacterial biomass for food supplements, biofuels, and biofertilizers for agriculture (Benson et al. 2014; Yamaguchi 1996).

These microorganisms play an important role in sustainable agriculture as they help in achieving food security without creating environmental hazards. Inoculation of fields with beneficial soil microbes including cyanobacteria is a new trend as they not only enhance the agricultural productivity but also alleviate greenhouse gas emissions (Singh 2011; Singh et al. 2011a). Application of various cyanobacteria and microalgae, for example, *Anabaena* sp. PCC 7120, *Chlorella* sp., and *Microcystis aeruginosa*, can aid in the growth, development, and metabolic activity of crops. Few cyanobacteria have the extraordinary ability to form close symbiotic associations with various plant groups. They are also an important component of diverse ecosystems that play a significant role in carbon and nitrogen cycling. The production of ammonia and indolic compounds can be enhanced by using cyanobacteria like *Anabaena* sp. and *Calothrix* sp. (Dhar et al. 2015). Lately, it has also been proposed that cyanobacteria can act as important organisms for the reclamation of degraded land (Singh 2014).

Cyanobacteria have the ability to convert insoluble organic phosphates with the help of phosphatase enzymes into mobilizing usable form (Bose and Nagpal 1971; Dorich et al. 1985). Cynaobacteria are also very important because of their ability to fix the atmospheric nitrogen and simultaneously perform oxygenic photosynthesis, thus allowing them to dwell and supply fixed nitrogen and organic carbon in a variety of habitats and provide nutrients to crop plants (Kannaiyan et al. 1994; Thomas et al. 1991).

Besides being a great source of biofertilizers, they have been also found to produce metabolites that have an impact on the growth of plant and development. Various studies have reported that cyanobacteria produce growth-promoting regulators which resemble auxin, gibberellins, cytokinin, and ABA; few vitamins such as vitamin B, polypeptides, amino acids, and exopolysaccharides are also produced which are antimicrobial in nature and toxin-like substances (Ahmad and Winter 1968a; Grieco and Desrochers 1978; Kulik 1995; Maršálek et al. 1992; Rodgers et al. 1979; Singh and Trehan 1973; Vorontsova et al. 1988). Growth-promoting effects of cyanobacterial inoculation have been reported for various crops such as rice, oat (*Avena sativa* L.), cotton (*Gossypium hirsutum* L.), chili (*Capsicum annuum* L.), tomato (*Solanum lycopersicum* L.), radish (*Raphanus sativus* L.), lettuce (*Lactuca sativa* L.), maize (*Zea mays* L.), bean (*Phaseolus vulgaris* L.), muskmelon (*Cucumis melo* L.), sugarcane (*Saccharum* sp.), soybean (*Glycine max* L. Merr.), and wheat (*Triticum aestivum* L.) (Arif et al. 2008; Karthikeyan et al. 2007; Maqubela et al. 2009; Rodgers et al. 1979; Saadatnia and Riahi 2009; Venkataraman 1972).

Since the issue of the plant development and nutrition is related to the biological component of the soils where microbes play a critical role in improving nutrient availability to the plants and, thus, enhancing the nutrient use efficiency of the plants especially in the nutrient-poor soil, it is therefore thought worthwhile to discuss the diversity and dynamics of microbes and especially cyanobacteria in the rhizosphere (Fig. 4.1).

4.2 Rhizospheric Diversity of Cyanobacteria

Rhizosphere is the hotspot for microbial abundance, dynamics, and diversity as they provide niche for the microbes and their metabolic activities which help in proper growth and development of the plants associated (Roger et al. 1993). Watanabe and Yamamoto (1971) reported the occurrence of cyanobacteria in only 5% of 911 soil samples, and 33% in 2213 samples was reported by Venkataraman, whereas Okuda (1952) reported the presence of cyanobacteria in 71% of Japanese soil samples. Thirty-eight soil samples from 11 districts of Dhaka (Bangladesh) for cyanobacterial flora were reported by Khan et al. (1994). Out of the documented 84 species, 50% were heterocystous diazotrophic cyanobacteria, chiefly *Fischerella*, *Nostoc*, and *Calothrix*, occurring in about 53%, 47%, and 26% of the soil samples, respectively. *Anabaena*, *Calothrix*, *Cylindrospermum*, *Nostoc*, *Gloeotrichia*, and *Scytonema* are reported dominant heterocystous species in many paddy fields (Roger et al. 1993).

Prasanna et al. (2009) carried out an investigation to study the abundance and diversity of cyanobacteria in the paddy fields' rhizosphere from diverse regions of India. Maximum populations of 9.1×10^4 and 1×10^6 were recorded in rice cultivars in nitrogen-deficient and nitrogen-supplemented media, respectively. *Nostoc* and *Anabaena* compromise 80% of the rhizospheric isolates. In addition to

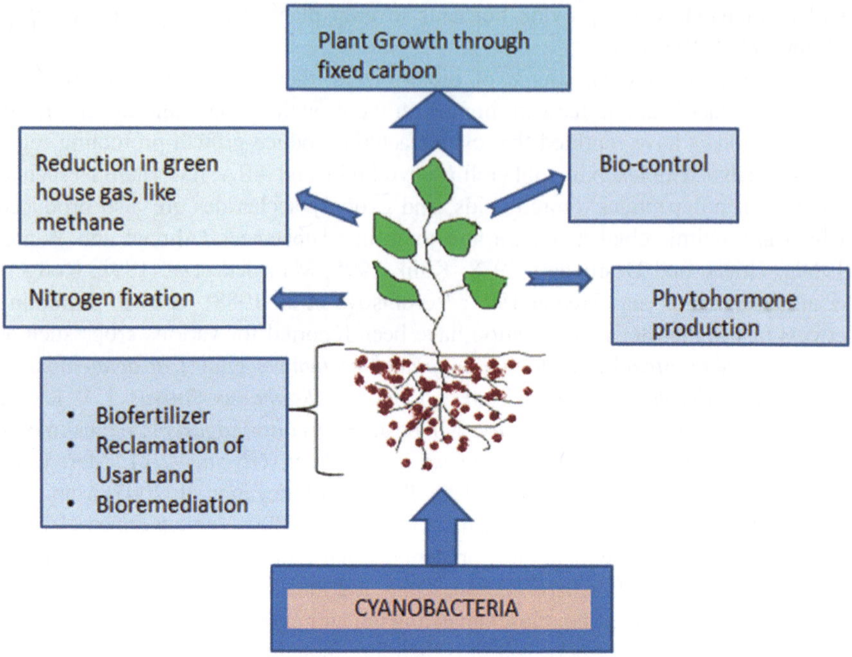

Fig. 4.1 Hypothetical diagram showing the application of cyanobacteria in various forms

Nostoc and *Anabaena*, *Hapalosiphon*, *Westiellopsis*, and *Calothrix* were the other heterocystous cyanobacteria that were isolated from the rhizosphere. Apart from heterocystous cyanobacteria mentioned above, non-heterocystous cyanobacteria, like *Oscillatoria* and *Phormidium* occurred in the rhizosphere; *Scytonema* were observed to predominate in saline soil samples.

Anabaena, *Aulosira*, *Cylindrospermum*, *Nostoc*, *Westiellopsis*, *Tolypothrix*, *Scytonema*, and several other genera are dominant in Indian paddy fields and are significant contributor of soil fertility in India (Swarnalakshmi et al. 2006). Venkataraman (1975) reported that in some of the eastern and southern states, cyanobacteria comprises 50% of the total algal population.

Highest diversity of cyanobacteria consisting of 20 cyanobacterial form across nine genera was isolated from soil samples from Jeypore, Odisha. *Anabaena* and *Nostoc* were dominant, in respect to abundance (Prasanna and Kaushik 2006; Prasanna and Nayak 2007).

Earlier studies regarding the distribution pattern of cyanobacteria in soils of various northern states of India have shown localized distribution of cyanobacteria depending upon various factors like the pH of the soil, electrical conductivity, and exchangeable sodium. *Calothrix*, *Hapalosiphon*, *Nostoc*, *Scytonema*, and *Westiellopsis* were predominant species and appeared to be tolerant to salt fluctuations (Kaushik 1961). *Calothrix* and *Nostoc* were predominant in salt-affected soils of Maharashtra. Mucilaginous cyanobacterial species of *Lyngbya*, *Scytonema*, and

Tolypothrix are also common. Singh (1961) observed in sugarcane that successive cultivation of BGA favors the environment as it may facilitate the production of good yield of crops after few years.

Analysis of the rhizosphere of rice and wheat demonstrated the morphological and functional diversity of the facultative prokaryotes. *Anabaena* and *Nostoc* were found to be predominant, and many strains exhibited potential of nitrogen fixation and production of plant growth promoter (Karthikeyan et al. 2009, 2007; Prasanna et al. 2009).

4.3 Biofertilizer

Non-judicious application of chemical fertilizers after the advent of modern agriculture has led to the pollution and contamination of the soil and water basins and has destroyed beneficial microbes and insects, thus making the crop more susceptible to diseases by reducing soil fertility. Apart from pollution, increasing cost of fuels and fertilizers are becoming burdensome for small farmers, thereby pushing them to adopt agricultural practices which are further detrimental for soil fertility. Further, there is a growing concern about environmental hazards, increasing threat to sustainable agriculture. Hence, the replacement of synthetic fertilizers by biofertilizers is considered as an eco-friendly practice (Rai et al. 2019).

Application of biofertilizer improves the soil quality and decreases our dependency on chemical fertilizer and pesticides. The diazotrophic cyanobacteria which are capable of growing at the expense of light, air, and water have been suggested to be an appropriate system for the source of fixed nitrogen in agricultural and non-agricultural ecosystems (Rai et al. 2019). Biofertilizers are defined as eco-friendly formulation of living or latent microorganisms which have the capability to promote plants' growth by increasing the ease of nutrients absorption by plants.

The majority of cyanobacteria are well-established fixers of molecular nitrogen. Free-living cyanobacteria function as a significant source of biofertilizer in rice fields of tropical countries (Singh 1961; Venkataraman 1972; Watanabe 1956), and their potential in temperate agricultural soils has also been identified (Jenkinson 1977).

4.3.1 Free-Living Cyanobacteria

Cyanobacteria belonging to various morphological domains are capable of fixing atmospheric nitrogen. They are classified into various filamentous heterocystous and non-heterocystous forms (Vaishampayan et al. 2001). Non-heterocystous forms have evolved to fix nitrogen under anaerobic conditions, while heterocystous forms fix nitrogen under aerobic conditions. Heterocyst is a thick walled cell which provides protection to nitrogenase, which is a nitrogen-fixing enzyme and is extremely oxygen-labile protein (Bergman et al. 1997; Rai et al. 1992a, b). Heterocysts are the site of nitrogen fixation. Several groups have performed comprehensive studies for the deeper understanding of nitrogen fixation functioning in bacteria and cyanobacteria (Brill 1983; Kim and Rees 1992; Kirn and Rees 1992; Mulligan and Haselkorn 1989).

Nitrogen fixed by cyanobacteria is assimilated by glutamine synthetase-glutamate synthase (GS-GOGAT) pathway (Stewart and Singh 1975; Wolk et al. 1976). Cyanobacteria help in seed germination and promote seedling growth (Gupta and Lata 1964). *Senna notabilis* and *Acacia hilliana* seeds were bio-primed with cyanobacteria genera *Microcoleus* and *Nostoc*, and seed germination and seedling growth were measured which demonstrated a positive effect (Muñoz-Rojas et al. 2018).

Cyanobacteria contribution toward total nitrogen content of paddy fields depends on their nitrogen fixation activity which is controlled mainly by various physiochemical and biotic factors. Estimates of the addition of fixed N by cyanobacteria in rice fields are reported to be 18–45 kg N/ha (Watanabe and Cholitkul 1978), 90 kg N/ha (Metting 1981), and 20–30 kg N/ha (Issa et al. 2014).

Cyanobacteria and maize hybrid interaction has shown to be a promising combination for improved yield (Prasanna et al. 2016b). *Anabaena-Trichoderma* biofilm is a cyanobacteria-based bioinoculant, which has shown to be the most effective in influencing soil fertility by increasing nutrient quotient of soil, thereby improving plant height and overall crop yield. Moreover, it was also demonstrated that *Anabaena torulosa-Trichoderma viride* biofilm formulations cause enhanced yields and appeared not only a promising plant growth promoter but also a disease-suppressing agent (Prasanna et al. 2016c). Likewise, consortium of *Anabaena-Trichoderma* biofilm along with *Chrysanthemum* also displayed promising positive effects in improving soil fertility (Prasanna et al. 2016a).

*4.3.2 Symbiotic System: Cyanobacteria-*Azolla

Cyanobacteria form symbiotic association with fungi, liverworts, ferns, and some plants, but the most common symbiotic associations for fixation of nitrogen are *Anabaena azollae* and free-floating aquatic fern, the *Azolla*. *Azolla*, belonging to the family Salviniaceae, is very effective biofertilizer and is widely distributed in freshwater habitats of temperate and tropical climates. The biofertilizer property is due to the presence of cyanobacteria inhabiting the leaf cavities present on the dorsal leaf lobe of *Azolla*. Cyanobacteria have enormous biological N fixation ability of up to 30–100 kg N/ha/crop through symbiotic association and thus appeared as a valuable source of fixed nitrogen for paddy crop (Ito and Watanabe 1985; Roy et al. 2016; Singh and Singh 1987).

The factor that is important in using *Azolla* as biofertilizer for rice crop is its quick decomposition in the soil and efficient availability of its nitrogen to rice plants. Besides N fixation, these bio-manures also contribute significant amounts of phosphorous, potassium, sulfur, zinc, iron, molybdenum, and other micronutrients to the plant.

The indiscriminate application of chemical fertilizer has led to environmental concern; therefore, *Azolla* is found to be an eco-friendly substitute for wetland paddy field in China, India, and Vietnam (Singh and Singh 1987). One of the most commonly propagated species of *Azolla* is *Azolla pinnata* (Mazid and Khan 2015) that has been found effective in enhancing nitrogen content of Indian paddy fields.

Therefore, *Azolla* biofertilizer is an eco-friendly and cost-effective option. Apart from nitrogen fixation, it can be used effectively for the following: (1) bioremediation of industrial effluents and sewage wastewater (Sood et al. 2012), (2) as animal feed, (3) human food, and (4) medicine (Wagner 1997).

4.4 Cyanobacterium Role in Bioremediation

4.4.1 Reclamation of Usar Land

Usar soils are saline or alkaline patch of land with high concentration of any kind of salt, electrical conductivity (EC) of soil saturation extract more than 4 dS/cm, and pH between 7.5 and 8.5. Usar soils are less productive, as the presence of excessive salts in the upper layers makes rigid soils impermeable to water. Owing to its character of high alkalinity, osmotic pressure, and impermeability, usar soils are inappropriate for agricultural applications (Rai et al. 2019).

Many Indian states, for example, Rajasthan (12.14 lakh hectares) and Uttar Pradesh (12.95 lakh hectares), have huge areas of unproductive usar lands. Hence, there is a need for the reclamation of such unfertile lands to meet the increasing demand for food for the increasing population. Different methods of reclamation have been adopted by farmers that reduce the salt content in soil, which includes irrigation with clean water, treatment with gypsum and sulfur (Dhar and Mukherji 1936), and cultivation of salt-tolerant crops, but these methods are neither cost-effective nor eco-friendly.

It was proposed (Singh 1961) that cyanobacteria could be useful in reclamation of usar soils as they are able to form a thick stratum on the soil surface and further help in conserving the organic carbon, nitrogen, phosphorous, moisture and they can also convert Na^+ clay to Ca^+ clay. Cyanobacteria help in binding of usar soil by adding organic matter and N, thus improving soil permeability and aeration (Singh 1961). Further it was found that cyanobacteria bind to the soil particles and at depth cause entangling of the soil particles as they form a superficial network of the trichomes on the soil (Nisha et al. 2007).

Kaushik (1994) reported that when cyanobacterial inoculation was used in paddy fields, requirement of gypsum for amelioration of sodic soils was reduced. Successive inoculation of cyanobacteria makes the environment more favorable for the crop cultivation, and it was observed by Singh (1961) in sugarcane that after 3 years of reclamation with cyanobacteria, enhanced yield of crops was produced.

Camptylonema (*C. lahorense*), *Cylindrospermum* (*C. licheniforme*, *C. muscicola*), *Microcoleus* (*M. chthonoplastes*, *M. vaginatus*), *Nostoc* (*N. commune*, *N. muscorum*, *N. punctiforme*), *Porphyrosiphon* (*P. notarisii*), and *Scytonema* (*S. ocellatum*, *S. javanicum*) are components of a comprehensive list of cyanobacteria which are capable of growing in alkaline soil, and when the soil became waterlogged at later stage, forms like *Aulosira fertilissima* and various species of *Anabaena*, *Cylindrospermum gorakhporense*, and *Wollea bharadwajae* appears. *Nostoc commune* dominated these populations, and hence it was assigned more credit in bio-amelioration (Singh 1950).

4.4.2 Cyanobacterial Bioremediation

Cyanobacteria have been found to be useful in removal of many kinds of environmental contaminants like pesticides, catechol, crude oil, phenanthrene, naphthalene, phenol, xenobiotics, heavy metals, etc. employing either accumulation or degradation techniques (Singh et al. 2011b; Megharaj et al. 1987, 1994; Al-Hasan et al. 1998, 2001; Sorkhoh et al. 1992; Cerniglia et al. 1980a, b; Shashirekha et al. 1997; Narro et al. 1992).

Cyanobacteria have been found to be useful in the detoxification of various industrial effluents such as from brewery and distilleries, dye and pharmaceutical industries, oil refinery, paper mill, and sugar mill due to their ability to absorb metal at high multiplication rate. Tertiary treatment of urban, agro-industrial effluents can also be performed by cyanobacteria, leading to alleviation in eutrophication and metal toxicity (Vilchez et al. 1997).

Presently, they are being used as bioremediating agents for treatment of nitrogen- and phosphorous-rich dairy wastewaters, hence converting these nutrients into biomass (Lincoln et al. 1996; Radwan and Al-Hasan 2000; Singh et al. 2011a). Several studies showed that cyanobacteria like *Oscillatoria salina*, *Plectonema terebrans*, *Aphanocapsa* sp., and *Synechococcus* sp. can degrade crude oil and other complex organic compounds like surfactants of oil spills that have been reported from the different parts of the world (Cohen 2002; Mansy and El-Bestawy 2002; Radwan and Al-Hasan 2000). Further, it is also reported that *Synechocystis* sp. successfully mineralized the anilofos herbicide and used the product as phosphate source (Lipok et al. 2007).

It is also reported (Cerniglia et al. 1979, 1980a, b) that *Oscillatoria* sp. and *Agmenellum* sp. oxidize naphthalene to 1-naphthol; *Oscillatoria* sp. oxidizes biphenyl to 4-hydroxybiphenyl, and *Agmenellum* sp. metabolizes phenanthrene into trans-9,10-dihydroxy-9,10-dihydrophenanthrene, and 1-methoxy-phenanthrene.

Thus, it can be said that cyanobacteria in wastewater lagoons have great potential to degrade the pollutants and pesticides and further can help in reducing the pollution load and support the growth of other microbial populations for reductions in the BOD and COD.

4.5 The Role of Cyanobacteria as Plant Growth Promoter

Cyanobacteria produce plant growth-promoting substances like auxin, gibberellins, cytokinin, ABA, few vitamins and amino acids (Ahmad and Winter 1968b; Grieco and Desrochers 1978; Maršálek et al. 1992; Rodgers et al. 1979; Singh and Trehan 1973; Vorontsova et al. 1988), antibiotics, and toxins. Studies relating to abundance and diversity have revealed about the dominance of heterocystous cyanobacteria *Nostoc* and *Anabaena* in many cropping lands in East and North India (Rogers and Burns 1994). It has been found that cyanobacteria showed efficiency in enhancing the germination and growth of rice and wheat seeds and increased the production of indole acetic acid (IAA) and proteins.

Cyanobacteria develop network of their filaments on the soil that enmeshes soil particles at depth (Nisha et al. 2007). In addition to this, they produce extracellular polysaccharides (EPS) that help binding soil particles together leading to improvement of soil quality as they are hygroscopic in nature (Flaibani et al. 1989). The studies portraying cyanobacterial inoculation as plant growth promoter to rice crop showed that it aided in rice seed germination and also helped in growth of roots and shoots (Misra and Kaushik 1989a, b; Nain et al. 2010; Obreht et al. 1993). The positive impact of cyanobacterial inoculation and colonization in roots of wheat plant was reported in terms of plant growth and enhancement of root dry weight and chlorophyll which can be attributed to the production of extracellular substances released by cyanobacteria (Gantar et al. 1995a, b).

Shariatmadari et al. (2013) worked on cucumber, tomato, and squash and have reported significant influence of cyanobacteria on plants in terms of plants physical attributes. Prasanna et al. (2014) have reported preparation of cyanobacterial formulations and biofilmed inoculants for leguminous crops and studied various parameters of growth like root and shoot length and weight, increase in N_2-fixing ability, microbial biomass and carbon of soil samples. Kumar et al. (2013) have reported the potential of two cyanobacteria along with eight thermotolerant bacteria as plant growth-promoting (PGP) agents for spice crops such as coriander, fennel, and cumin. Cyanobacteria also have the ability of solubilization and mobilization of insoluble organic phosphates, hence improving the bioavailability of phosphorus to plants (Bose and Nagpal 1971; Dorich et al. 1985).

Due to their capability of growing in diverse habitat and simple nutritional requirements, there is a scope of cyanobacterial species for the commercial application as PGPs (Ruffing 2011) (Table 4.1).

Table 4.1 Important phytohormones produced by cyanobacteria

Phytohormones	Cyanobacteria	Functions	Reference
Auxin	*Synechocystis, Chroococcidiopsis, Calothrix, Cylindrospermum, Glactothece, Plectonema, Anabaena, Anabaenopsis, Phormidium, Oscillatoria, Nostoc*	Increases growth level, biomass production, stress tolerance, oil content	Hussain et al. (2010), Mazhar et al. (2013), Sergeeva et al. (2002), Singh et al. (2016)
Gibberellins	*Anabaena, Anabaenopsis, Cylindrospermum, Phormidium*	Boost growth rate and biomass production	Gupta and Agarwal (1973), Singh et al. (2016), Tsavkelova et al. (2006a, b)
Cytokinin	*Synechocystis, Chroococcidiopsis, Anabaena, Phormidium, Oscillatoria, Calothrix, Chlorogloeopsis, Cylindrospermum, Rhodospirillum*	Enhances growth rate, oil content, and stress tolerance	Hussain et al. (2010), Singh et al. (2016), Tsavkelova et al. (2006a, b)
Abscisic acid	*Nostoc, Anabaena, Synechococcus, Trichormus*	Known to impart stress tolerance	Hartung (2010), Maršálek et al. (1992), Zahradníčková et al. (1991)
Ethylene	*Nostoc, Anabaena, Calothrix, Cylindrospermum, Scytonema, Synechococcus*	May be involved in programmed cell death, improved growth rate, and biomass production	Tsavkelova et al. (2006a, b)

4.6 Plant Growth Promotion Through Direct Transfer of Fixed Carbon

In bipartite lichen, cyanobiont symbiosis plays a role as the provider of fixed nitrogen and carbon. Lichens are composite organisms made up of a cyanobacteria or algae in a symbiotic relation with fungus, generally an ascomycete. The common cyanobacterial genus in lichen symbioses is *Nostoc*; apart from *Nostoc*, several other genera, like *Chroococcidiopsis*, *Gloeocapsa*, *Sertonema*, and *Stigonema*, are also known to associate in forming different lichens. Bipartite lichen where cyanobacterium is a photobiont is known as cyanolichen.

Since mycobiont (fungal partner) is photosynthetically inactive, the entire provision of both fixed carbon and nitrogen is the sole responsibility of the cyanobiont partner. Hence these cyanobionts are photosynthetically active and with the help of C3 pathway they fix CO_2, and around 70–80% of total fixed CO_2 produced by the cyanobiont is released to the mycobiont. The transfer of fixed carbon generally occurs in light and in the form of glucose (Rai and Bergman 2002). A unique feature of cyanobiont of lichen is its role as provider of fixed C and fixed N, which is transferred to its fungal partner for maintenance of symbiotic association.

4.7 Cyanobacterium Role as Biocontrol Agent

Cyanobacteria have been found as a potential source of bioactive compounds of antimicrobial in nature (Teuscher et al. 1992). Studies revealed that they can control the following diseases: (1) the incidence of *Botrytis cinerea* on strawberries, (2) *Erysiphe polygoni* on turnips, and (3) damping-off disease in tomato seedlings (Kulik 1995). Several compounds like fischerellin from *Fischerella muscicola* are antifungal in nature and act against several plant pathogenic fungi, for example, *Erysiphe graminis* (powdery mildew), *Uromyces appendiculatus* (brown rust), *Pyricularia oryzae* (rice blast), and *Phytophthora infestans* (Hagmann and Jüttner 1996; Papke et al. 1997).

Nostoc muscorum has been found to show antifungal property against soil fungi (De Caire et al. 1990). *N. muscorum* extracts prevent the in vitro growth of the fungal plant pathogens like *Rhizoctonia solani* and *S. sclerotiorum* (Kulik 1995). *Nostoc* sp. produces cryptophycin which acts against the fungi, insects, and nematodes (Biondi et al. 2004).

Therefore, it can be concluded that cyanobacteria efficiently serve the purpose of biocontrol agents by inhibiting the growth of plant pathogens. New assays are being developed for the commercial use of the cyanobacterial metabolites for the development of sustainable agriculture. However, studies on the biocontrols show that most experiments have been conducted in laboratory conditions. Therefore, extensive research should be carried out to find out the feasibility of application of cyanobacteria as the convincing biocontrol agents against variety of plant diseases.

4.8 Cyanobacterium Role in Reduction of Methane Gas

Methane (CH_4) has negative impact on atmosphere as it is a potent greenhouse gas with the impact approximately 20 times greater than that of CO_2 (Singh et al. 2011a). Apart from natural and anthropogenic sources, flooded paddy fields constitute one of the major contributors to atmospheric methane. Methanogenesis in anaerobic flooded paddy soils results in production of methane gas. The increasing population will not only lead to generation of huge amount of waste but also increased use of fossil fuels which will ultimately cause increase in concentration of methane. Therefore, a viable and eco-friendly tool is required to mitigate the ever-increasing concentration of methane gas.

The global warming problem caused by the greenhouse gas generated from various activities could be overcome by the use of cyanobacteria (Cuellar-Bermudez et al. 2015). Further, bio-agents like methanotrophs (Tiwari et al. 2015) in association with cyanobacteria can play a very important role in removing significant amount of the greenhouse gas like methane (Singh 2013; Singh and Pandey 2013; Singh and Singh 2013). There is lack of information on interaction between cyanobacteria and methanotrophs regarding methane flux regulation in rice fields (Kaushik and Venkataraman 1983). Therefore, assumption has been made that enhancement of the concentration of oxygen by cyanobacteria in the rhizosphere of paddy field may also consequently enhance the methane uptake activity of methanotrophs. The oxygen released during photosynthesis by cyanobacteria into the flooded soils, can create anaerobic environment, which is antagonistic for methane genesis; however it can boost up the population and activity of methanotrophs leading to augmentation of CH_4 oxidation (Prasanna et al. 2002; Singh and Pandey 2013).

Hence, it can be concluded that increasing diversity of cyanobacteria and methanotrophs in paddy field can act as potent strategy to curb the methane emission from agricultural field and can also improve crop productivity (Singh 2014; Singh and Singh 2012).

4.9 Conclusion

Indiscriminate use of chemical fertilizers and various anthropogenic activities have led to various environmental concerns like depletion of soil fertility, loss of fertile lands, pathogens being resistant to various synthetic drugs, and emission of various greenhouse gases leading to an increase in the rise of average global temperature. Further, the increasing population needs the urgent demand for food production in order to feed them. To counteract these and various other problems, cost-effective, environmentally friendly, and sustainable approaches need to be discovered and adopted.

Cyanobacteria have recently emerged as a great source to mitigate various environmental problems. Due to its ability to thrive in the extremes of the environment and the ability to grow in a simple nutrient medium, they form an inexpensive farm-grown input active against various environmental problems. Among biofertilizers, cyanobacteria constitute one of the most important inputs in rice cultivation, which are now also gaining importance in other crops like wheat. There is a definite need to utilize these biofertilizers along with organic composts as an eco-friendly approach. Further, cyanobacteria can be very effective for enhancing soil organic elements and making phosphorus available to the plants. They act as an excellent biocontrol agent controlling various pathogens of the economically important plants. They also help in bioremediation by accumulating or degrading various chemicals, pesticides, heavy metals, and oil-containing compounds. They have also been found great in the reclamation of usar land by converting it into fertile land for crop production. Cyanobacteria have also been

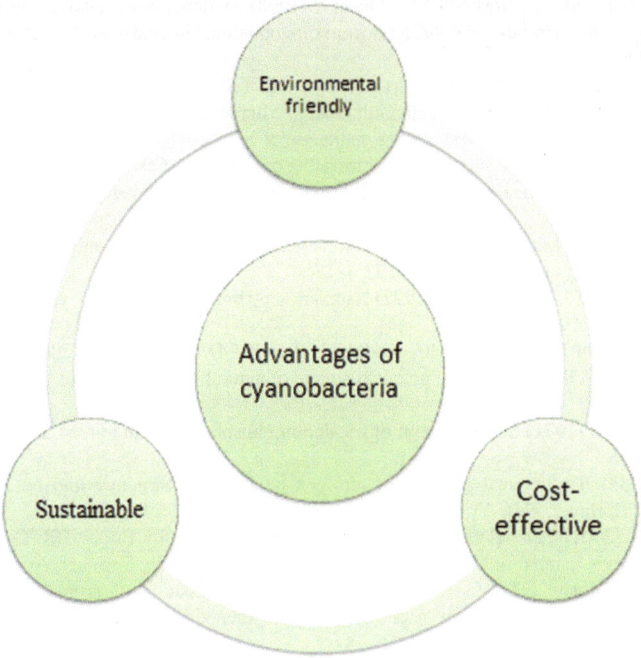

Fig. 4.2 Hypothetical diagram showing the advantages of using cyanobacteria

studied for its role in inhibiting the emission of methane gas, a greenhouse gas, in association with methanotrophs (Fig. 4.2).

However, it is necessary to carry out further investigation to utilize their full potential in order to combat various environmental concerns. Being multifunctional, cyanobacteria have emerged as bio-agents not only for safe and eco-friendly agriculture but also for environmental sustainability. Thus there is an urgent need of exploiting cyanobacteria, the better way. In addition to product developments, future research must take into account the improvement of strain of useful cyanobacteria in order to obtain high-quality food and fuel. Besides these, extensive field-level trial should be taken into account, and emphasis should be given on practical aspects pertaining to the method of application, economic benefits, environmental benefits, and long-term and short-term gains.

References

Ahmad MR, Winter A (1968a) Studies on the hormonal relationships of algae in pure culture. Planta 78:277–286

Ahmad MR, Winter A (1968b) Studies on the hormonal relationships of algae in pure culture: II. The effect of potential precursors of indole-3-acetic acid on the growth of several freshwater blue-green algae. Planta 81:16–27

Al-Hasan R, Al-Bader D, Sorkhoh N, Radwan S (1998) Evidence for n-alkane consumption and oxidation by filamentous cyanobacteria from oil-contaminated coasts of the Arabian Gulf. Mar Biol 130:521–527

Al-Hasan R, Khanafer M, Eliyas M, Radwan S (2001) Hydrocarbon accumulation by picocyanobacteria from the Arabian Gulf. J Appl Microbiol 91:533–540

Arif M, Gupta RK, Joshi M (2008) Studies on the use of cyanobacteria as biofertilizer for vegetable cultivation in hydroponic system. In: Schirmacher oasis region, East Antarctica. Eleventh Indian expedition to Antarctica scientific report. Department of ocean development. Technical publication, vol 9, p 243

Benson D, Kerry K, Malin G (2014) Algal biofuels: impact significance and implications for EU multi-level governance. J Clean Prod 72:4–13

Bergman B, Gallon J, Rai A, Stal L (1997) N_2 fixation by non-heterocystous cyanobacteria. FEMS Microbiol Rev 19:139–185

Biondi N, Piccardi R, Margheri MC, Rodolfi L, Smith GD, Tredici MR (2004) Evaluation of Nostoc strain ATCC 53789 as a potential source of natural pesticides. Appl Environ Microbiol 70:3313–3320

Bose P, Nagpal S (1971) Solubilization of tricalcium phosphate by blue-green algae. Curr Sci 40 (7):165–166

Brill WJ (1983) Biochemical genetics of nitrogen fixation. In: Structure and function of plant genomes. Springer, New York, NY, pp 231–237

Castenholz R (2001) General characteristics of the cyanobacteria. In: Boone DR, Castenholz RW (eds) Bergey's manual of systematic bacteriology, vol 1, 2nd edn. Springer, New York, NY

Cerniglia CE, Gibson DT, Van Baalen C (1979) Algal oxidation of aromatic hydrocarbons: formation of 1-naphthol from naphthalene by Agmenellum quadruplicatum, strain PR-6. Biochem Biophys Res Commun 88:50–58

Cerniglia CE, Gibson DT, Van Baalen C (1980a) Oxidation of naphthalene by cyanobacteria and microalgae. Microbiology 116:495–500

Cerniglia CE, Van Baalen C, Gibson DT (1980b) Oxidation of biphenyl by the cyanobacterium, Oscillatoria sp., strain JCM. Arch Microbiol 125:203–207

Cohen Y (2002) Bioremediation of oil by marine microbial mats. Int Microbiol 5:189–193

Cuellar-Bermudez SP, Garcia-Perez JS, Rittmann BE, Parra-Saldivar R (2015) Photosynthetic bioenergy utilizing CO2: an approach on flue gases utilization for third generation biofuels. J Clean Prod 98:53–65

De Caire G, De Cano M, De Mule M, De Halperin D (1990) Antimycotic products from the cyanobacterium Nostoc muscorum against Rhizoctonia solani. Phyton, Bueños Aires 51:1–4

Dhar N, Mukherji S (1936) Alkali soils and their reclamation. Proc Natl Acad Sci Ind 6:136–148

Dhar DW, Prasanna R, Pabbi S, Vishwakarma R (2015) Significance of cyanobacteria as inoculants in agriculture. In: Algal biorefinery: an integrated approach. Springer, New York, NY, pp 339–374

Dorich R, Nelson D, Sommers L (1985) Estimating algal available phosphorus in suspended sediments by chemical extraction. J Environ Qual 14:400–405

Flaibani A, Olsen Y, Painter TJ (1989) Polysaccharides in desert reclamation: compositions of exocellular proteoglycan complexes produced by filamentous blue-green and unicellular green edaphic algae. Carbohydr Res 190:235–248

Gantar M, Kerby N, Rowell P, Obreht Z, Scrimgeour C (1995a) Colonization of wheat (Triticum vulgare L.) by N_2-fixing cyanobacteria: IV. Dark nitrogenase activity and effects of cyanobacteria on natural 15N abundance in the plants. New Phytol 129:337–343

Gantar M, Rowell P, Kerby NW, Sutherland IW (1995b) Role of extracellular polysaccharide in the colonization of wheat (Triticum vulgare L.) roots by N 2-fixing cyanobacteria. Biol Fertil Soils 19:41–48

Grieco E, Desrochers R (1978) Production de vitamine B12 par une algue bleue. Can J Microbiol 24:1562–1566

Gupta A, Agarwal P (1973) Extraction, isolation, and bioassay of a gibberellin-like substance from Phormidium foveolarum. Ann Bot 37:737–741

Gupta A, Lata K (1964) Effect of algal growth hormones on the germination of paddy seeds. Hydrobiologia 24:430–434

Hagmann L, Jüttner F (1996) Fischerellin A, a novel photosystem-II-inhibiting allelochemical of the cyanobacterium Fischerella muscicola with antifungal and herbicidal activity. Tetrahedron Lett 37:6539–6542

Hartung W (2010) The evolution of abscisic acid (ABA) and ABA function in lower plants, fungi and lichen. Funct Plant Biol 37:806–812

Hussain A, Krischke M, Roitsch T, Hasnain S (2010) Rapid determination of cytokinins and auxin in cyanobacteria. Curr Microbiol 61:361–369

Irisarri P, Gonnet S, Monza J (2001) Cyanobacteria in Uruguayan rice fields: diversity, nitrogen fixing ability and tolerance to herbicides and combined nitrogen. J Biotechnol 91:95–103

Issa AA, Abd-Alla MH, Ohyama T (2014) Nitrogen fixing cyanobacteria: future prospect. Adv Biol Ecol Nitrogen Fix 2:24–48

Ito O, Watanabe I (1985) Availability to rice plants of nitrogen fixed by Azolla. Soil Sci Plant Nutr 31:91–104

Jenkinson D (1977) Nitrogen economy of the Broadbalk experiments. I. Nitrogen balance in the experiments. Rothamsted Experimental Station Report for 1976. Rothamsted Research, Harpenden

Kannaiyan S, Rao K, Hall D (1994) Immobilization of Anabaena azollae from Azolla filiculoides in polyvinyl foam for ammonia production in a photobioreactor system. World J Microbiol Biotechnol 10:55–58

Karthikeyan N, Prasanna R, Nain L, Kaushik BD (2007) Evaluating the potential of plant growth promoting cyanobacteria as inoculants for wheat. Eur J Soil Biol 43:23–30

Karthikeyan N, Prasanna R, Sood A, Jaiswal P, Nayak S, Kaushik B (2009) Physiological characterization and electron microscopic investigation of cyanobacteria associated with wheat rhizosphere. Folia Microbiol (Praha) 54:43–51

Kaushik B (1961) Blue-green algae (Cyanobacteria) in Agriculture. Adv Microbiol Ind Agric Res Inst 2004:19–34

Kaushik B (1994) Algalization of rice in salt-affected soils. Ann Agric Res 14:105–106

Kaushik B (1998) Use of cyanobacterial biofertilizers in rice cultivation: a technology improvement. Cyanobact Biotechnol:211–222

Kaushik B, Venkataraman G (1983) Response of cyanobacterial nitrogen fixation to insecticides. Curr Sci 52:321–323

Khan Z, Begum ZT, Mandal R, Hossain M (1994) Cyanobacteria in rice soils. World J Microbiol Biotechnol 10:296–298

Kim J, Rees D (1992) Structural models for the metal centers in the nitrogenase molybdenum-iron protein. Science 257:1677–1682

Kirn J, Rees D (1992) Crystallographic structure and functional implications of the nitrogenase molybdenum–iron protein from Azotobacter vinelandii. Nature 360:553–560

Koller M, Salerno A, Tuffner P, Koinigg M, Böchzelt H, Schober S, Pieber S, Schnitzer H, Mittelbach M, Braunegg G (2012) Characteristics and potential of micro algal cultivation strategies: a review. J Clean Prod 37:377–388

Kulik MM (1995) The potential for using cyanobacteria (blue-green algae) and algae in the biological control of plant pathogenic bacteria and fungi. Eur J Plant Pathol 101:585–599

Kumar M, Prasanna R, Bidyarani N, Babu S, Mishra BK, Kumar A, Adak A, Jauhari S, Yadav K, Singh R (2013) Evaluating the plant growth promoting ability of thermotolerant bacteria and cyanobacteria and their interactions with seed spice crops. Sci Hortic 164:94–101

Lincoln E, Wilkie A, French B (1996) Cyanobacterial process for renovating dairy wastewater. Biomass Bioenergy 10:63–68

Lipok J, Owsiak T, Młynarz P, Forlani G, Kafarski P (2007) Phosphorus NMR as a tool to study mineralization of organophosphonates—the ability of Spirulina spp. to degrade glyphosate. Enzym Microb Technol 41:286–291

Mansy AE-R, El-Bestawy E (2002) Toxicity and biodegradation of fluometuron by selected cyanobacterial species. World J Microbiol Biotechnol 18:125–131

Maqubela M, Mnkeni P, Issa OM, Pardo M, D'acqui L (2009) Nostoc cyanobacterial inoculation in South African agricultural soils enhances soil structure, fertility, and maize growth. Plant Soil 315:79–92

Maršálek B, Zahradníčková H, Hronková M (1992) Extracellular abscisic acid produced by cyanobacteria under salt stress. J Plant Physiol 139:506–508

Mazhar S, Cohen JD, Hasnain S (2013) Auxin producing non-heterocystous Cyanobacteria and their impact on the growth and endogenous auxin homeostasis of wheat. J Basic Microbiol 53:996–1003

Mazid M, Khan TA (2015) Future of bio-fertilizers in Indian agriculture: an overview. Int J Agric Food Res 3:10

Megharaj M, Venkateswarlu K, Rao A (1987) Metabolism of monocrotophos and quinalphos by algae isolated from soil. Bull Environ Contam Toxicol 39:251–256

Megharaj M, Madhavi D, Sreenivasulu C, Umamaheswari A, Venkateswarlu K (1994) Biodegradation of methyl parathion by soil isolates of microalgae and cyanobacteria. Bull Environ Contam Toxicol 53:292–297

Metting B (1981) The systematics and ecology of soil algae. Bot Rev 47:195–312

Misra S, Kaushik B (1989a) Growth promoting substances of cyanobacteria II. Detection of amino acids, sugars and auxins. Proc Indian Natl Sci Acad B55:499–504

Misra S, Kaushik B (1989b) Growth promoting substances of cyanobacteria. I: vitamins and their influence on rice plant. Proc Ind Natl Sci Acad B Biol Sci 55:295–300

Mulligan M, Haselkorn R (1989) Nitrogen fixation (nif) genes of the cyanobacterium Anabaena species strain PCC 7120. J Biol Chem 264:19200–19207

Muñoz-Rojas M, Chilton A, Liyanage G, Erickson T, Merritt D, Neilan B, Ooi M (2018) Effects of indigenous soil cyanobacteria on seed germination and seedling growth of arid species used in restoration. Plant Soil 429:91–100

Nain L, Rana A, Joshi M, Jadhav SD, Kumar D, Shivay Y, Paul S, Prasanna R (2010) Evaluation of synergistic effects of bacterial and cyanobacterial strains as biofertilizers for wheat. Plant Soil 331:217–230

Narro ML, Cerniglia C, Van Baalen C, Gibson D (1992) Metabolism of phenanthrene by the marine cyanobacterium Agmenellum quadruplicatum PR-6. Appl Environ Microbiol 58:1351–1359

Nisha R, Kaushik A, Kaushik C (2007) Effect of indigenous cyanobacterial application on structural stability and productivity of an organically poor semi-arid soil. Geoderma 138:49–56

Obreht Z, Kerby NW, Gantar M, Rowell P (1993) Effects of root-associated N 2-fixing cyanobacteria on the growth and nitrogen content of wheat (Triticum vulgare L.) seedlings. Biol Fertil Soils 15:68–72

Okuda A (1952) Algae and atmospheric nitrogen fixation in paddy soils. II. Relation between the growth of blue-green algae and physical or chemical properties of soils and effect of soil treatments and inoculation on the nitrogen fixation. Mem Res Inst Food Sci 4:1–11

Papke U, Gross EM, Francke W (1997) Isolation, identification and determination of the absolute configuration of Fischerellin B. A new algicide from the freshwater cyanobacterium Fischerella muscicola (Thuret). Tetrahedron Lett 38:379–382

Prasanna R, Kaushik B (2006) Cyanobacteria in soil health and sustainable agriculture. Health Environ 3:91–105

Prasanna R, Nayak S (2007) Influence of diverse rice soil ecologies on cyanobacterial diversity and abundance. Wetl Ecol Manag 15:127–134

Prasanna R, Kumar V, Kumar S, Yadav AK, Tripathi U, Singh AK, Jain M, Gupta P, Singh P, Sethunathan N (2002) Methane production in rice soil is inhibited by cyanobacteria. Microbiol Res 157:1–6

Prasanna R, Jaiswal P, Nayak S, Sood A, Kaushik BD (2009) Cyanobacterial diversity in the rhizosphere of rice and its ecological significance. Indian J Microbiol 49:89–97

Prasanna R, Triveni S, Bidyarani N, Babu S, Yadav K, Adak A, Khetarpal S, Pal M, Shivay YS, Saxena AK (2014) Evaluating the efficacy of cyanobacterial formulations and biofilmed inoculants for leguminous crops. Arch Agron Soil Sci 60:349–366

Prasanna R, Kanchan A, Kaur S, Ramakrishnan B, Ranjan K, Singh MC, Hasan M, Saxena AK, Shivay YS (2016a) Chrysanthemum growth gains from beneficial microbial interactions and fertility improvements in soil under protected cultivation. Hortic Plant J 2:229–239

Prasanna R, Kanchan A, Ramakrishnan B, Ranjan K, Venkatachalam S, Hossain F, Shivay YS, Krishnan P, Nain L (2016b) Cyanobacteria-based bioinoculants influence growth and yields by modulating the microbial communities favourably in the rhizospheres of maize hybrids. Eur J Soil Biol 75:15–23

Prasanna R, Ramakrishnan B, Ranjan K, Venkatachalam S, Kanchan A, Solanki P, Monga D, Shivay YS, Kranthi S (2016c) Microbial inoculants with multifaceted traits suppress Rhizoctonia populations and promote plant growth in cotton. J Phytopathol 164:1030–1042

Radwan SS, Al-Hasan RH (2000) Oil pollution and cyanobacteria. In: The ecology of cyanobacteria. Springer, New York, NY, pp 307–319

Rai AN, Bergman B (2002) Cyanolichens. In: Biology and environment: Proceedings of the royal irish academy, pp 19–22. JSTOR

Rai A, Borthakur M, Bergman B (1992a) Nitrogenase derepresssion, its regulation and metabolic changes associated with diazotrophy in the non-heterocystous cyanobacterium Plectonema boryanum PCC 73110. Microbiology 138:481–491

Rai A, Borthakur M, Soderback E, Bergman B (1992b) Immunogold localization of hydrogenase in the cyanobacterial-plant symbioses Peltigera canina, Anthoceros punctatus and Gunnera magellanica. Symbiosis 12:131

Rai A, Singh A, Syiem M (2019) Plant growth-promoting abilities in cyanobacteria. In: Cyanobacteria. Elsevier, Amsterdam, pp 459–476

Rodgers G, Bergman B, Henriksson E, Udris M (1979) Utilisation of blue-green algae as biofertilisers. Plant Soil 52:99–107

Roger PA, Zimmerman WJ, Lumpkin TA (1993) Microbiological management of wetland rice fields. In: Soil microbial ecology: applications in agricultural and environmental management. Marcel Dekker, New York, NY, pp 417–455

Rogers SL, Burns RG (1994) Changes in aggregate stability, nutrient status, indigenous microbial populations, and seedling emergence, following inoculation of soil withNostoc muscorum. Biol Fertil Soils 18:209–215

Roy D, Pakhira M, Bera S (2016) A review on biology, cultivation and utilization of Azolla. Adv Life Sci 5:11–15

Ruffing AM (2011) Engineered cyanobacteria: teaching an old bug new tricks. Bioeng Bugs 2:136–149

Saadatnia H, Riahi H (2009) Cyanobacteria from paddy fields in Iran as a biofertilizer in rice plants. Plant Soil Environ 55:207–212

Sergeeva E, Liaimer A, Bergman B (2002) Evidence for production of the phytohormone indole-3-acetic acid by cyanobacteria. Planta 215:229–238

Shariatmadari Z, Riahi H, Seyed Hashtroudi M, Ghassempour A, Aghashariatmadary Z (2013) Plant growth promoting cyanobacteria and their distribution in terrestrial habitats of Iran. Soil Sci Plant Nutr 59:535–547

Shashirekha S, Uma L, Subramanian G (1997) Phenol degradation by the marine cyanobacterium Phormidium valderianum BDU 30501. J Ind Microbiol Biotechnol 19:130–133

Singh RN (1950) Reclamation of 'Usar'lands in India through blue-green algae. Nature 165:325–326

Singh RN (1961) Role of blue-green algae in nitrogen economy of Indian agriculture. Indian Council of Agricultural Research, New Delhi

Singh JS (2011) Methanotrophs: the potential biological sink to mitigate the global methane load. Curr Sci 100:29–30
Singh JS (2013) Anticipated effects of climate change on methanotrophic methane oxidation. Clim Change Environ Sustain 1:20–24
Singh JS (2014) Cyanobacteria: a vital bio-agent in eco-restoration of degraded lands and sustainable agriculture. Clim Change Environ Sustain 2:133–137
Singh JS, Pandey VC (2013) Fly ash application in nutrient poor agriculture soils: impact on methanotrophs population dynamics and paddy yields. Ecotoxicol Environ Saf 89:43–51
Singh A, Singh P (1987) Influence of Azolla management on the growth, yield of rice and soil fertility: II. N and P contents of plant and soil. Plant Soil 102:49–54
Singh JS, Singh DP (2012) Reforestation: a potential approach to mitigate excess atmospheric CH4 build-up. Ecol Manag Restor 13:245–248
Singh S, Singh D (2013) Impact of anthropogenic disturbances on methanotrophs abundance in dry tropical forest ecosystems, India. Expert Opin Environ Biol 2(3):2
Singh V, Trehan K (1973) Effect of extracellular products ofAulosira fertilissima on the growth of rice seedlings. Plant Soil 38:457–464
Singh JS, Pandey VC, Singh D (2011a) Efficient soil microorganisms: a new dimension for sustainable agriculture and environmental development. Agric Ecosyst Environ 140:339–353
Singh JS, Singh D, Dixit S (2011b) Cyanobacteria: an agent of heavy metal removal. Bioremediation of pollutants. IK International Publisher, New Delhi, pp 223–243
Singh JS, Kumar A, Rai AN, Singh DP (2016) Cyanobacteria: a precious bio-resource in agriculture, ecosystem, and environmental sustainability. Front Microbiol 7:529
Sood A, Uniyal PL, Prasanna R, Ahluwalia AS (2012) Phytoremediation potential of aquatic macrophyte, Azolla. Ambio 41:122–137
Sorkhoh N, Al-Hasan R, Radwan S, Höpner T (1992) Self-cleaning of the Gulf. Nature 359:109–109
Stewart W, Singh H (1975) Transfer of nitrogen-fixing (NIF) genes in the blue-green alga nostocmuscorum. Biochem Biophys Res Commun 62:62–69
Swarnalakshmi K, Dhar D, Singh P (2006) Blue green algae: a potential biofertilizer for sustainable rice cultivation. Proc Ind Natl Sci Acad 72:167
Teuscher E, Lindequist U, Mundt S (1992) Blue-green algae, sources of natural active agents. Pharm Ztg Wiss 137:57–69
Thomas SP, Zaritsky A, Boussiba S (1991) Ammonium excretion by a mutant of the nitrogen-fixing cyanobacterium anabaena. Bioresour Technol 38:161–166
Tiwari S, Singh JS, Singh DP (2015) Methanotrophs and CH4 sink: effect of human activity and ecological perturbations. Clim Change Environ Sustain 3:35–50
Tsavkelova E, Klimova SY, Cherdyntseva T, Netrusov A (2006a) Hormones and hormone-like substances of microorganisms: a review. Appl Biochem Microbiol 42:229–235
Tsavkelova E, Klimova SY, Cherdyntseva T, Netrusov A (2006b) Microbial producers of plant growth stimulators and their practical use: a review. Appl Biochem Microbiol 42:117–126
Vaishampayan A, Sinha RP, Hader D-P, Dey T, Gupta A, Bhan U, Rao A (2001) Cyanobacterial biofertilizers in rice agriculture. Bot Rev 67:453–516
Venkataraman GS (1972) Algal biofertilizers and rice cultivation. Today & Tommorrow's Printers & Publishers, New Delhi
Venkataraman G (1975) The role of cyanobacteria in tropical rice cultivation. In: Nitrogen fixation by free living microorganisms. Cambridge University Press, Cambridge, pp 207–218
Vilchez C, Garbayo I, Lobato MV, Vega J (1997) Microalgae-mediated chemicals production and wastes removal. Enzym Microb Technol 20:562–572
Vorontsova G, Romanova N, Postnova T, Selyakh I, Gusev M (1988) Biostimulating effect of cyanobacteria and ways to increase it. I. Use of mutants, superproducers of amino acids. Moscow Univ Biol Sci Bull (USA) 43:14
Wagner GM (1997) Azolla: a review of its biology and utilization. Bot Rev 63:1–26

Watanabe A (1956) On the effect of the atmospheric nitrogen-fixing blue-green algae on the yield of rice. Bot Mag (Tokyo) 69:530–535

Watanabe I, Cholitkul W (1978) Field studies on nitrogen fixation in paddy [rice] soils [of Thailand]. In: Nitrogen and Rice Symposium College, Laguna (Philippines) 18-21 Sep 1978

Watanabe A, Yamamoto Y (1971) Algal nitrogen fixation in the tropics. Plant Soil 35:403–413

Woese CR (1987) Bacterial evolution. Microbiol Rev 51:221

Wolk CP, Thomas J, Shaffer P, Austin S, Galonsky A (1976) Pathway of nitrogen metabolism after fixation of 13N-labeled nitrogen gas by the cyanobacterium, Anabaena cylindrica. J Biol Chem 251:5027–5034

Yamaguchi K (1996) Recent advances in microalgal bioscience in Japan, with special reference to utilization of biomass and metabolites: a review. J Appl Phycol 8:487–502

Zahradníčková H, Maršálek B, Polišenská M (1991) High-performance thin-layer chromatographic and high-performance liquid chromatographic determination of abscisic acid produced by cyanobacteria. J Chromatogr A 555:239–245

Chapter 5
Effects of Herbicides on Soil Enzymes and Their Regulatory Factors in Agroecosystem: A Review

Laliteshwari Bhardwaj, Jitendra Pandey, and Suresh Kumar Dubey

Abstract Modern agriculture is heavily reliant upon herbicide application to control weeds for increasing crop productivity to meet the need of growing population and for economic benefits. However, such benefits bear high environmental cost including loss of soil fertility. An indispensable role is played by soil enzymes in the decomposition of xenobiotic and mineralization of organic compounds, and they are considered to be the best soil fertility indicators. Therefore, soil fertility sustenance and crop productivity maintenance demand a better understanding of response of soil enzymes to application of herbicides. The present chapter has made an attempt to present a comprehensive account on response of soil enzymes to different classes and types of herbicides under variable soil environment. Efforts were made to address the production and consumption of herbicides, types of regulatory determinants, and fate of herbicide-enzyme interaction. A critical analysis of in situ and controlled experiments suggests that herbicides applied individually or in combinations influence soil enzymes differently. Although the response shows dose dependence, a number of edaphic and climatic factors also play a significant role in regulating herbicide-enzyme cause-and-effect relationships. This has relevance for mechanistic understanding of enzyme-herbicide interaction and exploring strategies of soil management.

Keywords Herbicides · Soil enzymes · Herbicide-enzyme interaction · Soil fertility · Influencing factors

L. Bhardwaj · J. Pandey · S. K. Dubey (✉)
Molecular Ecology Laboratory, Centre of Advanced Study in Botany, Institute of Science, Banaras Hindu University, Varanasi, India
e-mail: skdubey@bhu.ac.in

S. K. Dubey, S. K. Verma (eds.), *Plant, Soil and Microbes in Tropical Ecosystems*, Rhizosphere Biology, https://doi.org/10.1007/978-981-16-3364-5_5

5.1 Introduction

Agriculture, which had been a major sector in Indian economy, now contributes to only 17% of gross domestic product (GDP) (Economic Survey Report, 2017–2018). Rapidly growing demand of population in developing countries has led to massive intensification of agricultural system. Agricultural weeds are important interspecies competitors of crops, leading to a sharp decline of about 29% and 47% in wheat and rice crops, respectively (Oerke 2005). This has forced the indiscriminate and tremendous application of pesticides mainly herbicides in the agricultural field (Nonga et al. 2011). Herbicides are toxic agrochemicals used against weeds and undesirable vegetation in the agricultural farms and gardens. Herbicide consumption accounts to 47.5%, insecticides 29.5%, fungicides 17.5%, and others only 5.5% on the global scale of pesticide expenditure. In India, insecticides hold 80% utilization, herbicides 15%, fungicide 1.46%, and others below 3%. The herbicide application has descending trend as wheat (44%), followed by rice (31%), plantation crop (10%), soybean (4%), and other crops (11%) (Sondhia 2014). Herbicides are biologically active eco-toxic compounds that may cause unexpected repercussions by influencing microbial populations, soil enzyme activities and therefore, the overall status of the soil because microbial communities are the key determinants of carbon flow, litter decomposition, and nutrient cycling. Such impacts reduce soil fertility and agricultural productivity in the long run (Tripathi et al. 2005; Pandey et al. 2007a, b).

Knowledge about the effect of herbicides, herbicidal efficacy, and consequential yield effects of the herbicide application either alone or amended with other agrochemicals and under organic or inorganic treatments, is important in crop management and long-term sustainable crop production strategies (Borowik et al. 2016). Considering all these issues, several research studies have been already conducted and are continued especially in the context of soil biological system and enzyme activities. The effects of herbicides butachlor, 2,4-D, and oxyfluorfen on dehydrogenase and urease activity have been studied by Baruah and Mishra (1986). The effects of glyphosate and diflufenican applied alone or in combination on soil biological properties have been examined by Tejada (2009). Du et al. (2018) illustrated the dose dependence of mesotrione on soil enzyme activities and microbial communities. A large number of reviews are available focusing on the effect of herbicide contamination on soil quality attributes. Recently, Raj and Syriac (2017) reviewed the dependency of herbicides on soil type, characteristic and concentration of herbicide, vulnerability of nontarget organisms, and climatic conditions in assessing the impact of soil health status. Riah et al. (2014) reviewed the effect of pesticides on soil enzymes. Most of the recent reviews have emphasized the context of weed control strategies. Mauprivez et al. (2019) explored the herbicidal effect on nontarget organisms. Very recently Macías et al. (2019) reviewed the advancement in allelopathy from knowledge to application to overcome the problems of weeds. Least attention has been paid to review the effects of herbicide mixtures in comparison to individual agrochemical on soil biological parameters although it seems to be more effective in weed killing and nontarget effects. Opportunities and challenges regarding interactions between different categories of herbicides and different classes of soil enzymes and their consequential dose-response relationship have not received sufficient attention so far, irrespective of the fact that soil enzymes quickly response to herbicides and also are the best indicators of soil fertility. The centerpiece of the present review is to precisely enumerate the available scientific literature related to the common soil enzymes, monitored by different herbicidal treatments, in various dose and on discrete soil types in either way as observed in field experiments or laboratory incubation studies, and it is also planned to summarize the knowledge base of factors altering the influence of herbicides on soil enzymes (Table 5.1). Our in-depth analysis of available literature shows that herbicide-soil enzyme interaction follows the "dose-response relationship." At higher doses, herbicides inhibit enzyme activity, while relatively lower concentration acts as stimulator. Further, a complex set of ecological factors like soil moisture, endogenous residues, soil type, herbicide quality, etc. influence the overall interaction between herbicide and soil enzymes. Therefore, extensive investigations to establish a mechanistic link between herbicide and soil enzyme and their regulatory factors, seem to be imperative.

5.2 Overall Production and Consumption of Herbicides

Weeds being the major impediment in crop productivity have caused a phenomenal growth in the application of herbicides for limiting and eliminating the weed population. Herbicide holds the highest position of global pesticide sale which accounts 47% making it a major class of pesticides followed by 29.4% insecticides, 17.5% fungicides, and remaining 5.5% only sold by others (Shea 1985). Herbicides alone share 47.5% of total 2 million tons of global annual pesticide consumption (Gupta 2004). As per the report of Sondhia (2014), herbicides account for 44% of total global annual pesticide consumption (share followed by insecticides (22%), fungicides (27%), and others (7%) at global forum of pesticide consumption). Herbicides consumption is around 60% of total pesticide at global level (Sondhia 2018). The derivatives of chlorophenoxy acid, 2,4-D, triazines having three heterocyclic N atoms in ring structure (atrazines), urea derivatives, substituted chloroacetanilides (propachlor) and sulfonylurea substituted (amidosulfuron and nicosulfuron). Glyphosate undoubtedly holds number one position, whereas paraquat ranks second in terms of worldwide sale (Woodburn 2000). Glyphosate (GP), 2,4-D, atrazine, metolachlor, diuron, imazapyr, pendimethalin, paraquat, and clodinafop propargyl (CF) are the most commonly applied herbicides (Singh and Singh 2014).

Table 5.1 An overview of herbicidal effects on soil enzymes and regulatory determinants

SN	Soil type, crop, study site	Experimental setup (field/incubated), herbicide dose, and mode of application (MA)	Enzymes studied	Observations/findings	References
1	Clay soil	Soil (500 g, 0–20 cm depth) incubated for 10 weeks at 28 °C.	Cellulase	Cellulase activity decreased, and enhancement was recorded once by low or high rates of Brominal, 1 week after soil treatment. Activity of AP inhibited at high dose and stimulated at low doses of herbicide. ALP accelerated with herbicide application low or high	Omar and Abdel-Sater (2001)
	Rice crop	Brominal (bromoxynil, 24%) was added in 1 L per feddan (0.6 g a.i. dws)	Acid phosphatase (AP)		
	Asyut, Egypt	MA: spray	Alkaline phosphatase (ALP)		
2	Fluvo-aquic soil	Soil (5 g, 3–15 cm depth) laboratory incubation for 24 h	Dehydrogenase	DHA increased gradually with increasing dose and showed the highest activity on the 16th Day after application of 22.0 mg g^{-1} dws of butachlor. Hydrogen peroxidase increased on lower dose of butachlor	Min et al. (2001)
	Paddy rice crop	Butachlor @ 5.5μg/g, 11.0 mg g^{-1} and 22.0 mg g^{-1} dws	Hydrogen peroxidase		
	China, Hangzhou				
3	Clay soil (Phaeozem)	Tested soil samples (0–20 cm in depth) preincubated at 25 °C for 7 days.	Urease	The activities of urease ranged from 75.7% to 120% and phosphatase from 41.2% to 136.4% considering control as 100%	Wang et al. (2006)
	Shanghai, China	Butachlor, MA: solution-based application	Phosphatase		
4	Loamy sand	Soil (8 g at 15 cm depth) incubated for 1–3 weeks at 28 °C, atrazine, butylate, ethalfluralin, imazethapyr, linuron, metolachlor, metriburin, and trifluralin (rate 10μg g^{-1} dws)	Dehydrogenase	No effect on amylase activity of all eight herbicidal treatments except ethalfluralin inhibited its activity. All the chemicals significantly inhibited the activity of phosphatase	Tu (1992)
	London, Ontario	MA: spray	Amylase		
5	Clayey texture soil and sandy loam texture soil	Soil (100 g) incubated for 180 days at 25 °C, 0.52 g of glyphosate, 2.08 g of diflufenican, and combination of both	Dehydrogenase	Activity of all the enzymes got inhibited, and more pronounced inhibition was observed in sandy loam soil than clay loam soil	Tejada (2009)

		0.52 g of glyphosate and 2.08 g of diflufenican			
	Seville, Spain	MA: the herbicides were applied using a machine of laboratory treatments equipped with fuzes of flat fan Teejet 80.02 E.VS, to a pressure of 3 kg cm^{-2} and 300 L ha^{-1} of application.	Urease		
		Applied in 1 L water solution	β-glucosidase		
			Phosphatase		
			Arylsulfatase		
6	Cassava farms, woody perennial plants (famine or drought crop)	Topsoil (up to 5 cm depth) treated for a period of 6 weeks, 4 L h^{-1} for paraquat, glyphosate, and primeextra, while recommended rate of 3 kg h^{-1} atrazine powder was used for atrazine treatment.	Dehydrogenase	In the treatment of 6-week period, soil treated with atrazine recorded highest DHA of 14.32μg g^{-1} min^{-1} after fourth week of treatment, while primeextra treatment had the lowest record of DHA 9.02, 12.55, and 16.09μg (g^{-1} min^{-1}) after second, fourth, and sixth week treatment	Sebiomo et al. (2011)
	Nigeria, Ogun State	MA: spray			
7	Sandy clay loam	Field experiment, soil samples (0–15 cm) were collected at 0, 15, 30, and 45 days after spraying of pre-emergent herbicides	Acid phosphatase	During 45-day observations, in lower doses, activity of all the enzymes increased except urease which got inhibited compared to the control	Sireesha et al. (2012)
	Radish crop	Pendimethalin (0.5/0.75 kg a.i. ha^{-1}), oxyfluorfen (0.1/0.15 kg a.i. ha^{-1})	Alkaline phosphatase		
	India, Hyderabad	MA: spraying	Dehydrogenase		
			Urease		
8	Inceptisol (sandy loam)	Field experiment, soil samples (0–10 cm depth) for 2 annual cycles	β-glucosidase	Activities of all the enzymes increased with respect to control	Singh and Ghoshal (2013)
	Rice and wheat	Butachlor (2 kg a.i. ha^{-1})	Alkaline phosphatase		

(continued)

Table 5.1 (continued)

SN	Soil type, crop, study site	Experimental setup (field/incubated), herbicide dose, and mode of application (MA)	Enzymes studied	Observations/findings	References
	India, BHU	MA: spraying	Urease		
9	Clay loam	Field experiment, pre-emergent application of oxyfluorfen @ 0.10 and 0.15 kg ha^{-1}, pretilachlor @ 1.00 and 1.50 kg ha^{-1}, pendimethalin @ 0.675 and 1.00 kg ha^{-1}, atrazine @ 1.25 kg ha^{-1}	Dehydrogenase	Dehydrogenase activity decreased with higher doses of herbicides.	Nadiger et al. (2013)
	Maize crop	MA: spraying		Enzyme activity was reduced till 20–30 DAS. At 40 DAS, the DHA in the soil was reduced in all the treatments compared to 20 DAS. However, at later stages of the crop growth (60, 80, and 100 DAS), there was a drastic increase in the activity of DHA in the plots treated with pretilachlor, oxyfluorfen, pendimethalin (at both applied doses), and atrazine	
	Dharwad, India				
10	Clay and loamy soil	Soil (0–20 cm depth, 80 g) incubated for 60 days at 25 °C, fomesafen (50––420μg kg^{-1} conc. @ 180–375 g a. i. ha^{-1})	Acid phosphatase (AP)	AP increased with dose. Urease activity inhibited first and then increased. DHA found most sensitive and showed enhanced activity in response to herbicide	Zhang et al. (2014)
	China, Qingdao	MA: spray	Alkaline phosphatase (ALP)		
			Urease		
			Dehydrogenase (DHA)		

11	Sandy loam soil	Pot culture experiment, soil (5 kg), sampling was undertaken at 0, 30, 60, 90, and 120 DAS (days after sowing), Pendimethalin at three different rates (500, 1000, and 1500 g a.i.) were applied.	FDAH (fluorescein diacetate hydrolysis)	Dehydrogenase and FDAH were the least tolerant to the effect of the herbicide, whereas alkaline phosphatase was the most tolerant one. Higher dose was more deleterious than the lower doses of pendimethalin	Singh (2014)
	Aligarh Muslim University, Aligarh, India	(MA): spray	Dehydrogenase		
			Acid and alkaline phosphatase		
12	Sandy loam and loamy soil	Soil (1 kg, 0–10 cm depth) incubated for 30 min at 20 °C nicosulfuron @ 0.3,0.6, 3.0, and 30.0 mg a.i. kg^{-1} of soil	β-glucosidase	The herbicide was found to stimulate β-glucosidase and protease activity in both types of the soil. Enzyme activity increased after treatment with nicosulfuron. Protease activity stimulated in both soil types on herbicide application	Santric et al. (2014)
	Belgrade, Serbia	MA: spray	Protease		
13	Clay loam	A field experiment was conducted for 2 years in randomized complete block design pattern. Buctril super (bromoxynil) herbicide was applied at 375, 750, 1500, and 22,500 mL ha^{-1}.	Dehydrogenase (DHA)	Dehydrogenase activity declined	Abbas et al. (2015)
	Wheat (variety Chakwal-50)	MA: spray	Urease	The highest dehydrogenase activity was found in control, followed by in 375 mL ha^{-1} treatment and least in 2250 mL ha^{-1} treatment. The highest urease activity in 375 mL ha^{-1} and lowest in 2250 mL ha^{-1} treatment were reported.	
	Rawalpindi, Pakistan		Alkaline phosphatase	30% and 31% reduction in urease activity, 36% inhibition in dehydrogenase activity, and 34% and 31% decline in alkaline phosphatase activity were recorded in two seasons	

(continued)

Table 5.1 (continued)

SN	Soil type, crop, study site	Experimental setup (field/incubated), herbicide dose, and mode of application (MA)	Enzymes studied	Observations/findings	References
14	Sandy clay loam	Soil (4 kg), incubated for 0 (before spray), 5, 10, 15, 20, 25, 30, and 45 days	Dehydrogenase	DHA at 20 days after pretilachlor application was inhibited by 27%, 28%, and 40% compared to initial values for RD, 2 RD, and 10 RD treatments, respectively. DHA was found elevated in control sample.	Sahoo et al. (2016)
	Rice crop (var. Naveen, Indica-type)	Pretilachlor @ recommended dose (RD), 600 g a.i. ha^{-1} @ 1200 g a.i. ha^{-1} (2 RD), and @ 6000 g a.i. ha^{-1} (10 RD) and control with no herbicide.	Fluorescein diacetate (FDA)	FDA activity was reported with an increase of 29%, 36%, 10%, and 36% for RD, 2 RD, 10 RD, and control treatments after 25 days of pretilachlor application, respectively.	
	Cuttack, India	MA: spray as pre-emergent herbicide	β-glucosidase	There was an increase of 29%, 36%, 10%, and 36% FDA activity from initial values for RD, 2 RD, 10 RD, and control treatments after 25 days of pretilachlor application, respectively. β-glucosidase and urease activity inhibited, compared to control, 5 days after herbicide application	
			Urease		
15	Eutric Cambisols	Soil (3 kg, 0–20 cm depth) incubated for 60 days	Dehydrogenase (DHA)	0.73 mg P + T kg^{-1} destabilized activity of all enzymes, and 14.63–468.16 mg P + T kg^{-1} strongly inhibited all the enzymes' activity. Inhibition trend was recorded as DHA > AP > U > Al. P > βG > ArS > C	Wyszkowska et al. (2016)
	Maize crop	Enzyme assay on Day 30 and 60 of the experiment, pethoxamid (P) (300 g dm^{-3}), terbuthylazine (T) (250 g dm^{-3})	Catalase (C)		
	Tomaszkowo, Poland		Urease (U)		
			Alkaline phosphatase (ALP)		
			Acid phosphatase (AP)		
			Arylsulfatase (ArS)		
			β-glucosidase (βG)		

16	Silty clay loam Maize Palampur, India	Field exp., soil samples were collected continued with an interval of 20 days till harvest. Tembotrione (110 g ha^{-1}), atrazine (1500 g ha^{-1} as pre-emergent) MA: spray fitted	Dehydrogenase	Enzyme activity decreased at higher doses from 2 to 60 days after sowing and increased 60–100 days in all treatments	Kumar et al. (2017)
17	Haplic Luvisol Lublin, Poland	Field experiment, soil (0–20 cm) collected in the end of May, each year. Azoprim (atrazine) @ 3 kg ha^{-1}, Roundup Ultra (glyphosate) 4 dm^{-3} ha^{-1} plus Chwastox Extra (MC PA) 2 dm^{-3} ha^{-1}, Azotop (simazine) 4 kg ha^{-1}. Roundup 3 dm^{-3} ha^{-1} plus ammonium sulfate approximately 12 kg ha^{-1}, and Dual 720 EC (metolachlorine) 1 dm^{-3} ha^{-1}	Dehydrogenases Phosphatase Urease Protease	Among the enzymes analyzed, the activity of phosphatases was the most sensitive indicator for soil contamination with triazine herbicides. The application of herbicide-triggered fallow land and a high level of mineral fertilization effected in the lowering of the enzymatic activity of the soil over the years. DHA reduced by 44.8%, phosphatase by 58%, urease by 46%, and protease by 43%	Bielińska and Pranagal (1994)
18	Soil Endocalcaric Cambisols, with sandy loam texture	Soil (1.5 g cm^{-3}, from 0 to 20 cm depth terbuthylazine (T), mesotrione (M), and S-metolachlor (S)). One cubic decimeter of the herbicide contains 187.5 g of T, 37.5 g of M, and 312.5 g of S.	Dehydrogenase	Activity decreased strongly dependent to dose in response to mixture of three Hbs.	Borowik et al. (2016)
	Maize crop Olsztyn, Poland	MA: applied to the soil in the form of aqueous suspension	Catalase (C) Urease (U) Arylsulfatase B-glucosidase (βG) Acid and alkaline phosphatase (AP and ALP)	AP 27%, C by 43%, ArS by 52%, ALP by 57%, DHA by 83%, U by 89%, and βG by 92%	

(continued)

Table 5.1 (continued)

SN	Soil type, crop, study site	Experimental setup (field/incubated), herbicide dose, and mode of application (MA)	Enzymes studied	Observations/findings	References
19	Sandy loam	Field experiment, pre-emergent herbicides (atrazine 50% WP @ 1.0 kg a.i. ha^{-1}, pendimethalin 30% EC @ 1.0 kg a.i. ha^{-1})	Urease	Activity decreased in pendimethalin-treated soil more than atrazine	Kumari et al. (2018)
	Maize	Post-emergent herbicides (pendimethalin 30% EC @ 1.0 kg a.i ha^{-1}, topramezone 42% SC @ 105 g a.i. ha^{-1}, tembotrione 42% SC @ 105 g a.i. ha^{-1})			
	India	Topramezone + atrazine @ 25.2 + 250 g a.i. ha^{-1} and tembotrione + atrazine @ 105 + 250 g kg a.i. ha^{-1}			
		MA: the pre-emergent herbicides were sprayed at zero days after sowing where post-emergent herbicides were sprayed after 15 days after sowing			
20	Taian brunisolic soil	Soil (50 g, 0–20 cm) incubated for 2 weeks at 25 °C, mesotrione exposure at doses of 0.1, 1.0, and 5.0 mg kg^{-1}	B-glucosidase (βG)	U and AP activity is found relatively stable in mesotrione-treated soil compared to control, while βG activity was reduced in the 5.0 mg kg^{-1} treatment of mesotrione application	Du et al. (2018)
	China, Taian	MA: solution-based application	Urease (U)		
			Acid phosphatase (AP)		

5.3 Classification of Herbicides

The understanding and management of herbicide resistance equally demand the classification of herbicides in order to overcome the continuous problems in sustainable agricultural management (Sherwani et al. 2015). Herbicides are categorized into diverse groups based on their chemical families, method of application, mode of action, target site, timing of application, target specificity, selectivity, and translocation (Sherwani et al. 2015; Vats 2014).

5.3.1 Based on Mode of Application

Singh and Singh (2014) have described the herbicide application methodologies. This includes foliar spray, soil application, and broadcasting, either covering complete regime or spot spray masking the specified area. Sherwani et al. (2015) have advocated that herbicide mixing with soil is a traditional approach, while weed-specialized eco-friendly herbicide spraying is a modern practice followed in advanced agricultural sector. The soil-applied herbicides such as fluchloralin in contrast to the foliar spray-applied herbicides such as glyphosate and paraquat, act primarily on the plant foliage. Soil-applied herbicides often leave a greater amount of residual herbicide (Sopeña et al. 2009). Along with the weed specificity, planned emplacement of herbicides at an appropriate rate is pivotal. The higher the rate of absorption and retention, the lesser will be the volume of herbicides required, and the lesser will be the potency compared to their counterparts (Sherwani et al. 2015).

5.3.2 Based on Formulation

Herbicides are generally not applied in the form they get synthesized. The basic idea of different types of herbicide formulations with their pros and cons is to make their handling easier in terms of effectiveness, safety, adverse impact minimization on non-target organisms, stability, management, and application.

Herbicide formulations are usually prepared for commercial purpose in which active ingredients are supplemented with the adjuvants and surfactants to meet regulatory standards without compromising the potency of the active ingredients (Sopeña et al. 2009).

5.3.3 Based on Translocation

Based on translocation, the herbicides can broadly be classified into three major classes:

- Symplastically translocated (source to sink capable of downward movement), e.g., glyphosate, 2,4-D, sulfonylureas
- Apoplastically translocated (capable of only upward movement), e.g., glyphosate
- Contact herbicide, those which do not move appreciably (kill very quickly), e.g., paraquat

5.3.4 Based on Application Time

Preplant herbicides are mechanically incorporated into the soil before planting is done. Pre-emergent herbicides such as dithiopyr and pendimethalin are introduced into the soil prior to weed seedling emergence. Post-emergent herbicides are subjected to the soil only after emergence of weed seedlings through the soil and require multiple applications. For example, 2,4-D is a selective, systemic, foliar-absorbed post-emergent herbicide (Vats 2014).

5.3.5 Based on Mode of Action

Herbicides belonging to the same chemical family, tend to share similar mode of actions, although a few, assigned to different chemical class, depict the same mode of actions. Some of the common groups of herbicide and their mode of action are described below (Sherwani et al. 2015):

1. Group 1. Lipid biosynthesis inhibitors (fluazifop-*p*-butyl and sethoxydim) inhibit acetyl-CoA carboxylase, the enzyme required for biosynthesis of phospholipid bilayer which results into disruption of structural and functional integrity of the cell membrane.
2. Group 2. Amino acid biosynthesis inhibitors or acetolactate synthase (ALS), the largest group of inhibitors (imidazolinones, pyrimidinyl thiobenzoates, sulfonyl glyphosate, imazapyr, and imazapic), which prevent protein synthesis by inhibiting branch chain amino acids, causing plant wilting and ultimately death.
3. Group 3: Root growth inhibitors (benzamide, benzoic acid, dinitroaniline, phosphoramidate, and pyridine), which inhibit the cell division and ultimately check the root extension and growth.
4. Group 4. Synthetic auxins or plant growth regulators (2,4-D, clopyralid, picloram, and triclopyr) which mimic indole acetic acid (IAA), thus increasing the transcription, translation, and protein biosynthesis within the cell leading to

uncontrolled disorganized vascular growth, causing cell bursts and ultimately cell and plant death.

5. Group 5, 6, and 7. Photosynthesis inhibitors (hexazinone, triazine, triazinone, nitriles, benzothiadiazinones, paraquat, phenyl urea, and amides), which cause disruption of photosynthetic pathway, especially PSII.

5.4 Fate of Herbicides After Application in the Soil

Apart from the very small fraction of herbicides reaching the target organisms (Pimentel 1995), a large proportion of residual herbicides end up into the soil, water, and atmosphere or in the harvested produce, posing a potential threat to nontarget organisms, including crop produce and health of consumers (Kudsk and Streibig 2003; Singh and Singh 2014; Zabaloy et al. 2011). Once introduced into the soil, herbicides simultaneously dissipate and degrade, resulting into redistribution or transformation into other metabolites. Dissipation mechanism comprises of multiple complex processes such as volatilization, soil adsorption, runoff, and downward leaching. On the other hand, degradation constitutes three main processes, photodegradation, chemical degradation, and microbial degradation leading to partial or total degradation of herbicide (Gianfreda and Rao 2008; Rana et al. 2016). Large and repeated dose of herbicide application in modern agricultural practices is a major concern. It may cause risk to soil microbial diversity, alter soil enzymes and overall performance of soil microflora (Kumari et al. 2018).

5.5 Soil Enzymes

5.5.1 An Indicator of Soil Health

Soil quality is evaluated in terms of microbial diversity, activity, bulk density, porosity, stability, texture, infiltration, governing water and solute flow, buffering capacity, and carbon and nutrient cycling (Dexter 2004; USDA 2015).

Soil enzymes are among the most important soil biological indicators driving mineralization of organic matter and release of nutrients for plant and microbial growth (Jimenez De la Paz et al. 2002; Kızılkaya et al. 2004; Khan et al. 2009; Buturugă et al. 2016). Quick response to soil management changes and environmental factors also high sensitivity towards agrochemicals especially herbicides, make soil enzymes as healthy indicators. They can be measured using cost-effective simple methods based on short-term laboratory incubations. These attributes make soil enzymes more suitable soil health detector and indicator compared to other determinants (Nannipieri et al. 2002, 2012; Gianfreda and Ruggiero 2006).

5.5.2 *Sources and Status of Enzymes in the Soil*

Living and dead microorganisms are primary source of soil enzymes. Additionally, plant roots also contribute a small share to overall enzyme pool (Infinita Biotech 2019). Enzymes occur either accompanying viable microbes or soil fauna, the *biotic* form or as excreted enzymes, linked to nonviable cells or amalgamated with mineral colloids in *abiotic* manner. The latter class is also known as "soil-bound enzymes" or "naturally immobilized enzymes" (Dick et al. 2011; Gianfreda and Bollag 1996).

5.5.3 *Indispensable Soil Enzymes*

Soil is a dynamic resource with unprecedented treasures of enzymes such as oxidoreductases, hydrolases, isomerases, lyases, and ligases, catalyzing enumerable reactions related to energy and material conversion (Gu et al. 2009).

5.5.3.1 Oxidoreductase

The class comprises a cluster of enzymes (dehydrogenase, catalase, and peroxidase) involved in catalyzing oxidation reaction in the cell with the help of cofactors NAD^+/NADH and flavins ($FAD/FADH_2$).

Dehydrogenase (EC 1.1.1.). It is the most important enzyme found in all living microorganisms intracellularly (Moeskops et al. 2010; Zhao et al. 2010; Yuan and Yue 2012), and is used to assess the overall microbial activity in the soil (Quilchano and Marañón 2002; Gu et al. 2009; Salazar et al. 2011; Dotaniya et al. 2019). These enzymes transfer H^+ ions and electron on either the nicotinamide adenine dinucleotide or nicotinamide adenine dinucleotide phosphate (Gianfreda and Rao 2014) and thus play a major role in biological oxidation of soil organic matter (Sebiomo et al. 2011). Assessment of immediate soil microbial metabolic activities can easily be represented by measuring dehydrogenase activity (Nannipieri et al. 2002).

Catalase (EC1.11.1.6). An enzymatic antioxidant, capable of breaking down H_20_2 into water and O_2 without generating free radical. These enzymes play a key role in soil fertility (Shiyin et al. 2004; Trasar-Cepeda et al. 2008).

Peroxidases (EC 1.11.1). Act as biological catalysts, mediated by free radical species generated while using H_2O_2 as an electron acceptor (Passardi et al. 2007). These serve as an important factor in biogeochemical processes, lignin degradation, H_2O_2 removal, oxidation of toxic substances, carbon mineralization sequestrations, and dissolved organic C export (Erman and Vitello 2002; Bach et al. 2013).

5.5.3.2 Hydrolases

A dominant class of extracellular enzymes (cellulases, glucosidases, phosphoesterases, sulfatases, amidases, urease, etc.) which mediate hydrolytic cleavage of complex macromolecules such as cellobiose, urea, and organophosphorus to provide smaller utilizable forms. *Cellulase* (endocellulase and exocellulase) hydrolyzes the glycosidic bonds of cellulose into simple, reasonable, and soluble sugar (Alvarez et al. 2013; Dotaniya et al. 2019).

β-glucosidase (EC 3.2.1.21). Catalyzes cellulose degradation, commenced with the breakdown of complex cellulose chain into smaller units involving endo-1,4-ß-glucanase (EC 3.1.2.4), followed by cellobiohydrolase (EC 3.1.2.91). The hydrolytic process is accomplished by the enzymatic action of β-glucosidase where 2 mol of glucose are extracted per mole of cellobiose (Turner et al. 2002). Thus it plays a vital role in C-cycle.

Urease (EC 3.5.1.5). Urease is produced by all the groups of microorganisms (Follmer 2008) that exist both as extracellular and intracellular forms (Mobley and Hausinger 1989). This enzyme hydrolyzes urea into ammonium and carbon dioxide (Byrnes and Amberger 1989; Mohammadi 2011; Fazekasova 2012; Zhang et al. 2014). This enzyme regulates N-cycle.

Phosphatase *alkaline* (EC 3.1.3.1) and acid phosphatase (EC3.1.3.2) hydrolyze ester-phosphate bonds of organic phosphorus and anhydrides of phosphoric acid into inorganic phosphorus accessible to plant and microbes and necessary for P cycling in P-deficient soil (Mohammadi 2011; Quiquampoix and Mousain 2005). It plays a major role in P-cycle.

5.6 Soil Enzyme and Herbicide Interaction

Floch et al. (2011) proposed soil enzyme activity as a sustainable indicator of pesticide effects on the soil. Herbicides are applied in agro-ecosystems to hit the target weeds and increase the harvest of desired crop but pose simultaneously a great threat to soil microbial community which eventually leads to the decline of the fertility of soils in agroecosystems. Herbicides may modify the interrelationships between different groups of organisms, thus making an impact on the amount and type of enzymes produced (Tripathi et al. 2005; Pandey et al. 2007a, b). Latha and Gopal (2010) reported a decline in the activity of enzymes when treated with substituted urea herbicides. While studying the dose response, Sireesha et al. (2012) examined increased enzyme activity at lower dose of herbicide application. Singh (2014) observed that overdose of pendimethalin was detrimental for soil enzymes as compared to low or medium dosages. Phenomenal changes in both qualitative and quantitative attributes of soil enzymes in response to herbicidal effects have been observed by many investigators (Sebiomo et al. 2011; Xia et al.

2011; Nikoloff et al. 2013). A number of similar studies consolidated that herbicides behave as enzyme inhibitors (Tejada 2009; Sofo et al. 2012; Vlădoiu et al. 2015).

5.6.1 Dehydrogenase

Baruah and Mishra (1986) conducted incubation studies to examine the influence of three post-emergent herbicides, namely, 2,4-D, butachlor, and oxyfluorfen on dehydrogenase activity with recommended doses in paddy field that constitutes sandy loam soil. They observed that peak rate of dehydrogenase activity followed a trend as follows: 2,4-D > oxyfluorfen > butachlor. Dehydrogenase activity increased with time for the first 7 days and then decreased in subsequent days. Abbas et al. (2015) noted 36% decline in dehydrogenase activity subjected to bromoxynil. Baboo et al. (2013) studied transitory impacts on types and rate of herbicides such as butachlor, pyrazosulfuron, paraquat, and glyphosate on microbial populations and dehydrogenase. Sireesha et al. (2012) conducted a field study for two seasons and found strong link between herbicide treatments and period of their interaction influencing soil enzymes. They reported that with the application of pendimethalin and oxyfluorfen, dehydrogenase activities increased and attained their peak at 30 DAA. They also noted that lower doses of herbicides enhanced the dehydrogenase activity. Tu (1992) conducted laboratory experiment using atrazine, butylate, ethalfluralin, imazethapyr, linuron, metolachlor, metribuzin, and trifluralin, applied to a loamy sand at a rate of 10μg g^{-1}, and reported that the soil dehydrogenase activities were lowered by ethalfluralin application for 1 week. Min et al. (2001) observed gradual increase in dehydrogenase activity in butachlor-treated fluvo-aquic paddy soil, and the enzyme activity showed linearity and attained the maxima on Day 16th following exposure to 22.0 mg g^{-1} butachlor. Zhang et al. (2014) in 60-days incubation experiment with clay and loamy soils showed dehydrogenase activity to be more sensitive to fomesafen compared to acid and alkaline phosphatase and urease. Dehydrogenase activity increased appreciably on Day 10th after herbicide application. Juan et al. (2015) measured the response of soil microbial biomass and enzyme activity to mesotrione, a triketone herbicide. When applied at 50 mg/kg, it escalated soil biomass but reduced the dehydrogenase activity. Dehydrogenases which generally do not accumulate in the extracellular environment received more attention of researchers in response to mesotrione. The activity drops initially but get stimulated in due course of time. Hang et al. (2001), Crouzet et al. (2010), Kaczynska et al. (2015), P. Juan et al. (2015), and Kaczynski et al. (2016) observed dose dependence of dehydrogenase and butachlor. Vandana et al. (2012) in a field experiment reported that butachlor and cyhalofop-butyl when applied at the rate of 1 kg ha^{-1} at 30, 45, and 60 days after transplanting (DAT) enhanced the dehydrogenase activity. Nadiger et al. (2013) also showed dehydrogenase activity at 20 and 40 days after sowing (DAS) in response to pendimethalin and oxyfluorfen when applied at the rate of 0.1 kg ha^{-1}, respectively. Borowik et al. (2016) performed a pot culture experiment using a mixture of three active ingredients of herbicide, Lumax 537.5 SE:

terbuthylazine (T), mesotrione (M), and S-metolachlor (S), using 2,3,5-triphenyltetrazolium chloride (TTC) as a substrate for dehydrogenase. The mixture did show largest variability (83%) in dehydrogenase activity on Day 60. Baćmaga et al. (2014) reported that metazachlor negatively influences dehydrogenases, catalase, urease, acid and alkaline phosphatase, arylsulfatase, and β-glucosidase. Similarly Muñoz-Leoz et al. (2011) found β-glucosidase activity to be negatively influenced by tebuconazole. Contrary to this, stimulating effect on β-glucosidase activity in response to chloroacetanilide herbicides (alachlor, butachlor, and pretilachlor) has been observed by Saha et al. (2012). Wyszkowska et al. (2016), using Eutric Cambisols-filled pot culture experiment, demonstrated dose dependence and persistence effect of pethoxamid (P) and terbuthylazine (T) mixture, with the half-life of 6.1–14.2 days and 5–116 days, respectively, on dehydrogenase activity. Even the smallest dose (0.73 mg P + T kg^{-1}) of soil destabilized enzyme. Higher doses (14.63–468.16 mg P + T kg^{-1}) inhibited the activity by 90.56%. Sebiomo et al. (2011) conducted incubation studies for dehydrogenase responses to four herbicides (atrazine, primeextra, paraquat, and glyphosate). A significant decrease in DHA was observed with values being lowest at 9.02μg (g^{-1} min^{-1}), 12.55μg (g^{-1} min^{-1}), and 16.09μg (g^{-1} min^{-1}) in response to primeextra after second, fourth, and sixth week of treatment, respectively. The highest DHA of 14.32μg (g^{-1} min^{-1}) was recorded after fourth week compared to other treatments. The enzyme exposed to glyphosate was found to be the highest 20.16μg (g^{-1} min^{-1}) after sixth week. A. Kumar et al. (2017) explained the effect of post-emergent herbicide tembotrione soil dehydrogenase. They observed a decrease in DHA at higher doses from 20 to 60 DAS. This was followed by a drastic increase on 60th to 100th Day in all the treatments. Tejada (2009) studied the effects of glyphosate, diflufenican, and a combination of these on dehydrogenase activity. He observed that all the three treatments declined the enzyme activity. The highest decline (37.3%) was recorded in respect of herbicide mixture followed by 35.7% for diflufenican and 32.2% for glyphosate.

5.6.2 Urease

Baruah and Mishra (1986), in an incubation study, examined the influence of recommended doses of three post-emergent herbicides, namely, 2,4-D, butachlor, and oxyfluorfen on urease activity and found no significant effect. Abbas et al. (2015) noticed a 30% decline in urease activity subjected to bromoxynil exposure. Baboo et al. (2013) established a transient effect of types and dose of herbicides butachlor, pyrazosulfuron, paraquat, and glyphosate on microbial populations and urease activity. Kumari et al. (2018) remarked a decline in urease activity on treatment with pre-emergent herbicides atrazine and pendimethalin. The effect was more severe due to pendimethalin in a 60-day incubation experiment. Zhang et al. (2014), unlike positive response on phosphatase and dehydrogenase, showed a remarkable decline in urease activity on Day 10th in response to fomesafen. Du et al. (2018), in another incubation experiment to study the effect of mesotrione exposure, found no effect on urease activity except a mild initial increase. Borowik et al. (2016) observed the largest variability (89%) in urease activity on Day 60 under exposure of herbicide mixture. They reported over 50% decrease in urease activity at 53.768 mg T + M + S. Wyszkowska et al. (2016) showed adverse effect of a mixture of pethoxamid (P) and terbuthylazine (T) on urease activity. They found that even small dose 0.73 mg P + T kg^{-1} could influence the enzyme activity and higher doses (14.63–468.16 mg P + T kg^{-1}) significantly inhibited the activity. Tejada (2009) noted 83.4%, 67.1%, and 58.2% decline in urease activity in response to glyphosate + diflufenican, diflufenican, and glyphosate.

5.6.3 Phosphatase

Bromoxynil application causes a decline in microbial population and consequently 34% reduction in alkaline phosphatase activity (Abbas et al. 2015). Sireesha et al. (2012) used reddish crop to establish connections between herbicide treatments and period of their interaction. They observed that application of pendimethalin and oxyfluorfen causes a decline in acid and alkaline phosphatase activities. Zhang et al. (2014), to show the response of acid and alkaline phosphatase against fomesafen (a diphenyl ether herbicide), conducted a laboratory experiment using clay and loamy soil. Both acid and alkaline phosphatase activities increased significantly on Day 10th after fomesafen treatments although the effect on alkaline phosphatase was relatively mild. Du et al. (2018) in their 20-day laboratory experiment determined the impact of mesotrione on acid phosphatase. They did not observe significant effect at experimental concentrations. Similarly, Aurora 40 WG (carfentrazone-ethyl) did not show negative effect on acid phosphatase (Baćmaga et al. 2014). Rao et al. (2012) showed the response of phosphatase to oxadiargyl, the activity being highest at 0.75 kg ha^{-1} and lowest at 1.5 kg ha^{-1}. Some investigators (Sukul 2006; Yu et al. 2006) unanimously believe a decline of acid phosphatase activity on herbicide application. Majumdar et al. (2010) showed that manual weed control promotes acid phosphatase activity. Borowik et al. (2016) in their pot culture experiment used 4-nitrophenyl phosphate disodium PNPP as a substrate for phosphatase to assess the effect in response to soil contamination with a mixture of three active ingredients of the herbicide Lumax 537.5 SE: terbuthylazine (T), mesotrione (M), and S-metolachlor (S). On Day 30, they observed highest decrease in alkaline phosphatase and acid phosphatase. In another pot culture experiment, the activities of alkaline and acid phosphatase declined by the P + T mixture where the duration of persistence brought 0.54% and 25.99% variability in alkaline and acid phosphatase activity, respectively (Wyszkowska et al. 2016).

5.6.4 *β-glucosidase*

β-glucosidase activity in the soil is sensitive to herbicide and varies with concentration and incubation period and soil status prior to, during, and post-application period (Hussain et al. 2009). Saha et al. (2012) noted higher β-glucosidase activity in the soil treated with pre-emergent herbicides, butachlor and pretilachlor. Latha and Gopal (2010) observed that soil application of pyrazosulfuron, butachlor, and pretilachlor at a rate 100 times the field rate inhibited β-glucosidase by 16.21%, 21.32%, and 10.09%, respectively. At the field rate, the respective decline was only 5.64%, 7.47%, and 3.59%. On the contrary, Sofo et al. (2012) found increased activity of β-glucosidase in response to triasulfuron applied at tenfold higher than the field rate. Santric et al. (2014) observed 5.6–29.4% rise in the response of β-glucosidase activity to nicosulfuron, a sulfonyl urea herbicide, at two elevated doses (3.0 and 30.0 mg) after 7–14 days of exposure. Borowik et al. (2016) conducted a pot culture experiment using *p*-nitrophenyl-β-D-glucopyranoside (PNG) as a substrate to find β-glucosidase activity. The activity reduced by 92% in response to terbuthylazine (T), mesotrione (M), and S-metolachlor (S). Kucharski et al. (2016) showed that dehydrogenase, catalase, urease, arylsulfatase, and β-glucosidase activities declined with soil application of Boreal 58 WG 40 mg kg^{-1}. In a pot culture experiment, Wyszkowska et al. (2016) used Cambisols soil and concluded that the sensitivity of enzymes can be ranked as dehydrogenases > acid phosphatase > urease > alkaline phosphatase > -β-glucosidase > arylsulfatase > catalase. Tejada (2009) used two soil types (Vertic Chromoxerert and Typic Haploxeralf with 575 g kg^{-1} and 161 g kg^{-1} clay content, respectively) to study the effect of herbicides on enzyme activity. Soil treatments with glyphosate + diflufenican, diflufenican, and glyphosate reduced enzyme activity by 7.2%, 5.8%, and 4.6%, respectively.

5.6.5 *Catalase*

Wyszkowska et al. (2016) noted a 21% decrease in catalase activity in response to 468.16 mg kg^{-1} dose of a mixture of pethoxamid (P) and terbuthylazine (T). Borowik et al. (2016) tested the effect of three active ingredients of herbicide Lumax 537.5 SE on the activity of catalase in maize crop and found that the mixture inhibited the activity strongly. About 43% variability in the activity was observed depending on the dose of mixture applied.

Perucci and Scarponi (1994) investigated the effects of imazethapyr, an imidazolinone derivative, on catalase where they observed no adverse effect in the activity at field rate (50 g a.i. ha^{-1}) for soybean weeding. The laboratory treatment at 10-fold and 100-fold higher than the field rates, catalase activity increased.

5.6.6 *Arylsulfatase*

This enzyme hydrolyzes sulfate ester bonds in the extracellular soil environment (Kertesz and Mirleau 2004). Wyszkowska et al. (2016) reported 14.95% decline in arylsulfatase which is relatively less compared to 90.56% decline in urease activity in response to pethoxamid (P) and terbuthylazine (T). Tejada (2009) observed a decreasing trend in the inhibition of arylsulfatase in response to glyphosate + diflufenican followed by diflufenican and glyphosate.

5.7 Factors Affecting Soil Enzyme-Herbicide Interactions

Soil microbial community, soil enzyme activity, and many soil physical chemical properties are influenced by the concentrations and toxicological response variability of herbicides and factors such as climatic variables, soil organic matter, soil texture, temperature, available soil moisture, and pH (Haney et al. 2000; Schreffler and Sharpe 2003). Management practices such as crop type, cultivation system and fertilization, or pesticide application also influence enzyme-herbicide interactions.

5.7.1 *Temperature*

Response of soil dehydrogenase activity (DHA) to temperature has been explored by a large number of researchers. Wolińska and Stępniewska (2012) have reported that dehydrogenase activity increases with increase in temperature unless it reaches to the level of denaturation. Brzezińska et al. (1998) propounded similar results about soil DHA stating that the enzyme activity can be optimized at 28–30 °C under laboratory conditions. Kumari et al. (2018) using Alfisols and Vertisols, incubated at different temperatures ranging from 20 to 70 °C, demonstrated temperature maxima of 70 °C for urease activity. They further studied Alfisols and Vertisols, in temperature ranging from 20 to 90 °C, and observed that acid phosphatase activity increased in temperature range of 20–70 °C and thereafter declined on further rise in temperature. Steinweg et al. (2012) found that β-glucosidase activity remained stable at 15, 25, and 35 °C. Herbicide application with highest efficacy and appropriate temperature and timing favorably influence absorption, translocation, and metabolic degradation of herbicides. Thus a combination of optimum temperature ranges and weed size synergistically influences the herbicide performance. Studies show very obvious effect of growth temperature before, during, and after herbicide application. Ganie et al. (2017) have illustrated that 2,4-D and glyphosate should be applied during warmer days (>20 or ≈29 °C) for better efficiency. According to "Leaders in Farming Technology (2020)," temperature drop is an important issue for weeds to absorb herbicides, very similar to plants facing difficulty in nutrient mobilization at low temperature. Atienza et al. (2001) reported that with a rise in temperature from 5 to 25 °C, the extent of triallate, a pre-emergent selective herbicide dissipation, increases from 14% to 60% in sandy soil and 5–25 °C in loamy soil. Thus, temperature is an important regulator of condition that determines herbicide sensitivity of soil enzymes.

5.7.2 Soil Moisture

Baldrian et al. (2011) observed strong correlation between acid phosphatase activity and soil moisture in horizons L and H both during spring and late summer. However, for other extracellular enzymes such as laccase, Mn-peroxidase, endo-1,4-ß-glucanase, endo-1,4-ß-xylanase, cellobiohydrolase, ß-glucosidase, ß-xylosidase, and chitinase. The correlations were case specific. Sardans and Penuelas (2005) found diminished soil enzyme activities together with fewer microbial biomasses during dry periods in forest soils. Criquet et al. (2000) and Criquet et al. (2004) observed that phenoloxidase, glucosidase, acid phosphatase, urease, and protease activities declined in dry seasons, and that was later endorsed by Sardans and Penuelas (2005). Criquet et al. (2000) found Mn-peroxidase activity in evergreen oak litter during moist season only. Steinweg et al. (2012) observed increased sensitivity of soil moisture to β-glucosidase in drought-treated plot. Zhang et al. (2001) have shown that efficiency of preplant-incorporated (PPI) imazethapyr (a broad-spectrum herbicide) on barnyard grass and red rice was reduced in response to high soil moisture condition, although post-emergent imazethapyr efficacy remained unaltered. Upchurch (1957) analyzed the response of cotton to diuron, DNBP, and CIPC herbicides under variable soil moisture conditions. He concluded that soil moisture had no absolute effect but a large relative effect on phytotoxic properties of diuron. Geisseler et al. (2011) reported that enzyme activity declines on reduction of soil moisture potential. Quilchano and Marañón (2002) did show that soil moisture content is positively correlated with dehydrogenase activity.

5.7.3 Soil Organic Matter

There exists very intimate relationship among soil enzyme activities, microbial population, and soil organic matter content. Bhavya et al. (2017) experimented with different cropping systems, namely, mango, cashew, vegetables, rose, and medicinal and aromatic plants at varying soil depths (0–15, 15–30, 30–50, 50–100 cm) in sandy loam setup. The highest organic carbon content (OCC) was found to be 6500.00 mg kg^{-1} at 0–15 cm, and with the increase in depth, OCCs decreased by 6316.00 mg kg^{-1}, 5846.00 mg kg^{-1}, and 4611.00 mg kg^{-1} at 15–30 cm, 30–50 cm, and 50–100 cm, respectively, obtained in mango orchard followed by cashew orchard. Medicinal and aromatic plant soil held less OCC as

4300.00 mg kg^{-1}, 3916.00 mg kg^{-1}, 3834.00 mg kg^{-1}, and 3786.00 mg kg^{-1} at 0–15 cm, 15–30 cm, 30–50 cm, and 50–100 cm, respectively. The highest dehydrogenase and urease activity 650.84μg g^{-1} soil triphenyl formazan (TPF) and 1230μg g^{-1} soil *p*-nitrophenol (PNP), respectively, was recorded in mango orchard, followed by cashew orchard (9624.64μg TPF g^{-1} soil and 1246μg PNP g^{-1} soil), rose (426.48μg TPF g^{-1} soil and 840.34μg PNP g^{-1} soil), vegetables (421.44μg TPF g^{-1} soil and 821μg PNP g^{-1} soil), and medicinal and aromatic block (418.14μg TPF g^{-1} soil and 800μg PNP g^{-1} soil). Dehydrogenase and urease activity varied with soil depth. The topsoil layer (0–15 cm) was richest in dehydrogenase and urease enzyme activity with the increase in depths; enzyme activities declined irrespective of crop systems. Sondhia (2005) elucidated that butachlor with half-life of 18.1–23.0 days rapidly dissipated under field condition under the influence of soil organic matter and moisture. Sondhia (2014) showed that physical, chemical, and biological properties of soil are influenced by organic manuring, which, in turn, determines the fate of herbicides.

5.7.4 Soil pH

Martínez and Tabatabai (2000) observed a proportional increase in all the 13 study enzymes with a rise in soil pH except acid phosphatase which showed a declining trend. The sensitivity of enzymes to soil pH did appear in the following order: L-glutaminase > alkaline phosphatase > phosphodiesterase > β-glucosidase > acid phosphatase > L-asparaginase > amidase > arylsulfatase > arylamidase > -β-galactosidase > urease > α-galactosidase > α-glucosidase > L-aspartate. Shuler and Kargi (2010) conceptualized that pH influences soil enzymes either by modifying their 3-D shape, altering substrate-enzyme affinity, or by changing active sites. Quilchano and Marañón (2002) and Moeskops et al. (2010) contemplated pH as an important factor influencing soil enzymes. Włodarczyk et al. (2002) observed pH 6.6–7.2 to be the optimum range for dehydrogenase activity.

5.7.5 Soil Texture, Type, and Depth Profile

Stotzky (1985) affirmed that soil textural property can be a key determinant of microbial ecology. Microbial biomass and activity regulating soil moisture content, nutrient translocation, and soil pH are affected by soil texture (Gorres et al. 1998; Leirós et al. 2000). Roy and Singh (2006) described residue retention of clodinafop (0.093–0.081μg g^{-1}) in alluvial, red, and black soil. Martínez et al. (2003) studied the effect of texture of Amarillo soil, Estacado loam, Acuff soil, and Patricia soil containing different ratios of clay, silt, and sand at various soil depth on activities of arylsulfatase, β-glucosidase, β-glucosaminidase, phosphodiesterase, arylamidase, and acid and alkaline phosphatase. The lowest enzyme activity was recorded in

Patricia soil, containing 85% sand and 10% clay, whereas the highest activity was recorded in Estacado loam containing 21% clay and 59% sand. In general, the enzyme activities declined with depth, and the effect was more pronounced in β-glucosidase and arylamidase. Landgraf and Klose (2002) stated that enzyme activities were 1.5-fold higher at 0.5 cm depth than those at 15–30 cm. Quilchano and Marañón (2002) did show positive correlation of clay content with DHA. Clay is fine microporous textured soil that harbors and protects mineralizing microbes from grazers. Therefore, it supports high microbial biomass and higher enzyme activity. Wolińska and Stępniewska (2012) incubated soil samples enriched with glyphosate to see the effect of Mollic Gleysol, Eutric Fluvisol, and Terric Histosol on DHA. The enzyme activity declined in response to pesticides in both the soil samples. At 10μg g^{-1} of glyphosate, the enzyme activity declined by 33–47% in Eutric Fluvisol and Terric Histosol. Tejada (2009) studied dehydrogenase, urease, phosphatase, β-glucosidase, and arylsulfatase in response to glyphosate, diflufenican, and in combination of these. All the enzymes responded negatively to these treatments, and the effects were severe in Typic Haploxeralf soil relative to Vertic Chromoxerert having 161 g kg^{-1} and 575 g kg^{-1} clay content, respectively.

5.7.6 Heavy Metal Amendment

Chemical contaminants pollute soil in complex mixtures rather than as an individual. The abundance, diversity, and distribution of soil organisms are affected by heavy metals. Earthworms in the soil are more sensitive to heavy metals compared to other terrestrial organisms. Uwizeyimana et al. (2017) studied the response of earthworms to pesticides and heavy metals. Pesticides such as atrazine exaggerated the toxic effects of Cd on earthworm. It is supposed that soil fertility is reduced with the decreased number of earthworms as they are assumed to be the key determinant of soil fertility. More than 50% surveyed literatures show synergistic effects of pesticides and heavy metals at higher concentrations.

5.7.7 Cultivation System

Martínez et al. (2003) evaluated the response of acid and alkaline phosphatase activities under four cultivation practices, namely, conservation reserve program (CRP), native rangeland (NR), cotton-cotton conventional tillage (Cv), and cotton-wheat conservation tillage (Cs). The authors observed three to five times higher microbial biomass and enzyme activities under CRP and NR compared to Cv probably due to scarcity of residues during spring and winter season. Other studies reveals that crop rotation promotes enzyme activity under CPR and NR much higher than conventional tillage (Ekenler and Tabatabi 2002; Martínez et al. 2003). Reduced tillage cultivated under various crop and rotation systems consolidate greater diversity of aerobic microbes, facultative anaerobes, and denitrifiers (Franzluebbers 1996; Angers et al. 1997). This probably supports greater microbial biomass responsible for increased soil enzyme activity.

5.7.8 Fertilizer and Pesticide Treatment

Martínez and Tabatabai (2000) explored the impact of lime application on soil pH and enzyme activities. The activity of all the 14 enzymes (α- and β-glucosidases, α- and β-galactosidases, amidase, arylamidase, urease, L-glutaminase, L-asparaginase, L-aspartate, acid and alkaline phosphatases, phosphodiesterase, and arylsulfatase.) increased from 4.9- to 6.9-fold after 7 years of lime application on Kenyon loam soil. Geisseler et al. (2011) concluded that organic residues play important role in regulating extracellular enzymes. Mohammadi (2011) monitored changes in the activities of soil dehydrogenase, acid and alkaline phosphatase, and urease in response to different farmyard manure (N1), compost (N2), and chemical fertilizers (N3);[(N4) = N1 + N2]; [(N5) = N1 + N2 + N3]. All the treatments enhanced the enzyme activities with values being the highest in N4 treatment and lowest in the N1-treated cropland. Singh and Ghoshal (2013) in 2 years of study evaluated the effect of butachlor independently or in combination with soil amendments on β-glucosidase, alkaline phosphatase, and urease in a rice-wheat summer unplowed crop-rotated agroecosystem. β-glucosidase and phosphatase activities were recorded highest under a combination of HC + wheat straw, followed by HC + FYM, HC + sesbania shoot, HC + chemical fertilizer, and HC + control. The urease activity declined under all the treatment mixtures excluding herbicide + wheat straw.

5.8 Conclusions

Soil application of herbicides has dramatically increased the crop yields by eliminating the weeds. However, it has levied a high environmental cost in terms of damages to water and soil environment. Soil enzymes, the major drivers of soil fertility, despite being a nontarget group, are invariably influenced by soil-herbicide interactions. A critical analysis of available literatures shows that although herbicides' interaction with certain enzymes may render stimulatory effects, most of the soil enzymes respond negatively. Here, we identify soil dehydrogenases and urease with strong negative effect of herbicides at higher dose. Enzymes such as acid and alkaline phosphatase, protease, and catalase are least affected due to herbicides' application. The magnitude of these responses, however, differs subject to edaphic and climatic variables that influence microbial communities in the soil. Here, we conclude that because the enzymes are intrinsic attributes of soil fertility, there is need to minimize the negative influence of herbicides on soil enzymes. Therefore, further studies need to be oriented to explore herbicide-specific changes in microbial community structure and function in the soil. This will help screening novel agronomic practices that can support desired microbial communities for maintaining soil fertility under case-specific herbicidal treatments.

Acknowledgment One of the authors, Laliteshwari Bhardwaj, is grateful to Joint-CSIR-UGC for financial assistance in the form of Junior Research Fellow (NTA Ref. No.: 19162004435).

References

Abbas Z, Akmal M, Khan KS, Hassan F (2015) Response of soil microorganisms and enzymes activity to application of buctril super (Bromoxynil) under rainfed conditions. Int J Agric Biol 17:305–312

Alvarez TM, Paiva JH, Ruiz DM, Cairo JPLF, Pereira IO, Paixão DAA, de Almeida RF, Tonoli CCC, Ruller R, Santos CR, Squina FM, Murakami MT (2013) Structure and function of a novel cellulase 5 from sugarcane soil metagenome. PLoS One 8:e83635

Angers D, Bolinder M, Carter M, Gregorich E, Drury C, Liang B et al (1997) Impact of tillage practices on organic carbon and nitrogen storage in cool, humid soils of eastern Canada. Soil Tillage Res 41:191–201

Atienza J, Tabernero MT, Álvarez-Benedí J, Sanz M (2001) Volatilisation of triallate as affected by soil texture and air velocity. Chemosphere 42:257–261

Baboo M, Pasayat M, Samal A, Kujur M, Maharana JK, Patel AK (2013) Effect of four herbicides on soil organic carbon, microbial biomass-C, enzyme activity and microbial populations in agricultural soil. Int J Res Environ Sci Technol 3:100–112

Bach CE, Warnock DD, Van Horn DJ, Weintraub MN, Sinsabaugh RL, Allison SD, German DP (2013) Measuring phenol oxidase and peroxidase activities with pyrogallol, l-DOPA, and ABTS: effect of assay conditions and soil type. Soil Biol Biochem 67:183–191

Baćmaga M, Kucharski J, Wyszkowska J, Borowik A, Tomkiel M (2014) Responses of microorganisms and enzymes to soil contamination with metazachlor. Environ Earth Sci 72 (7):2251–2262

Baldrian P, Kolařík M, Štursová M, Kopecký J, Valášková V, Větrovský TŠ, Žifčáková L, Šnajdr J, Rídl J, Vlček Č, Voříšková J (2011) Active and total microbial communities in forest soil are largely different and highly stratified during decomposition. ISME J 6:248–258

Baruah M, Mishra RR (1986) Effect of herbicide butachlor, 2,4-D and oxyfluorfen on enzyme activities and CO_2 evolution in submerged paddy field soil. Plant Soil 96:287–291

Bhavya VP, Anil Kumar S, Shivanna M, Shivakumar KM, Alur A (2017) Effect of organic matter on soil enzyme activity, organic carbon and microbial activity under different land use systems. Int J Chem Stud 5:301–305

Bielińska EJ, Pranagal L (1994) Effects of the herbicide imazethapyr on soil microbial biomass and various soil enzyme activities. Biol Fertil Soils 17:237–240

Borowik A, Wyszkowska J, Kucharski J, Bacmaga M, Tomkiel M (2016) Response of microorganisms and enzymes to soil contamination with a mixture of terbuthylazine, mesotrione, and S-metolachlor. Environ Sci Pollut Res 24:1910–1925

Brzezińska M, Stępniewska Z, Stępniewski W (1998) Soil oxygen status and dehydrogenase activity. Soil Biol Biochem 30:1783–1790

Buturugă MD, Stefanic G, Săndoiu DI, Bădulescu L (2016) Ecological methods of pedo-enzymatical analysis for soil fertility control. Rom Biotechnol Lett 22:12274–12280

Byrnes BH, Amberger A (1989) Fate of broadcast urea in a flooded soil when treated with N-(n-butyl) thiophosphoric triamide, a urease inhibitor. Fertil Res 18:221–231

Criquet S, Farnet A, Tagger S, Le Petit J (2000) Annual variations of phenoloxidase activities in an Evergreen oak litter: influence of certain biotic and abiotic factors. Soil Biol Biochem 32:11–12

Criquet S, Ferre E, Farnet AM, Le Petit J (2004) Annual dynamics of phosphatase activities in an evergreen oak litter: influence of biotic and abiotic factors. Soil Biol Biochem 36:1111–1118

Crouzet O, Batisson I, Besse-Hoggan P, Bonnemoy F, Bardot C, Poly F, Bohatier J, Mallet C (2010) Response of soil microbial communities to the herbicide mesotrione: a dose-effect microcosm approach. Soil Biol Biochem 42:193–202

Dexter AR (2004) Soil physical quality. Part I. Theory, effects of soil texture, density, and organic matter, and effects on root growth. Geoderma 120:201–214

Dick RP, Burns RG, Dick RP (2011) A brief history of soil enzymology research. In: Dick RP (ed) Methods of soil enzymology. Book Series, no. 9. Wiley, New York, NY

Dotaniya ML, Aparna K, Dotaniya CK, Mahendra S, Regar K (2019) Role of soil enzymes in sustainable crop production. In: Enzymes in food biotechnology. Academic Press, New York, NY, pp 569–589

Du Z, Zhu Y, Zhu L, Zhang J, Li B, Wang J, Wang J, Zhang C, Cheng C (2018) Effects of the herbicide mesotrione on soil enzyme activity and microbial communities. Ecotoxicol Environ Saf 164:571–578

Ekenler M, Tabatabi M (2002) β-Glucosaminidase activity of soils: effect of cropping systems and its relationship to nitrogen mineralization. Biol Fertil Soils 36:367–376

Erman JE, Vitello LB (2002) Yeast cytochrome c peroxidase: mechanistic studies via protein engineering. Biochim Biophys Acta Protein Struct Mol Enzymol 1597:193–220

Fazekasova D (2012) Evaluation of soil quality parameters development in terms of sustainable land use. In: Curkovic S (ed) Sustainable development - authoritative and leading edge content for environmental management. Intech Open, London

Floch C, Chevermont A, Joanico K, Capowiez Y, Criquest S (2011) Indicators of pesticide contamination: soil enzyme compared to functional diversity of bacterial communities via biolog ecoplates. Eur J Soil Biol 47:256

Follmer C (2008) Insights into the role and structure of plant ureases. Phytochemistry 69:18–28

Franzluebbers A (1996) Soil organic matter pools with conventional and zero tillage in a cold, semiarid climate. Soil Tillage Res 39:1–11

Ganie ZA, Lindquist JL, Mithila J, Kruger GR, Marx DB, Jhala AJ (2017) An integrated approach to control glyphosate-resistant giant ragweed (Ambrosia trifida) with preplant tillage and herbicides in glyphosate-resistant corn. Weed Res 57:112–122

Geisseler D, Horwath W, Scow K (2011) Soil moisture and plant residue addition interact in their effect on extracellular enzyme activity. Pedobiologia 54:71–78

Gianfreda L, Bollag J-M (1996) Influence of natural and anthropogenic factors on enzyme activity in soil. In: Stotzky G, Bollag J-M (eds) Soil biochemistry, vol 9, pp 123–194

Gianfreda L, Rao MA (2008) Interaction between xenobiotics and microbial and enzymatic soil activity. Crit Rev Environ Sci Technol 38:269–310

Gianfreda L, Rao AA (2014) Enzymes in agricultural sciences. OMICS Group e Books, Foster City, CA

Gianfreda L, Ruggiero P (2006) Enzyme activities in soil. In: Nannipieri P, Smalla K (eds) Nucleic acids and proteins in soil, vol 8. Springer, New York, NY, pp 257–311

Gorres JH, Dichiaro MJ, Lyons JB, Amador JA (1998) Spatial and temporal patterns of soil biology activity in a forest and an old field. Soil Biol Biochem 30:219–230

Gu Y, Wang P, Kong CH (2009) Urease, invertase, dehydrogenase and polyphenoloxidase activities in paddy soil influenced by allelopathic rice variety. Eur J Soil Biol 45:436–441

Gupta PK (2004) Pesticide exposure—Indian scene. Toxicology 198:83–90

Haney RL, Senseman SA, Hons FM, Zuberer DA (2000) Effect of glyphosate on soil microbial activity and biomass. Weed Sci 48:89–93

Hang M, Yangfang Y, Zhongyun C, Wei-xiang W, Yunfeng D (2001) Effects of butachlor on microbial populations and enzyme activities in paddy soil. J Environ Sci Health B 36:581–595

Hussain S, Siddique T, Saleem M, Arshad M, Khalid A (2009) Impact of pesticides on soil microbial diversity, enzymes, and biochemical reactions. Adv Agron 102:159–200

Infinita Biotech (2019). https://infinitabiotech.com/blog/category/enzymes

Jimenez De la Paz MED, Horra de la A, Pruzzo L, Palma M (2002) Soil quality: a new index based on microbiological and biochemical parameters. Biol Fertil Soils 35:302–306

Juan PE, Sánchez-Martín MJ, Herrero-Hernández E, Rodríguez-Cruz MS (2015) Application of mesotrione at different doses in an amended soil: dissipation and effect on the soil microbial biomass and activity. Sci Total Environ 536:31–38

Kaczynska G, Borowik A, Wyszkowska J (2015) Soil Dehydrogenases as an indicator of contamination of the environment with petroleum products. Water Air Soil Pollut 226(11):372

Kaczynski P, Lozowicka B, Hrynko I, Wolejko E (2016) Behaviour of mesotrione in maize and soil system and its influence on soil dehydrogenase activity. Sci Total Environ 571:1079–1088

Kertesz MA, Mirleau P (2004) The role of soil microbes in plant sulphur nutrition. J Exp Bot 55:1939–1945

Khan S, El-Latif Hesham A, Qiao M, Rehman S, He J-Z (2009) Effects of Cd and Pb on soil microbial community structure and activities. Environ Sci Pollut Res 17:288–296

Kızılkaya R, Aşkın T, Bayraklı B, Sağlam M (2004) Microbiological characteristics of soils contaminated with heavy metals. Eur J Soil Biol 40:95–102

Kucharski J, Tomkiel M, Baćmaga M, Borowik A, Wyszkowska J (2016) Enzyme activity and microorganisms diversity in soil contaminated with the Boreal 58 WG herbicide. J Environ Sci Health B 51:446–454

Kudsk P, Streibig JC (2003) Herbicides – a two-edged sword. Eur Weed Res Soc Weed Res 43:90–102

Kumar A, Rana M, Sharma N, Rana S (2017) Effect of post-emergence herbicide -tembotrione on yield, soil dehydrogenase activity and its phytotoxicity on maize (Zea mays L.) under mid hill conditions of Himachal Pradesh, India. Int J Curr Microbiol App Sci 6:2297–2303

Kumari JA, Rao PC, Madhavi M, Padmaja G (2018) Effect of herbicides on the activity of soil enzymes urease in maize crop. Ind J Agric Res 52:300

Landgraf D, Klose S (2002) Mobile and readily available C and N fractions and their relationship to microbial biomass and selected enzyme activities in a sandy soil under different management systems. J Plant Nutr Soil Sci 165:9

Latha PC, Gopal H (2010) Effect of herbicides on soil microorganisms. Indian J Weed Sci 42:217–222

Leirós M, Trasar-Cepeda C, Seoane S, Gil-Sotres F (2000) Biochemical properties of acid soils under climax vegetation (Atlantic oakwood) in an area of the European temperate–humid zone (Galicia, NW Spain): general parameters. Soil Biol Biochem 32:733–745

Macías FA, Mejías FJR, Molinillo JMG (2019) Recent advances in allelopathy for weed control: from knowledge to applications. Pest Manag Sci 75:2413–2436

Majumdar B, Saha AR, Sarkar S, Maji B, Mahapatra BS (2010) Effect of herbicides and fungicides application on fibre yield and nutrient uptake by jute (Corchorus olitorius), residual nutrient status and soil quality. Indian J Agric Sci 80:878

Martínez VA, Tabatabai MA (2000) Enzyme activities in a limed agricultural soil. Biol Fertil Soils 31:85–91

Martínez VA, Klose S, Zobeck TM (2003) Enzyme activities in semiarid soils under conservation reserve program, native grassland and cropland. J Plant Nutr Soil Sci 166:699–707

Mauprivez CT, Laurent FM, Calvayrac C, Barthelmebs L (2019) Effects of herbicide on non-target microorganisms: towards a new class of biomarkers? Sci Total Environ 684:314–325

Min H, Ye Y, Chen Z, Wu W, Yufeng D (2001) Effects of butachlor on microbial populations and enzyme activities in paddy soil. J Environ Sci Health 36:581–595

Mobley HL, Hausinger RP (1989) Microbial ureases: significance, regulation, and molecular characterization. Microbiol Rev 53:85–108

Moeskops B, Buchan D, Slutel S, Herawaty L, Husen E, Saraswati R, Setyorini D, De Neve S (2010) Soil microbial communities and activities under intensive organic and conventional vegetable farming in West Java, Indonesia. Appl Soil Ecol 45:112–120

Mohammadi K (2011) Effect of different fertilization methods on soil biological indexes, World Academy of Science, Engineering and Technology. Int J Agric Biosyst Eng 5:329

Muñoz-Leoz B, Ruiz-Romera E, Antigüedad I, Garbisu C (2011) Tebuconazole application decreases soil microbial biomass and activity. Soil Biol Biochem 43:2176–2183

Nadiger S, Babu R, Kumar BN (2013) Bioefficacy of pre-emergence herbicides on weed management in maize. J Agric Sci 26:17–19

Nannipieri P, Kandeler E, Ruggiero P (2002) Enzyme activities and microbiological and biochemical processes. In: Burns RG, Dick RP (eds) Enzymes in the environment. Activity, ecology and applications. Marcel Dekker, New York, NY, pp 1–3

Nannipieri P, Giagnoni L, Renella G, Puglisi E, Ceccanti B, Masciandaro G, Foranasier F, Moscatelli MC, Marinari S (2012) Soil enzymology: classical and molecular approaches. Biol Fertil Soils 48:743–762

Nikoloff N, Escobar L, Soloneski S, Larramendy ML (2013) Comparative study of cytotoxic and genotoxic effects induced by herbicide S-metolachlor and its commercial formulation Twin Pack Gold® in human hepatoma (HepG2) cells. Food Chem Toxicol 62:777–781

Nonga HE, Mdegela RH, Lie E, Sandvik M, Skaare JU (2011) Assessment of farming practices and uses of agrochemicals in Lake Manyara basin, Tanzania. Afr J Agric Res 6:2216–2230

Oerke EC (2005) Crop losses to pests. J Agric Sci 144:31–43

Omar SA, Abdel-Sater MA (2001) microbial populations and enzyme activities in soil treated with pesticides. Water Air Soil Pollut 127:49–63

Pandey RR, Sharma G, Tripathi SK, Singh AK (2007a) Litterfall, litter decomposition and nutrient dynamics in a subtropical natural oak forest and managed plantation in northeastern India. For Ecol Manag 240:96–104

Pandey S, Bhandari H, Ding S, Prapertchob P, Sharan R, Naik D, Taunk SK, Sastri A (2007b) Coping with drought in rice farming in Asia: insights from a cross-country comparative study. Agric Econ 37:2134–2224

Passardi F, Zamocky M, Favet J, Jakopitsch C, Penel C, Obinger C, Dunand C (2007) Phylogenetic distribution of catalase-peroxidases: are there patches of order in chaos? Gene 397:101–113

Perucci P, Scarponi L (1994) Effects of the herbicide imazethapyr on soil microbial biomass and various soil enzyme activities. Biol Fertil Soils 17:237–240

Pimentel D (1995) Amounts of pesticides reaching target pests: environmental impacts and ethics. J Agric Environ Ethics 8:17–29

Quilchano C, Marañón T (2002) Dehydrogenase activity in Mediterranean forest soils. Biol Fertil Soils 35:102–107

Quiquampoix H, Mousain D (2005) Enzymatic hydrolysis of organic phosphorus. In: Turner BL, Frossard E, Baldwin D (eds) Organic phosphorus in the environment. CAB International, Wallingford, pp 89–112

Raj SK, Syriac EK (2017) Herbicidal effect on the bio-indicators of soil health - a review. J Appl Nat Sci 9:2438–2448

Rana SS, Kumar R, Sharma N, Badiyala D (2016) Effect of continuous use of herbicides on weed shifts in rice-wheat system: technical Bulletin No-1. Department of Agronomy, College of Agriculture, CSK Himachal Pradesh Krishi Vishvavidyalaya, Palampur, HP

Rao PC, Lakshmi C, Sireesha A, Madhavi M, Swapna G (2012) Effect of oxadiargyl on soil enzyme activity. J Crop Weed 8:52–56

Riah W, Laval K, Laroche-Ajzenberg E, Mougin C, Latour X, Trinsoutrot-Gattin I (2014) Effects of pesticides on soil enzymes: a review. Environ Chem Lett 12:257–273

Roy S, Singh SB (2006) Effect of soil type, soil pH, and microbial activity on persistence of clodinafop herbicide. Bull Environ Contam Toxicol 77:260–266

Saha S, Dutta D, Karmakar R, Ray DP (2012) Structure–toxicity relationship of chloroacetanilide herbicides: relative impact on soil microorganisms. Environ Toxicol Pharmacol 34:307–314

Sahoo S, Adak T, Bagchi T, Kumar U, Munda S, Saha S, Berliner J, Jena M, Mishra B (2016) Effect of pretilachlor on soil enzyme activities in tropical rice soil. Bull Environ Contam Toxicol 98:439–445

Salazar S, Sánchez L, Alvarez J, Valverde A, Galindo P, Igual J et al (2011) Correlation among soil enzyme activities under different forest system management practices. Ecol Eng 37:1123–1131

Santric L, Radivojevic L, Umiljendic G, Djurovic-Pejcev R, Saric-Krsmanovic M (2014) Assessment of microbial activity and biomass in different soils exposed to nicosulfuron. Pesticidi i fitomedicina 29:213–219

Sardans J, Penuelas J (2005) Drought decreases soil enzyme activity in a Mediterranean holm oak forest. Soil Biol Biochem 37:455–461

Schreffler AM, Sharpe WE (2003) Effects of lime, fertilizer, and herbicide on forest soil and soil solution chemistry, hardwood regeneration, and hardwood growth following shelterwood harvest. For Ecol Manag 177:471–484

Sebiomo A, Ogundero VW, Bankole SA (2011) Effect of four herbicides on microbial population, soil organic matter and dehydrogenase activity. Afr J Biotechnol 10:770–778

Shea P (1985) Detoxification of herbicide residues in soil. Weed Sci 33:33–41

Sherwani SI, Arif IA, Khan HA (2015) Modes of action of different classes of herbicides. In: Herbicides, physiology of action and safety. IntechOpen, London

Shiyin L, Lixiao N, Panying P, Cheng S, Liansheng W (2004) Effects of pesticides and their hydrolysates on catalase activity in soil. Bull Environ Contam Toxicol 72(3):600–606

Shuler M, Kargi F (2010) Bioprocess engineering basic concepts. Prentice Hall Incorporation, Englewood Cliffs, New Yersey, USA. ISBN: 10:0130819085

Singh R (2014) Influence of pendimethalin application with different fertilizers on soil enzyme activities in bulandshahr soil. Int J Sci Res Publ 4:1

Singh A, Ghoshal N (2013) Impact of herbicide and various soil amendments on soil enzymes activities in tropical rainfed agroecosystem. Eur J Soil Biol 54:56–62

Singh B, Singh K (2014) Microbial degradation of herbicides. Crit Rev Microbiol 42:245–261

Sireesha A, Rao PC, Ramalaxmi CS, Swapna G (2012) Effect of pendimethalin and oxyfluorfen on soil enzyme activity. J Crop Weed 8:124–128

Sofo A, Scopa A, Dumontet S, Mazzatura A, Pasquale V (2012) Toxic effects of four sulphonylureas herbicides on soil microbial biomass. J Environ Sci Health 47:653–659

Sondhia S (2005) Phytotoxicity and persistence of metribuzin residues in black soils. Toxicol Environ Chem 87:389–397

Sondhia S (2014) Herbicides residues in soil, water, plants and non-targeted organisms and human health implications: an Indian perspective. Ind J Weed Sci 46:66–85

Sondhia S (2018) Environmental fate of herbicide use in Central India. Herb Resid Res India:29–104

Sopeña F, Maqueda C, Morillo E (2009) Controlled release formulations of herbicides based on micro-encapsulation. Cien Inv Agr 35(1):27–42

Steinweg JM, Dukes JS, Wallenstein MD (2012) Modeling the effects of temperature and moisture on soil enzyme activity: linking laboratory assays to continuous field data. Soil Biol Biochem 55:85–92

Stotzky G (1985) Mechanisms of adhesion to clays, with reference to soil systems. In: Bacterial adhesion. Plenum Press, New York, NY, pp 195–253

Sukul P (2006) Enzymatic activities and microbial biomass in soil as influenced by metalaxyl residues. Soil Biol Biochem 38:320–326

Tejada M (2009) Evolution of soil biological properties after addition of glyphosate, diflufenican and glyphosate + diflufenican herbicides. Chemosphere 76:365–373

Trasar-Cepeda C, Gil-Sotres F, Leirós MC (2008) Corrigendum to "Thermodynamic parameters of enzymes in grassland soils from Galicia, NW Spain". Soil Biol Biochem 40:1–269

Tripathi S, Kumari S, Chakraborty A, Gupta A, Chakrabarti K, Bandyapadhyay BK (2005) Microbial biomass and its activities in salt-affected coastal soils. Biol Fertil Soils 42:273–277

Tu CM (1992) Effect of some herbicides on activities of microorganisms and enzymes in soil. J Environ Sci Health 27:695–702

Turner BL, Hopkins DW, Haygarth PM, Ostle N (2002) β-Glucosidase activity in pasture soils. Appl Soil Ecol 20:157–162

Upchurch RP (1957) The influence of soil moisture content on the response of cotton to herbicides. Weeds 5:112

USDA Natural Resources Conservation Service (2015) Soil quality indicators physical, chemical, and biological indicators for soil quality assessment and management

Uwizeyimana H, Wang M, Chen W, Khan K (2017) The eco-toxic effects of pesticide and heavy metal mixtures towards earthworms in soil. Environ Toxicol Pharmacol 55:20–29

Vandana LJ, Rao PC, Padmaja G (2012) Effect of herbicides and nutrient management on soil enzyme activity. J Rice Res 5:1–2

Vats S (2014) Herbicides: history, classification and genetic manipulation of plants for herbicide resistance. In: Sustainable agriculture reviews. Springer, Cham, pp 153–192

Vlădoiu DL, Filimon MN, Ostafe V, Isvoran A (2015) Assessment of pesticides interactions with bacillus pasteurii urease. A computational study, Paper presented at the 14th International Balkan Workshop on Applied Physics, July 2-4, 2014, Constanta, Romania. Rom J Phys 60:583–592

Wang J, Lu Y, Shen G (2006) Combined effects of cadmium and butachlor on soil enzyme activities and microbial community structure. Environ Geol 51:1221–1228

Włodarczyk T, Stępniewski W, Brzezińska M (2002) Dehydrogenase activity, redox potential, and emissions of carbon dioxide and nitrous oxide from Cambisols under flooding conditions. Biol Fertil Soils 36:200–206

Wolińska A, Stępniewska Z (2012) Dehydrogenase activity in the soil environment. In: Canuto RA (ed) Dehydrogenases. IntechOpen, London

Woodburn AT (2000) Glyphosate: production, pricing and use worldwide. Pest Manag Sci 56:309–312

Wyszkowska J, Tomkiel M, Bacmaga M, Borowik A, Kucharski J (2016) Response of microorganisms and enzymes to soil contamination with a mixture of pethoxamid terbuthylazine. Environ Earth Sci 75:1285

Xia XM, Zhao M, Wang HY, Ma H (2011) Influence of butachlor on soil enzymes and Microbial growth. J Food Agric Environ 9:753–756

Yu YL, Shan M, Fang H, Wang X, Chu XQ (2006) Responses of soil microorganisms and enzymes to repeated applications of chlorothalonil. J Agric Food Chem 54:10070–10075

Yuan BC, Yue DX (2012) Soil microbial and enzymatic activities across a chronosequence of Chinese pine plantation development on the loess plateau of China. Pedosphere 22:1–12

Zabaloy MC, Zanini GP, Bianchinotti V, Gomez MA, Garland JL (2011) Herbicides in the soil environment: linkage between bioavailability and microbial ecology. In: Larramendy M (ed) Herbicides, theory and applications. Intech, London. ISBN: 978-953-307-975-2

Zhang W, Webster EP, Selim HM (2001) Effect of soil moisture on efficacy of imazethapyr in green house. Weed Technol 12:355–359

Zhang Q, Zhu L, Wang J, Xie H, Wang J, Wang F, Sun F (2014) Effects of fomesafen on soil enzyme activity, microbial; population, and bacterial community composition. Environ Monit Assess 186:2801–2812

Zhao B, Chen J, Zhang J, Qin S (2010) Soil Microbial biomass and activity response to repeated drying-rewetting cycles along a soil fertility gradient modified by long-term fertilization management practices. Geoderma 160:218–224

Chapter 6
Diversity of Pathogenic Fungi in Agricultural Crops

Shivannegowda Mahadevakumar and Kandikere Ramaiah Sridhar

Abstract Fungi constitute an important group of organisms that possess beneficial as well as negative traits against plants and animals. Association of fungi with plants is mostly saprotrophic and involves in decomposition. However, a multitude of fungal species are widely recognized as plant pathogens owing to many diseases in crops like potato, paddy, wheat, maize, pulses, oil-yielding plants, floricultural crops, horticultural crops, plantation crops, and so on. Over 70% percent of plant disease is due to fungal pathogens, and they are usually parasitic and exhibit disease symptoms. Biotrophic fungal pathogens exhibit long-term establishment by obtaining nutrients from live host tissues via specialized cells "haustoria" that develop inside the host. Necrotrophic pathogens fetch nutrients from the dead host tissues by killing the tissues with toxins or enzymes, whereas biotrophs have a narrow host range. However, necrotrophs are generalists with a wide host range or specialized with a narrow host range for their survival. Recent advances in molecular biology and sequencing platforms enable the exploration of diverse plant pathogenic fungi associated with crop plants. This chapter intends to summarize the diversity of plant pathogenic fungi on selected agriculturally important crops. It includes the detailed comprehension of plant disease concepts, classification of plant pathogenic fungi based on their lifestyle, fungal diseases of historical records, major fungal diseases of crop plants (rice, maize, and vegetables), and global perspectives of major pathogenic genera.

Keywords Cereals · Millets · Oilseeds · Pulses · Timbers · Vegetables · *Colletotrichum* · *Diaporthe* · *Fusarium* · *Pestalotiopsis* · *Phytophthora* · *Sclerotium rolfsii*

S. Mahadevakumar
Department of Studies in Botany, University of Mysore, Mysore, Karnataka, India

K. R. Sridhar (✉)
Department of Biosciences, Mangalore University, Mangalore, Karnataka, India

Centre for Environmental Studies, Yenepoya (Deemed to be University), Mangalore, Karnataka, India

S. K. Dubey, S. K. Verma (eds.), *Plant, Soil and Microbes in Tropical Ecosystems*, Rhizosphere Biology, https://doi.org/10.1007/978-981-16-3364-5_6

6.1 Introduction

India is known for diverse agro-climatic zones such as tropical, subtropical, and temperate climates ranging from average to high in temperature, humidity, and rainfall. Thus, the Indian subcontinent is the home to rich flora and fauna in the world. India ranks second in the output of fruits, vegetables, cereals, pulses, oilseeds, fiber crops, sugarcane, spices, and ornamentals (Anonymous 2009). About 40% of the geographical area of the Indian subcontinent is utilized for agriculture, thus playing a crucial role in comprehensive socioeconomic development. India is a major producer of several crops; the major crops could be placed in four different categories:

1. Food grains (maize, millets, pulses, rice, and wheat)
2. Cash crops (fiber-yielding crops, cotton and jute; sugar-yielding crops, sugarcane and tobacco; oilseeds, soybean, castor, sesame, and others)
3. Plantations (coffee, coconut, rubber, and tea)
4. Horticultural crops (fruits and vegetables)

Even though India is producing a large number of agricultural products, it succumbs to substantial loss of production due to diseases caused by various biotic factors. These include diseases caused by bacteria, fungi, insect pests, nematodes, phytoplasma, viroids, viruses, and others. The extent of losses caused by such biological agents differs from crops as well as seasons. As per the data released by the Parliament on Agriculture and Farmers Welfare, loss of crop production up to 15–20% is due to diseases. Thus, India raises agricultural production to ensure food security and nutrition for the teeming population. A total of 68% net loss of global agricultural production is due to various pests and diseases (microbial diseases, 16%; animals and pests, 18%; weeds, 34%; pathogenic fungi, 70–80%) (Oerke 2006).

The overall loss of yield occurs by various pests, diseases, and weeds during growth and post-harvest are of paramount significance for raise in loss up to 10–30%. The economic loss could be up to 15.4 billion US$. The average crop loss (20%) by pests and diseases was up to 1.4 billion US$ (Kumar and Gupta 2012). The current situation is further worsening due to the appearance of new diseases, pathotypes, and variants (pathovars) of pathogens, as they can adapt in varying climates. The extent of losses incurring due to pests and diseases is higher than the extent of production via innovative programs (Kumar and Saxena 2009). Agricultural loss due to plant diseases may also be attributed to other direct or indirect economic failures by various factors: (1) reduced quality as well as quantity of crop production; (2) increased cost of production; (3) threat on animal health and environment; (4) limiting the type of crops/varieties grown; (5) loss of natural resources; and (6) less remunerative alternatives (Kumar 2014).

Fungi are well known for agricultural diseases; being eukaryotes, fungi are capable to fetch nutrients externally and absorb them through their cell walls. The majority of fungi reproduce via spores and possess thallus constituting microscopic tubular hyphae. Fungi as heterotrophs obtain carbon and energy through live organisms or detritus. Nutrition derivation by fungi from living hosts is referred to as biotrophs, while nutrition from dead plants or animals is called as saprotrophs. The process of infecting a living host and killing host cells for nutrition purposes are referred to as necrotrophs. However, organisms conventionally classified as kingdom "fungi" are divided into three unrelated groups such as true fungi (eumycota), the oomycetes, and the slime molds. As per the classification proposed by Alexopoulos et al. (1997) and subsequent literature, there are four major groups of true fungi, viz., *Ascomycota*, *Basidiomycota*, *Chytridiomycota*, and *Zygomycota* (Webster and Weber 2007). Members having different cell wall compositions and flagellated zoospores are placed under Oomycota. Recent inferences suggest the addition of *Glomeromycota* as a phylum, under *Zygomycota*, which is known to develop an association with the roots of most of the plant species as arbuscular mycorrhizae. To date, it is unclear how many species of fungi exist globally. A conservative estimate based on the angiosperm/fungus ratio will be between 2.2 and 3.8 million species, and so far only 3.7–6.4% of fungi were recorded (Hawksworth and Lücking 2017). This chapter addresses the diversity of plant pathogenic fungi in crops with major emphasis on diverse symptoms, diverse groups of pathogenic fungi, diversity among the individual groups, and current developments in the evaluation of the diversity of plant pathogenic fungi in crop plants.

6.2 Concepts of Plant Disease

Plants make up the majority of the earth's living environment and provide necessary nutrition to humans as well as animals. Plants convert energy from sunlight into stored exploitable chemical energy, viz., carbohydrates, proteins, and lipids. Animals depend on plant substances for their survival (Agrios 2005). Plant growth is dependent on soil providing adequate moisture and nutrients and sufficient light reaching leaves, and temperature remains within the normal range. Sickness in plants leads to poor growth as well as poor production. The agents that cause infection in plants include bacteria, fungi, nematodes, protozoa, and viruses. Plants are also known to suffer from the competition with other plants (weeds) and also damages caused by insects.

Fungal plant pathogens cause a wide array of diseases in major crops globally and cause substantial loss of yield (Anderson et al. 2004; Strange and Scott 2005; Rossman et al. 2014). Many phytopathogenic fungi have devastating threats in the history of agriculture. So far, ancient knowledge recognized as major of rust infection in crops and also the smuts infecting monocot crops such as barley (*Ustilago hordei*), wheat (*Ustilago tritici*), and maize (*Ustilago maydis*). A historically known threatened fungal plant pathogen is *Claviceps purpurea* (ergot of rye, barley, oats, and wheat). The sclerotia of *C. purpurea* are known to have a broad range of toxic alkaloids and replace kernels in the heads of crops, thereby causing contamination of harvested grains and flours.

6.3 Diversity Based on Lifestyle

Fungal pathogens are collectively referred to as those that derive nutrients from plants and induce a negative impact on plants' health. Some pathogens completely depend on their host (as obligate parasites), and some others as facultative exist in close association with the hosts to compete with its life cycle (phenology). The facultative pathogens are capable to prosper in varied environments; thus, they are capable to cause diseases in several hosts successfully. The traditional classification of plant pathogens (necrotrophs, biotrophs, and hemibiotrophs) differentiates fungi based on their lifestyle and the strategies of dependence on host. The recent reports suggest that such division is less stringent than previously realized; however, such categorization defines basic denominators as those that are common to all in each class and simplifies the lifestyle of pathogens.

Plant pathogenic fungi are ubiquitous and exhibit varied lifestyles. Many of them display a range of lifestyles from biotrophy through necrotrophy and ultimately to saprotrophy. Biotrophism is dependent on host plant, whereas hemibiotrophs shift from the initial biotrophic phase to the necrotrophic phase. The necrotrophic life cycle involves the active killing of host cells by secretion of cell wall-degrading phytotoxins and enzymes. The biotrophic fungi develop a close relationship with the host by haustorium, a specialized structure for assimilation of nutrition. It is proposed that biotrophy could essentially modulate plant defense mechanisms. Therefore, biotrophs adopt diverse strategies to counter the host plant defenses (De Silva et al. 2016). Specific examples of biotrophic fungi as well as oomycetes and other lifestyles are presented in Table 6.1. The discussion proceeds on plant pathogenic fungi under three subheadings based on their lifestyles (biotrophs, necrotrophs, and hemibiotrophs).

6.3.1 Biotrophs

Biotrophic fungi are those dependent on a narrow range of hosts for deriving nutrition from host's living cells leading to the damage of host tissue. They produce structures such as haustoria and appressoria to penetrate and acquire nutrients from the host (De Silva et al. 2016). Plant pathogenic fungi adapted to biotrophic mode of lifestyle are either obligate or non-obligate parasites. The powdery mildews (*Ascomycota*) and rusts (*Basidiomycota*) are the best representatives of obligate biotrophs causing diseases on various economically important crops like cereals, millets, vegetables, and horticultural crops. The downy mildews and white rusts (except white rust on *Chrysanthemum*) are also obligate biotrophs that belonged to Oomycota (protozoa) (Schulze-Lefert and Panstruga 2003).

The obligate biotrophs like rusts and powdery mildews have developed their lifestyle to match with the phenology of the hosts to complete the life cycle. Specifically, rusts possess five varied spore stages (aeciospores, pycniospores,

Table 6.1 Examples of fungi with different lifestyles (including oomycetes)

Lifestyle	Fungal pathogen	Host	Reference
Biotrophs	Rusts:		Oliver and Ipcho (2004), Meinhardt et al. (2014)
	Puccinia arachidis	Groundnut	
	Puccinia graminis	Wheat	
	Puccinia horiana	*Chrysanthemum*	
	Uromyces appendiculatus	Cowpea	
	U. fabae	Common beans	
	Smuts:		Mendgen and Hahn (2002), Latijnhouwers et al. (2003)
	Sphacelotheca sorghi	*Sorghum*	
	Ustilago maydis	Corn/maize	
	Ustilaginoidea virens	Rice	
	Powdery mildews:		Oliver and Ipcho (2004), Delaye et al. (2013)
	Blumeria graminis	Barley	
	Erysiphe cichoracearum	Cucurbits	
	Leveillula taurica	Chili and tomato	
	Podosphaera oxyacanthae	–	
	Podosphaera xanthii	Brinjal	
	Sphaerotheca mors-uvae	Gooseberry	
	Sphaerotheca pannosa	Rose	
Biotrophic oomycetes	*Albugo candida*	Crucifers	Latijnhouwers et al. (2003), Figueiredo et al. (2015)
	Peronospora parasitica	–	
	Plasmopara viticola	Grapes	
Hemibiotrophs	*Colletotrichum destructivum*	–	Damm et al. (2014)
	C. lindemuthianum	Common bean	Mendgen and Hahn (2002)
	Fusarium oxysporum	–	Krola et al. (2015)
	Gibberella zeae	Maize/corn	Kabbage et al. (2015)
	Magnaporthe oryzae	Rice	Oliver and Ipcho (2004), Kankanala et al. (2007), Kabbage et al. (2015)
	Moniliophthora roreri	–	Meinhardt et al. (2014)
	Mycosphaerella graminicola	–	Spanu (2012)
Hemibiotrophic oomycetes	*Phytophthora capsici*, *Phytophthora* sp., and *Pythium* sp.	Potato, tomato, beans, and vegetables	Meadows (2011)
Necrotrophs	*Alternaria brassicicola*	Crucifers	Spanu (2012), Pandey et al. (2016)
	Botrytis cinerea	Fruit crops	
	Leptosphaeria maculans	–	
	Sclerotinia sclerotiorum	Cabbage	

urediniospores, teliospores, and basidiospores) developed in specialized fruit bodies (aecidium, pycnium, uredenium, telium, and basidium); thus, they are capable to infect alternate hosts to complete the life cycle. But not all rusts have such mechanisms of completion of life cycles (Petersen 1974). Rust diseases are due to members of basidiomycetes of the order *Pucciniales* (earlier considered under the order *Uredinales*) (Duplessis et al. 2011). Usually, the urediniospores (dikaryotic) germinate on the leaf surface (on primary host) and produce penetration plug or haustorium, which invades the mesophyll tissue through stomata. The haustoria further differentiate to form sub-stomatal vesicles to develop the hyphae within the host tissue. On the contrary, the powdery mildews are caused by filamentous ascomycetous fungi (haploids) belonging to the order *Erysiphales*. The germination of conidia of these fungi occurs on the leaf surface, and appressoria helps in the penetration of the epidermal tissues (Spanu et al. 2010; Hückelhoven and Panstruga 2011). Following the penetration from the surface, the rusts cross the wall of mesophyll, and powdery mildews cross the wall of the epidermis and differentiate into haustoria, which are exclusive infection erections that help to survive inside the host tissues through the acquisition of nutrients. Usually, the haustoria are formed behind the plant cell wall without disrupting the cell membrane, and they push and invaginate the cell membrane establishing a maintainable "cell within cell" complex (Heath and Skalamera 1997). Once the fungi establish inside the host tissues, the haustoria secrete a broad varied array of transporters to derive the nutrition from the host (surrounding the living tissues where the obligate pathogen entered) into the haustorium which, in turn, nourishes the fungi to successfully colonize the tissues and extends its territory inside the host (Struck 2015; Voegele and Mendgen 2011; Voegele et al. 2001). They also synthesize effector molecules to suppress the host defense and keep the cells alive (Kemen et al. 2005, 2013; Petre et al. 2016). The obligate biotrophic fungi fully depend on their host for energy, and they are aptly designated as obligate parasites or energy parasites (Schulze-Lefert and Panstruga 2003).

Non-obligate biotrophs are capable to survive as true biotrophs in living tissues/host and capable to grow and survive without the presence of a host. The ergot disease caused by *Claviceps purpurea* acts as a true biotroph in the host, and it can also be grown in axenic culture (Tudzynski and Scheffer 2004). These non-obligate biotrophic fungal pathogens are taxonomically diversified throughout a wide range of genera, and important ones are the smuts belonging to the order *Ustilaginales* (*Basidiomycota*) and certain species in *Claviceps* (*Ascomycota*, *Clavicipitaceae*). In ergot disease, the wind-borne ascospores germinate on the pistil surface during anthesis and penetrate through the stigmatic hairs and colonize the ovarian tissue and launch a specific and persisting host-pathogen interface. A mycelial stroma develops in the ovary with the production of mass of conidiospores, and they exude sugar-rich fluid from the phloem sap. Sclerotia (overwintering structures) are formed after 10–15 days of post-infection. *Claviceps purpurea* does not produce classical haustoria and intracellular hyphae; instead, this fungus is completely covered by the host plasma membrane (Tudzynski and Tenberge 2003). The smut pathogen in maize *U. maydis* stands out among other smuts due to its acquiescence to molecular genetic manipulation and its small genome size (Brefort et al. 2009).

6.3.2 Necrotrophs

The necrotrophic fungi are those which feed on dead plant tissues (by killing the healthy tissues). However, there are two terminologies such as true necrotrophs and secondary necrotrophs. The true necrotrophs attack and kill the healthy plants, while the secondary necrotrophs are saprophytic but may occasionally infect the plants that have been previously weakened (Doehlemann et al. 2017). The basic definition of necrotrophy is "the mode of infection in which the pathogen kills the tissue before colonization" (Oliver and Solomon 2010; Doehlemann et al. 2017). This statement contradicts the fact since the initial contact of the pathogen is with a living tissue. There are two early stages following the first contact of the pathogen with the host. To survive, the pathogen needs to subvert the host defense and generate a necrotic zone where the pathogen can survive from the host defense spread the necrosis around the initial zone (Doehlemann et al. 2017). There will be initial and late reactions in biotrophic fungi which exist in two-phase strategy "survive or die," but the early stage has no direct fight with the host (defense), but it follows easy/simple "sneaking in" strategy, and further stages necessitate close contact with the living host, hence resulting in continuous conflict with the host defense. In biotrophic fungal pathogens, they observe a "survive or die" strategy at the first meeting with the plant defense. Further, biotrophs prosper to keep the host tissue alive, and host defense strategy at this stage is a morbid in the form of a hypersensitive response (HR), and the pathogen strategy is deterrence of this response (Doehlemann et al. 2017). The necrotrophic pathogens have to deal with the plant defense during the first contact and overcome the initiation of infection. Thus, the initial phase of necrotrophic fungi uses an array of effector molecules to cope with and operate the host defense during infection (Choquer et al. 2007). The true necrotrophic pathogens include several species that belong to diverse genera. It is convenient to divide all necrotrophic pathogens into narrow-host-range and broad-host-range species (Mengiste 2012). The host specificity of necrotrophs is due to the synthesis of host-specific toxins (HST); these pathogenicity factors are crucial for compatible host. For example, T-toxin (*Cochliobolus heterostrophus*), HSTs (*Pyrenophora tritici-repentis* and *Parastagonospora nodorum*), and HSTs from *Cochliobolus* sp. include HC-toxin (*Cochliobolus carbonum*), victorin (*Cochliobolus victoriae*), and ToxA toxins and specify the host range in spot and blotch diseases (Faris et al. 2010). These HSTs which interact with a specific gene from a host (similar to that of "Avr" or "effector proteins") interact with resistance proteins (R proteins). The HSTs are regarded as effectors as they share many of the characters of the avirulence gene. The HSTs share many characteristics with avirulence gene products similar to primary virulence function, they are specifically recognized by the host resistance counterparts, and they can be recognized by the immune system of plant as virulence factors.

The broad-host-range necrotrophs lack HSTs and may attack several host plants across the families. The best-represented broad-host-range necrotrophs is *Botrytis cinerea*, which is closely related to *Sclerotinia sclerotiorum*. Both of them have more than 300–500 host plants, and each causes severe economic losses pre- and post-harvest annual crop worldwide (Bolton et al. 2006; Williamson et al. 2007). The melanized sclerotia have a crucial role in germination (vegetatively or carpogenically) as well as production of apothecia to release ascospores (Doehlemann et al. 2017). The genetic basis of resistance against broad-host-range necrotrophic pathogens is more complex and typically quantitative (Oliver and Solomon 2010; Mengiste 2012). This is in contrast to R-gene-mediated resistance or HST-blocking genes, which provide complete resistance. For this reason, it is difficult to control broad-host-range necrotrophic pathogens, which might partially explain their growing economic importance.

Originally, the necrotrophy was inferred as toxin-assisted maceration of the tissue of host, and several studies showed the use of hydrolytic enzymes by necrotrophic pathogens (Smith 1900; Cole 1956). However, the recent literature and developments reveal that these enzymes do not function alone and they might have additional roles other than sheer hydrolysis of plant polymers. The large set of genes coding for cell wall-degrading and other hydrolytic enzymes present in the genomes of necrotrophic fungi support this strategy (Soanes et al. 2008; Amselem et al. 2011). The high redundancy of many of these enzymes (which is unusual in fungi) supports necrotrophic fungi. Recent transcriptome and secretome studies revealed that horde of these enzymes are produced and secreted at different stages of the infection (González et al. 2016; Kim et al. 2016; McCotter et al. 2016).

6.3.3 Hemibiotrophs

Hemibiotrophic pathogens are those that have combined biotrophic and necrotrophic lifestyles. Their life cycle has an initial biotrophic phase followed by the necrotrophic phase (Oliver and Ipcho 2004; Divon and Fluhr 2006). Hemibiotrophs are defined as species that have a flexible length of initial biotrophic stage before switching over to necrotrophy (Perfect et al. 1998; O'Connell et al. 2012; Yi and Valent 2013). This definition involves an initial true biotrophic phase mediated by special biotrophic organs. In the beginning, fungal pathogens secrete effectors to suppress the plant defense, and later (at the end of the brief biotrophic stage), the fungus undergoes a substantial developmental change that facilitates the transition from a biotrophic to a necrotrophic mode.

Two examples that fit very well as hemibiotrophic lifestyle are the rice blast fungus *M. oryzae* and species under the genus *Colletotrichum* (Yi and Valent 2013). Upon penetration into the sub-epidermal or epidermal cells, they develop specialized hyphae that establish close contact with the host and invaginate the host cell membrane, leading to true (temporary) biotrophic interphase. The true biotrophic phase may last from one to several days and then switches to the necrotrophic mode. This transition differentiates new types of hyphae, secretions (enzymes and toxins), and delivery of specialized effectors.

Colletotrichum and *Magnaporthe* are important plant pathogenic fungi, which could be cultured and amenable to genetic manipulations owing to their conventional hemibiotrophic mode of lifestyle. Other fungi that follow a hemibiotrophic mode of lifestyle include *Fusarium*, *Mycosphaerella*, *Verticillium*, and many others (Fradin and Thomma 2006; Churchill 2011; Goodwin et al. 2011; Ma et al. 2013; Ploetz 2015). The most common feature of these species is a latent stage that varies in length, but in most instances, they do not differentiate into the typical biotrophic specialized organs and do not form a close contact with the host cell, but remain in the apoplastic or intracellular sphere. However, the pathogenic lifestyle of these species includes symptomless, quiescent, latent, or endophytic stages, but they do not meet the criteria of hemibiotrophy as defined above and therefore should not be treated in the same fashion. To support, a list of representative fungi showing different lifestyles are given in Table 6.1 along with diseases caused by those fungi exhibiting different lifestyles (rusts, powdery mildews, *Choanephora* species, sooty bolds, and others) in Fig. 6.1.

6.4 Historically Known Major Fungal Diseases

Phytopathogenic fungi have caused devastating diseases on various crop plants throughout the history of agriculture. The Theophrastus botanical studies contributed for the first time a relevant scientific explanation to the knowledge on plant disease. So far, ancient knowledge recognized is rust infection in crops and also the smuts infecting monocot crops such as barley, wheat, and/or maize by *Ustilago hordei*, *U. tritici*, and *U. maydis*, respectively. Similarly, the rusts were also traditionally recognized as serious threats in agriculture. A fungal plant pathogen with vast agrarian and cultural influence in human history was *Claviceps purpurea* (associated with ergot of rye, barley, oats, and wheat). The sclerotia of *C. purpurea* are known to possess a broad range of toxic alkaloids and replace kernels in the heads of crops and thereby contaminate the harvested grains. Further, there are numerous epidemics of the plant diseases affected human life by causing diseases such as “devil’s curse” or “holy fire” (Agrios 2005).

In the history of plant pathology, the development of fungal diseases left bitter experiences on human life and was responsible for a large number of deaths. Important catastrophic events include (1) the Irish famine by *Phytophthora infestans*, which caused late blight of potato (1840–1845); (2) the Ceylon rust caused by *Hemileia vastatrix* which destroyed the coffee plantations in Sri Lanka; and (3) the Bengal famine contributed by *Helminthosporium oryzae* in rice. The impact caused by the occurrence of late blight of potato was the reason to initiate a new

Fig. 6.1 Representative images of fungal diseases which explicate different mode of lifestyles: biotrophic lifestyle and groundnut rust (*Puccinia arachidis*) (**a**, **b**); rust on *Pongamia* tree (*Ravenelia* sp.) (**c**, **d**); powdery mildew of cucurbits (**e**–**g**); powdery mildew on cowpea (**h**); bud rot of okra by *Choanephora* sp. (**i**, **j**); dieback of sandalwood (**k**, **l**); and sooty mold on sandalwood (**m**)

discipline, the "plant pathology." The following sections briefly discuss the three diseases which left their bitter impressions in the human history.

6.4.1 Phytophthora Infestans

Phytophthora infestans, the causative fungus of late blight of potato (Irish famine), occurs on many varieties of Solanaceae crops (potato, brinjal, tomato, and many other hosts) around the world. The genus *Phytophthora* placed under Oomycota has been a causative agent for diseases like blights, wilting, damping-off, chlorosis, root rot, and the rotting plant organs. The late blight of potato in the field is presented in Fig. 6.2. Although several species of *Phytophthora* are identified in different hosts causing different diseases, *P. infestans* find a place in the history for its infamous Irish potato famine in the 1840s (Fry 2008). At present, *P. infestans*, besides infecting potato, causes severe damage to other important Solanaceae crops like tomato (*Solanum lycopersicum*), brinjal (*Solanum melongena*), chili (*Capsicum annuum*), and many other vegetable crops.

The genus *Phytophthora* is capable to cause destructive and epiphytic diseases like blights, foot rots, wilts, cankers, seedling blights, damping-off, gummosis, and various rots in field and storage conditions. Therefore, it has the implicit role of a virulent pathogen owing to its capacity of zoosporangia to germinate by liberating zoospores as well as by germ tube. Diversity of *Phytophthora* spp. associated with crops is presented in Table 6.2 along with their hosts and other details. It is evident from the recent studies the existence of a complex nature or diverse population among the species *Phytophthora* or altogether a new species is present in the current agro-ecosystem, which needs to be deciphered by advanced molecular tools. Recently, Scanu et al. (2015) reported the occurrence of nine species of *Phytophthora* on the decline of Mediterranean maquis vegetation (a scrubland vegetation of the Mediterranean region) using multi-locus barcoding and phylogenomic analysis of the population. Thus, advanced molecular tools play a crucial role in deciphering the hidden diversity of *Phytophthora* (morphologically difficult to differentiate many of these species) and revealed for the first time the involvement of highly invasive pathogen with a wide host range *P. cinnamomi* and several species of *Phytophthora* (Scanu et al. 2015). The subsequent conventional and sequence analysis (ITS and Cox1 gene regions) revealed association of multiple *Phytophthora* spp. with two new species (Jung et al. 2011). Based on the morphological characters and molecular sequence analysis, the isolates were identified as *P. asparagi*, *P. bilorbang*, *P. cryptogea*, *P. cinnamomi*, *P. gonapodyides*, *P. melonis*, *P. syringae*, and the two new species (*P. crassamura* and *P. ornamentata*).

The genus *Phytophthora* is diversified, and variations are observed on mating types. The mating types vary from country to country and between the hosts. Different management strategies have been developed due to unsuccessful attempts of eradication. Even after 172 years of famine, the problem persists. Further, the

Fig. 6.2 *Phytophthora* blight observed in potato and tomato from Karnataka: field view of potato (**a**); leaf blight caused due to *P. infestans* (**b–e**); severe late blight affected tomato field view from Mysore region, Karnataka (**f**, **g**); and blight symptoms on individual leaves showing presence of sporangiospores on lower leaf surface of tomato (**h–j**)

Table 6.2 Diversity of *Phytophthora* species associated with diseases of crops

Phytophthora species	Crop plants and their disease	Reference
P. alni	Collar rot of alder	Hansen (2015)
P. boehmeriae	Wide host range (forest tree pathogen) and cotton	Elena and Paplomatas (1998)
P. cactorum	*Rhododendron*	Yang et al. (2018)
P. cambivora	Ink disease on chest nut	Vannini and Vettraino (2011)
P. capsici	Bell pepper, chili, eggplant, and tomato	Granke et al. (2012)
P. cinnamomi	Dieback of *Eucalyptus*	Hansen (2015)
P. citricola	Citrus orchards	Jung and Burgess (2009)
P. citrophthora	Citrus	Jung and Burgess (2009)
P. colocasiae	Black pepper and *Piper betel*	Shrestha et al. (2017)
P. cryptogea	Several floricultural crops	Ampuero et al. (2008)
P. erythroseptica	Potato (pink rot)	Jiang et al. (2019)
P. europea	European oak	Vettraino et al. (2005)
P. fragariae	Raspberry	Koprivica et al. (2009)
P. fragariae rubi	Root rot of red raspberry	Koprivica et al. (2009)
P. ilicis	Holly (ilicis leaf blight and spots)	Scanu et al. (2014)
P. infestans	Brinjal (eggplant), chili, potato, and tomato	Dey et al. (2018)
P. inundata	Tree pathogen	Brasier et al. (2003)
P. kernoviae	European beach and *Rhododendron*	Brasier et al. (2005)
P. lateralis	Cedar root disease	Hansen (2015)
P. medicaginis	Root rot of *alfalfa* and chickpea	Vandemark and Barker (2003)
P. melonis	Fruit rot of cucurbits	Guharoy et al. (2006)
P. nemorosa	Foliar and bole pathogen of various evergreen hard-wood trees	Hansen et al. (2003)
P. nicotianae	*Citrus*, pepper, *Piper betel*, and tobacco	Meng et al. (2014)
P. palmivora	Bud rot of palms, fruit rot (or koleroga) of coconut and areca nut	Carella et al. (2018)
P. parasitica	Eggplant and tomato	Meng et al. (2014)
P. pinifolia	Radiata pine	Hansen (2015)
P. plurivora		

(continued)

Table 6.2 (continued)

Phytophthora species	Crop plants and their disease	Reference
	Infecting roots of: *Acer platanoides*, *Aesculus hippocastanum*, *Alnus glutinosa*, *Fagus sylvatica*, *Quercus robur*, *Tilia* spp., and conifer species	Jung and Burgess (2009)
P. psis	Causes root rot in: *Cicer arietinum* (chickpea) and *Lens culinaris* (lentil), *Lathyrus* spp. (pea), *Pisum sativum*; *Vicia faba*, *V. sativa* (garden vetch), and *V. benghalensis* (purple vetch)	Heyman et al. (2013)
P. ramorum	Sudden oak death and *Phytophthora ramorum* blight (also infects members of Ericaceae, Fagaceae, Larix, and others)	Elliott et al. (2009), Hansen (2015)

mystery surrounding the Irish famine was resolved by Yoshida et al. (2013). Until the 1970s, there was very low diversity of *P. infestans* confined only to Mexico and the USA. But, it later dominated the globe for a period of 150 years. However, Yoshida et al. (2013) concluded that the nineteenth-century epidemic was due to the HERB-1 unique genotype which persisted over 150 years. The HERB-1 genotype was distinct from modern strains, but closely related to the US-1 genotype that was replaced outside the Mexico in the twentieth century. It was proposed that HERB-1 and US-1 could have emerged from metapopulation established in the early 1800s outside of epicenter (Yoshida et al. 2013). This scenario holds good for many species of *Phytophthora*, which are causing devastating diseases in horticultural, floricultural, and other vegetable crops.

6.4.2 Hemileia vastatrix

One of the historical examples of disease is the coffee rust (or Ceylon coffee rust) caused by *Hemileia vastatrix*. Until the 1870s, Sri Lanka was one of the world's greatest coffee producers. This dramatically changed after *H. vastatrix* reached Ceylon in 1875. From 1870 to 1885, coffee production drastically dropped to 95%, the fungus destroyed the coffee plantations, and now Sri Lanka is known mainly for tea. Even today, the coffee rust is a significant threat to coffee productions with recent outbreaks in Central and South American regions (Avelino et al. 2015). The coffee leaf rust threat by *H. vastatrix* is one of the major diseases occurring in coffee plantations affecting commercial coffee species *Coffea arabica* (arabica coffee) and *Coffea canephora* (robusta coffee). This rust caused multiple outbreaks in several coffee-growing regions resulting in heavy loss of yield. The new races are constantly evolving as evidenced by the presence of fungus in plants that were previously resistant (Mahadevakumar and Sridhar 2020).

H. vastatrix is a hemicyclic fungus, and its source of inoculum is by the urediniosporic life cycle. The disease appears as chlorotic spots initially (can be visualized through naked eyes), followed by the development of differentiation of suprastomatal, bouquet-shaped, and orange-colored uredinia (Fig. 6.3). The disease has resulted in 35% loss of yield and caused polyetic epidemiological impact. Although *H. vastatrix* is the only fungus causing rust on coffee, there exists variation in population structure and genotype composition, which plays an important role in determining the virulence of the isolate/pathogen at different ecological and environmental conditions (Talhnihas et al. 2017). Despite exhibiting low genetic polymorphism, the large genomes of *H. vastatrix* (c797 Mbp) cover up great pathological diversity (>50 physiological races). The gene expression studies conducted, which suggested the activation of signaling pathways for the production of putative effectors, suggest the plant-fungus dialogue starts as early as the germ tube stage, which provides clues for the identification of avr genes (Talhnihas et al. 2017).

6.4.3 Helminthosporium oryzae

Pathogenic fungi associated with rice are important historically and economically as the diseases caused by them lead to severe economic loss as well as an acute shortage of staple food. Historically, the rice brown spot disease is very important as it witnessed famine in two instances. The first famine was during 1769–1770, and the second was during 1943 (called the great Bengal famine). The 1943 famine resulted in mortality of more than two million people due to 10–58% seedling mortality. The diseases are associated with all stages of growth and development of rice (from seedbed to harvest and post-harvest).

6.5 Global Perspectives of Major Fungal Diseases

The main sources of staple food for the majority of the globe include maize, rice, and wheat. These crops are not only serving as staple diets for humans but also used as livestock feeds, thus indirectly contributing toward the production of meat, dairy, and other animal-derived products. Soya bean is the fourth important crop, which is grown primarily as feed for livestock. The trade-in four crops constitute a major share of food system of 7% (rice), 12% (maize), 19% (wheat), and 30% (soya beans) being traded internationally between 1995 and 2010 (Dowlah 2015). All these four crops are susceptible to fungal infections. In a recent review, Fisher et al. (2012) detailed the major fungal pathogens of each species responsible for the reduction in yield.

The rice blast (*Magnaporthe oryzae*) is a widely distributed disease, potentially found wherever rice is grown (Fig. 6.2). This blast can cause up to 10–35% loss depending on crop variety and environmental conditions (Talbot 2003). Infection in wheat is associated with the stem rust *Puccinia graminis* (and other *Puccinia* spp.), and *Puccinia tritici* cause crop loss up to 70% (Leonard and Szabo 2005). The

Fig. 6.3 Coffee rust disease caused by *Hemileia vastatrix*: field view of the coffee plantation from Balehonnur region, Karnataka (**a**); individual leaves showing rust pustules in adaxial and abaxial surfaces (**b**, **c**); and immature and mature rust pustules due to colonization of *H. vastatrix* (**d**, **e**)

resistant wheat cultivars developed in the past decades have shown good control; however, an emergence of a new virulent strain was seen in Uganda during 1999 (UG99 or TTKSK) (Singh et al. 2011). The major fungal pathogen associated with maize is corn smut by *Ustilago maydis*, a causative agent of galling and other damages. It is native to the central and southern Americas and spread over to most of the maize-growing areas and caused a 20% loss of crop (Brefort et al. 2009). The soya bean is known for attack by the rust *Phakopsora pachyrhizi*, which causes up to 70% of the loss. It is originated in Asia and spread over to most areas where soya bean is grown. Besides soya bean, the rust also attacks other plants of the family Fabaceae, which can serve as a reservoir of agricultural infections (Hartman et al. 2011).

6.5.1 Diversity of Fungal Diseases of Rice

Rice (*Oryza sativa*) is indeed life for most people in Asia, and scarcity in production and availability could lead to a severe food crisis. Considering the significance of rice globally and also in human life history, the United Nations celebrated the year 2004 as the International Year of Rice. Rice has been recognized as an important crop globally, and it is the main source of energy for the majority of the world's population. It is a staple food for people living in the rural and urban areas of humid and sub-humid Asia responsible for 30–50% of agricultural production (Hossain and Fischer 1995). Rice provides national food security and generates employment and income for the low-income groups. However, its production is influenced by various diseases by fungi, bacteria, viruses, and others. The following sections provide diverse fungal diseases associated with rice in major rice-growing countries.

The Bengal famine in 1942 was in part attributed to brown spot disease in rice (Padmanabhan 1973). The rice blast epidemics in the 1970s in Korea led to a major food crisis (Ou 1985) due to loss of yield up to 10–50 billion US$. Thus, minimizing or managing the disease epidemics and reduction of loss are crucial in sustained rice production. To achieve this goal, it is important to understand the extent of damage brought about by the fungal diseases and to identify shifting disease problems associated with technological advances.

Figure 6.4 presents a comprehensive picture of fungal disease associated with rice in major rice-growing countries. In a nutshell, the major five fungal diseases posed a challenge to world food security. These diseases include (1) rice blast caused by *Magnaporthe oryzae*; (2) sheath blight and sheath rot caused by *Rhizoctonia solani*; and (3) bacterial blight and viral disease tungro. However, the brown spot disease by *Helminthosporium oryzae* is also one of the historically important diseases caused by significant loss of production in the past as well as present. If suitable management practices are not followed, the world has to face a shortage in rice production leading to jeopardy in food security. The post-harvest diseases of rice are also playing a significant role and lead to decline in production due to fungal infestation during storage.

Fig. 6.4 Major fungal diseases recorded from principal rice-growing countries

In the process of increasing the rice production, rice seeds appear to be vulnerable to infection by many pathogens. The sheath rot complex and grain discoloration is an important problem faced by the rice growers. This syndrome involves a characteristic browning discoloration or rotting of the flag leaf sheath and discoloration of the grain. The syndrome is widespread in tropical Asia since the introduction of modern semi-dwarf and photoperiod-insensitive rice cultivars. It is more prevalent in the rainy season than in the dry season. In the literature, the causal agent of sheath rot is always associated with *Sarocladium oryzae*. However, many reports projected that sheath rot is a complex problem caused by bacteria as well as fungi. Sheath rot and seed discoloration pathogen includes several fluorescent and non-fluorescent pseudomonads: *Pseudomonas glumae* (syn. *Burkholderia glumae*), *P. fuscovaginae*, and other non-pathogenic bacteria (Cottyn et al. 1996a, b). Similarly, along with *S. oryzae* and others pathogens viz., *Bipolaris oryzae* and Fusarium spp. were also isolated from discolored seeds showed the frequency of *S. oryzae* was lower than 10% (Lee et al. 1986).

The false smut caused by *Ustilaginoidea virens* has long been considered a minor problem in the global rice production. But, reports of severe damage caused by this disease in tropical and temperate regions in Africa, Asia, and America appear to be increasing in recent years. This disease was recently reviewed by Biswas (2001); however, the reduction in yield associated with false smut remains unclear. Similarly, the epidemiology of the disease in association with modern rice production is not well understood. A few reports showed that the high incidence and severity of false smut are correlated with an increase in some parameters involved in modern rice production systems.

6.5.2 Diversity of Fungal Diseases of Maize

Maize is an important cereal crop cultivated globally, and the USA is a major producer fulfilling nearly 35% of the global demand. The USA has the highest productivity (>9.6 tons/ha), and it is twice the global average (4.92 tons/ha). In India, the average maize production is around 2.43 tons/ha. The maize is the third most important food crop in India after rice and wheat. According to an estimate, its production was 22.23 metric tons (2012–2013) mainly during Kharif season, which occupies 80% of the area under cultivation. Maize contributes nearly 9% of the national food production in India. Maize serves as a staple food for humans and livestock and also as a basic raw material for several industrial products (alcoholic beverages, cosmetics, film, food sweeteners, gum, oil, package protein, paper industries, pharmaceuticals, starch, and textile).

Maize is affected by various diseases caused by bacteria, fungi, rusts, smuts, root rots, and ear rots leading to a severe reduction in the yield as well as quality. There are more than 25 fungal diseases that cause significant economic loss in the production of maize. The stalk rot of maize is a major threat in terms of crop loss and seed quality. Important fungal diseases of maize include root disease, foliar diseases, stalk rots, kernel rots, and ear rots. Along with the pathogenic fungi in maize, the secondary metabolites (mainly toxins) produced by these fungi cause major threat to human and livestock health. The major fungal diseases associated with maize at various stages of its growth and development are presented in Fig. 6.5.

Fig. 6.5 Major fungal diseases associated with maize in principal maize-growing countries

The earliest report of maize disease is that of head smut caused by *Sphacelotheca reiliana* recorded by Cooke (1876). Butler (1918) wrote that this disease was severe in the Kashmir, Himalayas, and other regions of India. In southwest Rajasthan, this disease appears sporadically, and the incidence became as high as 50% in certain fields cultivated by the tribals. In 1893, Watt in his classic book *Dictionary of the Economic Products of India* stated that "It is well known that smut and rust which do so much damage in other parts of the world, also occur in India," which signifies that smut and rust diseases have appeared on corn long back. Later, Butler (1918) recorded maize diseases:

1. Downy mildew caused by *Peronosclerospora philippinensis*
2. Common smut caused by *Ustilago maydis*
3. Head smut caused by *Sphacelotheca reiliana*
4. Common rust caused by *Puccinia sorghi*
5. *Turcicum* leaf blight caused by *Exserohilum turcicum* (Dharanendraswamy 2020)

The stalk rot caused by *Fusarium* spp. is a devastating infection affecting global maize-growing regions. It is a complex disease caused by several fungal pathogens, and it varies from region to region and comprehensively reduces the crop yield by interfering absorption and translocation of water and nutrients leading to premature death (Shan et al. 2017). This serious disease (root and stalk) was first reported from the USA by Pammel (1914) and in India by Arya and Jain (1964) for the first time from Rajasthan. The stalk rot of maize causes premature wilting or drying of plant and finally lodging. The typical symptoms observed during the early phase are premature drying of bottom leaves, eventually leading to death. The diseased stalks lose firmness, and the interior cells of the stalk dissolve (Fig. 6.6). The microscopic observations of stalks suggested softening and reddish coloration, and the pith appeared to be soft, disintegrating, and becoming light-brown to reddish. The infection of stalk was seen up to three inter-nodal regions from the stem-soil interface as reported by Dharanendraswamy et al. (2019a, b, 2020a, b).

Although *Fusarium verticillioides* is the major causative agent of stalk rot, there are several fusaria that cause stalk rots: *F. acuminatum*, *F. avenaceum*, *F. merismoides*, *F. nivale*, *F. subglutinans* (*F. semitectum*), *F. roseum*, *F. solani*, and *F. sulphurcum* (Rintelen 1965, Kommedahal et al. 1972; Dorn et al. 2009; Nur-Ain-Izzati et al. 2011). However, in India, only *F. moniliforme* and *F. semitectum* are the causative agents of stalk rot of maize (Lal and Dwivedi 1983; Khokhar et al. 2013, 2014). Recently, association of *F. equiseti* and *Lasiodiplodia pseudotheobromae* with post-flowering stalk rot of maize was also reported by Dharanendraswamy et al. (2020a, b).

6.5.3 *Diversity of Fungal Diseases of Vegetables*

Vegetables are an important source of regular diet, and India produces a significant quantity of vegetables annually. The fungal diseases associated with the vegetables are also causing significant loss of food leading to severe economic loss. A list of common fungal diseases associated with vegetables is provided in Table 6.3. Fungi damage the hosts by killing cells and causing plant stress. The fungal infection can be through infected seeds, soil, crop debris, nearby crops, and weeds. Some of the fungal diseases that occur on different vegetables include anthracnose and rots caused by various species of *Colletotrichum*, *Botrytis*, downy mildews, *Fusarium*,

Fig. 6.6 (**a**–**j**) Stalk rot of maize: It is one of the complex diseases caused by several fungal species

powdery mildews, rusts, *Rhizoctonia*, *Sclerotinia*, *Sclerotium*, *Phytophthora*, and others. Some of the other diseases which are specific to a particular crop include clubroot (*Plasmodiophora brassicae*) in brassicas, leaf blight (*Alternaria* sp.) in carrots, and red root complex in beans (*Gibberella* sp.). Some fungi that cause highly prevalent foliar diseases include downy mildews, powdery mildews, and white blister. Similarly, soil-borne diseases like clubroot and other diseases are caused by the species of *Fusarium*, *Pythium*, *Rhizoctonia*, *Sclerotinia*, and *Sclerotium*.

The following sections deal with the diversity of fungal diseases associated with a few vegetable crops (cowpea, common bean, brinjal, and tomato) cultivated in Karnataka state, southern India. Cowpea is an important pulses crop and is the livelihood of millions of people in the tropics (Quin 1997). The crop provides food and animal feed and strengthens the economy of the rural population. There are new emerging diseases due to fungi and fungi-like organisms being increasingly reported in many regions (Farr and Rossman 2018). The new fungal diseases on cowpea caused by various fungal pathogens are becoming the major constraints to the cowpea production. Recently, various workers have reported the occurrence of root rot and dry root rot disease caused by *F. equiseti* (Li et al. 2018), *F. oxysporum* (Shrestha et al. 2016a), and *F. proliferatum* (Shrestha et al. 2016b) from the USA; target leaf spot disease caused by *Corynespora cassiicola* from China (Li et al. 2014); and leaf spot disease caused by *Pestalotiopsis* sp. (Mahadevakumar and Janardhana 2014) and *Dactuliophora* sp. (Mahadevakumar and Janardhana 2012) and collar rot caused by *Aplosporella hesperidica* (Deepika et al. 2020) from India.

In recent past, common bean (*Phaseolus vulgaris*) production is limited due to various plant diseases caused by bacteria, fungi, viruses, phytoplasma, and other biotic factors. Studies have been carried out throughout the world on the fungal diseases of common bean and other leguminous plants. Stem rot (*Sclerotium rolfsii*), root rot (*Pythium* and *Rhizoctonia solani*), charcoal rot (*Macrophomina phaseolina*), wilt (*Fusarium oxysporum*), southern blight and leaf spot (*S. rolfsii*, *Alternaria*), powdery mildew (*Erysiphe polygoni*), ashy stem blight (*M. phaseolina*), rust (*Uromyces phaseoli*), anthracnose (*Colletotrichum lindemuthianum*), and many more fungal diseases have been recorded on bean (Hagedorn and Inglis 1986; Abawi and Pastor Corrales 1990; Allen et al. 1996; Mahadevakumar et al. 2015a, b, c). *Fusarium* also causes different diseases in beans like root rot, wilt, decline and damping-off, and so on. Cramer et al. (1996) characterized the *Fusarium* isolates causing wilt disease in and around the central plains of the USA, while Roman-Aviles et al. (2003) described the root rot of common beans caused by the *Fusarium solani* in Michigan. The common bean decline is also reported to be caused by *F. solani*, *F. oxysporum*, *F. sambucinum*, *R. solani*, and *Pythium debaryanum* (Saremi et al. 2011). The fungal diseases of common bean in subterranean regions are *Aphanomyces* root rot, black root rot, *Fusarium* root rot, *Fusarium* yellows (wilt), *Phymatotrichum* root rot, *Pythium* root rot, *Rhizoctonia* root rot, and southern blight and stem rot. Similarly, the fungal diseases of aerial parts include *Alternaria* leaf and pod spot, angular leaf spot, anthracnose, *Ascochyta* leaf spot, ashy stem blight, *Cercospora* leaf spot, *Chaetoseptoria* leaf spot, *Diaporthe* pod

Table 6.3 Some of the common fungal diseases associated with vegetable crops

Disease	Causal organism	Crops affected	Symptoms
Anthracnose	*Colletotrichum* spp.	Wide range of crops: cereals, fruits, oil-yielding crops, pulses, and vegetables	Typical symptoms begin with sunken and water-soaked spots appearing on leaves, stems, and/or fruit
Black root rot		Beans, cucurbits, lettuce, and other vegetable crops	Blackening of roots; stunted plants
Botrytis rots—for example, gray mold	*Botrytis cinerea*	Beans, brassicas, capsicum, celery, cucumber, lettuce, and tomato	Softening of plant tissues in the presence of gray fungal growth
Charcoal rot	*Macrophomina phaseolina*	>500 host range reports	–
Club root	*Plasmodiophora brassicae*	Members of Brassicaceae (cabbage, cauliflower, radish, and others)	Plants are yellow and stunted and may wilt in hotter parts of the day; large malformed club roots
Damping-off	*Aphanomyces*, *Fusarium*, *Pythium*, *Phytophthora*, and *Rhizoctonia*	Many vegetable crops: beans, beetroot, brassicas, carrots, coriander, cucurbits, eggplant, leafy vegetables, spring onions, and tomato	Young seedlings have necrotic stems or roots; seedlings die or show a reduction in growth
Downy mildews	Obligate parasites are host-specific except for a few species having a wide host range	Cucurbits, grapes, horticultural crops, maize, onions, pearl millet, rose, sorghum, and vegetables	Yellowish leaf spots and streaks then turn into brown
Fusarium wilts and rots	Various species of *Fusarium* (*F. oxysporum* and *F. solani* are frequent)	Wide host range: Beans, brassicas, carrots, cucurbits, herbs, onions, peas, potato, spring onions, and tomato	Causes severe root and crown rots or wilt diseases by attacking roots and basal stems
Powdery mildews (some species are restricted to particular crops or crop families)	Obligate parasites are host-specific except for a few species having a wide host range	Wide host range and very common especially in greenhouse crops: cucurbits and vegetable crops	Small, white, powdery patches on most aboveground surfaces; usually observed first on undersides of leaves but eventually cover both surfaces
Pythium rots	*Pythium aphanidermatum* and other species	A wide range of vegetable crops and horticultural crops	Usually infects at the early seedling stages and kills the seedlings or may infect at any

(continued)

Table 6.3 (continued)

Disease	Causal organism	Crops affected	Symptoms
			stage of its growth and development
Rhizoctonia rots	*Rhizoctonia solani*	Wide host range: Beans, beets, brassicas, capsicum, carrots, cucurbits, lettuce, peas, potato, and tomato	Range of symptoms depending on the crop being grown, but can affect fruits, roots, leaves, stems, and tubers; plants wilt and may collapse and die
Rust diseases	Obligate biotrophic organism in some cases exhibits host specificity	Coffee, cowpea, groundnut, maize, sorghum, soybean, wheat, yard long bean, and others	Small, red, or reddish-brown pustules that form on the underside of the leaves and sometimes on the pods as well
Sclerotinia rots	*Sclerotinia sclerotiorum* and *Sclerotinia minor*	Most vegetable crops, pulses, and cereals	Water-soaked rotting of stems, leaves, and sometimes fruit; followed by a fluffy, white, and cottony fungal growth, which contain hard black pebble-like sclerotia
Sclerotium rots	*Sclerotium rolfsii* and *S. cepivorum*	*S. rolfsii* has a wide host range (>500 hosts are reported); *S. cepivorum* affects only onions, garlic, and related *Allium* (leeks, shallots, and spring onions)	*S. rolfsii* affects lower stem and root causing rots; coarse threads of white fungal growth surround the diseased areas; *S. cepivorum* produce yellowing and wilting and fluffy fungal growth containing black sclerotia forms at the bases of bulbs
White blister/white rust	*Albugo candida*	Members of Brassicaceae (cabbage, cauliflower, radish, and others)	Produces white blisters and swellings on leaves of affected plants

blight, downy mildew, *Entyloma* leaf smut, gray leaf spot, scab, rust, web blight, powdery mildew, and others (Schwartz et al. 2005).

Brinjal (*Solanum melongena*) is another essential vegetable crop cultivated in tropics and subtropics and grown extensively in China, India, Bangladesh, Pakistan, and the Philippines. It is also cultivated in America, Europe, and other parts of Asia. In India, brinjal is one of the most important vegetable crops (Zeven and Zhukovsky 1975; Rashid 1976; Sekara et al. 2007). It is susceptible to various biotic and abiotic stresses during its growth and development. Along with biotic stress, fungal pathogen infections such as leaf blight and fruit rot (*Phomopsis vexans*), damping-off (*Pythium aphanidermatum*), wilt (*Verticillium dahliae*), leaf spots (*Alternaria melongenae* and *Cercospora melongenae*), and root rot (*Sclerotinia sclerotiorum*) are associated with brinjal (Shivaprakasam and Soumini 1974; Iqbal et al. 2003). Fruit rot and leaf blight caused by *Phomopsis vexans* are major threats decreasing yield as well as market value (20–30%) (Beura et al. 2008; Pandey 2010). Diseases caused by *Phomopsis vexans* have been reported from Assam, Jammu, Karnataka, and Pantnagar (Srinivasa et al. 2005; Thippeswamy et al. 2005; Akhtar and Chaube 2006; Muneeshwar et al. 2011; Das and Sarma 2012; Jayaramaiah and Janardhana 2011; Mahadevakumar 2016; Mahadevakumar and Janardhana 2016c; Mahadevakumar et al. 2017).

6.6 Top Ten Fungal Diseases

Emerging infectious diseases caused by plant pathogens could lead to unexpected and serious epidemics. Farmers spend billions of dollars on disease management without adequate technical support; thus, the disease devastates natural ecosystems and causes habitat loss (Bellard et al. 2012). Yield loss causes hunger and starvation especially in underdeveloped countries, due to limited access to disease control, which causes annual losses of common major crops up to 30–50%. In some years, the losses are much severe, and the results were catastrophic for those who extensively depend only on the food crop (Flood 2010). The major food disease outbreaks have caused devastating famines and mass migrations of population throughout the history.

In 2012, a team of scientists surveyed to gather information on the most influential fungal pathogen or top ten fungal pathogens (Dean et al. 2012). These fungal pathogens were considered very important from the scientific and economic point of view. Accordingly, among the top ten most important fungal pathogens, *M. oryzae* stands as the topmost fungal pathogen causing blast disease in *Oryza sativa*. Table 6.4 lists rice blast pathogen, and most of the serious plant pathogens come across in many crop plants including vegetables, cereals, millets, horticultural crops, forest trees, and other economically important plant species. The genus *Botrytis* having more than 200 hosts causing significant economic damage is considered in the second position; *Puccinia* sp. causing wheat rust, a serious disease in major wheat-growing regions, is in the third position; and *Fusarium* spp. associated with more than 500 plant species causing diseases (in the field and post-harvest conditions) occupied the fourth and fifth positions in the list of pathogens. *Fusarium oxysporum* and *Blumeria graminis* occupied the sixth and seventh positions as the most important fungal pathogens in the world, respectively. Members of basidiomycetes including smut (*U. maydis* causing corn smut disease) and rust (*M. lini*) associated with flax rust were placed in the ninth and tenth positions, respectively. The genus *Colletotrichum* has occupied the eighth position and was well known to cause many diseases (anthracnose, blights, dieback, and others) in various crop

Table 6.4 Major fungal diseases in the world (top ten fungal pathogens)

Top ten	Fungal pathogen	Disease	Host
1	*Magnaporthe oryzae*	Blast disease	*Oryza sativa*
2	*Botrytis cinerea*	Fruit rot and gray mold	>200 host species
3	*Puccinia* spp.	Rust on wheat	*Triticum aestivum*
4	*Fusarium graminearum*	Head blight	*Zea mays*
5	*Fusarium oxysporum*	Vascular wilt	Wide host range
6	*Blumeria graminis*	Powdery mildew	*Hordeum vulgare*
7	*Mycosphaerella graminicola*	Septoria leaf blotch	*Triticum aestivum*
8	*Colletotrichum* species	Anthracnose, fruit rots, and dieback	>500 host species
9	*Ustilago maydis*	Corn smut	*Zea mays*
10	*Melampsora lini*	Flax rust	*Linum usitatissimum*

plants. However, among *Colletotrichum*, there are over 1000 species; most of them are regarded as pathogenic, and some of them are also endophytes. It has been widely considered as a model fungus to study the lifestyle behavior of hemibiotrophs. Although the list provides the most significant fungal pathogens, there are several serious plant pathogens, which play a pivotal role in agricultural production, and those that are not included in the list are *Phytophthora* spp. (late blight and damping-off), *Sclerotium rolfsii* (southern blight and foot rot), *Puccinia arachidis* (groundnut rust), downy mildew diseases, *Rhizoctonia solani* (damping-off), *Phakopsora pachyrhizi* (Asian soybean rust), and *Diaporthe* spp. (associated with many economically important crop plants causing a wide range of disease). Further, it necessitates that this top ten list of fungal pathogens needs revision periodically to focus on control measures of diseases.

6.7 Diversity of Major Disease-Causing Fungal Genera

6.7.1 Colletotrichum

The genus *Colletotrichum* has a wide number of species affecting economically important crops. This genus primarily occurs abundantly in tropical and subtropical regions, but there are some important/novel species causing diseases in crops grown in temperate regions too. Diseases caused by *Colletotrichum* species lead to severe loss of production of various agricultural commodities in the field as well as post-harvest stages (Dean et al. 2012).

In temperate regions, fruit productions of high-value crops like strawberry, mango, citrus, avocado, banana, and others are severely affected. In Africa, species of *Colletotrichum* are known to cause devastating disease of coffee berries, which also causes significant economic loss to cereal growers, and also affect the important crops like maize, sugarcane, and sorghum. In the top ten fungal pathogens of the world, the genus *Colletotrichum* is voted to the eighth rank based on the perceived scientific and economic importance (Canon et al. 2012; Dean et al. 2012). *Colletotrichum* often cause anthracnose and other diseases which include red rot disease sugarcane, coffee berry disease, crown rot of strawberry, and banana and brown blotch of cowpea (Lenné 2002; Canon et al. 2012). Further, many *Colletotrichum* spp. are latent plant pathogens, and some of them are recorded as endophytes and saprobes, and they can switch to a pathogenic lifestyle when host plants are subjected to stress conditions or during post-harvest storage (Crous et al. 2016). The germinating conidia develop the appressorium through which the infection initiates by penetrating the cuticle of fresh tissues or occasionally through the epidermal cells via hyphal structures (Bailey and Jeger 1992; Deising et al. 2000).

The typical symptoms of anthracnose by *Colletotrichum* include the development of necrotic lesions on leaves, stems, flowers and fruits, and crown, and it also causes stem rots, seedling blight, and so on (Waller et al. 2002; Agrios 2005). The disease symptoms associated with *Colletotrichum* recorded from Karnataka have been represented in Fig. 6.7. Many species are seed-borne, dwell in soil saprobically on dead plant debris, and may spread disease through conidial dispersal by water splash as well as air transmission of ascospores from the sexual morph (Nicholson and Moraes 1980).

The genus *Colletotrichum* is a major threat among pathogenic fungi as it can thrive on a wide host range in warmer and humid environments and present globally (Ford et al. 2004; Shenoy et al. 2007; Damm et al. 2009; Diao et al. 2014; He et al. 2016; De-Silva et al. 2017). The anthracnose in several vegetables, fruits, and other crops is associated with *Colletotrichum* infections (Hyde et al. 2009). In pepper, tomato, potato, cabbage, and papaya, anthracnose is a destructive disease responsible for significant yield loss (Than et al. 2008; Hyde et al. 2009; Liu et al. 2016; He et al. 2016; Torres-Calzada et al. 2018). In India, this pathogen has been reported in chili, tomato, and garlic (Saxena et al. 2014; Saini et al. 2017; Salunkhe et al. 2018). Identification of *Colletotrichum* spp. is solely relied on morphological characteristics (Sutton 1992). Some of the species of *Colletotrichum* exhibit conidial overlapping; thus, molecular identification tools are being employed recently (Sherriff et al. 1995; Hyde et al. 2009; Canon et al. 2012; He et al. 2016). The important crop plants affected by *Colletotrichum* anthracnose are presented in Table 6.5 and Fig. 6.7.

6.7.2 Diaporthe

The genus *Diaporthe* belongs to *Diaporthaceae* under *Sphaeropsidales* of mitosporic fungi. It is typified by *Diaporthe eres*, an asexual form referred to as *Phomopsis* (Dissanayake et al. 2017a, b; Senanayake et al. 2017). The nomenclature is considered for genus name *Diaporthe* over *Phomopsis* based on the priority of publication (Rossman et al. 2014). Earlier, the identification of various *Diaporthe*

Fig. 6.7 Anthracnose disease caused by various species of *Colletotrichum*: eggplant fruit rot (**a**); caused by *C. parasitica* on eggplant fruits (**b**, **c**); on *C. capsici* (**d**, **e**); on bell pepper caused by *C. capsici* (**f**); on *Polianthes tuberosa* caused by *C. truncatum* (**g–i**); and on beans caused by *C. lindemuthianum* (**j**, **k**)

Table 6.5 Diversity of *Colletotrichum* spp. associated with agriculturally important crops

Host	Disease	Causal organism	Reference
Apple	Fruit rot	*C. acutatum*, *C. fioriniae*, *C. fructicola*, *C. gloeosporioides*, *C. karstii*, *C. nymphaeae*, *C. siamense*, and *C. theobromicola*	Velho et al. (2015), Munir et al. (2016), Park et al. (2018)
Banana	Anthracnose	*C. gloeosporioides*, *C. karstii*, *C. musae*, *C. paxtonii*, *C. scovillei*, *C. siamense*, *C. tropicale*, and *C. theobromicola*	Vieira et al. (2017), Zhou et al. (2017)
Bell pepper	Anthracnose and fruit rot	*C. brevisporum*, *C. fructicola*, *C. scovillei*, *C. siamense*, *C. sichuanensis*, and *C. truncatum*	Ramdial and Rampersad (2015)
Brinjal		*C. fioriniae*	Xu et al. (2018)
Butter fruit/ avocado (*Persea americana*)	Anthracnose	*C. alienum*, *C. boninense*, *C. fructicola*, *C. gloeosporioides*, and *C. karstii*	Giblin et al. (2018), Kimaru et al. (2018)
Chickpea (*Cicer arietinum*)		*C. dematium* and *C. truncatum*	Nene et al. (2012), Mahmodi et al. (2013)
Chili (*Capsicum annum*)	Anthracnose, fruit rot, and stem blight	*C. cairnsense*, *C. cliviae*, *C. gloeosporioides*, *C. queenslandicum*, *C. siamense*, *C. simmondsii*, and *C. truncatum* (*capsici*)	Than et al. (2008), Saxena et al. (2014), De-Silva et al. (2017), Saini et al. (2017)
Citrus	Anthracnose, post-bloom fruit drop, and stem-end rot on fruit	*C. boninense*, *C. catinaense*, *C. gloeosporioides*, *C. helleniense*, *C. hystricis*, *C. karstii*, *C. limonicola*, and *C. novae-zelandiae*	Guarnaccia et al. (2017)
Coffee (*Coffea arabica* and *C. robusta*)		*C. acutatum*, *C. asianum*, *C. boninense*, *C. capsici*, *C. fragariae*, *C. fructicola*, *C. gloeosporioides*, *C. kahawae* subsp. *kahawae*, and *C. siamense*	Prihastuti et al. (2009), Nguyen et al. (2010), Canon et al. (2012), Silva et al. (2012)
Cotton (*Gossypium hirsutum*)	Anthracnose	*C. gossypii* var. *cephalosporioides*	Moreno-Moran and Burbano-Figueroa (2016)
Cowpea	–	*C. gloeosporioides* and *C. lindemuthianum*	
Mango	Anthracnose	*C. asianum*, *C. cliviicola*, *C. cordylinicola*, *C. endophytica*, *C. fructicola*, *C. gigasporum*, *C. gloeosporioides*, *C. karstii*,	Mo et al. (2018), Li et al. (2019)

(continued)

Table 6.5 (continued)

Host	Disease	Causal organism	Reference
		C. liaoningense, *C. musae*, *C. scovillei*, *C. siamense*, and *C. tropicale*	
Mung bean	–	*C. acutatum*, *C. lindemuthianum*, and *C. truncatum*	Shen et al. (2010), Roopadevi and Jamadar (2015)
Murraya koenigii	Anthracnose and leaf spot	*C. gloeosporioides*, *C. karstii*, *C. siamense*, and *C. simmondsii*	Padman and Janardhana (2012), Guarnaccia et al. (2017)
Papaya	Anthracnose	*C. acutatum*, *C. capsici*, *C. gloeosporioides*, and *C. truncatum*	Torres-Calzada et al. (2013)
Pigeon pea (*Cajanus cajan*)	–	*C. truncatum*	Khan and Singh (1975)
Pomegranate	–	*C. acutatum*, *C. fioriniae*, *C. gloeosporioides*, *C. nymphaeae*, *C. simmondsii*, *C. theobromicola*, and *C. siamense*	Jayalakshmi et al. (2015), Xavier et al. (2019)
Pyrus species	Anthracnose and leaf blight	*C. aenigma*, *C. citricola*, *C. conoides*, *C. fioriniae*, *C. fructicola*, *C. gloeosporioides*, *C. jinshuiense*, *C. karstii*, *C. plurivorum*, *C. pyrifolia*, *C. siamense*, and *C. wuxiense*	Fu et al. (2019)
Tomato	Anthracnose	*C. acutatum*, *C. coccodes*, *C. dematium*, and *C. gloeosporioides*	Dillard (1989), Byrne et al. (1997), Sanogo et al. (1997), LeBoeuf (2007)
Tuberose (*Polianthes tuberosa*)	Anthracnose	*C. truncatum*	Mahadevakumar et al. (2019)

species was based on morphological characteristics and also on host details (Brayford 1990; Rehner and Uecker 1994). With the advancement in sequencing platforms, the species diversity associated with a particular crop plant was discarded, and identity and assignment of species were considered over by multi-locus barcoding including internal transcribed spacer ribosomal DNA (ITS-rDNA), elongation factor-1a (EF-1α), β-tubulin, partial histone H3 (HIS), and calmodulin (CAL) of DNA sequences along with morphological characteristics (Udayanga et al. 2011; Gomes et al. 2013; Gao et al. 2017; Guarnaccia et al. 2018; Yang et al. 2018). In the recent studies on systematics, pathology, and environmental microbiology/ecology (endophytes), *Diaporthe* is primarily based on the usage of multi-locus approach, and it helped to resolve the species boundaries of *Diaporthe*/*Phomopsis* genus

(Udayanga et al. 2011, 2014a, b; Gao et al. 2017; Marin-Felix et al. 2019). Various species under the genus *Diaporthe* have been reported as endophytes (e.g., in a large number of medicinal plants), many of them are pathogenic causing severe damage to crops (leaf blight and fruit rot of eggplant; dieback of citrus), and some of the species are also regarded as saprobic on a wide range of hosts worldwide (Mahadevakumar et al. 2014, 2017; Liu et al. 2015; Hyde et al. 2016; Marin-Felix et al. 2019). Common diseases are dieback in forest trees (Yang et al. 2018); leaf and pod blights and seed decay in soybean (Udayanga et al. 2015); leaf spots in tea (Guarnaccia and Crous 2017); melanose, stem-end rot, and gummosis in *Citrus* spp. (Mondal et al. 2007; Udayanga et al. 2014a; Mahadevakumar et al. 2014; Guarnaccia and Crous 2017, 2018); and stem canker in sunflower (Muntañola-Cvetković et al. 1981; Thompson et al. 2011).

Phomopsis cane and leaf spot caused by *Diaporthe* species on the grapevine is one of the most complex grapevine trunk diseases worldwide (Úrbez-Torres et al. 2013; Dissanayake et al. 2015; Guarnaccia et al. 2018). The symptoms include breakage of shoots, stunting, dieback, loss of vigor, reduced bunch set, and fruit rot (Pine 1958, 1959; Pscheidt and Pearson 1989; Pearson and Goheen 1994; Wilcox et al. 2015). On the infected stem, brown to black necrotic irregular lesions could be seen. Once the clusters are infected, rachis necrosis and brown and shriveled berries during harvest time could be seen (Pearson and Goheen 1994). More than one *Diaporthe* species is frequently reported as causative agents from one geographical region (Dissanayake et al. 2015; Guarnaccia et al. 2018). Earlier, grapevine trunk disease was known to be caused by *Phomopsis viticola*, but the current knowledge on trunk diseases gives a different picture. Advancement in the field of molecular biology and techniques (sequencing platforms) revealed the occurrence of high diversity of pathogenic *Diaporthe* species associated with grapevine.

There are about 33 *Diaporthe* spp. known to cause dieback in grape-producing countries (Table 6.6) (Mostert et al. 2001; Van Niekerk et al. 2005; Udayanga et al. 2011, 2014a, b; White et al. 2011; Baumgartner et al. 2013; Úrbez-Torres et al. 2013; Hyde et al. 2014; Dissanayake et al. 2015; Guarnaccia et al. 2018; Lesuthu et al. 2019). All these species are associated with one disease; they differ in their symptoms, aggressiveness, and virulence, which differs from region to region and the variety of grape. In general, *D. ampelina* has a long history as the most common and severe pathogenic species together with *D. amygdali* (Mostert et al. 2001; Van Niekerk et al. 2005). *Diaporthe perjuncta* and *D. ampelina* cause cane bleaching (Kajitani and Kanematsu 2000; Mostert et al. 2001; Van Niekerk et al. 2005; Rawnsley et al. 2006). In South Africa, *D. ampelina*, *D. nebulae*, and *D. novem* have been reported to be most virulent species associated with grapevines. Further, *D. eres* was reported as a weak to moderate pathogen in many regions (Kaliterna et al. 2012; Baumgartner et al. 2013). In China, so far, four *Diaporthe* spp. are reported to cause grapevine dieback disease (*D. eres*, *D. hongkongensis*, *D. phaseolorum*, and *D. sojae*) (Dissanayake et al. 2015). These results specify the intricacy and high species richness of *Diaporthe* associated with the grapevines (Lesuthu et al. 2019).

Table 6.6 Diversity of *Diaporthe* spp. associated with grapevine trunk disease

Host	*Diaporthe* species	Reference
Grapes (*Vitis vinifera*): Associated with grapevine dieback	*D. eres*	Manawasinghe et al. (2019)
	D. gulyae	
	D. hubeiensis	
	D. pescicola	
	D. sojae	
	D. unshiuensis	
	D. vinifera	
Grapes: Grapevine swelling arm	*D. ambigua*	Dissanayake et al. (2017a, b)
	D. ampelina	Úrbez-Torres et al. (2013), Lawrence et al. (2015)
	D. amygdali	Gomes et al. (2013), Guarnaccia et al. (2018)
	D. australafricana	Gomes et al. (2013)
	D. baccae	Guarnaccia et al. (2018)
	D. bohemiae	
	D. celeris	
	D. chamaeropis	Lawrence et al. (2015)
	D. cytosporella	Lawrence et al. (2015), Dissanayake et al. (2017a, b), Guarnaccia et al. (2018), Farr and Rossman (2019)
	D. eres	
	D. foeniculina	
	D. helianthi	
	D. hispaniae	Dissanayake et al. (2017a, b)
	D. hongkongensis	
	D. hungariae	Guarnaccia et al. (2018)
	D. kyushuensis	Kajitani and Kanematsu (2000)
	D. nebulae	Lesuthu et al. (2019)
	D. neotheicola	Úrbez-Torres et al. (2013)
	D. nobilis	Lawrence et al. (2015), Dissanayake et al. (2017a, b)
	D. novem	Lawrence et al. (2015)
	D. perjuncta	Mostert et al. (2001)
	D. perniciosa	Stoykow and Denchev (2006)
	D. phaseolorum	Dissanayake et al. (2017a, b)
	D. rudis	Guarnaccia et al. (2018)
	D. serafiniae	Lesuthu et al. (2019)
	D. sojae	Dissanayake et al. (2017a, b)
Cane and leaf spot	*Phomopsis viticola*	Pscheidt and Pearson (1989)

6.7.3 Sclerotium rolfsii

Sclerotium rolfsii (or *Athelia rolfsii*) is a potent fungal pathogen causing diseases on a wide variety of plants including cereals, vegetables, fruits, ornamentals, and turfs at various stages of growth (Aycock 1966; Punja 1985; Smith et al. 1989; Mullen 2001). This pathogen is known to persist in the soil for 2–3 years and capable to cause infection when the new crop comes up (Smith et al. 1989). The disease has been named as southern blight or southern stem blight. The pathogen is known to cause infection in all stages of plant tissues although it is known to generally infect the lower part of the stem at the soil-air interface (Mullen 2001). This pathogen is also known to attack seedlings, herbaceous plants, woody plants, fleshy roots, bulbs, and fruits (Mullen 2001). The most important crop plants associated with southern blight and leaf spot diseases include southern blight of common bean (*Phaseolus vulgaris*), leaf spot of Indian jasmine (*Jasminum multiflorum*), boll rot of cotton (*Gossypium hirsutum*), fruit rot of pumpkin (*Cucurbita maxima*), and southern blight of wild coffee (*Psychotria nervosa*) (Mahadevakumar et al. 2015a, b, c; Mahadevakumar and Janardhana 2014, 2016a, b; Mahadevakumar et al. 2018).

6.8 Diversity of Emerging Fungal Pathogens in Agro-Ecosystem

The global increase in virulent infectious diseases of natural populations and managed landscapes are mainly due to unprecedented fungal diseases leading to severe economic loss and threat to food security (Fisher et al. 2012). The plant pathogens could enter agricultural ecosystems by several mechanisms like host tracking, host jumps, hybridization, and horizontal gene transfer. Agro-ecosystem is defined as "the ecosystem that develops on farmed land, which includes both the crop species and its associated micro- and macro-organisms" (Stukenbrock and McDonald 2008). A long timescale is necessary for the development of complex biochemical machinery of pathogen attack and plant defense in pathogen-plant interactions, but the agriculture is fairly recent, and domestication faces the severity of acclimatized pathogens (Balter 2007; Stukenbrock and McDonald 2008). The development of new crop cultivars and agricultural practices has resulted in the emergence of new pathogens causing significant variation population of pathogens preexisting on the wild ancestors of the cultivated crops. The new agro-ecosystem provided a denser and genetically more uniform host population that enabled the pathogen transmission compared to the natural habitats (Stukenbrock and McDonald 2008).

The introduction of plants or pathogens into new environments could result in novel host-pathogen interactions, where pathogens cause severe damage in native host populations. For example, the introduction of the potato late blight pathogen *Phytophthora infestans* into Ireland in the mid-nineteenth century caused the Irish potato famine (Goodwin et al. 1994), and the movement of the wheat stripe rust pathogen *Puccinia striiformis* f. sp. graminis into the USA long ago caused severe economic losses (Carleton 1915). The host and the pathogen coevolved during the process of host domestication and the development of the agro-ecosystem specific to the host crop. During domestication, the selection and cultivation of desired host genotypes simultaneously select pathogen genotypes that are adapted to the selected individuals in a specific agro-ecosystem (Stukenbrock and McDonald 2008). However, nowadays, a greater number of new disease reports indicate either new or crossing the host barrier or crossing the geographic location and becoming more serious constraints to modern agriculture. Even though the western countries have adopted different strategies to oversee the emergence of new diseases (along with fungi, viruses, bacterial diseases, and others), new diseases are emerging. India is yet to adopt modern diagnostic techniques for disease diagnosis and to suggest solutions or management strategies. As a result, a wide range of new outbreaks are unnoticed or not recorded so far, or the existing host-pathogen system needs to be explored at the genetic level. This strategy supports breeding programs to develop new cultivars, hybrids, or varieties for improved and sustained agricultural production.

The most neglected part of understanding the fungal diseases and their impact on human life is storage diseases. Every agricultural product consumed directly or after harvest will be subjected for storage depending on the type of produce. The storage fungi produce diverse secondary metabolites, which are toxic to the human being as well as livestock. Once the stored agricultural produce is affected by storage fungi, they will be unfit for consumption. In vegetables, the moisture content favors the development of various molds that cause damage. In Fig. 6.8, some of the common storage fungi are presented which include association of *Aspergillus* on cucumber; *Rhizopus* on jackfruit; *Alternaria* on apple; and *Colletotrichum* on watermelons, mango, and others.

6.9 Perspectives and Future Outlook

Production of sufficient food with assured quality and quantity remains of paramount importance for the sustenance of quality life. Inadvertent introduction of pathogenic fungi has adverse consequences on the cultivated crops throughout the world. The economic concussions by such introductions result in loss of yield, increased cost of cultivation, and disease control. Fungi being a unique group of organisms that have the potential of earning billion-dollar profit as the source of a wonder drug have also incurred a billion-dollar loss to a nation by their virulence. Diverse pathogenic fungi are the sole reason for more than 80% of crop loss in the Indian subcontinent. The plant pathogens also play a crucial role in regulating host populations in the geographic and ecological setup of a natural ecosystem. As a result, they can distress the availability of food sources to other living systems (Lindahl and Grace 2015). The majority of diverse microbial pathogens exhibit a high genetic variability due to narrow generation time, maximum population size, and rapid adaptability to various environmental conditions (Alberts et al. 2002; Lindahl and Grace 2015). Therefore, it is necessary to understand the plant pathogens at the genetic and population level to develop sustainable management practices in agriculture.

Fig. 6.8 Fungal diseases associated with fruits and vegetables (storage/post-harvest): *Aspergillus* rot of cucurbits (storage fungal disease) (**a**); soft rot of jackfruit (*Rhizopus* sp.) (**b**); apple fruit rot (storage/post-harvest disease) (**c**, **d**); fruit rot of watermelon (*Colletotrichum* sp.) (**e**, **f**); and on mango fruits (*Colletotrichum* sp.) (**f–h**)

To combat the diseases caused by fungal plant pathogens globally, it is crucial to confirm whether the same species/genotypes are present in other countries, as each species/genotype can possess varied attack patterns and responses to fungicides as well as climatological conditions. It is also necessary to follow what are their host ranges and mating strategies to relate to different disease control mechanisms. The movement of agricultural and forestry produce is inextricably cross-linked between geographic regions, and in turn, it becomes a global concern. Knowledge on which pathogen occurs and its attack on crop facilitates to enhance the yield and reduce the economic loss. Systematic and extensive research on emerging diseases has not been attempted especially in India on various crop plants. In terms of intensive plant material exchange and climate change, result in new pathogens needs stringent quarantine measures. Future plant disease management should aim at improving the food safety for a growing population with scope for simultaneous attempts to conserve the ecosystem integrity. Insights into the alternate food crops, traditionally important plant resources, and collateral hosts are vital to control the impact of pathogens. The diversity of fungal pathogens associated with a crop provides necessary strategies to adapt for biological control methods to manage diseases.

Caution should be exercised to follow up on the diversity of new and emerging fungi detrimental to crop production and food preservation to fulfill the needs of the teeming population.

Acknowledgments The first author (SM) is grateful to the Council of Scientific and Industrial Research (CSIR), Government of India, New Delhi, for the award of Research Associate Fellowship and also acknowledges the support and encouragement by the Department of Studies in Botany, University of Mysore.

References

Abawi GS, Pastor Corrales MA (1990) Root rots of beans in Latin America and Africa – diagnosis, research methodologies and management strategies. Centro Internacional de Agricultura Tropical, Cali, Colombia, pp 7–47

Agrios GN (2005) Plant pathology, 5th edn. Academic press, London

Akhtar J, Chaube HS (2006) Variability in *Phomopsis* blight pathogen (*Phomopsis vexans* (Sacc. & Syd.) Harter). Ind Phytopathol 59:439–444

Alberts B, Johnson A, Lewis J, Raff M, Roberts K, Walter P (2002) Introduction to pathogens in molecular biology of the cell, 4th edn. Garland Science, New York. https://www.ncbi.nlm.nih.gov/books/NBK26917

Alexopoulos CJ, Mims CW, Blackwell M (1997) Introductory mycology. John Wiley & Sons, New York, NY

Allen DJ, Ampofo JKO, Wortmann CS (1996) Pests, diseases, and nutritional disorders of the common bean in Africa: a field guide, vol 260. Centro Internacional de Agricultura Tropical, Cali, Colombia, pp 49–74

Ampuero J, Latorre BA, Torres R, Chávez ER (2008) Identification of *Phytophthora cryptogea* as the cause of rapid decline of petunia (*Petunia* × *hybrida*) in Chile. Plant Dis 92:1529–1536

Amselem J, Cuomo CA, van Kan JAL, Viaud M, Benito EP et al (2011) Genomic analysis of the necrotrophic fungal pathogens *Sclerotinia sclerotiorum* and *Botrytis cinerea*. PLoS Genet 7: e1002230. https://doi.org/10.1371/journal.pgen.1002230

Anderson PK, Cunningham AA, Patel NG, Morales FJ, Epstein PR, Daszak P (2004) Emerging infectious diseases of plants: pathogen pollution, climate change and agro technology drivers. Trends Ecol Evol 19:535–544

Anonymous (2009) Statistical data on horticultural crops in Karnataka state: Department of Horticulture. Lalbagh, Bangalore, Karnataka

Arya HC, Jain BL (1964) *Fusarium* seedling blight of maize in Rajasthan. Ind Phytopathol 17:51–57

Avelino J, Cristancho M, Georgiou S, Imbach P, Aguilar L et al (2015) The coffee rust crises in Colombia and Central America (2008–2013): impacts, plausible causes and proposed solutions. Food Sec 7:303–321

Aycock R (1966). Stem rot and other diseases caused by *Sclerotium rolfsii* or the status of Rolfs' fungus after 70 years. North Carolina Agricultural Experimental Station Technical Bulletin # 174 Raleigh : North Carolina State University

Bailey JA, Jeger MJ (1992) *Colletotrichum* biology, pathology and control. CABI, Wallingford

Balter M (2007) Seeking agriculture's ancient roots. Science 316:1830–1835

Baumgartner K, Fujiyoshi F, Travadon R, Castlebury LA, Wilcox WF, Rolshausen PE (2013) Characterization of species of *Diaporthe* from wood cankers of grape in Eastern North American vineyards. Plant Dis 97:912–920

Bellard C, Bertelsmeier C, Leadley P, Thuiller W, Courchamp F (2012) Impacts of climate change on the future of biodiversity. Ecol Lett 15:365–377. https://doi.org/10.1111/j.1461-0248.2011.01736.x

Beura SK, Mahanta IC, Mahapatra KB (2008) Economics and chemical control of *Phomopsis* twig blight and fruit rot of brinjal. J Mycopathol Res 46:73–76

Biswas A (2001) False smut disease of rice: a review. Environ Ecol 19:67–83

Bolton MD, Thomma BP, Nelson BD (2006) *Sclerotinia sclerotiorum* (Lib.) de Bary: biology and molecular traits of a cosmopolitan pathogen. Mol Plant Pathol 7:1–16. https://doi.org/10.1111/j.1364-3703.2005.00316.x

Brasier CM, Sanchez-Hernandez E, Kirk SA (2003) *Phytophthora inundata* sp. nov., a part heterothallic pathogen of trees and shrubs in wet or flooded soils. Mycol Res 107:477–484. https://doi.org/10.1017/S0953756203007548

Brasier CM, Beales PA, Kirk SA, Denman S, Rose J (2005) *Phytophthora kernoviae* sp. nov., an invasive pathogen causing bleeding stem lesions on forest trees and foliar necrosis of ornamentals in the UK. Mycol Res 109:853–859. https://doi.org/10.1017/S0953756205003357

Brayford D (1990) Vegetative incompatibility in Phomopsis from elm. Mycol Res 94:745–752. https://doi.org/10.1016/S0953-7562(09)81373-7

Brefort T, Doehlemann G, Mendoza-Mendoza A, Reissmann S, Djamei A, Kahmann R (2009) *Ustilago maydis* as a pathogen. Annu Rev Phytopathol 47:423–445. https://doi.org/10.1146/annurev-phyto-080508-081923

Butler EJ (1918) Fungi and disease in plants. Thacker, Spink & Co, Calcutta, West Bengal

Byrne JM, Hausbeck MK, Hammerschmidt R (1997) Conidial germination and appressorium formation of *Colletotrichum coccodes* on tomato foliage. Plant Dis 81:715–718

Canon PF, Damm U, Johnston PR, Weir BS (2012) *Colletotrichum* - current status and future directions. Stud Mycol 73:181–213

Carella P, Gogleva A, Tomaselli M, Alfs C, Schornack S (2018) *Phytophthora palmivora* establishes tissue-specific intracellular infection structures in the earliest divergent land plant lineage. Proc Natl Acad Sci 115:E3846–E3855. https://doi.org/10.1073/pnas.1717900115

Carleton MA (1915) A serious new wheat rust in this country. Science 42:58–59

Choquer M, Fournier E, Kunz C, Levis C, Pradier JM et al (2007) *Botrytis cinerea* virulence factors: new insights into a necrotrophic and polyphageous pathogen. FEMS Microbiol Lett 2778:1–10. https://doi.org/10.1126/science.1164627

Churchill AC (2011) *Mycosphaerella fijiensis*, the black leaf streak pathogen of banana: progress towards understanding pathogen biology and detection, disease development, and the challenges of control. Mol Plant Pathol 12:307–328. https://doi.org/10.1111/j.1364-3703.2010.00672.x

Cole JS (1956) Studies in the physiology of parasitism XX. The pathogenicity of *Botrytis cinerea, Sclerotinia fructigena*, and *Sclerotinia laxa*, with special reference to the part played by pectolytic enzymes. Ann Bot 20:15–38

Cooke MC (1876) Some Indian fungi. Grevillea 4:115

Cottyn B, Cerez MT, Van Outryve MF, Barroga J, Swings J, Mew TW (1996a) Bacterial diseases of rice.1. Pathogenic bacteria associated with sheath rot complex and grain discoloration of rice in the Philippines. Plant Dis 80:429–437

Cottyn B, Van Outryve MF, Cerez MT, De Cleene M, Swings J, Mew TW (1996b) Bacterial diseases of Rice II. Characterization of pathogenic bacteria associated with sheath rot complex and grain discoloration of rice in Philippines. Plant Dis 80:438–445

Cramer RA, Brick MA, Byrne PF, Schwartz HF, Wickliffe E (1996) Characterization of *Fusarium* wilt isolates collected in the central high plains. Department of Bio-Agricultural Sciences and Pest Management, Colorado State University, Ft. Collins, CO

Crous PW, Groenewald JZ, Slippers B, Wingfield MJ (2016) Global food and fibre security threatened by current inefficiencies in fungal identification. Phil Trans R Soc B Biol Sci 371:20160024. https://doi.org/10.1098/rstb.2016.0024

Damm U, Woudenberg J, Cannon PF, Crous PW (2009) *Colletotrichum* species with curved conidia from herbaceous hosts. Fungal Divers 39:45–87

Damm U, O'Connell RJ, Groenewald JZ, Crous PW (2014) The *Colletotrichum destructivum* species complex - hemibiotrophic pathogens of forage and field crops. Stud Mycol 79:49–84

Das SN, Sarma TC (2012) Some micro-fungi and their association on the incidence of diseases of brinjal (*Solanum melongena* L.) in Western Assam. Bioscan 1:303–306

De Silva NI, Lumyong S, Hyde KD, Bulgakov T, Phillips AJL, Yan JY (2016) Mycosphere essays 9: defining biotrophs and hemibiotrophs. Mycosphere 7:545–559. https://doi.org/10.5943/mycosphere/7/5/2

Dean R, Van Kan JAL, Pretorius ZA, Hammond-Kosack KE, Pietro AD et al (2012) Top 10 fungal pathogens in molecular plant pathology. Mol Plant Pathol 13:414–430. https://doi.org/10.1111/J.1364-3703.2011.00783.X

Deepika YS, Mahadevakumar S, Amruthesh KN, Lakshmidevi N (2020) A new collar rot disease of cowpea (*Vigna unguiculata*) caused by *Aplosporella hesperidica* in India. Lett Appl Microbiol 71:154. https://doi.org/10.1111/lam.13293

Deising HB, Werner S, Wernitz M (2000) The role of fungal appressoria in plant infection. Microbes Infect 2:1631–1641. https://doi.org/10.1016/s1286-4579(00)01319-8

Delaye L, Guzmán GG, Heil M (2013) Endophytes versus biotrophic and necrotrophic pathogens are fungal life-styles evolutionarily stable traits? Fungal Divers 60:125–135

De-Silva DD, Ades PK, Crous PW, Taylor PWJ (2017) *Colletotrichum* species associated with chili anthracnose in Australia. Plant Pathol 66:254–267. https://doi.org/10.1111/ppa.12572

Dey T, Saville A, Myers K, Tewari S, Cooke DE et al (2018) Large sub-clonal variation in Phytophthora infestans from recent severe late blight epidemics in India. Sci Rep 8:4429. https://doi.org/10.1038/s41598-018-22192-1

Dharanendraswamy S (2020) Studies on disease incidence and severity of stalk rot of maize caused by *Fusarium moniliforme* and *Cephalosporium acremonium* in Karnataka and its management. PhD Thesis, Department of Microbiology, University of Mysore, India

Dharanendraswamy S, Mahadevakumar S, Amruthesh KN, Swapnil R, Lakshmidevi N (2019a) Survey and surveillance for disease prevalence and incidence of stalk rot of maize in Karnataka. Int J Agric Sci 11:8633–8637

Dharanendraswamy S, Mahadevakumar S, Hemareddy HB, Amruthesh KN, Mamatha S et al (2019b) Morphological and Molecular characterization of *Fusarium verticillioides* (*F. moniliforme*) associated with post-flowering stalk rot (PSFR) of maize in Karnataka. KAVAKA 53:106–116. https://doi.org/10.36460/Kavaka/53/2019/106-116

Dharanendraswamy S, Mahadevakumar S, Hemareddy HB, Amruthesh KN, Mamatha S et al (2020a) First report of *Fusarium equiseti* associated with post flowering stalk rot of maize (*Zea mays* L.) in India. Crop Prot 129:1–7. https://doi.org/10.1016/j.cropro.2019.105035

Dharanendraswamy S, Mahadevakumar S, Vasanthakumar T, Amruthesh KN, Hemareddy HB et al (2020b) First report of *Lasiodiplodia pseudotheobromae* associated with post flowering stalk rot of maize (*Zea mays* L.) from India. Plant Dis 104:2524. https://doi.org/10.1094/PDIS-11-19-2418-PDN

Diao YZ, Zhang C, Lin D, Liu XL (2014) First report of *Colletotrichum truncatum* causing anthracnose of tomato in China. Plant Dis 98:687

Dillard HR (1989) Effect of temperature, wetness duration, and inoculums density on infection and lesion development of *Colletotrichum coccodes* on tomato fruit. Phytopathology 79:1063–1066

Dissanayake AJ, Liu M, Zhang W, Chen Z, Udayanga D et al (2015) Morphological and molecular characterization of Diaporthe species associated with grapevine trunk disease in China. Fungal Biol 119:283–294. https://doi.org/10.1016/j.funbio.2014.11.003

Dissanayake AJ, Phillips AJL, Hyde KD, Yan JY, Li XH (2017a) The current status of species in *Diaporthe*. Mycosphere 8:1106–1156

Dissanayake AJ, Zhang W, Liu M, Hyde KD, Zhao WS et al (2017b) *Diaporthe* species associated with peach tree dieback in Hubei, China. Mycosphere 8:512–528. https://doi.org/10.5943/mycosphere/8/5/2

Divon HH, Fluhr R (2006) Nutrition acquisition strategies during fungal infection of plants. FEMS Microbiol Lett 266:65–74. https://doi.org/10.1111/j.1574-6968.2006.00504.x

Doehlemann G, Ökmen B, Zhu W, Sharon A (2017) Plant pathogenic fungi. Microbiol Spectrum 5: FUNK-0023-2016

Dorn B, Forrer HR, Schurch S, Vogelgsang S (2009) *Fusarium* species complex on maize in Switzerland: occurrence, prevalence, impact and mycotoxin in commercial hybrids under natural infection. Eur J Plant Pathol 125:51–61

Dowlah CAF (2015) International trade, competitive advantage and developing economies. Routledge, London

Duplessis S, Cuomo CA, Lin YC, Aerts A, Tisserant E et al (2011) Obligate biotrophy features unraveled by the genomic analysis of rust fungi. Proc Natl Acad Sci U S A 108:9166–9171. https://doi.org/10.1073/pnas.1019315108

Elena K, Paplomatas EJ (1998) *Phytophthora boehmeriae* boll rot: a new threat to cotton cultivation in the Mediterranean region. Phytoparasitica 26:20–26

Elliott M, Shamoun SF, Sumampong G, James D, Masri S, Varga A (2009) Evaluation of several commercial biocontrol products on European and North American populations of Phytophthora ramorum. Biocontrol Sci Tech 19:1007–1021. https://doi.org/10.1080/09583150903243870

Faris JD, Zhang Z, Lu H, Lu S, Reddy L et al (2010) A unique wheat disease resistance-like gene governs effector triggered susceptibility to necrotrophic pathogens. Proc Natl Acad Sci 107:13544–13549. https://doi.org/10.1073/pnas.1004090107

Farr DF, Rossman AY (2018) Fungal databases, systematic mycology and microbiology laboratory. ARS, USDA, Washinton, DC. http://nt.ars-grin.gov/fungaldatabase/

Farr DF, Rossman AY (2019) Fungal databases - fungus-host distributions, systematic mycology and microbiology laboratory. ARS, USDA, Washinton, DC. http://nt.ars-grin.gov/fungaldatabase/

Figueiredo A, Monteiro F, Sebastiana M (2015) First clues on a jasmonic acid role in grapevine resistance against the biotrophic fungus *Plasmopara viticola*. Eur J Plant Pathol 142:645–652

Fisher MC, Henk DA, Briggs CJ, Brownstein JS, Madoff LC et al (2012) Emerging fungal threats to animal, plant and ecosystem health. Nature 484:186–194. https://doi.org/10.1038/nature10947

Flood JL (2010) The importance of plant health to food security. Food Sec 2:215–231

Ford R, Banniza S, Photita W, Taylor PWJ (2004) Morphological and molecular discrimination of Colletotrichum truncatum causing anthracnose on lentil in Canada. Aust Plant Pathol 33:559–569

Fradin EF, Thomma BP (2006) Physiology and molecular aspects of *Verticillium* wilt diseases caused by *V. dahliae* and *V. albo-atrum*. Mol Plant Pathol 7:71–86. https://doi.org/10.1111/j.1364-3703.2006.00323.x

Fry W (2008) *Phytophthora infestans*: the plant (and R gene) destroyer. Mol Plant Pathol 9:385–402

Fu M, Crous PW, Bai Q, Zhang PF, Xiang J et al (2019) *Colletotrichum* species associated with anthracnose of *Pyrus* spp. in China. Persoonia 42:1–35. https://doi.org/10.3767/persoonia.2019.42.01

Gao Y, Liu F, Duan W, Crous PW, Cai L (2017) *Diaporthe* is paraphyletic. IMA Fungus 8:153–187. https://doi.org/10.5598/imafungus.2017.08.01.11

Giblin FR, Tan YP, Mitchell R, Coates LM, Irwin JAG, Shivas RG (2018) *Colletotrichum* species associated with pre-and post-harvest diseases of avocado and mango in eastern Australia. Aust Plant Pathol 47:269–276. https://doi.org/10.1007/s13313-018-0553-0

Gomes RR, Glienke C, Videira SIR, Lombard L, Groenewald JZ, Crous PW (2013) *Diaporthe*: a genus of endophytic, saprobic and plant pathogenic fungi. Persoonia 31:1–41. https://doi.org/10.3767/003158513X666844

González C, Brito N, Sharon A (2016) Infection process and fungal virulence factors. In: Fillinger S, Elad Y (eds) *Botrytis*: the fungus, the pathogen and its management in agricultural systems. Springer International, Heidelberg, pp 229–246. https://doi.org/10.1007/978-3-319-23371-0_12

Goodwin SB, Cohen BA, Fry WE (1994) Panglobal distribution of a single clonal lineage of the Irish potato famine fungus. Proc Natl Acad Sci U S A 91:11591–11595

Goodwin SB, M'Barek SB, Dhillon B, Wittenberg AHJ, Crane CF et al (2011) Finished genome of the fungal wheat pathogen *Mycosphaerella graminicola* reveals dispensome structure, chromosome plasticity, and stealth pathogenesis. PLoS Genet 7:e1002070. https://doi.org/10.1371/journal.pgen.1002070

Granke L, Quesada-Ocampo L, Lamour K, Hausbeck MK (2012) Advances in research on *Phytophthora capsici* on vegetable crops in the United States. Plant Dis 95:1588–1600

Guarnaccia V, Crous PW (2017) Emerging citrus diseases in Europe caused by species of Diaporthe. IMA Fungus 8:317–334. https://doi.org/10.5598/imafungus.2017.08.02.07

Guarnaccia V, Crous PW (2018) Species of *Diaporthe* on *Camellia* and *Citrus* in the Azores Islands. Phytopathol Mediterr 57:307. http://hdl.handle.net/10019.1/106732

Guarnaccia V, Groenewald JZ, Polizzi G, Crous PW (2017) High species diversity in *Colletotrichum* associated with citrus diseases in Europe. Persoonia 39:32–50. https://doi.org/10.3767/persoonia.2017.39.02

Guarnaccia V, Groenewald JZ, Woodhall J, Armengol J, Cinelli T, Eichmeier A (2018) *Diaporthe* diversity and pathogenicity revealed from a broad survey of grapevine diseases in Europe. Persoonia 40:135–153. https://doi.org/10.3767/persoonia.2018.40.06

Guharoy S, Bhattacharyya S, Mukherjee SK, Mandal N, Khatua DC (2006) Phytophthora melonis associated with fruit and vine rot disease of pointed gourd in india as revealed by RFLP and sequencing of ITS region. J Phytopathol 154:612–615. https://doi.org/10.1111/j.1439-0434.2006.01155.x

Hagedorn DJ, Inglis DA (1986) Handbook of bean diseases. University of Wisconsin- extension publication # 3374. University of Wisconsin--Extension, Madison, Wi

Hansen EM (2015) *Phytophthora* species emerging as pathogens of forest trees. Curr For Rep 1:16–24. https://doi.org/10.1007/s40725-015-0007-7

Hansen EM, Reeser PW, Davidson JM, Garbelotto M, Ivors K, Douhan L, Rizzo DM (2003) *Phytophthora nemorosa*, a new species causing cankers and leaf blight of forest trees in California and Oregon, USA. Mycotaxon 88:129–138

Hartman GL, West ED, Herman TK (2011) Crops that feed the world 2. Soybean—worldwide production, use, and constraints caused by pathogens and pests. Food Sci 3:5–17. https://doi.org/10.1007/s12571-010-0108-x

Hawksworth DL, Lücking R (2017) Fungal diversity revisited: 2.2 to 3.8 million species. Microbiol Spectrum 5:FUNK-0052-2016. https://doi.org/10.1128/microbiolspec.FUNK-0052-2016

He Y, Chen Q, Shu C, Yang M, Zhou E (2016) *Colletotrichum truncatum*, A new cause of anthracnose on Chinese flowering cabbage (*Brassica parachinensis*) in China. Trop Plant Pathol 4:183–192

Heath MC, Skalamera D (1997) Cellular interactions between plants and biotrophic fungal parasites. In: Andrews PR, Tommerup IC (eds) Advances in botanical research, vol 24. Academic Press, San Diego, CA, pp 195–225

Heyman F, Blair JE, Persson L, Wikström M (2013) Root rot of pea and faba bean in southern Sweden caused by *Phytophthora pisi* sp. nov. Plant Dis 97:461–471

Hossain M, Fischer KS (1995) Rice research for food security and sustainable agricultural development in Asia: achievements and future challenges. GeoJ 3:286–298

Hückelhoven R, Panstruga R (2011) Cell biology of the plant-powdery mildew interaction. Curr Opin Plant Biol 14:738–746. https://doi.org/10.1016/j.pbi.2011.08.002

Hyde KD, Cai L, McKenzie EHC, Yang YL, Zhang JZ, Prihastuti H (2009) *Colletotrichum*: a catalogue of confusion. Fungal Divers 39:1–17

Hyde KD, Nilsson RH, Alias SA, Ariyawansa HA, Blair JE et al (2014) One stop shop: backbone trees for important phytopathogenic genera - I. Fungal Divers 67:21–125. https://doi.org/10.1007/s13225-014-0298-1

Hyde KD, Hongsanan S, Jeewon R, Bhat DJ, McKenzie EHC et al (2016) Fungal diversity notes 367-491: taxonomic and phylogenetic contributions to fungal taxa. Fungal Divers 80:1–270. https://doi.org/10.1007/s13225-016-0373-x

Iqbal SM, Ghafoor A, Ahmad Z, Haqqani AM (2003) Pathogenicity and fungicidal efficacy for *Sclerotinia* rot brinjal. Int J Agric Biol 5:618–620

Jayalakshmi K, Nargund VB, Raju J, Benagi VI, Raghu S et al (2015) Pomegranate anthracnose caused by *Colletotrichum gloeosporioides*: a menace in quality fruit production. J Pure Appl Microbiol 9:3093–3097

Jayaramaiah KM, Janardhana GR (2011) Mycobiota and their relative per cent incidence on freshly harvested brinjal seeds (*Solanum melongena* L.) produced in Karnataka. Asian J Microbiol Biotechnol Environ Sci 13:281–283

Jiang N, Cui J, Shi Y, Yang G, Zhou X et al (2019) Tomato lncRNA23468 functions as a competing endogenous RNA to modulate NBS-LRR genes by decoying miR482b in the tomato - *Phytophthora infestans* interaction. Hortic Res 6:28. https://doi.org/10.1038/s41438-018-0096-0

Jung T, Burgess TI (2009) Re-evaluation of *Phytophthora citricola* isolates from multiple woody hosts in Europe and North America reveals a new species, *Phytophthora plurivora* sp. nov. Persoonia 22:95–110

Jung T, Stukely MJC, Hardy GESJ, White D, Paap T, Dunstan WA, Burgess TI (2011) Multiple new *Phytophthora* species from ITS Clade 6 associated with natural ecosystems in Australia: evolutionary and ecological implications. Persoonia 26:13–39

Kabbage M, Yarden O, Dickman MB (2015) Pathogenic attributes of Sclerotinia sclerotiorum: switching from a biotrophic to necrotrophic lifestyle. Plant Sci 233:53–60

Kajitani Y, Kanematsu S (2000) *Diaporthe kyushuensis* sp. nov. the teleomorph of the causal fungus of grapevine swelling arm in Japan, and its anamorph *Phomopsis vitimegaspora*. Mycoscience 41:111–114. https://doi.org/10.1007/BF02464318

Kaliterna J, Milicevic T, Cvjetkovic B (2012) Grapevine trunk diseases associated with fungi from the *Diaporthaceae* family in Croatian vineyards. Arch Ind Hyg Toxicol 63:471–478

Kankanala P, Czymmek K, Valen B (2007) Roles for rice membrane dynamics and plasmodesmata during biotrophic invasion by the blast fungus. Plant Cell 19:706–724

Kemen E, Kemen AC, Rafiqi M, Hempel U, Mendgen K, Hahn M, Voegele RT (2005) Identification of a protein from rust fungi transferred from haustoria into infected plant cells. Mol Plant-Microbe Interact 18:1130–1139. https://doi.org/10.1094/MPMI-18-1130

Kemen E, Kemen A, Ehlers A, Voegele R, Mendgen K (2013) A novel structural effector from rust fungi is capable of fibril formation. Plant J 75:767–780. https://doi.org/10.1111/tpj.12237

Khan MW, Singh RK (1975) A note on anthracnose of arhar (*Cajanus cajan*) incited by *Colletotrichum truncatum* (India). Ind Phytopathol 27:622–624

Khokhar MK, Hooda KS, Sharma SS, Singh V, Saini A (2013) *Fusarium* stalk rot: a major threat to maize production in India. Maize J 2:1–5

Khokhar MK, Hooda KS, Sharma SS, Singh V (2014) Post flowering stalk rot complex of maize - present status and future prospects. Maydica 59:226–242

Kim KT, Jeon J, Choi J, Cheong K, Song H, Choi G, Kang S, Lee YH (2016) Kingdom-wide analysis of fungal small secreted proteins (SSPs) reveals their potential role in host association. Front Plant Sci 7:186. https://doi.org/10.3389/fpls.2016.00186

Kimaru SK, Monda E, Cheruiyot RC, Mbaka J, Alakonya A (2018) Morphological and molecular identification of the causal agent of anthracnose disease of avocado in Kenya. Int J Microbiol 2018:4568520. https://doi.org/10.1155/2018/4568520

Kommedahal T, Windels CE, Stucker RE (1972) Occurrence of *Fusarium* species in root and stalks symptoms in corn plants during the growing season. Phytopathology 69:961–966

Koprivica M, Dulic-Markovic I, Jevtic R, Cooke DEL (2009) Methods for detection of *Phytophthora fragariae* var. *rubi* on raspberry. Pest Phytomed 24:17–184

Krola P, Igielski R, Pollmannb S, Kepczynska EK (2015) Priming of seeds with methyl jasmonate induced resistance to hemi-biotroph *Fusarium oxysporum* f. sp. *lycopersici* in tomato via

12-oxo-phytodienoic acid, salicylic acid, and flavonol accumulation. J Plant Physiol 179:122–132

Kumar S (2014) Plant disease management in India. Afr J Agric Res 9:1207–1217

Kumar S, Gupta O (2012) Expanding dimensions of plant pathology. JNKVV Res J 46:286–293

Kumar J, Saxena SC (2009) Proceedings of the 21st training on recent advances in plant disease management. GBPUA&T, Pantnagar, pp 1–3

Lal S, Dwivedi RR (1983) *Cephalosporium* and *Fusarium* stalk rots of maize. In: Husain A, Singh K, Singh BP, Agnihotri VP (eds) Recent advances in plant pathology, pp 344–360

Latijnhouwers M, de Wit PJGM, Govers F (2003) Oomycetes and fungi: similar weaponry to attack plants. Trends Microbiol 11:462–469

Lawrence DP, Travadon R, Baumgartner K (2015) Diversity of *Diaporthe* species associated with wood cankers of fruit and nut crops in northern California. Mycologia 107:926–940. https://doi.org/10.3852/14-353

LeBoeuf J (2007) Identifying the species causing anthracnose of pepper and tomato in Southern Ontario. OMAFRA. http://www.omafra.gov.on.ca/crops/html

Lee SC, Alvenda ME, Bonman JM, Heinrichs EA (1986) Insects and pathogens associated with rice grain discoloration and their relationship in the Philippines. Kor J Plant Prot 25:107–112

Lenné JM (2002) Some major plant diseases. In: Waller JM, Lenné JM, Waller SJ (eds) Plant pathologist's pocketbook, 3rd edn. CABI, Wallingford, pp 4–18

Leonard KJ, Szabo LJ (2005) Stem rust of small grains and grasses caused by *Puccinia graminis*. Mol Plant Pathol 6:99–111. https://doi.org/10.1111/j.1364-3703.2005.00273.x

Lesuthu P, Mostert L, Spies CF, Moyo P, Regnier T, Halleen F (2019) *Diaporthe nebulae* sp. nov. and first report of *D. cynaroidis*, *D. novem*, and *D. serafiniae* on grapevines in South Africa. Plant Dis 103:808–817. https://doi.org/10.1094/PDIS-03-18-0433-RE

Li JT, Mo SX, Fu HB (2014) First report of target leaf spot caused by *Corynespora cassiicola* on cowpea in China. Plant Dis 98:1427

Li YG, Zhang SQ, Sun LP, Li S, Ji P (2018) First report of root rot of cowpea caused by *Fusarium equiseti* in Georgia in the United States. Plant Dis 101:1674

Li Q, Bu J, Shu J, Yu Z, Tang L, Huang S, Guo T, Mo J, Luo S, Solangi GS, Hsiang T (2019) *Colletotrichum* species associated with mango in southern China. Sci Rep 9:18891. https://doi.org/10.1038/s41598-019-54809-4

Lindahl JF, Grace D (2015) The consequences of human actions on risks for infectious diseases: a review. Infect Ecol Epidemiol 5:30048. https://doi.org/10.3402/iee.v5.30048

Liu JK, Hyde KD, Jones EBG, Ariyawansa HA, Bhat DJ et al (2015) Fungal diversity notes 1–110: taxonomic and phylogenetic contributions to fungal species. Fungal Divers 72:1–197. https://doi.org/10.1007/s13225-015-0324-y

Liu F, Wang M, Damm U, Crous PW, Cai L (2016) Species boundaries in plant pathogenic fungi: a *Colletotrichum* case study. BMC Evol Biol 16:81

Ma LJ, Geiser DM, Proctor RH, Rooney AP, O'Donnell K et al (2013) *Fusarium* pathogenomics. Annu Rev Microbiol 67:399–416. https://doi.org/10.1146/annurev-micro-092412-155650

Mahadevakumar S (2016) Biochemical and molecular characterization of pathogenesis related proteins (PRP's) and their role in conferring resistance in *Solanum melongena* L. against Phomopsis vexans. PhD Thesis, Department of Botany, University of Mysore, India

Mahadevakumar S, Janardhana GR (2012) First report of *Dactuliophora* species causing leaf spot of cowpea in India. New Dis Rep 25:17

Mahadevakumar S, Janardhana GR (2014) First report of *Pestalotiopsis* species causing leaf spot of cowpea (*Vigna unguiculata* (L.) Walp) in India. Plant Dis 98:686

Mahadevakumar S, Janardhana GR (2016a) First report of leaf spot disease caused by *Sclerotium rolfsii* on *Jasminum multiflorum* in India. J Plant Pathol 98:355

Mahadevakumar S, Janardhana GR (2016b) Morphological and molecular characterization of southern leaf blight of *Psychotria nervosa* (wild coffee) incited by *Sclerotium rolfsii*. J Plant Pathol 98:351–354

Mahadevakumar S, Janardhana GR (2016c) The leaf blight and fruit rot disease of brinjal caused by *Phomopsis vexans* in six agro-ecological regions of South West India. Plant Pathol Quar 6:5–12

Mahadevakumar S, Sridhar KR (2020) Plant-microbes interaction: current developments and future challenges. In: Yadav N, Mishra S, Yadav N, Kumar A (eds) Advances in plant microbiomes and sustainable agriculture. Springer Nature, Singapore

Mahadevakumar S, Yadav V, Tejaswini GS, Sandeep SN, Janardhana GR (2014) First report of *Phomopsis citri* H.S. Fawc. associated with dieback of *Citrus lemon* (L.) Burm. f. in India. Plant Dis 98:1286. https://doi.org/10.1094/PDIS-04-14-0388-PDN

Mahadevakumar S, Tejaswini GS, Shilpa N, Janardhana GR, Dharanendraswamy S, Yadav V (2015a) First report of *Sclerotium rolfsii* associated with boll rot of cotton in India. Plant Dis 100:229

Mahadevakumar S, Tejaswini GS, Yadav V, Janardhana GR (2015b) First report of southern blight and leaf spot (*Sclerotium rolfsii*) on common bean (*Phaseolus vulgaris*) in India. Plant Dis 99:1280

Mahadevakumar S, Yadav V, Tejaswini GS, Janardhana GR (2015c) Morphological and molecular characterization of *Sclerotium rolfsii* associated with fruit rot of *Cucurbita maxima*. Eur J Plant Pathol 145:215–219

Mahadevakumar S, Amruthavalli C, Sridhar KR, Janardhana GR (2017) Prevalence, incidence and molecular characterization of *Phomopsis vexans* (*Diaporthe vexans*) causing leaf blight and fruit rot disease of brinjal in Karnataka (India). Plant Pathol Quar 7:29–46

Mahadevakumar S, Chandana C, Deepika YS, Sumashri KS, Yadav V, Janardhana GR (2018) Pathological studies on southern blight disease of China aster (*Callistephus chinensis*) caused by *Sclerotium rolfsii*. Eur J Plant Pathol 151:1081–1087

Mahadevakumar S, Chandana C, Janardhana GR (2019) Association of *Colletotrichum truncatum* with anthracnose disease of tuberose (*Polianthes tuberosa*) in India. Crop Prot 118:1–5

Mahmodi F, Kadir JB, Nasehi A, Puteh A, Soleimani N (2013) Occurrence of anthracnose caused by *Colletotrichum truncatum* on chickpea (*Cicer arietinum*) in Malaysia. Plant Dis 97:1507

Manawasinghe IS, Dissanayake AJ, Li X, Liu M, Wanasinghe DN et al (2019) High genetic diversity and species complexity of *Diaporthe* associated with grapevine dieback in China. Front Microbiol 10:1936. https://doi.org/10.3389/fmicb.2019.01936

Marin-Felix Y, Hernández-Restrepo M, Wingfield MJ, Akulov A, Carnegie AJ et al (2019) Genera of phytopathogenic fungi: GOPHY 2. Stud Mycol 92:47–133. https://doi.org/10.1016/j.simyco.2018.04.002

McCotter SW, Horianopoulos LC, Kronstad JW (2016) Regulation of the fungal secretome. Curr Genet 62:533–545. https://doi.org/10.1007/s00294-016-0578-2

Meadows R (2011) Why biotrophs can't live alone. PLoS Path 9:1–2

Meinhardt LW, Costa GGL, Thomazella DPT, Teixeira PJPL, Carazzolle MF et al (2014) Genome and secretome analysis of the hemibiotrophic fungal pathogen, *Moniliophthora roreri*, which causes frosty pod rot disease of cacao: mechanisms of the biotrophic and necrotrophic phases. BMC Genomics 15:164–189

Mendgen K, Hahn M (2002) Plant infection and the establishment of fungal biotrophy. Trends Plant Sci 7:352–356

Meng Y, Zhang Q, Ding W, Shan W (2014) *Phytophthora parasitica*: a model oomycete plant pathogen. Mycology 5:43–51. https://doi.org/10.1080/21501203.2014.917734

Mengiste T (2012) Plant immunity to necrotrophs. Annu Rev Phytopathol 50:267–294. https://doi.org/10.1146/annurev-phyto-081211-172955

Mo J, Zhao G, Li Q, Solangi GS, Tang L, Guo T, Huang S (2018) Identification and characterization of *Colletotrichum* species associated with mango anthracnose in Guangxi, China. Plant Dis 102:1283–1289. https://doi.org/10.1094/PDIS-09-17-1516-RE

Mondal SN, Vincent A, Reis RF, Timmer LW (2007) Saprophytic colonization of citrus twigs by *Diaporthe citri* and factors affecting pycnidial production and conidial survival. Plant Dis 91:387–392. https://doi.org/10.1094/PDIS-91-4-038

Moreno-Moran M, Burbano-Figueroa O (2016) First report of *Colletotrichum gossypii* var. *cephalosporioides* causing cotton ramulosis in Colombia. Plant Dis 100:653

Mostert L, Crous PW, Kang JC, Phillips AJ (2001) Species of *Phomopsis* and a *Libertella* sp. occurring on grapevines with specific reference to South Africa: morphological, cultural, molecular and pathological characterization. Mycologia 93:146–167. https://doi.org/10.1080/00275514.2001.12061286

Mullen J (2001) Southern blight, southern stem blight, white mold. Plant Health Instruct. https://doi.org/10.1094/PHI-I-2001-0104-01

Muneeshwar S, Razdan VK, Sourav G (2011) Occurrence of *Phomopsis* leaf blight and fruit rot of brinjal caused by *Phomopsis vexans* in Jammu. Ann Plant Prot Sci 19:396–399

Munir M, Amsden B, Dixon E, Vaillancourt L, Ward-Gauthier NA (2016) Characterization of *Colletotrichum* species causing bitter rot of apple in Kentucky orchards. Plant Dis 100:2194–2203

Muntañola-Cvetković M, Mihaljčević M, Petrov M (1981) On the identity of the causative agent of a serious *Phomopsis* - *Diaporthe* disease in sunflower plants. Nova Hedwigia 34:417–435

Nene YL, Reddy MV, Haware MP, Ghanekar AM, Amin KS et al (2012) Field diagnosis of chickpea diseases and their control. Information Bulletin # 28 (revised). International Crops Research Institute for the Semi-Arid Tropics, Andhra Pradesh

Nguyen PTH, Pettersson OV, Olsson P, Liljeroth E (2010) Identification of *Colletotrichum* species associated with anthracnose disease of coffee in Vietnam. Eur J Plant Pathol 127:73–87. https://doi.org/10.1007/s10658-009-9573-5

Nicholson RL, Moraes WRC (1980) Survival of *Colletotrichum graminicola*: importance of the spore matrix. Phytopathology 70:255–261

Nur-Ain-Izzati MZ, Azmi AR, Siti Nordahliawate MS, Norazlina J (2011) Contribution to knowledge of diversity of *Fusarium* associated with maize in Malaysia. Plant Prot Sci 47:20–24

O'Connell RJ, Thin MR, Hacquard S, Amyotte SG, Kleemann J et al (2012) Lifestyle transitions in plant pathogenic Colletotrichum fungi deciphered by genome and transcriptome analyses. Nat Genet 44:1060–1065. https://doi.org/10.1038/ng.2372

Oerke EC (2006) Crop losses to pests. J Agric Sci 144:31–43. https://doi.org/10.1017/S0021859605005708

Oliver RP, Ipcho SVS (2004) *Arabidopsis* pathology breathes new life into the necrotrophs vs. biotrophs classification of fungal pathogens. Mol Plant Pathol 5:347–352

Oliver RP, Solomon PS (2010) New developments in pathogenicity and virulence of necrotrophs. Curr Opin Plant Biol 13:415–419. https://doi.org/10.1016/j.pbi.2010.05.003

Ou SH (1985) Rice diseases, 2nd edn. Commonwealth Mycological Institute, Kew

Padman M, Janardhana GR (2012) PCR based detection of *Colletotrichum gloeosporioides* (Penz.) Penz. and Sacc. The causative agent of leaf spot disease of *Murraya Koenigii* L. As. J Microbiol Biotechnol Environ Sci 14:79–82

Padmanabhan SY (1973) The Great Bengal Famine. Annu Rev Phytopathol 11:11–26

Pammel LH (1914) Serious root and stalk diseases of corn. IOWA Agric 15:156–158

Pandey A (2010) Studies on fungal diseases of eggplant in relation to statistical analysis and making of disease calendar. Recent Res Sci Technol 2:1–3

Pandey D, Rajendran SRCK, Sajeesh MGPK, Kumar A (2016) Plant defense signaling and responses against necrotrophic fungal pathogens. J Plant Growth Reg 35:1–16

Park MS, Kim BR, Park IH, Hahm SS (2018) First Report of Two *Colletotrichum* species associated with bitter rot on apple fruit in Korea - *C. fructicola* and *C. siamense*. Mycobiology 46:154–158. https://doi.org/10.1080/12298093.2018.1478220

Pearson RC, Goheen C (1994) *Phomopsis* cane and leaf spot. In: Hewitt WB, Pearson RC (eds) Compendium of grape diseases. APS Press, St Paul, MI, pp 17–18

Perfect SE, O'Connell RJ, Green EF, Doering-Saad C, Green JR (1998) Expression cloning of a fungal proline-rich glycoprotein specific to the biotrophic interface formed in the Colletotrichum-bean interaction. Plant J 15:273–279. https://doi.org/10.1046/j.1365-313X.1998.00196.x

Petersen RH (1974) The rust fungus life cycle. Bot Rev 40:453–513. https://doi.org/10.1007/BF02860021

Petre B, Lorrain C, Saunders DG, Win J, Sklenar J et al (2016) Rust fungal effectors mimic host transit peptides to translocate into chloroplasts. Cell Microbiol 18:453–465. https://doi.org/10.1111/cmi.12530

Pine TS (1958) Etiology of the dead-arm of grapevines. Phytopathology 48:192–196

Pine TS (1959) Development of the grape dead-arm disease. Phytopathology 49:738–743

Ploetz RC (2015) Management of *Fusarium* wilt of banana: a review with special reference to tropical race 4. Crop Prot 73:7–15. https://doi.org/10.1016/j.cropro.2015.01.007

Prihastuti H, Cai L, Chen H, McKenzie EHC, Hyde KD (2009) Characterization of *Colletotrichum* species associated with coffee berries in northern Thailand. Fungal Divers 39:89–109

Pscheidt JW, Pearson RC (1989) Time of infection and control of *Phomopsis* fruit rot of Grape. Plant Dis 73:829–833. https://doi.org/10.1094/PD-73-0829

Punja ZK (1985) The biology, ecology and control of *Sclerotium rolfsii*. Annu Rev Phytopathol 23:97–127

Quin FM (1997) Advances in cowpea research. Colorcraft, Hong Kong

Ramdial H, Rampersad SN (2015) Characterization of *Colletotrichum* spp. causing anthracnose of bell pepper (*Capsicum annuum* L.) in Trinidad. Phytoparasitica 43:37–49. https://doi.org/10.1007/s12600-014-0428-z

Rashid MM (1976) Bangladesher Sabji, 1st edn. Bangla Academy, Dhaka

Rawnsley B, Wicks TJ, Stummer BE, Scott ES (2006) The role of *Diaporthe perjuncta*, in delayed budburst and death of grapevine buds. Aust Plant Pathol 35:265–273

Rehner SA, Uecker FA (1994) Nuclear ribosomal internal transcribed spacer phylogeny and host diversity in the Coelomycete *Phomopsis*. Can J Bot 72:1666–1674

Rintelen J (1965) *Fusarium culmorum* and *Fusarium* Artenals Erreger einer Stengelfaule an reifenden Maispflanzen. Z Pflanzenkr (Pflanzenpathol) Pflanzenschutz 72:89–91

Roman-Aviles B, Snapp SS, Kelly JD, Kirk WW (2003) *Fusarium* root rot of common beans. Michigan State University Extension Bulletin # E2876. Michigan State University Extension, Eat Lansing, MI

Roopadevi, Jamadar MM (2015) Physiological studies on anthracnose of green gram [*Vigna radiata* (L.) Wilczek] caused by *Colletotrichum truncatum* (Schw.) Andrus and Moore. Glob J Biol Agric Health Sci 4:9–12

Rossman A, Udayanga D, Castlebury LA, Hyde KD (2014) (2304) Proposal to conserve the name *Diaporthe eres* against twenty-one competing names (Ascomycota: Diaporthales: Diaporthaceae). Taxon 63:934–935. https://doi.org/10.12705/634.23

Saini TJ, Gupta SG, Anandalakshmi R (2017) Detection of chilli anthracnose caused by *Colletotrichum cliviae* in India. Aust Plant Dis Notes 12:33

Salunkhe O, Khare PK, Kumari R, Khan ML (2018) A systematic review on the aboveground biomass and carbon stocks of Indian forest ecosystems. Ecol Process 17:7. https://doi.org/10.1186/s13717-018-0130-z

Sanogo S, Pennypacker SP, Stevenson RE, MacNab AA (1997) Weather variables associated with infection of tomato fruit by *Colletotrichum coccodes*. Plant Dis 81:753–756

Saremi H, Amiri ME, Ashrafi J (2011) Epidemiological aspects of bean decline disease caused by *Fusarium* species and evaluation of the bean resistant cultivars to disease in Northwest Iran. Afr J Biotechnol 10:14954–14961

Saxena A, Raghuvanshi R, Singh HB (2014) Molecular, phenotypic and pathogenic variability in *Colletotrichum* isolates of subtropical region in North Eastern India, causing fruit rot of chillies. J Appl Microbiol 117:1422–1434. https://doi.org/10.1111/jam.12607

Scanu B, Linaldeddu BT, Peréz-sierra A, Deidda A, Franceschini A (2014) *Phytophthora ilicis* as a leaf and stem pathogen of *Ilex aquifolium* in Mediterranean islands. Phytopathol Mediterr 53:480–490. https://doi.org/10.14601/Phytopathol_Mediterr-14048

Scanu B, Linaldeddu BT, Deidda A, Jung T (2015) Diversity of *Phytophthora* species from declining Mediterranean Maquis vegetation, including two new species, *Phytophthora crassamura* and *P. ornamentata* sp. nov. PLoS One 10:e0143234. https://doi.org/10.1371/journal.pone.0143234

Schulze-Lefert P, Panstruga R (2003) Establishment of biotrophy by parasitic fungi and reprogramming of host cells for disease resistance. Annu Rev Phytopathol 41:641–667. https://doi.org/10.1146/annurev.phyto.41.061002.083300

Schwartz HF, Steadman JR, Hall R, Forster RL (2005) Compendium of bean diseases, 2nd edn. American Phytopathological Society, St Paul, MN

Sekara A, Cebula S, Kunicki E (2007) Cultivated eggplants - origin, breeding objectives and genetic resources, a review. Folia Hortic 19:97–114

Senanayake IC, Crous PW, Groenewald JZ, Maharachchikumbura SS, Jeewon R et al (2017) Families of diaporthales based on morphological and phylogenetic evidence. Stud Mycol 86:217–296. https://doi.org/10.1016/j.simyco.2017.07.003

Shan LY, Cui WY, Zhang DD, Zhang J, Ma NN et al (2017) First report of *Fusarium brachygibbosum* causing maize stalk rot in China. Plant Dis 101:837

Shen YM, Liu HL, Chang ST, Chao CH (2010) First Report of anthracnose caused by *Colletotrichum acutatum* on mung bean sprouts in Taiwan. Plant Dis 94:131

Shenoy BD, Jeewon R, Lam WH, Bhat DJ, Than PP et al (2007) Morpho-molecular characterisation and epitypification of *Colletotrichum capsici* (Glomerellaceae, Sordariomycetes), the causative agent of anthracnose in chilli. Fungal Divers 27:197–211

Sherriff C, Whelan MJ, Arnold GM, Bailey JA (1995) rDNA sequence analysis confirms the distinction between *Colletotrichum graminicola* and *C. sublineolum*. Mycol Res 99:475–478

Shivaprakasam K, Soumini RCK (1974) Effect of nitrogen on the incidence of *Verticillium* wilt disease of eggplant caused by *Verticillium dahliae* Kleb. Plant Soil 40:217–220

Shrestha U, Ownley BH, Butler DM (2016a) First report of stem and root rot of cowpea caused by *Fusarium oxysporum* in Tennessee. Plant Dis 100:649

Shrestha U, Ownley BH, Butler DM (2016b) First report of dry root and stem rot of cowpea caused by *Fusarium proliferatum* in the United States. Plant Dis 100:860

Shrestha SK, Miyasaka SC, Shintaku M, Kelly H, Lamour K (2017) *Phytophthora colocasiae* from Vietnam, China, Hawaii and Nepal: intra- and inter-genomic variations in ploidy and a long-lived, diploid Hawaiian lineage. Mycol Prog 16:893–904. https://doi.org/10.1007/s11557-017-1323-z

Silva DN, Talhinhas P, Várzea V, Cai L, Paulo OS, Batista D (2012) Application of the Apn2/MAT locus to improve the systematics of the *Colletotrichum gloeosporioides* complex: an example from coffee (*Coffea* spp.) hosts. Mycologia 104:396–409. https://doi.org/10.3852/11-145

Singh RP, Hodson DP, Huerta-Espino J, Jin Y, Bhavani S et al (2011) The emergence of Ug99 races of the stem rust fungus is a threat to world wheat production. Annu Rev Phytopathol 49:465–481. https://doi.org/10.1146/annurev-phyto-072910-095423

Smith RE (1900) *Botrytis* and *Sclerotinia*: their relation to certain plant diseases and to each other. Bot Gaz 29:369–407. https://doi.org/10.1086/328001

Smith VL, Jenkins SF, Punja ZK, Benson DM (1989) Survival of sclerotia of *Sclerotium rolfsii*: influence of sclerotial treatment and depth of burial. Soil Biol Biochem 21:627–632

Soanes DM, Alam I, Cornell M, Wong HM, Hedeler C et al (2008) Comparative genome analysis of filamentous fungi reveals gene family expansions associated with fungal pathogenesis. PLoS One 3:e2300. https://doi.org/10.1371/journal.pone.0002300

Spanu PD (2012) The genomics of obligate (and nonobligate) biotrophs. Annu Rev Phytopathol 50:91–109

Spanu PD, Abbott JC, Amselem J, Burgis TA, Soanes DM et al (2010) Genome expansion and gene loss in powdery mildew fungi reveal tradeoffs in extreme parasitism. Science 330:1543–1546. https://doi.org/10.1126/science.1194573

Srinivasa C, Niranjana SR, Shetty HS (2005) Effect of bio-agents and fungicides against *Phomopsis vexans* and on seed quality of brinjal. Crop Improv 32:95–101

Stoykow DY, Denchev CM (2006) Current knowledge of Diaporthales (Ascomycota) in Bulgaria. Mycol Balc 3:179–185. https://doi.org/10.5281/zenodo.2547618

Strange RN, Scott PR (2005) Plant disease: a threat to global food security. Annu Rev Phytopathol 43:83–116. https://doi.org/10.1146/annurev.phyto.43.113004.133839

Struck C (2015) Amino acid uptake in rust fungi. Front Plant Sci 6:40. https://doi.org/10.3389/fpls.2015.00040

Stukenbrock EH, McDonald BA (2008) The origins of plant pathogens in agro-ecosystems. Annu Rev Phytopathol 46:75–100

Sutton BC (1992) The genus *Glomerella* and anamorph *Colletotrichum*. In: Bailey JA, Jear MJ (eds) *Colletotrichum*: biology, pathology and control. CABI, Wallingford, pp 1–26

Talbot NJ (2003) On the trail of a cereal killer: exploring the biology of *Magnaporthe grisea*. Annu Rev Microbiol 57:177–202. https://doi.org/10.1146/annurev.micro.57.030502.090957

Talhnihas P, Batista D, Diniz I, Vieira A, Silva DN et al (2017) The coffee leaf rust pathogen Hemileia vastatrix: one and a half centuries around the tropics. Mol Plant Pathol 18:1039–1051

Than PP, Jeewon R, Hyde KD, Pongsupasamit S, Mongkolporn O, Taylor PWJ (2008) Characterization and pathogenicity of *Colletotrichum* species associated with anthracnose disease on chilli (*Capsicum* spp.) in Thailand. Plant Pathol 57:562–572. https://doi.org/10.1111/j.1365-3059.2007.01782.x

Thippeswamy B, Krishnappa M, Chakravarthy CN (2005) Location and transmission of *Phomopsis vexans* and *Alternaria solani* in brinjal. Ind Phytopathol 58:410–413

Thompson SM, Tan YP, Young AJ, Neate SM, Aitken EAB, Shivas RG (2011) Stem cankers on sunflower (*Helianthus annuus*) in Australia reveal a complex of pathogenic Diaporthe (Phomopsis) species. Persoonia: Mol Phylo Evol Fungi 27:80. https://doi.org/10.3767/003158511X617110

Torres-Calzada C, Tapia-Tussell R, Higuera-Ciapara I, Perez-Brito D (2013) Morphological, pathological and genetic diversity of *Colletotrichum* species responsible for anthracnose in papaya (*Carica papaya* L). Eur J Plant Pathol 135:67–79. https://doi.org/10.1007/s10658-012-0065-7

Torres-Calzada C, Yapia-Tussel R, Higuera-Ciapara I, Huchin-Poot E, Martin-Mex R et al (2018) Characterization of *Colletotrichum truncatum* from papaya, pepper and physic nut based on phylogeny, morphology and pathogenicity. Plant Pathol 67:821–830

Tudzynski P, Scheffer J (2004) *Claviceps purpurea*: molecular aspects of a unique pathogenic lifestyle. Mol Plant Pathol 5:377–388. https://doi.org/10.1111/j.1364-3703.2004.00237.x

Tudzynski P, Tenberge KB (2003) Molecular aspects of host-pathogen interactions and ergot alkaloid biosynthesis in *Claviceps*. Mycol Ser 19:445–474

Udayanga D, Liu X, McKenzie EH, Chukeatirote E, Bahkali AH, Hyde KD (2011) The genus *Phomopsis*: biology, applications, species concepts and names of common pathogens. Fungal Divers 50:189–225. https://doi.org/10.1007/s13225-011-0126-9

Udayanga D, Castlebury LA, Rossman AY, Hyde KD (2014a) Species limits in *Diaporthe*: molecular re-assessment of *D citri*, *D cytosporella*, *D foeniculina* and *D rudis*. Persoonia 32:83–101. https://doi.org/10.3767/003158514X679984

Udayanga D, Castlebury LA, Rossman AY, Chukeatirote E, Hyde KD (2014b) Insights into the genus *Diaporthe*: phylogenetic species delimitation in the *D. eres* species complex. Fungal Divers 67:203–229. https://doi.org/10.1007/s13225-014-0297-2

Udayanga D, Castlebury LA, Rossman AY, Chukeatirote E, Hyde KD (2015) The *Diaporthe* sojae species complex: phylogenetic re-assessment of pathogens associated with soybean, cucurbits and other field crops. Fungal Biol 119:383–407. https://doi.org/10.1016/j.funbio.2014.10.009

Úrbez-Torres JR, Peduto F, Smith RJ, Gubler WD (2013) *Phomopsis* dieback: a grapevine trunk disease caused by *Phomopsis viticola* in California. Plant Dis 97:1571–1579. https://doi.org/10.1094/PDIS-11-12-1072-RE

Van Niekerk JM, Groenewald JZ, Farr DF, Fourie PH, Haleen F, Crous PW (2005) Reassessment of *Phomopsis* species on grapevines. Aust Plant Pathol 34:27–23

Vandemark GJ, Barker BM (2003) Quantifying *Phytophthora medicaginis* in susceptible and resistant Alfalfa with a real-time fluorescent PCR assay. J Phytopathol 151:577–583. https://doi.org/10.1046/j.0931-1785.2003.00768.x

Vannini A, Vettraino A (2011) *Phytophthora cambivora*. For Phytoph 1:1811. https://doi.org/10.5399/osu/fp.1.1.1811

Velho AC, Alaniz S, Casanova L, Mondino P, Stadnik MJ (2015) New insights into the characterization of *Colletotrichum* species associated with apple diseases in southern Brazil and Uruguay. Fungal Biol 119:229–244. https://doi.org/10.1016/j.funbio.2014.12.009

Vettraino AM, Morel O, Perlerou C, Robin C, Diamandis S, Vannini A (2005) Occurrence and distribution of *Phytophthora* species in European chestnut stands, and their association with Ink Disease and crown decline. Eur J Plant Pathol 111:169. https://doi.org/10.1007/s10658-004-1882-0

Vieira WAS, Lima WG, Nascimento ES, Michereff SJ, Câmara MPS, Doyle VP (2017) The impact of phenotypic and molecular data on the inference of *Colletotrichum* diversity associated with Musa. Mycologia 109:912–934. https://doi.org/10.1080/00275514.2017.1418577

Voegele RT, Mendgen KW (2011) Nutrient uptake in rust fungi: how sweet is parasitic life? Euphytica 179:41–55. https://doi.org/10.1007/s10681-011-0358-5

Voegele RT, Struck C, Hahn M, Mendgen K (2001) The role of haustoria in sugar supply during infection of broad bean by the rust fungus *Uromyces fabae*. Proc Natl Acad Sci U S A 98:8133–8138. https://doi.org/10.1073/pnas.131186798

Waller JM, Lenné JM, Waller SJ (2002) Plant pathologists's pocketbook. CABI, Wallingford

Watt GW (1893) Dictionary of the economic products of India, Volume 6, Part IV. Central Printing Office, Government of India, Calcutta, West Bengal

Webster J, Weber RWS (2007) Introduction to fungi. Cambridge University Press, New York, NY

White C, Halleen F, Fischer M, Mostert L (2011) Characterisation of the fungi associated with esca diseased grapevines in South Africa. Phytopathol Mediterr 50:204–223

Wilcox WA, Ellis MA, Ramsley B et al (2015) *Phomopsis cane* and leaf spot. In: Wilcox WF, Gubler WD, Uyemoto JK (eds) Compendium of grape diseases, disorders and pests, 2nd edn. APS, St Paul, MN

Williamson B, Tudzynski B, Tudzynski P, van Kan JA (2007) *Botrytis cinerea*: the cause of grey mould disease. Mol Plant Pathol 8:561–580. https://doi.org/10.1111/j.1364-3703.2007.00417.x

Xavier KV, Kc AN, Peres NA, Deng Z, Castle W, Lovett W, Vallad GE (2019) Characterization of *Colletotrichum* species Causing anthracnose of pomegranate in the Southeastern United States. Plant Dis 103:2771–2780. https://doi.org/10.1094/PDIS-03-19-0598-RE

Xu SJ, Aktaruzzaman M, Kim BS, Kim I, Shin J (2018) First report of anthracnose caused by *Colletotrichum fioriniae* on eggplant fruits in Korea. Plant Dis 102:2642

Yang Q, Fan XL, Guarnaccia V, Tian CM (2018) High diversity of *Diaporthe* species associated with twelve new species described. MycoKeys 39:97–149. https://doi.org/10.3897/mycokeys.39.26914

Yi M, Valent B (2013) Communication between filamentous pathogens and plants at the biotrophic interface. Annu Rev Phytopathol 51:587–611. https://doi.org/10.1146/annurev-phyto-081211-172916

Yoshida K, Schuenemann VJ, Cano LM, Pais M, Mishra B et al (2013) The rise and fall of the *Phytophthora infestans* lineage that triggered the Irish potato famine. elife 2:e00731. https://doi.org/10.7554/eLife.00731

Zeven AC, Zhukovsky PM (1975) Dictionary of cultivated plants and their centres of diversity. Centre for Agricultural Publishing and Documentation, Wageningen

Zhou Y, Huang JS, Yang LY, Wang GF, Li JQ (2017) First report of banana anthracnose caused by *Colletotrichum scovillei* in China. Plant Dis 101:381

Chapter 7
Application of Soil Microorganisms for Agricultural and Environmental Sustainability: A Review

Vivek Kumar Singh, Shraddha Rai, Deepti Singh, and R. S. Upadhyay

Abstract Microorganisms play a significant role in the edaphic ecosystem. Distribution and diversity of soil microorganisms such as bacteria, fungi, actinomycetes, algae, protozoans, and viruses are important to understand their functional significance at a given site of soil. In the edaphic ecosystem, microbial processes determine the exchange of matter and flow of energy between plant and soil which affect productivity and ecosystem stabilization. Thus, soil microorganisms show precise contributions to sustainable biosphere. They are also extremely important sources of food, feed, medicines, enzymes, and antimicrobial substances. More recently, their potential to serve in human and animal health applications, genetic engineering technology, environmental protection measures, agricultural biotechnology, and management of agricultural and municipal wastes has taken them in the category of "jewels of the environment." Their significance toward a prosperous environment helps them to be "jewels." Nowadays, genetically modified organisms are being used for applications in agriculture, bioremediation, industries, and human health. Many new methods and technologies have been added to understand the relationship between microbial diversity and its function in soil processes. Now, with technical improvements and focused researches, we can hypothesize the results from micro-scale to large-scale processes for the prediction of climate changes.

Keywords Biodiversity · Bioinformatics · Biopesticides · Bioremediation · Metagenomics · Metaproteomics

V. K. Singh (✉)
University Department of Botany, T.M. Bhagalpur University, Bhagalpur, India

S. Rai · D. Singh
T.D.P.G. College, V.B.S. Purvanchal University, Jaunpur, India

R. S. Upadhyay
Centre of Advanced Study in Botany, Banaras Hindu University, Varanasi, India

S. K. Dubey, S. K. Verma (eds.), *Plant, Soil and Microbes in Tropical Ecosystems*, Rhizosphere Biology, https://doi.org/10.1007/978-981-16-3364-5_7

7.1 Introduction

The microbial world comprises miscellaneous existing organisms in any ecosystem, and different organisms are discovered regularly. It is the largest unexplored reservoir of biodiversity on the earth. Microorganisms play a specific role in the maintenance and functioning of the ecosystems for preparing a sustainable biosphere (Nannipieri et al. 2002). Microbes are divided into six groups according to their distinct characteristics: prions, virus, bacteria, protozoan, unicellular algae, and fungi. Huge diversity is found within these groups. Soil is considered to be a complex and dynamic ecosystem as it is difficult to determine the microbial community composition in soil. Soil is a structured, heterogeneous, and discontinuous system and also a medium pulsating with life in the environment. Being a perfect culture medium for the growth and development of microbial communities, the soil is known as a complex microhabitat by having several unique properties (Nannipieri and Badalucco 2003).

Microbial diversity encompasses genetic as well as ecological diversity. Genetic diversity refers to the amount and distribution of genetic information within microbial communities, while ecological diversity portrays the structural variations in the communities, interaction complexities, number of trophic levels, and number of ecological guilds. Soil possesses several different groups of microorganisms among which bacteria are the most abundant in comparison to the other microorganisms. In the soil, microbes are found maximum in the upper portion (i.e., horizon A/topsoil) and decrease with the depth. Different soil organisms play a significant role in specific change/transformation occurring in the soil. The major role of microorganisms in the soil is to make the soil an excellent medium for the proper growth and development of higher plants. A huge diversity of microbes is observed not only in pristine soils but also in polluted soils and in most environments under extreme conditions (Guimaraes et al. 2010). Therefore, such contaminated soils should be conserved for their unique biopotentials and microbial diversity.

Soil microbial diversity is very crucial for life on earth. Various phenomena occurring above the ground are determined directly or indirectly by microbial processes in the soil (Wardle 2002; Bardgett and Bowman 2005). Structure and function of various organisms are regulated directly by soil microorganisms through the stimulation or inhibition of their growth and development. On the other hand, soil microorganisms play an important role in the regulation of aboveground communities indirectly by altering the nutrient dynamics (Van Der Putten 2003; Wardle et al. 2004).

Human health issues have provoked the awareness regarding soil ecosystem and geochemistry, while soil and water conservation problems are already becoming hot cakes in several parts of the world (Sparks 2001; Ward and Pulido-Velazquez 2008; Nowak 2013). Soil influences human health through contact with pathogens (Burras et al. 2013). The study of soil ecosystem is significant for global change and biodiversity preservation. The other facet of this study is that the impact of human activities on soil and water resources is increasing continuously with the growing population resulting in the loss of organic matter, fertility, erosion, pollution, losses of soil microbial diversity, and losses of soil functions. The present review could be helpful to open more opportunities for soil scientists, soil microbiologists, professionals of other related disciplines, and industrialists for obtaining a more comprehensive perceptiveness of the environment and sustainable development in the future.

7.2 Microbial Functions in the Soil

Microbial processes occurring in the soil are responsible for the structure and functioning of aboveground world. Soil microbes play a significant role in plant nutrition by organic material decomposition and increasing nutrient availability to the plants. Through nitrogen fixation, plants are benefitted by using an infinite source of nitrogen from the atmosphere, and this procedure concurrently increases soil fertility as dead plant root remains add some of the biologically available nitrogen to the soil. Some soil microorganisms act as determinants for the mineralogical properties of most soils and sediments. Microorganisms play an important role in weathering process which liberates many essential elements (C, S, N, P) from the lithospheric resources within which they are generally unavailable to many living organisms (Douglas and Beveridge 1998). Another important role of the microbes is biomineralization which supports soil structural characteristics (Wardle et al. 2004).

Some microbes develop mutual beneficial relationships with the plants. These microbes colonize plant roots and obtain nutrients from the soil. Soil microbes protect roots from pests and pathogens and also provide a greater root area for nutrient uptake. Along with the beneficial microbes, pathogenic microorganisms are also present in the soils which are involved in the pathogenesis in host plants. These pathogenic microorganisms infect the plant and kill living tissue, creating a weakened and diseased plant. High biodiversity in soil suppresses soil-borne pathogens and diseases. In suppression mechanisms, native microorganisms outcompete the pathogenic organisms, physically protect the roots, and provide better nutrition to the plants. Thus, soil microbiota performs various modification and biotransformation in the soil and provides better nutrition to the plants for proper growth and development in the soil. Some microbes execute important soil functions like nutrient cycling, disease suppression, and soil and water dynamics, all of which promote plants to become healthy, disease resistant, and vigorous.

7.3 Applications of Soil Microbes

The exceptionalities and biosynthetic capabilities of the soil microbes have made them the most desired *organisms* for overcoming some major *problems in the life sciences* and other relevant fields. The pivotal role of microorganisms in several areas such as genetic engineering, advanced medical technology, human and animal health, pharmaceutical drugs, enzyme *technology,* food processing, food safety and quality, environmental protection, agricultural biotechnology, and agricultural and municipal waste management has provided a most remarkable achievement. Major applications of soil microorganisms such as enhanced symbiotic or associative N_2 fixation (Alexander 1984; Stacey and Upchurch 1984), plant growth promotion (Burr and Caesar 1984; Gaskins et al. 1985), biological control of soil-borne plant pathogens (Watrud et al. 1985), degradation of xenobiotic compounds (Brunner et al. 1985), and exploitation of industrially important enzymes on commercial scale (Nigam and Singh 1995; Nigam 2013; Prasad et al. 2013) are deciphering their potentials. Nowadays, enormous scopes in the beneficial application of soil microorganisms and the potential for developing specific strains through genetic engineering and molecular techniques have definitely contributed to various fields.

7.3.1 Application of Soil Microbes in Agriculture

The farmers generally use synthetic chemical methods for increasing agricultural production, and these practices definitely enhance the crop yield. In turn, the random application of agrochemicals has resulted in environmental pollution and poor human and animal health. Thus, the alternative methods are needed in place of chemical-based conventional agriculture to reduce these problems. Soil-borne microbes are becoming very popular and beneficial as an additive to chemical fertilizers in improving the quality and yield of crops and are now applied in a wide variety of agricultural systems for better productivity and integrated pest management (Antoun and Prevost 2005). Regarding this, plant growth-promoting microorganisms (PGPM) are found as potential contributors in sustainable crop production (Shoebitz et al. 2009).

7.3.1.1 Soil Microbes as Biofertilizers

Rhizospheric soil of plants possesses *several* beneficial microorganisms (Kathiresan and Selvam 2006). Application of these beneficial soil microorganisms for improving the plant growth and productivity as "biofertilizer" has been intensively studied (Artursson et al. 2006; Berg 2009). An extensive variety of bacterial species are applied as biofertilizers in the plants. These bacteria include strains of *Azospirillum*, *Azotobacter*, *Bacillus*, *Burkholderia*, *Enterobacter*, *Erwinia*, *Flavobacterium*, *Pseudomonas*, and *Rhizobium* (Lagos et al. 2015) and are termed as plant growth-promoting rhizobacteria (PGPR) and act as biofertilizers (Burr and Caesar 1984; Podile and Kishore 2007). The *Bacilli* and *Pseudomonas* are the predominant genera among the diverse bacteria (Podile and Kishore 2007). These rhizobacteria improve plant growth by increasing photosynthetic capacity (Xie et al. 2009); synthesizing precursors of phytohormones (Ahmad et al. 2008), antibiotics, enzymes, vitamins, and siderophores (Burd et al. 2000); and inhibiting ethylene synthesis (Khan et al. 2009). In addition, the rhizobacterial strains can solubilize inorganic P (Khan et al. 2007), mineralize organic P (Ponmurugan and Gopi 2006), improve plant tolerance to salt and drought stress (Xie et al. 2009; Zhang et al. 2010), improve plant growth and plant nutrition, and provide *plant resistance* to *phytopathogenic organisms* (Avis et al. 2008; Hayat et al. 2010; Pii et al. 2015). Dai et al. (2016) conducted an experiment to show that pyrogenic organic matter addition in soil induced the root growth and several soil parameters more in rhizospheric soils in comparison to bulk soil. Thus, PGPR application as eco-friendly biofertilizer may facilitate in reducing the environmental problems caused by the excessive use and high production costs of fertilizers. The application of PGPR also improves the physicochemical properties of the soil which facilitate the growth and efficiency of symbiotic soil microbes such as nitrogen-fixing rhizobia and arbuscular mycorrhizal fungi.

Some rhizospheric fungi are also capable of promoting plant growth through root colonization like PGPR and are known as plant growth-promoting fungi (PGPF) such as *Trichoderma*, *Penicillium*, *Fusarium*, and *Phoma* (Hyakumachi 1994). Some species of PGPF are found to induce systemic resistance against several *pathogens* in cucumber plants (Shoresh et al. 2005). Being non-pathogenic soil-inhabiting saprophytes, PGPF have been reported as beneficial microbes for several crop plants with the properties of growth promotion and protection from several diseases (Shivanna et al. 1994). Among PGPF, some isolates of *Penicillium simplicissimum* and *Phoma* sp. were found effective against cucumber anthracnose caused by *Colletotrichum orbiculare* through the activation of systemic resistance (Koike et al. 2001). Generally, P-solubilizing ability of PGPF is greater than PGPR. Some PGPF genera like *Aspergillus*, *Penicillium*, and *Trichoderma* have been reported as efficient P-solubilizers (Altomare et al. 1999; Babana and Antoun 2005).

7.3.1.2 Soil Microbes as Biocontrol Agents

The extensive use of rhizobacteria and PGPF application for overcoming the soil-borne diseases is replacing the chemical pesticides, which is a major concern in inducing the environmental pollution and health hazards (Walsh et al. 2001). The most commonly used soil microorganisms as biopesticides include biofungicides (*Trichoderma* sp.), bioherbicides (*Phytophthora* sp.), and bioinsecticides (*Bacillus thuringiensis* and *B. sphaericus*). Several bacteria, particularly *Pseudomonas* and *Bacillus* strains, are capable of controlling the growth of various fungi. *Burkholderia* sp. was found to suppress the virulence factors (that normally activate immune response in several plants) by forming a biofilm at the root surface (Paungfoo-Lonhienne et al. 2016). Rhizobacteria act as biocontrol agent and protect the root surface from soil-borne pathogens. Being rhizosphere competent, they have the capacity to rapidly colonize the root surface and spread down the root after single seed treatment or drench application in the soil (Rangarajan et al. 2003). The biocontrol potential of *Bacillus* spp. was assessed in many crops including chickpea and found it as an important agent to resist root and soil-borne pathogens (Landa et al. 1997). The antagonistic actions of *Pseudomonas fluorescens* have been studied extensively against several plant pathogens (Saravanakumar and Samiyappan 2007) and also in diseases of crops grown in saline agricultural soils (Paul and Nair 2008). There are several PGPR that suppress diseases by releasing antimicrobial or antifungal compounds that prevent plant pathogens (Weller et al. 2002). Members of the genus *Trichoderma* were found very effective biocontrol agents against several soil-borne plant pathogens (Benitez et al. 2004). *Glomus fasciculatum* and *Gigaspora margarita* have been reported to suppress root rot diseases of asparagus caused by *Fusarium oxysporum* f. sp. asparagi (Matsubara et al. 2001) and *Glomus clarum* against root necrosis caused by *Rhizoctonia solani* in cowpea (Abdel-Fattah and Shabana 2002). The arbuscular mycorrhizal fungus *Glomus mosseae* was found to suppress "take-all" disease caused by *Gaeumannomyces graminis* var. tritici in barley (Al-Askar and Rashad 2010).

7.3.1.3 Soil Microbes in Saline Agricultural Soils

Soil salinity is a serious problem affecting the vegetables and crops causing growth inhibition particularly in plants of arid and semiarid areas (Parida and Das 2005). It has been reported that plant growth under salt stress can be improved by inoculation of PGPR and PGPF (Cho et al. 2006) and application of mycorrhizal fungi which promotes abiotic stress tolerance in host plants and plays a significant role in plant survival under different stress conditions (Rodriguez et al. 2009). Thus, selected PGPR, PGPF, and other microbes, particularly, AM fungi, could serve as a potential tool for alleviating salinity stress in salt-sensitive crop plants.

7.3.2 Applications of Microbes in Industries

Microorganisms are progressively more important to industry, where they are used in large-scale processes ranging from food production to soil/water treatment. The development of recombinant DNA technology brought many changes to industrial applications of microorganisms.

7.3.2.1 Enzyme Production

Majority of the industrial enzymes are of microbial origin. Enzymes from soil microorganisms are of *great* significance in various *industries* such as pharmaceutical, food, dairy, textile, leather, detergent, paper and pulp, animal feed, biosurfactants, bioplastics, natural bioproducts, cosmetics, etc., and their range of applications is gradually increasing. Soil microbes are used in the production of several enzymes such as cellulase, lipase, amylase, proteases, and pectinases.

Cellulase is produced by several fungi (such as *Aspergillus*, *Penicillium*, *Fusarium*, *Trichoderma*, *Chaetomium*, and *Phoma*), aerobic bacteria (such as *Bacillus*, *Acidothermus*, *Pseudomonas*, *Cellvibrio*, *Staphylococcus*, *Streptomyces*, and *Xanthomonas*), and anaerobic bacteria (such as *Acetivibrio*, *Bacteroides*, *Butyrivibrio*, *Clostridium*, *Erwinia*, *Eubacterium*, *Caldocellum*, *Pseudonocardia*, *Ruminococcus*, and *Thermoanaerobacter*) (Zhang et al. 2006). Crude enzymes produced by these microorganisms are commercially available for agricultural and industrial use. Commercial lipases are produced from *Rhizopus*, *Geotrichum*, *Rhizomucor*, *Aspergillus*, *Burkholderia cepacia*, *Candida antarctica*, *Candida rugosa*, *Pseudomonas alcaligenes*, and *Pseudomonas mendocina* (Jaeger and Reetz 1998). However, α-amylase-producing species are *Aspergillus niger*, *A. fumigatus*, *A. foetidus*, *A. terreus*, and *Rhizopus delemar* (Pandey et al. 2005). Proteases are produced by *Aspergillus niger*, *A. oryzae*, *Bacillus amyloliquefaciens*, *B. stearothermophilus*, *M. pusillus*, and *Mucor miehei*. However, pectinase producers are *Aspergillus*, *Bacillus*, *Trichoderma*, *Rhizopus*, *Pseudomonas*, *Penicillium*, *Fusarium*, *Kluyveromyces*, and *Erwinia* (De Gregorio et al. 2002). The fungi synthesizing pectinolytic enzymes such as *Aspergillus niger*, *Aspergillus carbonarius*, and *Lentinus edodes* are mostly preferred in industries. Specificity, thermostability, and pH response of the microbial enzymes are critical properties for the growing interest in soil microbial enzymes compared to chemical processes for their industrial use. This led to the search of new strains of soil microorganism, which can be used in the development of processes for producing such microbial enzymes on a commercial scale.

7.3.2.2 Triacylglycerol Production

The members of actinomycetes such as *Streptomyces*, *Mycobacterium*, *Nocardia*, *Rhodococcus*, *Dietzia*, and *Gordonia* produce the triacylglycerols efficiently. They produce variable amounts of neutral lipids on culture media containing different carbon sources. Eukaryotic microorganisms such as fungi and yeast also accumulate TAG during metabolic stress (Lemann 1997).

7.3.2.3 Biosurfactants

Microbial biosurfactants are useful biotechnological products with a broad range of applications in various industries (Mulligan 2009). Biosurfactants are an assorted group of surface active chemical compounds produced by a variety of soil microbes. These include bacteria, yeasts, and filamentous fungi (Mulligan 2005). Bacterial surfactant-producing members include *Pseudomonas aeruginosa* (mono- and di-rhamnolipid); *Corynebacterium*, *Rhodococcus*, and *Nocardia* (phospholipids, trehalose dimycolates/dicorynomycolates, glycolipids, etc.); *Arthrobacter paraffineus* (trehalose and sucrose lipids); *Bacillus subtilis* (surfactin); and *Bacillus licheniformis* (lipopeptide similar to surfactin). Fungi involved in surfactant production include yeasts such as *Candida* spp. (liposan, phospholipids) and *Torulopsis* spp. (sophorolipids). Several researches have demonstrated the increase in pollutant desorption and availability by application of biosurfactants (Oberbremer et al. 1990; Volkering et al. 1995). It was observed that a biosurfactant-producing species of *Burkholderia* isolated from oil-contaminated soil could be used for the bioremediation of various pesticide-contaminated sites (Wattanaphon et al. 2008). Hermane et al. (1995) suggested the application of biosurfactants in controlling the bioavailability of toxicants in soils and other environment due to their biodegradability. *Pseudomonas* produces biosurfactants which solubilize and degrade the polycyclic aromatic hydrocarbons (PAHs) such as phenanthrene (Burd and Ward 1996). Noordman et al. (2002) observed the effect of biosurfactant produced from *Pseudomonas aeruginosa* on hexadecane degradation. The biosurfactants are used extensively in agriculture for improvement of soil quality, plant growth promotion, enhanced biodegradation of pollutants, and protection from plant pathogens because they show antimicrobial activity and increase plant-microbe interactions which are beneficial to the crop plants (Dhara and Swaranjit 2013).

7.3.2.4 Food Industry

Application of soil microbes in the food industry has been used widely in the production of several commercially important foods such as yoghurt, cheese, pickles, brewing, winemaking industries, etc. *Saccharomyces cerevisiae* is extensively used in food industries. The microorganisms involved in the food biopreservation are especially lactic acid bacteria and some yeast such as *Acetobacter*, *Brevibacterium*, *Corynebacterium*, *Gluconobacter*, *Pseudomonas*, and *Erwinia* (Sugisawa et al. 1990; Sauer et al. 2004; Bremus et al. 2006). Several microbes are extensively used for vitamin production in food industry, for example, vitamin B12 is produced on an industrial scale by *Propionibacterium shermanii* or *Pseudomonas denitrificans* (Bremus et al. 2006). Microbial enzymes produced by microbial systems have extended application in food industries.

7.3.3 Pharmaceutical Applications

Soil microorganisms are also infinite source of some novel chemicals with various potential therapeutic applications. The members of *Actinomycetes* group isolated from soil serve as potential sources of antiinfection, antitumor, and antidiabetic compounds and also agents for the treatment of various neurodegenerative diseases (Thomashow et al. 1997). Antibiotics are one of the commercially exploited secondary metabolites produced by several microorganisms like bacteria and fungi. Approximately 80% of the world's antibiotics are known to be produced from actinomycetes, mostly from the genera *Streptomyces* and *Micromonospora*. The genus *Streptomyces* produces amphotericin, erythromycin, streptomycin, tetracycline, and rifamycin (Thomashow et al. 1997).

7.3.4 Environmental Applications

Some soil microorganism-based bioremediation techniques for controlling the environmental pollution have been developed in the recent years to utilize the potential of certain taxa to degrade and detoxify *the* contaminants (Lee et al. 1983; Guengerich 1990). Soil microorganisms have the potential to degrade various environmental pollutants without producing toxic and harmful compounds as byproducts (Kothe et al. 2005) and evolved multifaceted mechanisms to neutralize the toxic effects of pollutants (Silver and Phung 1996). These microbial systems are more cost-effective and help in the development of appropriate techniques for cleaning up soil-contaminated environments for environmental restoration and protection. Nowadays, several soil microbes are isolated from contaminated sites and are extensively used for the bioremediation of numerous environmental pollutants (Machado et al. 2008; Ray and Ray 2009; Ruta et al. 2010).

7.3.4.1 Bioremediation

Bioremediation is an eco-friendly technique which utilizes the microorganisms to reduce or neutralize pollutants present in the contaminated environments. Some soil microorganisms have the ability to decompose or transform the petroleum products. The bacterial groups such as *Arthrobacter*, *Achromobacter*, *Acinetobacter*, *Alcaligenes*, *Bacillus*, *Flavobacterium*, *Burkholderia*, *Nocardia*, and *Pseudomonas* sp. are used to degrade hydrocarbons in soil environments. The fungi and yeasts such as *Amorphotheca*, *Graphium*, *Neosartorya*, *Talaromyces*, *Candida*, *Yarrowia*, and *Pichia* isolated from petroleum-contaminated soil were found to be effective in hydrocarbon degradation (Chaillana et al. 2004). Singh (2006) also reported that *Aspergillus*, *Cephalosporium*, and *Penicillium* were potential degraders of crude oil hydrocarbons. The fungi *Rhodotorula*, *Sporobolomyces*, *Aspergillus*, and *Penicillium* possess biodegradation potential of oil. Applications of soil bacteria *Pseudomonas*, *Acinetobacter*, *Alcaligenes*, and *Arthrobacter* sp. are known for toxic waste management in polluted sites (Brunner et al. 1985; Nicholas 1987). PGPR has also been reported as an efficient remediator of contaminated soils (Zhuang et al. 2007).

7.3.4.2 Phytoremediation

Phytoremediation is the technique of cleanup of contaminants using green plants, and its efficiency is affected by the activity of a variety of rhizospheric microorganisms (Khan et al. 2009). Rhizospheric bacteria degrade and detoxify the toxic compounds (rhizodegradation) (Kuiper et al. 2004). The combined application of both plants and biodegradative bacteria is used to remove petroleum products (Alarcón et al. 2008), polycyclic hydrocarbons and other aromatic compounds (Daane et al. 2001), as well as a variety of halogenated compounds (Leigh et al. 2006) from contaminated soils. Rhizodegradation enhances the plants' yield in the polluted soils (Lucy et al. 2004), for example, the amendment of some PGPR (*Pseudomonas* and *Acinetobacter*) has been found to enhance the phytoremediation abilities of non-hyperaccumulating maize (*Zea mays* L.) plants by favoring their growth and biomass production (Lippmann et al. 1995).

7.3.5 Applications of Soil Microbes as Genetically Modified Microorganisms

Soil microbes are utilized in several aspects such as agriculture, human health, environmental protection, and industries (such as food, paper, pharmaceuticals, textiles, leather, etc.) after the development of molecular techniques and recombinant DNA technology. These modified microbes are termed as genetically modified microorganisms (GMMs). The applications of GMMs include enhancement of nitrogen fixation (Gerhold and Stacey 1990), fungal pathogen restriction (Howell 1990), insect pest control, or biodegradation of pesticide residues (Snow et al. 2005) and production of proteins (insulin, interferons, and interleukins) for therapeutic use. *Rhizobium* species have been genetically modified either to improve their nitrogen fixation efficiency (Cullen et al. 1998) or to enhance their survival by the application of marker genes (Mendum et al. 2001; Hirsch 2004). Genetic manipulation of phosphate-solubilizing bacteria has been made to enhance their ability to improve growth and productivity of plants (Rodríguez and Fraga 1999). Another important application of genetically modified microorganisms is as a sensor to assess biologically relevant concentrations of agrochemicals, petroleum products, heavy metals, and toxins in various environmental samples of contaminated sites (Belkin 2003).

7.4 Advances in Soil Microbial Ecology

The primary target of microbial ecology is to determine the position and number of microbes in the environment after the development of several modern molecular techniques (Brock 1996). Recent molecular methods have contributed in the knowledge of microbial diversity in soils and also the interactions between diversity and its function in soil processes. Recently, the interest toward the soil microbes and ecology has been increased after knowing their role in the maintenance of biosphere and environment. Now the world has started moving in task of preserving the environment and maintaining the sustainable land and exploitation of genetic resources. The recent advancements in the field of soil microbial ecology are offering fresh perspectives in the under-appreciated microbial world.

7.4.1 DNA Extraction, PCR, Cloning, and Sequencing Techniques

Nucleic acid isolation and characterization of microbes has revolutionized the microbial ecology (Nesme et al. 2016). DNA isolation is the primary and most essential step in the molecular studies of microbial ecology in which firstly DNA is recovered from the soil. The important task in the extraction is to isolate a sufficient amount of DNA without contamination, which inhibits the amplification of nucleic acid during PCR (Macrae 2000). PCR amplification of 16S rRNA genes (16S rDNA) using specific bacterial primers and separation of the resultant PCR amplicons either by cloning or by denaturing gradient gel electrophoresis (DGGE) or temperature gradient gel electrophoresis (TGGE) and sequencing are the most popular molecular techniques for the determination of soil bacterial ecology (Muyzer and Smalla 1998). The numbers of rRNA gene copies are related to the life strategy of bacteria, and species with lesser copy numbers inhabit low nutrient environment (Větrovský and Baldrian 2013). In the past few years, these molecular studies have been carried out in various diverse environments (Rheims et al. 1996; Duinveld et al. 1998). These studies have developed the ribosome-based sequences, and the environmental sequences deposited online are used for the design and application of oligonucleotide probes for the isolation, identification, and screening of several bacterial species in diverse environments (Busse et al. 1996). Another advantage of the extensive use of 16S rDNA techniques is to study bacterial diversity in geographically distinct soils (Ludwig et al. 1997). Therefore, differentiation in 16S rRNA gene sequences of different bacterial species has enormously improved our understanding about the ecological diversity of bacterial communities in soil.

7.4.2 Fungal PCR Primers

The bacterial species are identified as variation in the 16S rRNA gene, whereas taxonomic identification of fungi is based on 18S rRNA which is more challenging with identification usually restricted to family or genus level. The highest 18S rRNA sequence variation was observed between species belonging to phylum *Glomeromycota* (Schüßler et al. 2001). Therefore, 18S rDNA primers are used more commonly for symbiotic arbuscular mycorrhizal fungi as there is significant variation in 18S rRNA gene sequences of different fungal species to differentiate isolates to species level and below (Vandenkoornhuyse et al. 2002). White et al.

(1990) designed the first fungal PCR primers for the amplification of fungal 18S rDNA and ITS regions of the fungal DNA. Although these primers were designed with limited reference gene sequence informations, they have been proved to be very useful and powerful tools in genetic studies of fungi. These primers were generally used to amplify as broad taxonomic range as possible, and some of them were also used to amplify plant DNA from the mixed DNA samples of plant and fungi (Gardes and Bruns 1993; White et al. 1990). Such lack of specificity for fungal templates limits their effectiveness in mixed DNA samples especially where the ratio of fungal DNA to non-fungal DNA is low. Later, Gardes and Bruns (1993) designed ITS1F and ITS4B primers for the specific amplification of basidiomycetous fungal DNA from mixed DNA samples extracted from the colonized ectomycorrhizal (ECM) plant root tips. Subsequently, these fungal primers have been extensively used in ECM fungal researches and have increased our knowledge about ECM fungal communities and their ecology. Furthermore, ITS1F primer has been used in association with ITS4A primer, specifically to amplify templates from mixed DNA samples of fungal communities (Chen and Cairney 2002; Dickie et al. 2002; Lord et al. 2002; White et al. 1990), and with the ITS reverse primer ITS4A, especially for ascomycete fungal DNA (Larena et al. 1999). Thus, different fungal primers were designed for specific fungi.

7.4.3 Metagenomics, Metaproteomics, and Metatranscriptomics

Metagenomics involves the construction of DNA library followed by sequencing and functional analysis. Phylogenetics (based on the 16S rRNA/DNA) revolutionized the field of microbial ecology (Woese 1987). 16S rRNA gene analysis is very helpful in studying diversity and evolution of microbial populations. It has been reported that microbes with identical 16S rDNA sequences may have different overall genomes and show remarkably different physiologies and growth patterns (Jasper and Overmann 2004; Hahn and Pöckl 2005). Due to 16S rRNA gene analysis, soil is known as the most abundant diverse habitat for prokaryotes in earth which was not investigated by the cultivation-based methods. Nowadays, the goal of microbial ecology is to concern the identities of various microbes to the processes carried out by them in that environment, and this could be achieved using the 16S rDNA to identify clones belonging to specific microorganism and gene sequencing to gain information about the physiology of the microorganism. Fluorescent in situ hybridization (FISH) is a classical microbial technique which has been developed for the need of metagenomic research, using fluorescent probes to detect 16S rRNA.

Metatranscriptomics deals with the characterization of mRNA which provides the knowledge of metabolic phenomena of the microbial communities (Simon and Daniel 2011; deMenezes et al. 2012). Therefore, metatranscriptomics has the ability to find out novel genes and functions which allow the detection of active members in rhizospheric microbial communities correlated with their metabolic activities in soil (Kim et al. 2014).

Microbial functions generally refer to proteins, so the investigation of the microbial proteins is the most appropriate tool for confirming the potential activity of the microbial community (Myrold et al. 2013). Metatranscriptomics has certain limitations toward the study of indigenous microbial communities such as short half-life of RNA, differential transcriptional kinetics of similar genes present in different populations, and low correlation between RNA levels and corresponding protein synthesis (Hurt et al. 2001; Zhou and Thompson 2002), so these limitations have increased interest in metaproteomics. Wilmes and Bond (2004) studied the diversity in proteins of microbial communities present in activated sludges, and Schulze et al. (2004) characterized proteins from the samples taken from soil solutions, lake water, and soil particles by electrophoresis coupled with mass spectrometry (MS). Together with metagenomics and metatranscriptomics, there has been a steady evolution in the methodology for the extraction and analysis of proteins from soils (Bastida et al. 2009; Hettich et al. 2010; Siggins et al. 2012). In the last decades, the advances in proteomic technologies, in addition to the sequencing of various microorganisms, have enabled us to link phylogeny with the microbial functions. In this way, through metaproteomics study, several novel researches would be correlated with microbial ecology as a link between genetic and functional diversity in microbial communities and its relative contribution toward taxonomic and functional diversity for ecosystem stability.

7.4.4 *Community Profiling Techniques*

Community fingerprinting techniques are generally used for investigating different bacterial communities and have extensively improved our knowledge about their role and diversity in the soil (Johnsen et al. 2001; Ranjard et al. 2003). However, these techniques include denaturing gradient gel electrophoresis (DGGE), temperature gradient gel electrophoresis (TGGE), single-strand conformation polymorphism (SSCP), terminal restriction fragment length polymorphism (T-RFLP), amplified rDNA restriction analysis (ARDRA), amplified ribosomal intergenic spacer analysis (ARISA), and cloning which have recently been adopted and used successfully for the community study of soil fungi.

7.4.4.1 DGGE and TGGE

Genetic diversity of a microbial community can be determined by fingerprinting techniques. In the early decades, electrophoretic separation technique was used, but later on, DGGE and TGGE were introduced which have the potential to separate DNA fragments of the same length but with different sequences (Riesner et al. 1991;

Muyzer and Smalla 1998). Now, DGGE is more widely used to investigate community complexity, community changes, isolation of bacteria, monitoring of the enrichment, and detection of microheterogeneity in rRNA encoding genes. One of the major limitations with these techniques is the separation of only relatively small fragments (up to 500 base pairs) which shortens the amount of sequence information for phylogenetic inferences as well as for probe designing (Myers et al. 1995).

7.4.4.2 T-RFLP Analysis of 16SrDNA for Characterization of Microbial Communities

Terminal restriction fragment length polymorphism (T-RFLP) analysis of PCR-amplified genes is a well-known fingerprinting technique for profiling of microbial community structure and dynamics in natural habitats (Schütte et al. 2008). This analysis depends upon the restriction endonuclease-mediated digestion of fluorescently end-labeled PCR products. The digested products are firstly mixed with a DNA size standard (already labeled with a distinct fluorescent dye) and then after fragments are separated by capillary or gel electrophoresis using an automated sequencer. After analysis, only the terminal end-labeled restriction fragments are detected and recorded. An electropherogram is prepared at the end which shows a profile of microbial community as a series of peaks of varying heights. This technique has been extensively used in the examination of complex microbial environments and in the ecological study of bacterial, archaeal, and eukaryal populations growing in natural habitats (Singh et al. 2006).

7.4.4.3 SSCP Analysis for Microbial Characterization

SSCP is used to identify and characterize specific microorganisms from the microbial communities in soil samples. In this technique, double-strand DNA of each microorganism is firstly transformed to single strand and separated by polyacrylamide gel electrophoresis. SSCP analysis has the ability to differentiate small variations within same-length DNA of different microorganisms due to the presence of differences in retention time, temperature, ionic strength, and electrophoretic mobility of single-stranded DNA (Bharathi et al. 2016).

7.4.4.4 ARDRA and ARISA

Amplified ribosomal DNA restriction analysis (ARDRA) is a commonly used technique to study microbial diversity which relies on DNA polymorphism (Deng et al. 2008). In this technique, amplicons containing 16S rDNA gene fragments are firstly amplified and then digested by restriction endonucleases, followed by separation of the resulting fragments through high-density acrylamide gel electrophoresis. Amplified ribosomal intergenic spacer analysis (ARISA) is used to amplify both bacterial and fungal community in various soils. Various researches showed that ARISA is a high-resolution, high-reproducible, and vigorous technique to discriminate diverse microbial communities in soils (Ranjard et al. 2001).

7.4.5 *Microarray Technology*

Microarray is an extraordinary, precise, sophisticated, quantitative, and high-throughput technique used for the detection, identification, and characterization of microorganisms in the natural habitats. Due to swift advances in fingerprinting technology, microarrays contain hundreds to thousands of probes. Various studies have used microarray technology for investigating ecological problems. Some modern techniques such as PCR fingerprinting, real-time PCR, reverse transcriptase PCR, reporter genes, and fluorescence in situ hybridization (FISH) technology have developed to study the dynamics of simple microbial communities or small groups of dominant microbes in natural environments. Later, microarray technology has been predominantly developed to study gene expression profiling of pure cultures of diverse microorganisms; moreover, some major advances have been made regarding their efficient application to different environmental samples. Microarrays detect only the dominant populations of microorganisms in many environmental samples (Denef et al. 2003; Rhee et al. 2004). Different types of microarrays such as phylogenetic oligonucleotide arrays (POAs), functional gene arrays (FGAs), metagenomic arrays (MGAs), community genome arrays (CGAs), and whole-genome open reading frame arrays (WGAs) have been successfully used in microbial ecology research. These arrays are useful for functional genomic study of individual organisms and comparative genomic analyses and also for investigating the interactions of multiple organisms at the transcriptional level (Denef et al. 2003; Rhee et al. 2004).

7.5 Future Prospects of Soil Microbial Ecology

Soil microbiology is a very fast-growing area of research with many relevant topics regarding the development of model ecosystems and sustainable environmental management. For maintaining and protecting the life-supporting natural resources and soil biodiversity, it is essential to develop and standardize the methodology and to specify overall data collection and quality assurance techniques. It is also important to understand the spatial and temporal variation of soil microbiological characteristics for the successful execution of monitoring programs. Another challenge for future research is to be efficient to generalize the results from microscale to large-scale processes even for the prediction of global climate changes. Although reliable techniques are crucial, the quality of research results does not depend solely on technical improvements. The advancement of knowledge/technology needs skillful evaluation of the appropriate techniques and data analysis tools to be applied in each specific question regarding environmental sustainability. As per the perspectives of soil microbiologists, new molecular techniques offer new ways to explore community composition and processes of microbes on a microscale. In situ hybridization is able to explain where microorganisms exist and play an active role (Lübeck et al. 2000). The study of microbial hotspots occurring inside the guts of soil microfauna or around root surfaces is needed for future research. In microbial hotspots, various turnover processes occur, and microbial loops are formed (Clarholm 1994.). Sustainable management of soil ecosystem aims to establish desired microbial populations successfully. Such microorganisms may play an active role as degraders of xenobiotics, nitrogen fixers, or pathogen antagonists. In the future, alteration in a single key biological agent in the soil in a desired way would result in the alteration of soil functions for the benefit of human beings. Hopefully, such strategy would increase the agricultural sustainability and also help to remediate polluted soils and protect natural resources successfully.

Microarray technology and genome sequencing would have a major impact on our ecosystems. Microarray technology enables us to assess and analyze the community diversity in soils by directly expressing and hybridizing oligonucleotides fixed on specific membranes (Guschin et al. 1997; Ogram 2000). Another application of this technology to correlate community structure with community function by using mRNA and by combining with PCR amplification and/or rDNA would be possible (Gottschal et al. 1997). Using computational methods, it might be possible to describe a three-dimensional physical and functional model of a microbial niche. This goal of synthesis of complex information brings us closer to the computational sciences. Microbial model has many advantages over the macroecological models in our system. Many genes and traits will be potentially explorable at the genetic level by mutation. We will likely see the phylogenetic tree as a bush showing a continuum of many types of species.

Another emerging field is "bioinformatics" which is popular in almost all the branches of biological sciences. Bioinformatics converts complex biological information into the understandable model by using computer science and technology. Bioinformatics utilizes the integrated efficiency of the computational methods, simulation, analysis, and modelling to extract information and prediction of biological processes what exactly going on within a cell naturally (Altman and Klein 2002). Integration of genomic, proteomic, and metabolomic data sources will enable us to predict genetic mutations after the molecular analysis of disease symptoms and vice versa. The effects and outcomes of diseases and pests in agricultural systems can be predicted with the integration of GIS data like geographical mapping and weather systems with crop health and genotypic traits. Another challenging research area for bioinformatics is comparative genomic studies at large scale which could be achieved by the development of practical tools and techniques. The problems with digitization of phenotypic data such as complex behavior of microbes in diverse ecological niches correlated with crop or soil health offer future opportunities in the field of bioinformatics. Currently, there is quite a need to develop bioinformatics tools to open the hidden mystery of central dogma-based biological processes occurring within the tiny and often unseen microbial life forms. Thus, bioinformatics technique will enable us to understand the complex biological processes of any organism through the integration of informations obtained from these key biological processes within the cells.

The current microbiological researches are focusing only on pathophysiological mechanisms behind microbial diseases in plants rather than their management measures. For this purpose, a recently emerging field of "omics" era called metabolomics has the potential to find out solutions along with bioinformatics capabilities toward data integration, analysis, and management in biological studies. In the coming years, targets of microbial research and development such as molecular taxonomy, microbial mapping, identification of different agroecological sites using culture-dependent or metagenomics approaches, searching of potential genes and gene products for the microbial management of disturbed agricultural soils, bioprospecting for novel metabolites, enhancement of biotic and abiotic stress tolerance in crop plants (Tiwari et al. 2011), microbe-associated soil fertility and crop improvement programs, enhanced bioremediation efficiency, enhanced biofermentation capability, and development of next-generation microbial inoculants as biofertilizers and biopesticides (Singh et al. 2011) could not be achieved without the applications of bioinformatics (Wollenweber et al. 2005). Recent emerging fields like interactome, which includes sets of protein-protein interactions, and localizome, which deals with the subcellular localizations of proteins, will certainly play a significant role in future molecular researches. In the future, the ultimate goal of microbial biotechnology will be the integration of genetic resources and biological databases which would result in the computational representation of any aspect of biology of living cells and microorganisms.

7.6 Conclusion

In summary, it can be concluded that soil microorganisms play a pivotal role in the functioning of the ecosystems for maintaining a sustainable environment and productivity. Soil microorganisms have precise contributions to the nutrient cycles and as sources of useful chemicals. Soil microbial diversity plays a key role in human survival and economic development and also provides a major reservoir of natural resources which can be utilized for the betterment of human lives. Thus, the future of soil microbial ecology is bright because many major challenges of society have their root in it. Soil microbiology is the rapidly growing area which would greatly benefit agriculture, industries, environment, and human health through the application of advanced technologies, development of suitable ecosystem models and ecological theories, sustainable soil management, and realization of ecosystem stabilization and global changes. For prosperous environment, we should save such "jewels" and use them wisely.

Acknowledgments The authors gratefully acknowledge the Head, University Department of Botany, T.M. Bhagalpur University, Bhagalpur, India, for providing necessary facilities during the paper writing.

Authors' contributions: The author (VKS) designed and wrote the chapter. Another author (RSU) supervised the whole work. All authors have contributed, read, and approved the final manuscript.

References

Abdel-Fattah GM, Shabana YM (2002) Efficacy of the arbuscular mycorrhizal fungus *Glomus clarum* in protection of cowpea plants against root rot pathogen *Rhizoctonia solani*. J Plant Dis Prot 109:207–215

Ahmad F, Ahmad I, Khan MS (2008) Screening of free-living rhizospheric bacteria for their multiple plant growth promoting activities. Microbiol Res 163:173–181

Alarcón A, Davies FT Jr, Autenrieth RL, Zuberer DA (2008) Arbuscular mycorrhiza and petroleum-degrading microorganisms enhance phytoremediation of petroleum contaminated soil. Int J Phytoremediation 10:251–263

Al-Askar AA, Rashad YM (2010) Arbuscular mycorrhizal fungi: a biocontrol agent against common bean *Fusarium* root rots disease. Plant Pathol J 9(1):31–38

Alexander M (1984) Ecology of *rhizobium*. In: Alexander M (ed) Biological nitrogen fixation, ecology, technology and physiology. Plenum Press, New York, pp 39–50

Altman RB, Klein TE (2002) Challenges for biomedical informatics and pharmacogenomics. Annu Rev Pharmacol Toxicol 42:113–133

Altomare C, Norvell WA, Bjorkman T, Harman GE (1999) Solubilization of phosphates and micronutrients by the plant-growth-promoting and biocontrol fungus *Trichoderma harzianum* Rifai 1295-22. Appl Environ Microbiol 65:2926–2933

Antoun H, Prevost D (2005) Ecology of plant growth promoting rhizobacteria. In: Siddiqui ZA (ed) PGPR: biocontrol and biofertilization. Springer, Dordrecht, pp 1–38

Artursson V, Finlay R, Jansson J (2006) Interactions between arbuscular mycorrhizal fungi and bacteria and their potential for stimulating plant growth. Environ Microbiol 8:1–10

Avis TJ, Gravel V, Antoun H, Tweddell RJ (2008) Multifaceted beneficial effects of rhizosphere microorganisms on plant health and productivity. Soil Biol Biochem 40:1733–1740

Babana AH, Antoun H (2005) Effect of Tilemsi phosphate rock-solubilizing microorganisms on phosphorus uptake and yield of field-grown wheat (*Triticum aestivum* L.) in Mali. Plant Soil 287 (1–2):51–56

Bardgett RD, Bowman WD (2005) A temporal approach to linking aboveground and belowground ecology. Trends Ecol Evol 20(11):634–641

Bastida F, Moreno JL, Nicolas C, Hernandez T, Garcia C (2009) Soil metaproteomics: a review of an emerging environmental science, significance, methodology and perspectives. Eur J Soil Sci 60:845–859

Belkin S (2003) Microbial whole cell sensing systems of environmental pollutants. Curr Opin Microbiol 6:206–212

Benitez T, Rincon AM, Limonn MC, Codon A (2004) Biocontrol mechanisms of *Trichoderma* strains. Int Microbiol 7(4):249–260

Berg G (2009) Plant-microbe interactions promoting plant growth and health: perspectives for controlled use of microorganisms in agriculture. Appl Microbiol Biotechnol 84:11–48

Bharathi M, Balachandar D, Kumar K (2016) Single strand conformational polymorphism as a molecular marker for detection of variation between the small length ss DNA of *Azospirillum*, *Phosphobacteria* and *Pseudomonas* from the inoculated Bhendi field (COBH 1). IRA-Int J Appl Sci 4(1):200–210

Bremus C, Herrmann U, Bringer-Meyer S, Sahm H (2006) The use of microorganisms in L-ascorbic acid production. J Biotechnol 124:196–205

Brock TD (1996) Principles of microbial ecology. Englewood Cliffs, Prentice-Hall, p 19

Brunner W, Sutherland FH, Focht DD (1985) Enhanced biodegradation of polychlorinated biphenyls in soil by analog enrichment and bacterial inoculation. J Environ Qual 14:324–328

Burd G, Ward OP (1996) Bacterial degradation of polycyclic aromatic hydrocarbons on agar plates: the role of biosurfactants. Biotechnol Tech 10:371–374

Burd GI, Dixon DG, Glick BR (2000) Plant growth promoting bacteria that decrease heavy metal toxicity in plants. Can J Microbiol 46:237–245

Burr TJ, Caesar A (1984) Beneficial plant bacteria. Crit Rev Plant Sci 2:1–20

Burras CL, Nyasimi M, Butler L (2013) Soils, human health, and wealth: a complicated relationship. In: Brevik EC, Burgess LC, Boca Raton FL (eds) Soils and human health. CRC Press, Boca Raton, pp 215–226

Busse HJ, Denner EBM, Lubitz W (1996) Classification and identification of bacteria: current approaches to an old problem; overview of methods used in bacterial systematics. J Biotechnol 47:3–38

Chaillana F, Flècheb A, Burya E, Phantavonga Y, Saliot A, Oudot J (2004) Identification and biodegradation potential of tropical aerobic hydrocarbon-degrading microorganisms. Res Microbiol 155(7):587–595

Chen DM, Cairney JWG (2002) Investigation of the influence of prescribed burning on ITS profiles of ectomycorrhizal and other soil fungi at three Australian sclerophyll forest sites. Mycol Res 106:532–540

Cho K, Toler H, Lee J, Ownley B, Stutz JC, Moore JL (2006) Mycorrhizal symbiosis and response of *Sorghum* plants to combined drought and salinity stresses. J Plant Physiol 163:517–528

Clarholm M (1994) The microbial loop in soil. In: Ritz K, Dighton J, Giller KE (eds) Beyond the biomass. Wiley, New York, pp 221–230

Cullen DW, Nicholson PS, Mendum TA, Hirsch PR (1998) Monitoring genetically modified rhizobia in field soils using the polymerase chain reaction. J Appl Microbiol 84:1025–1034

Daane LL, Harjono I, Zylstra GJ, Häggblom MM (2001) Isolation and characterization of polycyclic aromatic hydrocarbon-degrading bacteria associated with the rhizosphere of salt marsh plants. Appl Environ Microbiol 67:2683–2691

Dai Z, Hu J, Xu X, Zhang L, Brookes PC, He Y, Jianming X (2016) Sensitive responders among bacterial and fungal microbiome to pyrogenic organic matter (biochar) addition differed greatly between rhizosphere and bulk soils. Sci Rep 6:36–101

De Gregorio A, Mandalani G, Arena N, Nucita F, Tripodo MM, Lo Curto RB (2002) SCP and crude pectinase production by slurry-state fermentation of lemon pulps. Bioresour Technol 83 (2):89–94

deMenezes A, Clipson N, Doyle E (2012) Comparative metatranscriptomics reveals widespread community responses during phenanthrene degradation in soil. Environ Microbiol 14:2577–2588

Denef VJ, Park J, Rodrigues JLM, Tsoi TV, Hashsham SA, Tiedje JM (2003) Validation of a more sensitive method for using spotted oligonucleotide DNA microarrays for functional genomics studies on bacterial communities. Environ Microbiol 5:933–943

Deng W, Xi D, Mao H, Wanapat M (2008) The use of molecular techniques based on ribosomal RNA and DNA for rumen microbial ecosystem studies: a review. Mol Biol Rep 35:265–274

Dhara PS, Swaranjit SC (2013) Biosurfactants in agriculture. Appl Microbiol Biotechnol 97:1005–1016

Dickie IA, Xu B, Koide RT (2002) Vertical distribution of ectomycorrhizal hyphae in soil as shown by T-RFLP analysis. New Phytol 156:527–535

Douglas S, Beveridge TJ (1998) Mineral formation by bacteria in natural microbial communities. FEMS Microbiol Ecol 26:79–88

Duinveld BM, Rosado AS, van Elsas JD, van Veen JA (1998) Analysis of the dynamics of bacterial communities in the rhizosphere of the *Chrysanthemum* via denaturing gradient gel electrophoresis and substrate utilisation patterns. Appl Environ Microbiol 64:4950–4957

Gardes M, Bruns TD (1993) ITS primers with enhanced specificity for basidiomycetes: application to the identification of mycorrhiza and rusts. Mol Ecol 2:113–118

Gaskins MH, Albrecht SL, Hubbell DH (1985) Rhizosphere bacteria and their use to increase plant productivity: a review. Agric Ecosyst Environ 12:99–116

Gerhold D, Stacey G (1990) Recent advances in molecular biology techniques for studying phytosymbiotic microbes. In: Nakas JP, Hagedorn C (eds) Biotechnology for plant-microbe interactions. McGraw-Hill, New York, pp 51–84

Gottschal JC, Meijer WG, Oda Y (1997) Use of molecular probing to assess microbial activities in natural ecosystems. In: Insam H, Rangger A (eds) Microbial communities. Springer, Heidelberg, pp 10–18

Guengerich FP (1990) Enzymatic oxidation of xenobiotic chemicals. Crit Rev Biochem Mol Biol 25:97–153

Guimaraes BCM, Arends JBA, van der Ha D, van der Wiele T, Boon N, Verstraete W (2010) Microbial services and their management: recent progress in soil bioremediation technology. Appl Soil Ecol 45(2):157–167

Guschin DY, Mobarry BK, Proudnikov D, Stahl DA, Rittmann BE, Mirzabekov AD (1997) Oligonucleotide microchips as genosensors for determinative and environmental studies in microbiology. Appl Environ Microbiol 63:2397–2402

Hahn MW, Pöckl M (2005) Ecotypes of planktonic actinobacteria with identical 16S rRNA genes adapted to thermal niches in temperate, subtropical and tropical freshwater habitats. Appl Environ Microbiol 71:766–773

Hayat R, Ali S, Amara U, Khalid R, Ahmed I (2010) Soil beneficial bacteria and their role in plant growth promotion: a review. Ann Microbiol 60:579–598

Hermane DC, Artiola JF, Miller RM (1995) Removal of cadmium, lead and zinc from soil by a rhamnolipid biosurfactant. Environ Sci Technol 29:2280–2285

Hettich RL, Chourey K, Jansson J, Ver Berkmoes N, Shah M, Chavarria KL, Tom LM, Brodie EL (2010) Direct cellular lysis/protein extraction protocol for soil metaproteomics. J Proteome Res 9:6615–6622

Hirsch PR (2004) Release of transgenic bacterial inoculants-rhizobia as a case study. Plant Soil 266:1–10

Howell CR (1990) Fungi as biological control agents. In: Nakas JP, Hagedorn C (eds) Biotechnology for plant-microbe interactions. McGraw-Hill, New York, pp 257–286

Hurt RA, Qiu X, Wu L, Roh Y, Palumbo AV, Tiedje JM, Zhou J (2001) Simultaneous recovery of RNA and DNA from soils and sediments. Appl Environ Microbiol 67:4495–4503

Hyakumachi M (1994) Plant growth promoting fungi from Turfgrass rhizosphere with potential for disease suppression. Soil Microorg 44:53–68

Jaeger KE, Reetz MT (1998) Microbial lipases form versatile tools for biotechnology. Trends Biotechnol 16:396–403

Jaspers E, Overmann J (2004) Ecological significance of microdiversity: identical 16S rRNA gene sequences can be found in bacteria with highly divergent genomes and ecophysiologies. Appl Environ Microbiol 70:4831–4839

Johnsen K, Jacobsen CS, Torsvik V, Sørensen J (2001) Pesticide effects on bacterial diversity in agricultural soils—a review. Biol Fertil Soils 33:443–453

Kathiresan K, Selvam MM (2006) Evaluation of beneficial bacteria from mangrove soil. Bot Mar 49 (1):86–88

Khan MS, Zaidi A, Wani PA (2007) Role of phosphate solubilizing microorganisms in sustainable agriculture: a review. Agro Sustain Dev 27:29–43

Khan MS, Zaidi A, Wani PA, Oves M (2009) Role of plant growth promoting rhizobacteria in the remediation of metal contaminated soils. Environ Chem Lett 7(1):1–19

Kim J, Chang JH, Kim EJ, Kim KJ (2014) Crystal structure of (R)-3-hydroxybutyryl-CoA dehydrogenase PhaB from Ralstonia eutropha. Biochem Biophys Res Commun 443:783–788

Koike N, Hyakumachi M, Kageyama K, Tsuyumu S, Doke N (2001) Induction of systemic resistance in cucumber against several diseases by plant growth promoting fungi: lignification and superoxide generation. Eur J Plant Pathol 107:523–533

Kothe E, Bergmann H, Büchel G (2005) Molecular mechanisms in bio-geo-interactions: from a case study to general mechanisms. Chem Erde 65:7–27

Kuiper I, Lagendijk EL, Bloemberg GV, Lugtenberg BJJ (2004) Rhizoremediation: a beneficial plant-microbe interaction. Mol Plant-Microbe Interact 17:6–15

Lagos ML, Maruyama F, Nannipieri P, Mora ML, Ogram A, Jorquera MA (2015) Current overview on the study of bacteria in the rhizosphere by modern molecular techniques: a mini-review. J Soil Sci Plant Nutr 15(2):504–523

Landa BB, Hervas A, Bethiol W, Jimenez-Diaz RM (1997) Antagonistic activity of bacteria from the chickpea rhizosphere against *Fusarium oxysporum* f. sp. *ciceris*. Phytoparasitica 25 (4):305–318

Larena I, Salazar O, González V, Julián MC, Rubio V (1999) Design of a primer for ribosomal DNA internal transcribed spacer with enhanced specificity for ascomycetes. J Biotechnol 75:187–194

Lee MD, Wilson JT, Ward CH (1983) Microbial degradation of selected aromatics in a hazardous waste site. Dev Ind Microbiol 25:557–565

Leigh MB, Prouzova P, Mackova M, Macek T, Nagle DO, Fletcher JS (2006) Polychlorinated biphenyl (PCB)-degrading bacteria associated with trees in a PCB-contaminated site. Appl Environ Microbiol 72:2331–2342

Lemann J (1997) Oleaginous microorganisms: an assessment of the potential. Adv Appl Microbiol 43:195–243

Lippmann B, Leinhos V, Bergmann H (1995) Influence of auxin producing rhizobacteria on root morphology and nutrient accumulation of crops: changes in root morphology and nutrient accumulation in maize (*Zea mays* L.) caused by inoculation with indole-3-acetic acid (IAA) producing *Pseudomonas* and *Acinetobacter* strains or IAA applied exogenously. Angew Bot 69:31–36

Lord NS, Kaplan CW, Shank P, Kitts CL, Elrod SL (2002) Assessment of fungal diversity using terminal restriction fragments (TRF) pattern analysis: comparison of 18S and ITS ribosomal regions. FEMS Microbiol Ecol 42:327–337

Lübeck PS, Hansen M, Sorensen J (2000) Simultaneous detection of the establishment of seed-inoculated *Pseudomonas fluorescens* strain DR54 and native soil bacteria on sugar beet root surface using fluorescence antibody and in situ hybridization techniques. FEMS Microbiol Ecol 33:11–19

Lucy M, Reed E, Glick BR (2004) Applications of free living plant growth-promoting rhizobacteria: review. Antonie Van Leeuwenhoek 86:1–25

Ludwig W, Bauer SH, Bauer M, Held I, Kirchhof G, Schulze R, Huber I, Spring S, Hartmann A, Schleifer K (1997) Detection and in situ identification of representatives of a widely distributed new bacterial phylum. FEMS Microbiol Lett 153:181–190

Machado MD, Santos MSF, Gouveia C, Soares HMVM, Soares EV (2008) Removal of heavy metal using a brewer's yeast strain of *Saccharomyces cerevisiae*: the flocculation as a separation process. Bioresour Technol 99:2107–2115

Macrae A (2000) The use of 16s rDNA methods in soil microbial ecology. Braz J Microbiol 31:77–82

Matsubara Y, Ohba N, Fukui H (2001) Effects of arbuscular mycorrhizal fungus infection on the incidence of fusarium root rot in *Asparagus* seedlings. J Jap Soc Hort Sci 70:202–206

Mendum TA, Clark IM, Hirsch PR (2001) Characterization of two novel *Rhizobium leguminosarum* bacteriophages from a field release site of genetically modified rhizobia. Antonie Van Leeuwenhoek 79:189–197

Mulligan CN (2005) Environmental application for biosurfactants. Environ Pollut 133:183–198

Mulligan CN (2009) Recent advances in the environmental applications of biosurfactants. Curr Opin Colloid Interface Sci 14:372–378

Muyzer G, Smalla K (1998) Application of denaturing gradient gel electrophoresis (DGGE) and temperature gradient gel electrophoresis (TGGE) in microbial ecology. Antonie Van Leeuwenhoek Int J Gen Mol Microbiol 73:127–141

Myers RM, Fischer SG, Lerman LS, Manialis T (1995) Nearby all base substitution in DNA fragments joined to a GC clamp can be detected by denaturing gradient gel electrophoresis. Nucleic Acids Res 13:3131–3145

Myrold DD, Zeglin LH, Jansson JK (2013) The potential of metagenomic approaches for understanding soil microbial processes. Soil Sci Soc Am J 78:3–10

Nannipieri P, Badalucco L (2003) Biological processes. In: Bembi DK, Nieder R (eds) Processes in the soil-plant system: modelling concepts and applications. The Haworth Press, Binghamton

Nannipieri P, Kandeler E, Ruggiero P (2002) Enzyme activities and microbiological and biochemical processes in soil. In: Burns RG, Dick R (eds) Enzymes in the environment. Marcel Dekker, New York, pp 1–33

Nesme J, Achouak W, Agathos SN et al (2016) Back to the future of soil metagenomics. Front Microbiol 7:73. https://doi.org/10.3389/fmicb.2016.00073

Nicholas RB (1987) Biotechnology in hazardous waste disposal: an unfulfilled promise. ASM News 53:138–142

Nigam P (2013) Microbial enzymes with special characteristics for biotechnological applications. Biomol Ther 3:597–611

Nigam P, Singh D (1995) Enzymes and microbial systems involved in starch processing. Enzyme Microb Technol 17:770–778

Noordman WH, Wachter JJJ, de Boer GJ, Janseen DB (2002) The enhancement by biosurfactants of hexadecane degradation by *Pseudomonas aeruginosa* varies with substrate availability. J Biotechnol 94:195–212

Nowak P (2013) Thinking about a future conservation agenda. J Soil Water Conserv 68(2):50A–52A

Oberbremer A, Muller-Hurtig R, Wagner F (1990) Effect of the addition of microbial surfactants on hydrocarbon degradation in a soil population in a stirred reactor. Appl Microbiol Biotechnol 32:485–489

Ogram A (2000) Soil molecular microbial ecology at age 20: methodological challenges for the future. Soil Biol Biochem 32:1499–1504

Pandey A, Webb C, Soccol CR, Larroche C (2005) Enzyme technology. Asia Tech Publishers, New Delhi

Parida AK, Das AB (2005) Salt tolerance and salinity effects on plants: a review. Ecotoxicol Environ Saf 60:324–349

Paul D, Nair S (2008) Stress adaptations in a plant growth promoting rhizobacterium (PGPR) with increasing salinity in the coastal agricultural soils. J Basic Microbiol 48:378–384

Paungfoo-Lonhienne C, Lonhienne TGA, Yeoh YK, Donose BC, Webb RI, Parsons J, Liao W, Sagulenko E, Lakshmanan P, Hugenholtz P, Schmidt S, Ragan MA (2016) Crosstalk between sugarcane and a plant-growth promoting *Burkholderia* species. Sci Rep 6:37389

Pii Y, Mimmo T, Tomasi N, Terzano R, Cesco S, Crecchio C (2015) Microbial interactions in the rhizosphere: beneficial influences of plant growth-promoting rhizobacteria on nutrient acquisition process: a review. Biol Fertil Soils 51:403–415

Podile AR, Kishore K (2007) Plant growth-promoting rhizobacteria. In: Plant associated bacteria. Springer, Dordrecht, pp 195–230

Ponmurugan P, Gopi C (2006) Distribution pattern and screening of phosphate solubilizing bacteria isolated from different food and forage crops. J Agron 5:600–604

Prasad V, Singh VK, Meena M, Tiwari A, Zehra A, Upadhyay RS (2013) Production and technological applications of enzymes from microbial sources. In: Gupta VK et al (eds) Applications of microbial genes in enzyme technology. Nova Science, New York, pp 175–204

Rangarajan S, Saleena LM, Vasudevan P, Nair S (2003) Biological suppression of rice diseases by *Pseudomonas* spp. under saline soil conditions. Plant Soil 251:73–82

Ranjard L, Poly F, Lata JC, Mougel C, Thioulouse J, Nazaret S (2001) Characterization of bacterial and fungal soil communities by automated ribosomal intergenic spacer analysis fingerprints: biological and methodological variability. Appl Environ Microbiol 67(10):4479–4487

Ranjard L, Lejon DPH, Mougel C, Scheher L, Merdinoglu D, Chaussod R (2003) Sampling strategy in molecular microbial ecology: influence of soil sample size on DNA fingerprinting of fungal and bacterial communities. Environ Microbiol 5:1111–1120

Ray SA, Ray MK (2009) Bioremediation of heavy metal toxicity with special reference to chromium. Al Ameen J Med Sci 2(2):57–63

Rhee SK, Liu X, Wu L, Chong SC, Wan X, Zhou J (2004) Detection of genes involved in biodegradation and biotransformation in microbial communities by using 50-mer oligonucleotide microarrays. Appl Microbiol Biotechnol 70:4303–4317

Rheims H, Rainey FA, Stackebrandt E (1996) A molecular approach to search for diversity among bacteria in the environment. J Ind Microbiol 17:159–169

Riesner D, Henco K, Steger G (1991) Temperature-gradient gel electrophoresis: a method for the analysis of conformational transitions and mutations in nucleic acids and proteins. In: Chrambach A, Dunn MJ, Radola BJ (eds) Advances in electrophoresis, vol 4. VCH, Weinheim, pp 169–250

Rodríguez H, Fraga R (1999) Phosphate solubilizing bacteria and their role in plant growth promotion. Biotechnol Adv 17:319–339

Rodriguez RJ, White JF Jr, Arnold AE, Redman RS (2009) Fungal endophytes: diversity and functional roles. New Phytol 182:314–330

Ruta L, Paraschivescu C, Matache M, Avramescu S, Farcasanu IC (2010) Removing heavy metals from synthetic effluents using "kamikaze" *Saccharomyces cerevisiae* cells. Appl Microbiol Biotechnol 85:763–771

Saravanakumar D, Samiyappan R (2007) ACC deaminase from *Pseudomonas fluorescens* mediated saline resistance in groundnut (*Arachis hypogea*) plants. J Appl Microbiol 102:1283–1292

Sauer M, Branduardi P, Valli M, Porro D (2004) Production of L-ascorbic acid by metabolically-engineered *Saccharomyces cerevisiae* and *Zygosaccharomyces bailii*. Appl Environ Microbiol 70(10):6086–6091

Schulze WX, Gleixner G, Kaiser K, Guggenberger G, Mann M, Schulze ED (2004) A proteomic fingerprint of dissolved organic carbon and of soil particles. Oecologia 142:335–343

Schüßler A, Schwarzott D, Walker C (2001) A new phylum, the *Glomeromycota*: phylogeny and evolution. Mycol Res 105:1413–1421

Schütte UME, Abdo Z, Bent SJ, Conrad S, Christopher JW, Jacob DP, Larry JF (2008) Advances in the use of terminal restriction fragment length polymorphism (T-RFLP) analysis of 16S rRNA genes to characterize microbial communities. Appl Microbiol Biotechnol 80:365–380

Shivanna MB, Meera MS, Hyakumachi M (1994) Sterile fungi from Zoysiagrass rhizosphere as plant growth promoters in spring wheat. Can J Microbiol 40:637–644

Shoebitz M, Ribaudo CM, Pardo MA, Cantore ML, Ciampi L, Curá JA (2009) Plant growth promoting properties of a strain of *Enterobacter ludwigii* isolated from *Lolium perenne* rhizosphere. Soil Biol Biochem 41(9):1768–1774

Shoresh M, Yedida I, Chat I (2005) Involvement of jasmonic acid/ethylene signaling pathway in the systemic resistance induced in cucumber by *Trichoderma asperellum* T203. Phytopathology 95:76–84

Siggins A, Gunnigle E, Abram F (2012) Exploring mixed microbial community functioning: recent advances in metaproteomics. FEMS Microbiol Ecol 80(2):265–280

Silver S, Phung LT (1996) Bacterial heavy metal resistance: new surprises. Annu Rev Microbiol 50:753–789

Simon C, Daniel R (2011) Metagenomics analyses: past and future trends. Appl Environ Microbiol 77:1153–1161

Singh H (2006) Mycoremediation: fungal bioremediation. Wiley Interscience, New York

Singh BK, Nazaries L, Munro S, Anderson IC, Campbell DC (2006) Use of multiplex terminal restriction fragment length polymorphism for rapid and simultaneous analysis of different components. Appl Environ Microbiol 72(11):7278–7285

Singh DP, Prabha R, Yandigeri MS, Arora DK (2011) Cyanobacteria-mediated phenylpropanoids and phytohormones in rice (*Oryza sativa*) enhance plant growth and stress tolerance. Antonie Van Leeuwenhoek 100:557–568

Snow AA, Andow DA, Gepts P, Hallerman EM, Power A, Tiedji JM, Wolfenbarger LL (2005) Genetically engineered organisms and the environment: current status and recommendations. Ecol Appl 15(2):377–404

Sparks DL (2001) Elucidating the fundamental chemistry of soils: past and recent achievements and future frontiers. Geoderma 100:303–319

Stacey G, Upchurch RG (1984) *Rhizobium* inoculation of legumes. Trends Biotechnol 2:65–69

Sugisawa T, Hoshino T, Masuda S, Nomura S, Setoguchi Y, Tazoe M, Shinjoh M, Someha S, Fujiwara A (1990) Microbial production of 2-keto-L-gulonic acid from L-sorbose and D-sorbitol by *Gluconobacter melanogenus*. Agric Biol Chem 54:1201–1209

Thomashow LS, Bonsall RF, Weller DM (1997) Antibiotic production by soil and rhizosphere microbes *in situ*. In: Hurst CJ, Knudson GR, Mclnerney MJ, Stetzenbach LD, Walter MV (eds) Manual of environmental microbiology. ASM Press, Washington, DC, pp 493–499

Tiwari S, Singh P, Tiwari R, Meena KK, Yandigeri M et al (2011) Salt-tolerant rhizobacteria mediated induced tolerance in wheat (*Triticum aestivum*) and chemical diversity in rhizosphere enhance plant growth. Biol Fertil Soils 47:907–916

Van der Putten WH (2003) Plant defence belowground and spatiotemporal processes in natural vegetation. Ecology 84(9):2269–2280

Vandenkoornhuyse P, Baldauf SL, Leyval C, Straczek J, Young JPW (2002) Extensive fungal diversity in plant roots. Science 295:20–51

Větrovský T, Baldrian P (2013) The variability of the 16S rRNA gene in bacterial genomes and its consequences for bacterial community analyses. PLoS One 8:1–10

Volkering F, Breure AM, van Andel JG, Rulkens WH (1995) Influence of nonionic surfactants on bioavailability and biodegradation of polycyclic aromatic hydrocarbons. Appl Environ Microbiol 61:1699–1705

Walsh UF, Morrisey JP, O'Gara F (2001) *Pseudomonas* for biocontrol of phytopathogens: from functional genomics to commercial exploitation. Curr Opin Biotechnol 12:289–295

Ward FA, Pulido-Velazquez M (2008) Water conservation in irrigation can increase water use. Proc Natl Acad Sci 105(47):18218–18220

Wardle DA (2002) Communities and ecosystems—linking the aboveground and belowground components. Princeton University Press, Princeton

Wardle DA, Bardgett RD et al (2004) Ecological linkages between above ground and below ground biota. Science 304(5677):1629–1633

Watrud LS, Perlak FJ, Tran MT et al (1985) Cloning of the *Bacillus thuringiensis* subsp. *Kurstaki* delta-endotoxin gene into *Pseudomonas fluorescens*: molecular biology and ecology of an engineered microbial pesticide. In: Halvorson HO, Pramer O, Rogul M (eds) Engineered organisms in the environment: scientific issues. American Society of Microbiology, Washington, DC, pp 40–46

Wattanaphon HT, Kerdsin A, Thammacharoen C, Sangvanich P, Vangnai AS (2008) A biosurfactant from *Burkholderia cenocepa*cia BSP3 and its enhancement of pesticide solubilization. J Appl Microbiol 105:416–423

Weller DM, Raaijmakers JM, Mcspadden BB, Thomashow LS (2002) Microbial populations responsible for specific soil suppressiveness to plant pathogens. Annu Rev Phytopathol 40:309–348

White TJ, Bruns TD, Lee S, Taylor J (1990) Analysis of phylogenetic relationships by amplification and direct sequencing of ribosomal RNA genes. In: Innis MA, Gelfand DH, Sninsky JJ, White TJ (eds) PCR protocols: a guide to methods and applications. Academic Press, New York, pp 315–322

Wilmes P, Bond PL (2004) The application of two-dimensional polyacrylamide gel electrophoresis and downstream analyses to a mixed community of prokaryotic microorganisms. Environ Microbiol 6:911–920

Woese CR (1987) Bacterial evolution. Microbiol Rev 51:221–271

Wollenweber B, Porter JR, Lubberstedt T (2005) Need for multidisciplinary research towards a second green revolution. Curr Opin Plant Biol 8:337–341

Xie X, Zhang H, Paré PW (2009) Sustained growth promotion in *Arabidopsis* with long-term exposure to the beneficial soil bacterium *Bacillus subtilis* (GB03). Plant Signal Behav 4:1–6

Zhang YH, Himmel ME, Mielenz JR (2006) Outlook for cellulase improvement: screening and selection strategies. Biotech Adv 24:452–481

Zhang H, Murzello C, Sun Y, Kim MS, Xie X et al (2010) Choline and osmotic-stress tolerance induced in *Arabidopsis* by the soil microbe *Bacillus subtilis* (GB03). Mol Plant-Microbe Interact 23:1097–1104

Zhou J, Thompson DK (2002) Challenges in applying the microarrays to environmental studies. Curr Opin Biotechnol 13:204–207

Zhuang XL, Chen J, Shim H, Bai Z (2007) New advances in plant growth-promoting rhizobacteria for bioremediation. Environ Int 33:406–413

Chapter 8
Biotic Constraints to Wheat Production in Tropics: Microbial Control Strategies and Mechanism

Vandana Jaggi and Manvika Sahgal

Abstract Wheat is the premier food crop worldwide. It is grown in a diverse climate. Phytopathogenic diseases are biotic constraints to its production. Biological control is an established method for plant disease management. It refers to the suppression of phytopathogens by beneficial organisms and their metabolites. However, one of the factors that obstruct the large-scale application of biocontrol is the lack of efficient, commercially available biocontrol agents (BCAs). The identification of novel BCAs, determining their modes, and mechanism of action is a critical step in the success of commercial biocontrol products. A robust screening of suitable candidates is required to develop potential BCAs. In this chapter, we present an overview of well-known wheat diseases and their biological control. Moreover, we also reviewed status of commercially available BCAs and summarize research of organizations working in the area of plant disease management.

Keywords Wheat · Disease managements · Biocontrol agents and secondary metabolites

8.1 Introduction

Cereals are the mainstay for food security in developing nations where the cereal-based production system is predominant (Nikos and Jelle 2012; Shiferaw et al. 2013). Three major cereals, namely, maize, rice, and wheat, contribute approximately 5%, 19%, and 19% of daily calories and 4%, 13%, and 21% of daily dietary protein, respectively (Ali et al. 2011), and are a major source of carbohydrates (Enghiad et al. 2017). Globally, wheat occupies the largest total cultivated area (38.8%) among the cereals. However, its production is significantly lower than rice and maize (FAO, IFAD, UNICEP, WFP and WHO 2018). The wheat demand

---

V. Jaggi · M. Sahgal (✉)
Department of Microbiology, G.B. Pant University of Agriculture & Technology, Pantnagar, Uttarakhand, India

S. K. Dubey, S. K. Verma (eds.), *Plant, Soil and Microbes in Tropical Ecosystems*, Rhizosphere Biology, https://doi.org/10.1007/978-981-16-3364-5_8

worldwide will increase by 60% in the next few decades (Aggarwal 2009). The global wheat production for the year 2018–2019 was 731 million tonnes, 32.19 million tonnes less than that in the year 2017–2018. During the year 2017–2018, wheat production was 763.19 million metric tonnes. This reduction was attributed to diseases. Short-duration winters and terminal temperature stress are reported to increase the spread of diseases and their severity (Langridge 2017). Every 1 °C increase in temperature will cause around a 10% reduction in world wheat production (Zayan 2019). Since the increase in disease severity due to climate change is creating a barrier to the global wheat production, therefore, it is necessary to pursue some goals to mitigate these discrepancies. The prominent goal is to minimize crop losses caused by diseases, which are estimated between 20 and 40%, through a sustainable approach with lesser harmful consequences for livelihoods, public health, and the environment (Newitt et al. 2019).

8.2 Wheat Production: Acreage and Yield

8.2.1 Global Scenario

Wheat is grown in diverse climatic types such as tropical, subtropical, and temperate (Curtis et al. 2019; Peleg et al. 2011; Raza et al. 2019; Salim and Raza 2020). Globally, wheat (*Triticum aestivum* L.) occupies around ~217 million hectares with an annual production of 731 million tonnes. India shares the maximum area (14%) under wheat cultivation followed by Russia (12.43%), China (11.14%), and the USA (6.90%) together accounting for 45% of the total global wheat acreage (https://www.statista.com/statistics/267268/production-of-wheat-worldwide-since-1990/).
Approximately, 58% (449 MT) of the global wheat production is from 7 countries, namely, China, India, Russia, the USA, Canada, Ukraine, and Pakistan. China tops the production at 136 million tonnes (MT) followed by India (98.51 MT), Russia (85), and the USA (47.35). However, the average wheat yield in major wheat-growing countries is significantly low. The wheat yield is maximum in China (5.48 tonnes/ha) followed by Ukraine, India, and the USA (Ramadas et al. 2019).

8.2.2 Indian Scenario

In India, the total wheat cultivated area is nearly 31.08 million hectares with an average production of ~98.51 million tonnes (MT) and productivity of 3.37 tonnes/ha during 2018–2019 (Ramadas et al. 2019). It is a Rabi crop growing from November to April when temperature ranges between 3 and 32 °C (Enghiad et al. 2017). It is cultivated in subtropical and tropical regions, between of 30°N to 60°N latitudes and 27°S to 40°S longitude, up to 3000 m above sea level (mabsl). The wheat cultivated area in India is classified into five agro-climatic zones (Table 8.1).

Table 8.1 Agro-climatic zones of wheat in India, regions covered, and area under cultivation (source https://nfsm.gov.in/StatusPaper/Wheat2016.pdf)

S. no.	Zones covered	States/regions	Area in cultivation during 2013–2014 (million ha)
1	Northern Hill Zone (NHZ)	Hilly areas of Jammu & Kashmir (except Jammu, Kathua and Samba districts), Himachal Pradesh (except Una & Paonta valley), Uttarakhand (excluding Tarai region) & Sikkim	0.8
2	North Western Plains Zone (NWPZ)	Punjab, Haryana, Western Uttar Pradesh (except Jhansi Div), Rajasthan (excluding Kota & Udaipur div), Delhi, Tarai region of Uttarakhand, Una & Paonta valley of Himachal Pradesh, Jammu, Samba & Kathua districts of Jammu & Kashmir and Chandigarh	11.5
3	North Eastern Plains Zone (NEPZ)	Eastern Uttar Pradesh (28 dist), Bihar, Jharkhand, West Bengal, Assam, Odisha and other North Eastern states (except Sikkim)	11.1
4	Central Zone	Madhya Pradesh, Gujarat, Chhattisgarh, Kota & Udaipur Div of Rajasthan & Jhansi Div of Uttar Pradesh	6.08
5	Peninsular Zone	Maharashtra, Tamil Nadu (except Nilgiris & Palani Hills), Karnataka & Andhra Pradesh	1.6
		Total	31.08

Three wheat-producing regions, namely, Gangetic Plains, Central, and Peninsular India, are the heat-stressed regions (Joshi et al. 2007). The increased frequency of high temperature results in increased prevalence of diseases. Variability in climate is a major threat to Indian wheat production by promoting the occurrence of plant diseases (Chakraborty and Pangga 2004; Garrett et al. 2009). The important diseases affecting wheat are rusts, powdery mildew, foliar blights, etc. discussed later in the chapter.

8.3 Biotic Constraints for Wheat Production

Wheat production world over is diminished mainly by abiotic (temperature, salt stress, heavy and unseasonal rains, drought) and biotic (diseases, pests, and insects) stresses. Of these, phytopathogenic diseases are the most important limiting factor of wheat production causing huge loss in yield and quality. It accounts for 30–40% loss of wheat globally (Singh et al. 2016; Serfling et al. 2016).

8.3.1 Fungal Diseases

The main fungal diseases of wheat are rusts, powdery mildew, blotches, blight, and blast. These disease-causing fungi are characterized as biotrophic, necrotrophic, and hemibiotrophic. Biotrophic fungi feed on living host tissue, necrotrophic kill host tissues and feed on them, and hemibiotrophic is an intermediate state characterized by temporal and/or spatial transitions between biotrophic and necrotrophic (Spanu and Panstruga 2017). Wheat yield and grain quality are affected more by fungal diseases than bacterial and viral diseases (Gautam et al. 2015).

8.3.1.1 Wheat Rust

There are three different types of wheat rusts: stripe, stem, and leaf rust. The stripe rust is caused by *Puccinia striiformis* f.sp. tritici, stem rust by *P. graminis Pers.* f.sp. tritici Eriks. & Henn., and leaf rust by *P. triticina Eriks.* Wheat rust disease in the country was first reported in the year 1922 (Mehta 1940). The causative fungus is a biotrophic and obligate parasite (Chen et al. 2014). The disease reduces the plant height and grain yield by affecting the photosynthetic ability of the plant (Roelfs 1992). Out of the three wheat rust diseases, stem rust is the most damaging (Gessese et al. 2019). It along with stripe rust affected ten million hectares of wheat cultivation in Northern India and seven million hectares in Central and Peninsular India. In contrast, leaf rust is less damaging than stem and stripe rust but is spread across all the wheat-growing regions of India. The infection of wheat rusts is caused byurediniospores spreading through wind and infects host plants several miles away. The rust fungus continuously evolves to new pathotypes. Therefore, the management of wheat rust pathogens is a challenging task (Bhardwaj et al. 2019).

8.3.1.2 Powdery Mildew

Powdery mildew is caused by an obligate biotrophic fungus, *Blumeria graminis* f.sp. tritici (Bgt) (Alam et al. 2013; Friebe et al. 1996; Shi et al. 1998). Bgt is distributed throughout the world especially in cool, warm, and humid areas (Priestley and Bayles 1988; Huang et al. 2004). In India, it is prevalent in Northern Hill Zone and North Western Plain Zone (Li et al. 2011; Royse et al. 1980). The disease symptoms mainly appear on leaves, but during severe cases, symptoms also appear on leaf sheath, stem, and ear. The fungus produces conidiospores, which disperse to long distances through air leading to an epidemic. Infection occurs during tillering, stem elongation, and booting stages. These result in up to 40% yield loss (Bowen et al. 1992; Kang et al. 2020).

8.3.1.3 Karnal Bunt

Karnal bunt (KB) is caused by the fungus *Tilletia indica* Mitra (syn. *Neovossia indica* (Mitra) Mundkur). The causative fungus belongs to the class *Basidiomycotina*. The disease was first reported from Karnal, Haryana, India (Mitra 1931). Initially, the disease was restricted to Northwestern India. Currently, the disease is worldwide in occurrence. It has been reported from Mexico (Duran 1972), Pakistan (Munjal 1975), Brazil (Da Luz et al. 1993), Nepal (Singh et al. 1995), the USA (APHIS 1996), Iran (Torarbi et al. 1996), and the Republic of South Africa (Crous et al. 2001). This disease mainly affects the wheat grain. The infected grain is converted into black powder of bunt spores. The disease is devastating because 1–4% of kernel infection is sufficient to make wheat grain unpalatable and 5% of kernel infection causes a distinct deterioration in flour quality (Rush et al. 2005; Ullah et al. 2012).

8.3.1.4 Fusarium Head Blight

Fusarium head blight (FHB) is caused by *Fusarium graminearum*, a hemibiotrophic fungus. The disease was first reported in England in 1884. Later on, it spread to humid and semi-humid wheat-growing areas of the world, especially Asia, Australia, Europe, and North and South America (Dickson 1942; Teli et al. 2016). In India, it is confined to Himachal Pradesh, Punjab, and Tamil Nadu. Recently, an increase in the FHB epidemic has been observed due to global warming (Saharan and Naef 2008; Shah et al. 2014). The fungus spends its asexual cycle on infested crop debris and the sexual cycle on living wheat tissues (Gunupuru et al. 2017). *Fusarium graminearum* releases the mycotoxins such as trichothecenes that make the grain unsafe for consumption. Hence, the disease incurs a huge economic loss to growers (Venkataramana et al. 2014). According to the International Maize and Wheat Improvement Center (CIMMYT), FHB has been considered as one of the most destructive wheat diseases impacting the production globally (Bottalico and Perrone 2002; McMullen et al. 1997; Yi et al. 2018).

8.3.1.5 Blotch

Spot blotch disease is caused by hemibiotrophic ascomycetous fungus *Bipolaris sorokiniana* (syn. *Drechslera prorokiniana* syn. *Helminthosporium sativum*, teleomorph *Cochliobolus sativus*). The disease was first reported in Europe by Mohy in 1914 (Mitra 1931). Since the past few decades, it has been seen as a serious constraint for wheat production worldwide. The disease affects an estimated 25 m ha of wheat-growing area in Africa, Asia, Australia, Canada, and South America (Duveiller et al. 2005; Van Ginkel and Rajaram 1998). In India alone, approximately 10 m ha in the wheat-growing belt is affected by the disease, out of which 9 m ha

Fig. 8.1 Wheat plants showing infection by different fungal pathogens. (**a**) Stem rust C.O *Puccinia graminis*, (**b**) stripe rust C.O. *Puccinia striiformis*, (**c**) leaf rust C.O. *Puccinia triticina*, (**d**) powdery mildew C.O. *Blumeria graminis*, (**e**) fusarium head blight C.O. *Fusarium* sp., (**f**) spot blotch C.O. *Bipolaris sorokiniana*, (**g**) leaf blight C.O. *Alternaria triticina,* (**h**) wheat blast C.O. *Magnaporthe oryzae Triticum*. (**Source: modified from** Langridge 2017)

area is in the Indo-Gangetic Plains (Nagarajan and Kumar 1998; Iftikhar et al. 2010). The average yield loss due to spot blotch has been estimated to be in the range of 15–30% (Gupta et al. 2018). Disease symptoms typically appear on the leaf, sheath, node, and glumes as small light brown lesions, mostly oval to oblong in shape (Fig. 8.1). Gradually they increase in size and coalesce to form larger necrotic patches (Viani et al. 2017).

8.3.1.6 Leaf Blight

Leaf blight is one of the most important foliar diseases of wheat. The causative fungus, *A. triticina*, is a hemibiotrophic fungus. In the Indian subcontinent, this disease is responsible for 18–22% loss to wheat yield (Bandyopadhyay et al. 2016). In India, it was first reported from Maharashtra (Kulkarni 1924). Presently, it is also prevalent in Himachal Pradesh, Punjab, Jammu and Kashmir, and Madhya Pradesh. The disease initially appears as small and irregularly scattered chlorotic lesions on the leaves. As the disease progresses, several spots coalesce and cover partial or whole leaf giving a burnt appearance to the leaf.

8.3.1.7 Wheat Blast

Wheat blast (WB) is caused by the hemibiotrophic fungus, *Magnaporthe oryzae* pathotype triticum (MoT). The disease was first reported in the Brazilian state of Paraná (Urashima et al. 1993). In 2016, the wheat blast outbreak was first reported outside of South America, in Bangladesh, South Asia (Callaway 2016; Malaker et al. 2016). In Bangladesh, nearly 15,000 ha (3.5% of the total 0.43 million ha of wheat area) of wheat cultivation was affected by the wheat blast and caused up to 51% yield loss (Ceresini et al. 2018).

8.3.2 Bacterial Diseases

The most frequently reported wheat bacterial diseases are bacterial leaf streak/black chaff caused by *Xanthomonas translucens* pv. undulosa (XTU), basal glume rot caused by *Pseudomonas syringae* pv. atrofaciens (PSA), and bacterial leaf blight caused by *P. syringae* pv. syringae (PSS) (Valencia-Botín and Cisneros-López 2012). These diseases are distributed in tropical and subtropical wheat-growing regions (Maraite et al. 2007). Despite numerous reports of bacterial diseases on wheat worldwide, the study of bacterial diseases on wheat has been limited, and quantitative information, for example, on crop losses and disease epidemiology, is rarely available (Duveiller et al. 2012).

8.4 Biological Control

Biological control includes the application of BCAs as inoculants, as well as active metabolites directly derived from natural origin (microorganisms) with a low or no impact on the environment and non-target microorganisms (Jacobsen et al. 2004). Firstly, William Roberts in 1874 reported the antagonistic action of bacteria against *Penicillium glaucum* in liquid culture and introduced the term "antagonism." However, the term biological control was coined for the first time by C. F. Von in 1914 as a feasible preposition of plant disease management. There is growing demands for biological control of diseases. In the past few years, commercial interest and innovation have been increased in biological control research and achieved a success to commercialize them. Moreover, to get more success in the future, it required extensive research through molecular techniques to find the abundance of microbial antagonists in natural environment and find the microbial ecology of pathogen and antagonists.

8.4.1 Type of Interactions Contributing to Biological Control

Biological control is the result of the complex interactions occurring among pathogen and biological control agents (BCAs). In all types of interactions, pathogens are inhibited or disintegrated through direct and indirect antagonism. Direct antagonism results from physical contact such as hyperparasitism. In contrast, indirect antagonism results from the suppression of the pathogens through the production of cell wall-degrading enzymes, antibiotics, competition for nutrients, and stimulation of plant host defense pathways by BCAs (Jamalizadeh et al. 2011; Köhl et al. 2019). Pathogen suppression in the natural environment is mediated via cell wall-degrading enzymes (CWDEs), antibiosis, competition for nutrients and space, and induced resistance in the host plant (Fig. 8.2).

8.4.1.1 Cell Wall-Degrading Enzymes (CWDEs) and Parasitism

Parasitism of fungi by various microorganisms relies on the production of fungal CWDEs (Raymaekers et al. 2020). Microorganisms secrete various extracellular hydrolytic enzymes like chitinase, protease, cellulase, and amylase. These enzymes interfere with pathogen growth by hydrolyzing a wide variety of polymeric compounds, including chitin, proteins, cellulose, and hemicelluloses. Various examples of parasitism involving CWDEs are hyperparasitism of powdery mildews by the *Coelomycetes* and *Ampelomyces quisqualis* (Sztejnberg et al. 1989); control of *Magnaporthe oryzae* triticum (MoT) by *Bacillus amyloliquefaciens* (BTLK6A), *B. subtilis* (BTS-3), and *B. amyloliquefaciens* (BTS-4) mediated by chitinase expression (Dutta et al. 2018); and chitinase-mediated growth inhibition of pathogenic *Aspergillus flavus*, *A. niger*, *A. terreus*, *Fusarium oxysporum*, *Ralstonia solanacearum*, and *Rhizopus* sp. by *B. thuringiensis* and *B. licheniformis* (Gomaa 2012). Köhl et al. (2019) demonstrated that biocontrol activity of *T. harzianum* T39 against biotrophic and necrotrophic foliar pathogens was due to the production of CWDEs. Hirpara et al. (2017) found that the application of *T. harzianum* inhibits the radial growth of *S. rolfsii by* 88%. Similarly, *T. viride* was found to be an antagonist to *Bipolaris sorokiniana*, the causal agent of root rot, seedling, and foliar blights of wheat (Prasad et al. 1978; Krivoshchekova and Mishchenk 1990). Recently, a technological advancement in biocontrol was established. The transgenic plants containing the gene for endochitinase from *T. harzianum* T39 with increased resistance against plant pathogenic fungi have been developed (Kannojia et al. 2019).

8.4.1.2 Antibiosis

Antibiotics are secondary metabolites of microbial origin, effective at low concentrations. In some instances, the antibiotics are particularly effective at suppressing

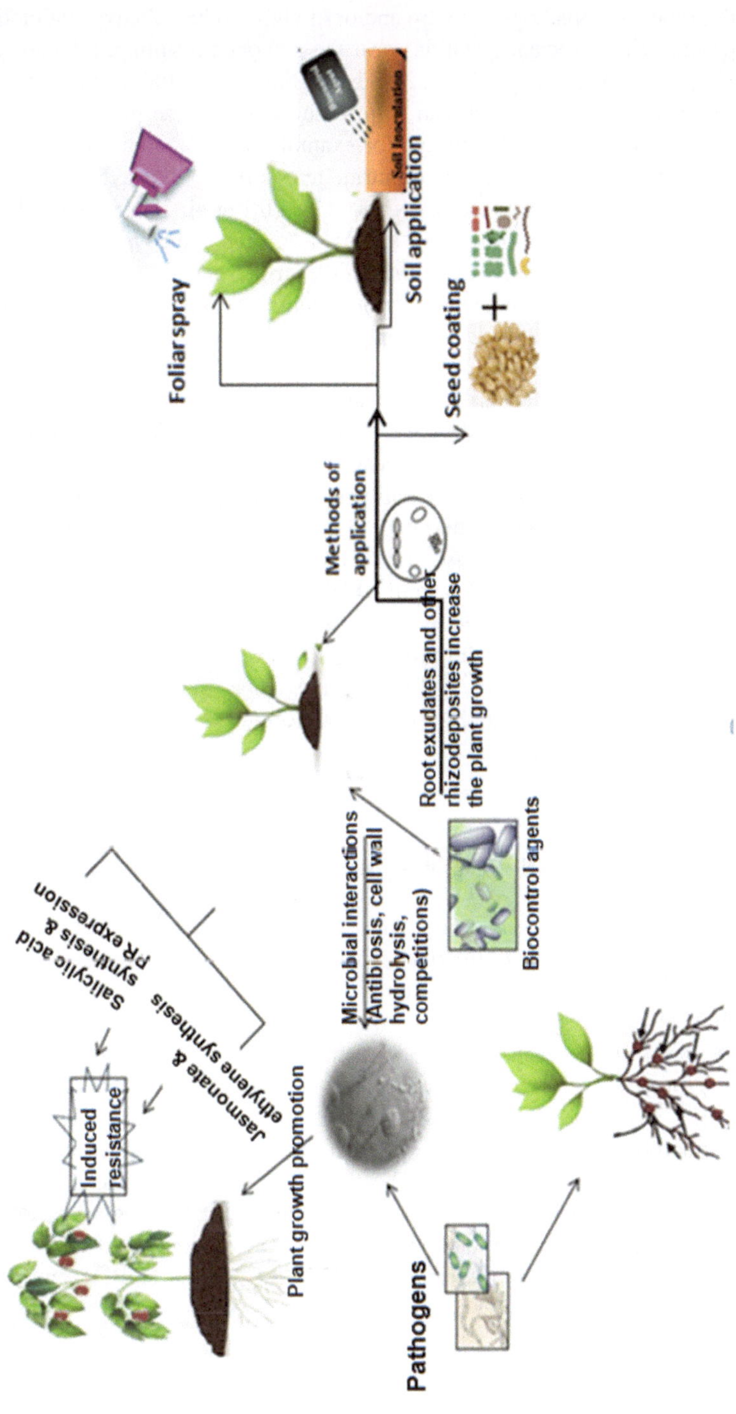

Fig. 8.2 Interaction within the plants, pathogenic microorganisms, and biocontrol agents. Methods of application of biocontrol agents (soil inoculation, seed application, and foliar spray). (**Modified:** Sharma et al. 2017; Law et al. 2017)

the growth of the target pathogen in vitro and/or in vivo. To be effective, antibiotics must be produced in sufficient quantities near the pathogen resulting in biocontrol (Weller et al. 2007; Mavrodi et al. 2012). Several biocontrol strains are known to produce multiple antibiotics which can help to suppress diverse plant pathogens, displaying enhanced biocontrol activity. For example, *Pseudomonas* spp., able to produce phenazines, pyoluteorin, pyrrolnitrin, and DAPG, displayed improved capacities to suppress plant diseases of wheat (Glandorf et al. 2001; Bakker et al. 2002). Moreover, phenazine-producing strain of *Pseudomonas fluorescens* was effective against *Gaeumannomyces graminis* var. tritici causing take-all disease (Thomashow et al. 1990), an endophytic *Pseudomonas aurantiaca* strain suppressed leaf rust caused by *Puccinia triticina* (Wang et al. 2012), and two *Pseudomonas putida* strains (JD204 and JC186) were found effective against stripe rust caused by *Puccinia striiformis* f.sp. tritici Eriks on wheat (Pang et al. 2016). Studies reported that *P. putida* (Flaishman et al. 1996) and *P. protegens* strain CHA0 (Bellameche et al. 2020) were capable of suppressing wheat leaf rust caused by *Puccinia triticina*. Several bacillus strains with biocontrol ability have been reported, namely, *Bacillus cereus* strain UW85 producing zwittermicin (Silo-Suh et al. 1994) and kanosamine (Milner et al. 1996), *Bacillus subtilis* strain E1R-j (Gao et al. 2015), and *Bacillus amyloliquefaciens* (formerly *subtilis*) strain QST 713 (Matzen et al. 2019) controlling wheat powdery mildew (caused by *Blumeria graminis*). The biocontrol potential of these strains was due to the production of several secondary metabolites including siderophores, antibiotics, and hydrogen cyanide. The secondary metabolites inhibited conidial germination, appressorial formation, and development of haustoria and extension of mycelia.

8.4.1.3 Competition for Nutrients and Space

In a microbial context, soils and living plant surfaces are nutrient-limited environments. To successfully colonize the rhizosphere, microorganisms must effectively compete for the available nutrients. Bacteria, yeasts, and filamentous fungi restrict the growth of phytopathogens by competing for nutrients such as carbon, nitrogen, and macro- and micro-elements (Elad and Freeman 2002). Competition for rare but essential micronutrients, such as iron, has been extensively studied. Iron is extremely limiting (10^{-18} M) in the rhizosphere. This concentration is too low to support the growth of microorganisms. To sustain in such an environment, microorganisms secrete siderophores to sequester iron. The majority of microorganisms found in the soil produce siderophores (Neilands 1981). For the first time, siderophore production as a mechanism for biological control was demonstrated in *Erwinia carotovora* by *Pseudomonas fluorescens* strains A1, BK1, TL3B1, and B10 (Kloepper et al. 1980; Keswani et al. 2019).

8.4.1.4 Induced Resistance in the Host Plants

Plants actively respond to a variety of chemical stimuli produced by soil and plant-associated microbes. Such stimuli often enhance resistance against pathogenic infection (Maurya 2020). Disease suppression through the induction of resistance in host is an alternative mode of action of BCAs. Induction of resistance results in the release of elicitors (proteins, antibiotics, and volatiles). The elicitors induce the expression of the genes involved in the salicyclic acid pathway or the jasmonic acid/ethylene pathway of host (Pieterse et al. 2014). Induced resistance is local or systemic representing two distinct pathways: systemic acquired resistance (SAR) and induced systemic resistance (ISR). SAR is typically induced by pathogens while ISR by non-pathogenic bacteria (Whipps 2001). SAR is mediated by salicylic acid that leads to the expression of pathogenesis-related (PR) proteins such as PR-1, PR-2, chitinases, and peroxidases (Kageyama and Nelson 2003; Park et al. 2000; Ramamoorthy et al. 2001; Choudhary and Johri 2009). In contrast, ISR is mediated by jasmonic acid (JA) and/or ethylene. During ISR, the host plant physiology and metabolic responses are altered, leading to an enhanced synthesis of plant defense chemicals upon pathogen challenge (Ramamoorthy et al. 2001; Nowak and Shulaev 2003).

8.5 BCAs: Methods of Delivery

Despite the well-known benefits of biocontrol agents in disease reduction, the delivery of microbial inoculum in agriculture crops frequently hampers its effect. Depending on the type of disease and the mode of action, microbial inoculants are typically applied as seed treatment, soil amendment, and foliar spray (Rocha et al. 2019). Seed treatment with microbial inoculants is one of the efficient delivery systems for the management of seed–/soil-borne diseases. Here, seed is coated with dry powder of bio-inoculants just before sowing. For commercial purpose, dry powder of antagonist is used @ 3–10 g/kg seed based on seed size (Mukhopadhyay et al. 1992; Das and Hazarika 2000; Puyam 2016). For soil treatment, 4 kg of the recommended biofertilizer is mixed in 200 kg of compost and kept overnight. This mixture is then incorporated in the soil at the time of sowing or planting. Bio-inoculants in the form of spray can be also applied more frequently depending on the disease. Successful spraying is typically achieved by timely administration of an efficient microbial formulation with proper spraying equipment (Preininger et al. 2018).

Insight into the research showed the successful application of bio-inoculants as seed priming. Hossain et al. (2015) observed that seed treatment with BAU-biofungicide (a *Trichoderma*-based preparation) significantly decreased the disease severity in wheat and resulted in 20.51% higher grain yield in both pot and field experiments. Foliar application of *Bacillus subtilis* strain E1R-j effectively controlled the induction of wheat powdery mildew and stripe rust in both greenhouse and field trial (Liu et al. 2009; Li et al. 2013). Similarly, the foliar application of antagonistic *B. velezensis* CC09 at two leaf stages was also found effective in reducing incidence of wheat powdery mildew disease by 31.21% (Cai et al. 2017).

8.6 Commercial BCAs for Wheat Diseases

Nowadays, farmers are interested in reducing dependence on chemical inputs, so use of BCAs instead of synthetic chemicals is gaining popularity in plant disease management. After a time span of almost 100 years, only a few BCAs are available for commercial application. The first BCA to be registered was *Agrobacterium radiobacter* strain K 84 for the control of crown gall in 1972. This was registered as Gallex or Galltrol with the United States Environmental Protection Agency (EPA) in 1979. The first fungus to be registered as BCA was *Trichoderma harzianum* ATCC 20476 in 1989 for the control of plant diseases (Fravel 2005). Till 2013, 12 fungi and 14 bacteria were registered with the EPA for the control of different plant diseases (Agrios 2005). These are Kodiak (*Bacillus subtilis* for the control of seed-borne diseases), bio-jet (*Pseudomonas aureofaciens* strain TX-1 effective against *Pythium* and *Rhizoctonia solani*), etc. (Table 8.2) (Junaid et al. 2013). However, presently, no effective BCAs are registered against the wheat pathogens. The technology of commercialization is still in its initial phase. Of all the organisms registered, 65% were registered within the past 10 years. The commercialization and application of BCAs have been slow due to their variable performances under different environmental conditions and soil types in the fields (Heydari and Pessarakli 2010). Many BCAs perform well in the laboratory and greenhouse conditions but fail to do so in the field. This problem can only be solved by better understanding of the environmental parameters that affect biocontrol agents (Wang et al. 2003; Fravel 2005). In addition, there has also been relatively little investment in the development and production of commercial formulation of BCAs probably due to the absence of the cost-effective process of development, testing, registration, and marketing (Ardakani et al. 2009).

8.7 Biological Control Research in India: Past to Present

In India, organized and systematic biological control research began with the establishment of the Indian Station of Commonwealth Institute of Biological Control (CIBC) in 1957 (Birthal and Sharma 2004). Later, other organizations come forward to fillip the workers in biocontrol research in the world. These organizations provide effective services to world agriculture to coordinate and administer agricultural informations, species identification, biocontrol programs, and other services.

Table 8.2 List of biocontrol products, targets, and manufacturer (modified: Junaid et al. 2013)

S. no.	Bio control agent	Product	Target disease/organism	Manufacturer
1	*Ageobacterium radiobacter* strain 84	Galtrol	*Agrobacterium tumefaciens*	AgBioChem, USA
2	*Ageobacterium radiobacter* strain K 1026	Nagol	*Agrobacterium tumefaciens*	Bio-care
3	*Bacillus subtilis* strain GB34	GB34	*Rhizoctonia, Fussarium*	Gustafon, USA
4	*Bacillus subtilis* strain GB03	Kodiac, companion	*Rhizoctonia, Aspergillus*	Growth prod-ucts,USA
5	*Pseudomonas aureofaciensstrain* TX-1	Bio–jet, spot less	*Pythium, Rhizoctonia solani*	Ingreen houses EcoSoil system
6	*Pseudomonas fluorescence* strain A506	Frostban	Fire blight, bunch rot	Plant Health Technologies
7	*Streptomyces griseoviridis*	Mycostop	Soil borne pathogens	Kemira Oy, Finland
8	*Trichoderma harzianum* T22	Root shield, plant shield	Soil borne pathogens	Bio works, USA
9	*Trichoderma harzianum T39*	Trichodex	*Botrytis cinerea*	Bio works, USA
10	*Ampelomyces qulsquallis* isolate M-10	AQ10	Powdery mildew	Ecogen, USA
11	*Aspergillus flavus* AF36	Alfa guard	*Aspergillus flavus*	Circleone globa, USA
12	*Gliocladium catenulatum strain JI446*	Prima stop soil guard	Soil borne pathogens	Kemira Ag
13	*B. subtilis*	Epic	*Rhizoctonia, Fusarium, Alternaria, Aspergillus and Pythium*	AgraQuest, Inc. Gustafson, Inc.
		Kodiak		
		MBI 600		
		Cillus		
		Green-all G		
		HiStick N/T[a]		
		Subtilex		
14	*B. subtilis FZB24*	Rhizo-Plus Konz	*Rhizoctonia, Fusarium, Alternaria, Verticillium* and *Streptomyces*	FZB Biotechnik, GmbH
15	*B. thuringiensis var. galleriae*	Spicturin	leaf folder of rice and diamondback moth on vegetables	ISCB, India

(continued)

Table 8.2 (continued)

S. no.	Bio control agent	Product	Target disease/organism	Manufacturer
16	*Pseudomonas* + *Azospirillum*	BioJet	Effective against brown patch and dollar spot soil pathogens	Eco-Soil
17	*Pseudomonas fluorescens*	Su-Mona	Bacterial wilt	PCI Pest Control Private Limited, India

[a]It is ready to apply biostacked co-inoculum of *B. subtilis* strain MBI 600 and *Bradyrhizobium japonicum*

8.7.1 Commonwealth Institute of Biological Control (CIBC)

Systematic biological control research in India started with the establishment of the Indian Station of Commonwealth Institute of Biological Control (CIBC) at Bengaluru in 1957. During the past seven decades, our knowledge of pests and weeds of crop plants and control methods have increased manifold (Barratt et al. 2017).

8.7.2 All India Coordinated Research Project on Biological Control of Crop Pests and Weeds (AICRP-BC&W)

Under the aegis of the Indian Council of Agricultural Research (ICAR), All India Coordinated Research Project on Biological Control of Crop Pests and Weeds (AICRP-BC&W) was launched in 1977. AICRP coordinates with many centers in different regions of the country. The details of the centers and salient achievements are provided in Table 8.3.

8.7.3 The Society for Biocontrol Advancement (SBA)

This was established in 1986 as an Indian Society for Biocontrol Advancement (ISBA) at the Tamil Nadu Agricultural University, Coimbatore, India. In 1996, it shifted at Project Directorate of Biological Control, Bangalore. This facility has been recently upgraded as National Bureau of Agriculturally Important Insects (NBAIR). The intent goal of SBA is to promote research and create awareness of biological control of pests, pathogens, and weeds in India. Additionally, it is also active in publication, conducting seminar/symposium/conference on biological control, and

Table 8.3 All India Coordinated Research Projects (AICRP) centers in different regions and their salient achievements

Regions	Institutes	Salient achievements in area of biological control (2012–2017)
Eastern region	Assam Agricultural University (AAU), Jorhat (Assam)	Foliar sprays of *Beauveria bassiana* Bb5a has minimized significantly the damage of sucking pests in hot chilli and improved the yield (51.29 q/ha).
Southern region	Central Tobacco Research Institute (CTRI), Rajahmundry (Andhra Pradesh)	
	Acharya N.G. Ranga Agricultural University (ANGRAU), Hyderabad (Andhra Pradesh)	Actively involved in the management of sugarcane borers.
	Indian Institute of Horticultural Research (IIHR), Bangalore (Karnataka)	
	Tamil Nadu Agricultural University (TNAU), Coimbatore (Tamil Nadu)	Talc formulation of *Metarhizium anisopliae* of IIHR strain @ 1 kg/100 L recorded 77.1% mortality of mango hoppers.
		Application of *Beauveria bassiana* (NBAIR formulation) at 5 g/L of water along with 6 releases of *Trichogramma chilonis* at 10 days interval from bud initiation stage suppressed the jasmine bud borer with minimum bud damage of 2.9%.
		Classical biological control of papaya mealybug, Paracoccus marginatus with mass multiplication and release of parasitoid, Acerophagus saved a crop loss of Rs. 435 crores (papaya, tapioca and mulberry) and input cost on pesticides to the tune of Rs. 244.5 crores annually in Tamil Nadu.
	Sugarcane Breeding Institute (SBI), Coimbatore (Tamil Nadu)	
	Kerala Agricultural University (KAU), Thrissur (Kerala)	The IPM practices (*Pseudomonas fluorescens* @ 10 g/kg of seed, followed by five releases of *Trichogramma japonicum* @ 1 lakh/ha starting from 20 days after transplanting or 40 days after sowing) recorded higher incidence of natural enemies and resulted in reduction in stem borer population in rice by 37%.
	Central Plantation Crops Research Institute (CPCRI), Kayangulam (Kerala)	
Western region	Mahatama Phule Krishi Vidyapeeth (MPKV), Rahuri, College of Agriculture, Pune (Maharashtra)	Three sprays of *Metarhizium anisopliae* @ 108 cfu/mL or six releases of Blastothecus. pallescens @ 20 nymphs/

(continued)

Table 8.3 (continued)

Regions	Institutes	Salient achievements in area of biological control (2012–2017)
		m rows reduced the thrips, *Thrips tabaci* population in onion .
	Gujarat Agricultural University (GAU), Anand (Gujarat)	
Northern region	Dr. Y.S. Parmar University of Horticulture & Forestry (YSPUH & F), Nauni-Solan (Himachal Pradesh)	Soil application of *Beauveria bassiana*, *Metarhizium anisopliae*, *Heterorhabditis indica* and *Steinernema carpocapsae* resulted in low potato tuber damage by the white grubs (31.4–38.0%) as compared to control (59.2%).
	Sher-e-Kashmir University of Agricultural Sciences & Technology (SKUAS & T), Srinagar (Jammu & Kashmir)	Root dip treatment of tomato seedlings with *Paecilomyces lilacinus* @ 2.0×10^8 spores/L of water 15 min before transplantation significantly decreased the soil population of root-knot nematodes, *Meloidogyne hapla* by 85% and increased the yield up to 84%.
		Two sequential releases of *Trichogramma* spp. @ 2500–3000 adult wasps/tree and twice use of pheromone traps @ 4 traps/orchard effectively suppressed codling moth, *Cydia pomonella.*
	Punjab Agricultural University (PAU), Ludhiana (Punjab)	Application of botanical (Neem baan 1% @ 1250 and 1500 mL/ha) and biopesticides (*Lecanicillium lecanii* 2% AS and *Metarhizium anisopliae* 1% WP @ 1200 mL/ha) was effective in suppressing cotton whitefly.
	Indian Institute of Sugarcane Research (IISR), Lucknow (Uttar Pradesh)	A field trial was conducted at IISR research farm on sugarcane variety CoLk8102. The treatments were: (a) release of *Trichogramma chilonis* @ 50,000 ha^{-1} from July to October at 10 days interval, (b) release of *Cotesia flavipes* @ 500 gravid females/ha from July to November at 7 days interval release of *Tetrastichus howardi* @ 5000 adults/ha at monthly interval from July to November. *T. chilonis* release plots recorded lowest internode borer incidence (5.6%) compared to *C. flavipes* or *T. howardi* released fields, but *C. flavipes* release plots recorded lowest incidence of stalk borer (5.6%) followed by *T. howardi* and *T. chilonis*. Highest yield was recorded in *T. chilonis* release plots

(continued)

Table 8.3 (continued)

Regions	Institutes	Salient achievements in area of biological control (2012–2017)
	Indian Agricultural Research Institute (IARI), New Delhi (Delhi)	
	G.B. Pant University of Agriculture & Technology (GBPUA&T), Pantnagar (Uttarakhand)	The invert emulsion based *Trichoderma* formulation (IEF2) was found most effective in managing seed and seedling mortality in chickpea due to wilt disease.
		Trichoderma isolates, Th-14, Th-89, Th-82, TCMS 43, TCMS 9 and TCMS 36 were found most effective in reducing rice sheath blight by 60% and brown spot disease by 50% in rice.
		Trichoderma harzianum (Th-3) and *Pichiaguillier mondii* (Y-12) isolate was effective in reducing fruit rot incidence and increasing yield of Chilli.
		By the rhizosphere biology research group a total of 45 rhizobacterial isolates were isolated from rhizosphere of wheat across India and evaluated for their antagonistic activity against foliar blight complex. In vitro and glasshouse studies revealed that *Bacillus* isolates were significantly more effective in reducing the disease severity and could be exploited in a better way for crop health management.

developing national and international cooperation with societies and organizations engaged in similar activities (https://nbair.res.in/SBA/activities.html). The salient achievements of the society are as follows:

- Development and promotion of multiple insecticide-tolerant strain (MITS) and high-temperature-tolerant strain (HTTS) of *Trichogramma chilonis* and *Chrysoperla carnea* for the management of stem borers and sucking pests on various crops.
- Over 1 lakh insect specimens and 239 type specimens are housed in the National Repository at NBAIR. Eleven open-access databases of agricultural insects are hosted on NBAIR website. Two thousand nine hundred fifty-three identifications were provided. One thousand three hundred twenty-nine insects and their resources were molecularly characterized and DNA available in the Genomics Repository.
- Management of sugarcane woolly aphid, *Ceratovacuna lanigera*, and *rugose spiralling* whitefly through conservation biocontrol and classical biological control of eucalyptus gall wasp *Leptocybe invasa* and *papaya mealybug*, *Paracoccus marginatus*. Development of cost-effective liquid formulation of *Bacillus thuringiensis* for the management of pod borers.

8.7.4 Project Directorate of Biological Control (PDBC)

In 1993, Project Directorate of Biological Control (PDBC) was established. It is the nodal agency in India to undertake research on biological control of pests of agricultural importance. It coordinates 16 centers spread across the country, which forms a strong network for field studies on biological control of pests and weeds of crops such as sugarcane, rice, cotton, pulses, oilseeds, tobacco, coconut, fruits, and vegetables. PDBC has a team of experienced scientists and trainers in all the disciplines of biological control programs, viz., biological control of crop pests and diseases, mass production of biocontrol agents, control of plant parasitic nematodes, entomopathogenic bacteria, and insect viruses. They help to provide the technological backstop for the establishment of biological control agent production centers to government and private entrepreneurs (Kumar 2015). For example, an evaluation was done by the scientist team of PDBC in a farmer's field on a 45-day-old ratoon crop of sugarcane. There has been a severe damage by early shoot borer with the symptoms of dead hearts in this field. *T. chilonis* releases were initiated at higher dosages of 40,000 per acre, and totally 14 releases were made. The pest incidence, intensity, and infestation index were significantly lower in the treatment plots in comparison to those in the control plot, and the final yield was recorded as 34.7 tonnes/acre.

Apart from this, other research institutes such as the Indian Institute of Wheat and Barley Research (IIWBR) Karnal, India, are engaged in the detection of new virulence or pathotypes and mycotoxin analysis of rusts, powdery mildew, and Karnal bunt diseases. It is also involved in the screening for biocontrol agents and genetics of disease resistance of the disease pathogens. Salient achievements are as follows:

- Seed treatment with carboxin (75 WP @ 2.5 g/kg seed) or carbendazim (50 WP @ 2.5 g/kg seed) or tebuconazole (2DS @ 1.25 g/kg seed) or a combination of a reduced dosage of carboxin (75 WP @ 1.25 g/kg seed) and a bioagent fungus *Trichoderma viride* (@ 4 g/kg seed) is recommended for the integrated management of loose smut.
- Identified 80 new pathotypes of wheat and barley rusts and sources conferring resistance to new pathotypes. Monitored pathotypic variability in wheat and barley rusts in India and neighboring countries. Maintained national repository of 127 pathotypes of rust pathogens. DNA fingerprinted and sequenced three wheat rust pathogens.

8.8 Future Outlook

Biological control research has been conducted over the past many years. There are several examples of ill-conceived and poorly targeted biocontrol research worldwide. Therefore, biocontrol researchers need to look forward to facilitate new biocontrol technologies and applications. Currently, advancements in molecular biology, computing, analytical chemistry, and statistics have led to new research aimed at characterizing the structure and functions of biocontrol agents, pathogens, and host plants at the molecular, cellular, and ecological levels. There is a need to enhance the research criteria that will advance our understanding of biological control and the conditions under which it can be most fruitfully applied. Since fungal plant pathogens are very diverse and their pathogenicity is different to the host plants, therefore it is very important to look for new and novel biocontrol agents with different mechanism of biocontrol. In this regard, there is a need to investigate for the most potent microbes as biological control agents, study the roles of genes and gene products involved in pathogen suppression and application of combinations in comparison with individual agents, and study on the signal molecules of plant and microbial origin which regulate the expression of biocontrol traits. Nevertheless, for attaining success in the commercialization of BCAs, surely we have to convince funding agencies and, more importantly, researchers/scientists to direct research resources toward more fruitful and targeted areas which collectively will advance our chances of attaining more commercial success.

References

Aggarwal PK (2009) Global climate change and Indian agriculture—case studies from the ICAR Network Project. Indian Council of Agricultural Research, New Delhi

Agrios GN (2005) Plant pathology, 5th edn. Elsevier Academic Press, Amsterdam, p 922

Alam MA, Wang C, Ji W (2013) Chromosomal location and SSR markers of a powdery mildew resistance gene in common wheat line N0308. Afr J Microbiol Res 7(6):477–482

Ali A, Syed AAW, Khaliq T, Asif M, Aziz M, Mubeen M (2011) Effects of nitrogen on growth and yield components of wheat (report). Biol Sci 3(6):1004–1005

APHIS (1996) Karnal bunt: situation report update. USDA-APHIS, plant protection and quarantine. http://www.aphis.usda.gov/oa/bunt

Ardakani SS, Heydari A, Khorasani NA, Arjmandi R, Ehteshami M (2009) Preparation of new biofungicides using antagonistic bacteria and mineral compounds for controlling cotton seedling damping-off disease. J Plant Prot Res 49(1):49–55

Bakker AB, Phillips JH, Lanier LL (2002) U.S. patent no. 6,416,973. U.S. Patent and Trademark Office, Washington, DC

Bandyopadhyay R, Ortega-Beltran A, Akande A, Mutegi C, Atehnkeng J, Kaptoge L, Cotty PJ (2016) Biological control of aflatoxins in Africa: current status and potential challenges in the face of climate change. World Mycotoxin J 9(5):771–789

Barratt BIP, Mason PG, Cock MJW, Klapwijk J, Van Lenteren JC, Brodeur J, Heimpel GE (2017) Access and benefit sharing: best practices for the use and exchange of invertebrate biological control agents. In: Symposium on biological control of arthropods, p 71

Bellameche F, Pedrazzini C, Mauch-Mani B et al (2020) Efficiency of biological and chemical inducers for controlling Septoria tritici leaf blotch (STB) on wheat (Triticum aestivum L.). Eur J Plant Pathol 158:99–109. https://doi.org/10.1007/s10658-020-02057-y

Bhardwaj SC, Singh GP, Gangwar OP, Prasad P, Kumar S (2019) Status of wheat rust research and progress in rust management-Indian context. Agronomy 9(12):892

Birthal PS, Sharma OP (2004) Integrated pest management for sustainable agriculture. In: Integrated pest management in Indian agriculture. NCAP and NCIPM, New Delhi, p 11

Bottalico A, Perrone G (2002) Toxigenic Fusarium species and mycotoxins associated with head blight in small-grain cereals in Europe. In: Mycotoxins in plant disease. Springer, Dordrecht, pp 611–624

Bowen P, Menzies J, Ehret D, Samuels L, Glass AD (1992) Soluble silicon sprays inhibit powdery mildew development on grape leaves. J Am Soc Hortic Sci 117(6):906–912

Cai XC, Liu CH, Wang BT, Xue YR (2017) Genomic and metabolic traits endow *Bacillus velezensis* CC09 with a potential biocontrol agent in control of wheat powdery mildew disease. Microbiol Res 196:89–94

Callaway E (2016) Devastating wheat fungus appears in Asia for first time. Nature News 532 (7600):421

Ceresini PC, Castroagudín VL, Rodrigues FÁ, Rios JA, Aucique-Pérez CE, Moreira SI, Maciel JLN (2018) Wheat blast: past, present, and future. Annu Rev Phytopathol 56:427–456

Chakraborty S, Pangga IB (2004) Plant disease and climate change. In: Gillings M, Holmes A (eds) Plant microbiology. BIOS Scientific, London, pp 163–180

Chen W, Wellings C, Chen X, Kang Z, Liu T (2014) Wheat stripe (yellow) rust caused by *Puccinia striiformis f. sp. tritici*. Mol Plant Pathol 15(5):433–446

Choudhary DK, Johri BN (2009) Interactions of *Bacillus spp.* and plants—with special reference to induced systemic resistance (ISR). Microbiol Res 164(5):493–513

Crous PW, Van Jaarsveld AB, Castlebury LA, Carris LM, Frederick RD, Pretorius ZA (2001) Karnal bunt of wheat newly reported from the African continent. Plant Dis 85(5):561–561

Curtis TY, Raffan S, Wan Y, King R, Gonzalez-Uriarte A, Halford NG (2019) Contrasting gene expression patterns in grain of high and low asparagine wheat genotypes in response to sulphur supply. BMC Genomics 20(1):628

Da Luz WC, Mendes MAS, Ferreira MASV, Urben AF (1993) *Tilletia indica* on wheat in the south of the state of Rio Grande do Sul, Brazil and measures for eradication. Fitopatol Bras 18:S329

Das BC, Hazarika DK (2000) Biological management of sheath blight of rice. Indian Phytopathol 53(4):433–435

Dickson JG (1942) Scab of wheat and barley and its control. USDA farmers' bull. no. 1599 (revised), pp 1–22

Duran R (1972) Aspects of teliospore germination in north American smut fungi. II. Can J Bot 50 (12):2569–2574

Dutta S, Surovy MZ, Gupta DR, Mahmud NU, Chanclud E, Win J, Islam T (2018) Genomic analyses reveal that biocontrol of wheat blast by Bacillus spp. may be linked with production of antimicrobial compounds and induced systemic resistance in host plants. Figshare 17:48

Duveiller E, Kandel YR, Sharma RC, Shrestha SM (2005) Epidemiology of foliar blights (spot blotch and tan spot) of wheat in the plains bordering the Himalayas. Phytopathology 95 (3):248–256

Duveiller E, Singh RP, Singh PK, Dababat AA, Mezzalama M (2012) Wheat diseases and pests: a guide for field identification. CIMMYT, Mexico

Elad Y, Freeman S (2002) Biological control of fungal plant pathogens. In: Agricultural applications. Springer, Berlin, pp 93–109

Enghiad A, Ufer D, Countryman AM, Thilmany DD (2017) An overview of global wheat market fundamentals in an era of climate concerns. Int J Agron 2017:3931897

FAO, IFAD, UNICEF, WFP, WHO (2018) The state of food security and nutrition in the world 2018. Building climate resilience for food security and nutrition. Rome, FAO. Licence: CC BY-NC-SA 3.0 IGO

Flaishman MA, Eyal Z, Zilberstein A, Voisard C, Haas D (1996) Suppression of Septoria tritici blotch and leaf rust of wheat by recombinant cyanide-producing strains of *Pseudomonas putida*. Mol Plant-Microbe Interact 9(7):642–645

Fravel DR (2005) Commercialization and implementation of biocontrol. Annu Rev Phytopathol 43:337–359

Friebe B, Jiang J, Raupp WJ, McIntosh RA, Gill BS (1996) Characterization of wheat-alien translocations conferring resistance to diseases and pests: current status. Euphytica 91(1):59–87

Gao X, Gong Y, Huo Y, Han Q, Kang Z, Huang L (2015) Endophytic *Bacillus subtilis* strain E1R-J is a promising biocontrol agent for wheat powdery mildew. Biomed Res Int 2015:462645

Garrett KA, Zúñiga LN, Roncal E, Forbes GA, Mundt CC, Su Z, Nelson RJ (2009) Intraspecific functional diversity in hosts and its effect on disease risk across a climatic gradient. Ecol Appl 19(7):1868–1883

Gautam RK, Singh PK, Sakthivel K, Srikumar M, Kumar N, Kumar K, Roy SD (2015) Analysis of pathogenic diversity of the rice bacterial blight pathogen (*Xanthomonas oryzae pv. oryzae*) in the Andaman Islands and identification of effective resistance genes. J Phytopathol 163 (6):423–432

Gessese M, Bariana H, Wong D, Hayden M, Bansal U (2019) Molecular mapping of stripe rust resistance gene Yr81 in a common wheat landrace Aus27430. Plant Dis 103(6):1166–1171

Glandorf DC, Verheggen P, Jansen T, Jorritsma JW, Smit E, Leeflang P, Bakker PA (2001) Effect of genetically modified *Pseudomonas putida* WCS358r on the fungal rhizosphere microflora of field-grown wheat. Appl Environ Microbiol 67(8):3371–3378

Gomaa EZ (2012) Chitinase production by *Bacillus thuringiensis* and *Bacillus licheniformis*: their potential in antifungal biocontrol. Res J Microbiol 50(1):103–111

Gunupuru LR, Perochon A, Doohan FM (2017) Deoxynivalenol resistance as a component of FHB resistance. Trop Plant Pathol 42(3):175–183

Gupta PK, Chand R, Vasistha NK, Pandey SP, Kumar U, Mishra VK, Joshi AK (2018) Spot blotch disease of wheat: the current status of research on genetics and breeding. Plant Pathol 67 (3):508–531

Heydari A, Pessarakli M (2010) A review on biological control of fungal plant pathogens using microbial antagonists. J Biol Sci 10(4):273–290

Hirpara DG, Gajera HP, Hirpara HZ, Golakiya BA (2017) Antipathy of *Trichoderma* against *Sclerotium rolfsii* Sacc.: evaluation of cell wall-degrading enzymatic activities and molecular diversity analysis of antagonists. J Mol Microbiol Biotechnol 27(1):22–28

Hossain MM, Hossain I, Khalequzzaman KM (2015) Effect of seed treatment with biological control agent against bipolaris leaf blight of wheat. Int J Sci Res Agric Sci 2(7):151–158

Huang XQ, Kempf H, Ganal MW, Röder MS (2004) Advanced backcross QTL analysis in progenies derived from a cross between a German elite winter wheat variety and a synthetic wheat (*Triticum aestivum* L.). Theor Appl Genet 109(5):933–943

Iftikhar S, Asad S, Anjum M (2010) Incidence of *Bipolaris sorokiniana* in Punjab and Khyber pakhtoon khawa, Pakistan. Pak J Phytopathol 22(2):95–97

Jacobsen BJ, Zidack NK, Larson BJ (2004) The role of Bacillus-based biological control agents in integrated pest management systems: plant diseases. Phytopathology 94(11):1272–1275

Jamalizadeh M, Etebarian HR, Aminian H, Alizadeh A (2011) A review of mechanisms of action of biological control organisms against post-harvest fruit spoilage. EPPO Bull 41(1):65–71

Joshi AK, Mishra B, Chatrath R, Ortiz Ferrara G, Singh RP (2007) Wheat improvement in India: present status, emerging challenges and future prospects. Euphytica 157:431–446

Junaid JM, Dar NA, Bhat TA, Bhat AH, Bhat MA (2013) Commercial biocontrol agents and their mechanism of action in the management of plant pathogens. Int J Modern Plant Anim Sci 1 (2):39–57

Kageyama K, Nelson EB (2003) Differential inactivation of seed exudates stimulation of *Pythium ultimum* sporangium germination by *Enterobacter cloacae* influences biological control efficacy on different plant species. Appl Environ Microbiol 69:1114–1120

Kang Y, Zhou M, Merry A, Barry K (2020) Mechanisms of powdery mildew resistance of wheat—a review of molecular breeding. Plant Pathol 69(4):601–617

Kannojia P, Choudhary KK, Srivastava AK, Singh AK (2019) PGPR bioelicitors: induced systemic resistance (ISR) and proteomic perspective on biocontrol. In: PGPR amelioration in sustainable agriculture. Woodhead Publishing, Sawston, pp 67–84

Keswani C, Prakash O, Bharti N, Vílchez JI, Sansinenea E, Lally RD, de Lima R (2019) Re-addressing the biosafety issues of plant growth promoting rhizobacteria. Sci Total Environ 690:841–852

Kloepper JW, Leong J, Teintze M, Schroth MN (1980) *Pseudomonas* siderophores: a mechanism explaining disease-suppressive soils. Curr Microbiol 4(5):317–320

Köhl J, Kolnaar R, Ravensberg WJ (2019) Mode of action of microbial biological control agents against plant diseases: relevance beyond efficacy. Front Plant Sci 10:845

Krivoshchekova TG, Mishchenk VS (1990) The effectiveness of trichodermin Zashchita Rastenii (Moskova) no 11, 22

Kulkarni GS (1924) Report of work done in the plant pathology section during the year 1922–23. CABI, Wallingford

Kumar S (2015) History, progress and prospects of classical biological control in India. Indian J Weed Sci 47(3):306–320

Langridge P (2017) Achieving sustainable cultivation of wheat. Volume 1: breeding, quality traits, pests and diseases. Burleigh Dodds Science Publishing Limited, Sawston

Law JWF, Ser HL, Khan TM, Chuah LH, Pusparajah P, Chan KG, Lee LH (2017) The potential of *Streptomyces* as biocontrol agents against the rice blast fungus, *Magnaporthe oryzae* (*Pyricularia oryzae*). Front Microbiol 8:3

Li A, Zhang R, Pan L, Tang L, Zhao G, Zhu M, Jia J (2011) Transcriptome analysis of H_2O_2-treated wheat seedlings reveals a H_2O_2-responsive fatty acid desaturase gene participating in powdery mildew resistance. PLoS One 6(12):e28810

Li H, Zhao J, Feng H, Huang L, Kang Z (2013) Biological control of wheat stripe rust by an endophytic *Bacillus subtilis* strain E1R-j in greenhouse and field trials. Crop Prot 43:201–206

Liu B, Qiao H, Huang L et al (2009) Biological control of take-all in wheat by endophytic *Bacillus subtilis* E1R-j and potential mode of action. Biol Control 49(3):277–285

Malaker PK, Barma NC, Tiwary TP, Collis WJ, Duveiller EP, Singh K, Pedley KF (2016) First report of wheat blast caused by *Magnaporthe oryzae pathotype triticum* in Bangladesh. American Phytopathological Society (APS), Saint Paul

Maraite H, Bragard C, Duveiller E (2007) The status of resistance to bacterial diseases of wheat. In: Wheat production in stressed environments. Springer, Dordrecht, pp 37–49

Matzen N, Heick TM, Jørgensen LN (2019) Control of powdery mildew (Blumeria graminis spp.) in cereals by Serenade® ASO (*Bacillus amyloliquefaciens* (former *subtilis*) strain QST 713). Biol Control 139:104067

Maurya S (2020) Biological control a sustainable approach for plant diseases management: a review. J Pharmacogn Phytochem 9(2):1514–1523

Mavrodi OV, Walter N, Elateek S, Taylor CG, Okubara PA (2012) Suppression of *Rhizoctonia* and pythium root rot of wheat by new strains of *Pseudomonas*. Biol Control 62(2):93–102

McMullen MP, Schatz B, Stover R, Gregoire T (1997) Studies of fungicide efficacy, application timing, and application technologies to reduce Fusarium head blight and deoxynivalenol. Cereal Res Commun 25(3):779–780

Mehta K (1940) Farther studies on cereal rusts in India. CABI, Wallingford

Milner JL, Silo-Suh LAURA, Lee JC, He H, Clardy J, Handelsman JO (1996) Production of kanosamine by *Bacillus cereus* UW85. Appl Environ Microbiol 62(8):3061–3065

Mitra M (1931) A new bunt of wheat in India. Ann Appl Biol 18:178–179

Mukhopadhyay AN, Shrestha SM, Mukherjee PK (1992) Biological seed treatment for control of soil-borne plant pathogens. Boletin Fitosanitario de la FAO (FAO) 40:21–30

Munjal RL (1975) Status of karnal bunt (*Neovossia indica*) of wheat in northern India during 1968-69 and 1969-70. Indian J Mycol Plant Pathol 5:185–187

Nagarajan S, Kumar J (1998) An overview of the increasing importance of research of foliar blights of wheat in India: germplasm improvement and future challenges towards a sustainable high yielding wheat production. Helminthosporium blight of wheat: spot blotch and tan spot. CIMMYT, Mexico, pp 52–58

Neilands JB (1981) Iron absorption and transport in microorganisms. Annu Rev Nutr 1(1):27–46

Newitt JT, Prudence SM, Hutchings MI, Worsley SF (2019) Biocontrol of cereal crop diseases using streptomycetes. Pathogens 8(2):78

Nikos A, Jelle B (2012) World agriculture towards 2030/2050: the 2012 revision. FAO Agricultural Development Economics Division, Rome

Nowak J, Shulaev J (2003) Priming for transplant stress resistance in in vitro propagation. In Vitro Cell Dev Biol Plant 39:107–124. https://doi.org/10.1079/IVP2002403

Pang F, Wang T, Zhao C, Tao A, Yu Z, Huang S, Yu G (2016) Novel bacterial endophytes isolated from winter wheat plants as biocontrol agent against stripe rust of wheat. Biol Control 61 (2):207–219

Park K, Park HJ, Jeun YC, Kim CH, Kloepper JW (2000) Induced systemic resistance against anthracnose in cucumber plant by a selected PGPR, *Bacillus amyloliquefaciens* and its mechanism of action. In: Proceedings of the fifth international PGPR workshop, Cordoba, Argentina. http://www.ag.auburn.edu/argentina

Peleg Z, Fahima T, Korol AB, Abbo S, Saranga Y (2011) Genetic analysis of wheat domestication and evolution under domestication. J Exp Bot 62(14):5051–5061

Pieterse CM, Zamioudis C, Berendsen RL, Weller DM, Van Wees SC, Bakker PA (2014) Induced systemic resistance by beneficial microbes. Annu Rev Phytopathol 52:347–375

Prasad KSK, Kulkarni S, Siddaramaiah AL (1978) Antagonistic action of *Trichoderma sp.* and *Streptomyces sp.* on *Drechslera sativum* (*Helminthosporium sativum* Pam, King and Bakke) Subram. and Jain. Curr Res 12:202–203

Preininger C, Sauer U, Bejarano A, Berninger T (2018) Concepts and applications of foliar spray for microbial inoculants. Appl Microbiol Biotechnol 102(17):7265–7282

Priestley RH, Bayles RA (1988) Contribution and value of resistant cultivars to disease control in cereals. In: Clifford BC, Laster E (eds) Control of plant diseases: costs and benefits. Blackwell Scientific Publications, Oxford

Puyam A (2016) Advent of Trichoderma as a bio-control agent—a review. J Nat Appl Sci 8 (2):1100–1109

Ramadas S, Kumar TK, Singh GP (2019) Wheat production in India: trends and prospects. In: Global wheat production. IntechOpen, London

Ramamoorthy V, Viswanathan R, Raguchander T, Prakasam V, Samiyappan R (2001) Induction of systemic resistance by plant growth promoting rhizobacteria in crop plants against pests and diseases. Crop Prot 20(1):1–11

Raymaekers K, Ponet L, Holtappels D, Berckmans B, Cammue BP (2020) Screening for novel biocontrol agents applicable in plant disease management—a review. Biol Control 144:104240

Raza MA, Feng LY, van der Wer W, Iqbal N, Khan I, Hassan MJ, Ahmed M (2019) Optimum leaf defoliation: a new agronomic approach for increasing nutrient uptake and land equivalent ratio of maize soybean relay intercropping system. Field Crops Res 244:107647

Rocha IDS, Ma Y, Souza-Alonso P, Vosátka M, Freitas H, Oliveira RS (2019) Seed coating: a tool for delivering beneficial microbes to agricultural crops. Front Plant Sci 10:1357

Roelfs AP (1992) Rust diseases of wheat: concepts and methods of disease management. CIMMYT, Mexico

Royse DJ, Gregory LV, Ayers JE, Cole H Jr (1980) Powdery mildew of wheat: relation of yield components to disease severity. Can J Plant Pathol 2(3):131–136

Rush CM, Stein JM, Bowden RL, Riemenschneider R, Boratynski T, Royer MH (2005) Status of Karnal bunt of wheat in the United States 1996 to 2004. Plant Dis 89(3):212–223

Saharan MS, Naef A (2008) Detection of genetic variation among Indian wheat head scab pathogens (*Fusarium spp.*/isolates) with microsatellite markers. Crop Prot 27(7):1148–1154

Salim N, Raza A (2020) Nutrient use efficiency (NUE) for sustainable wheat production: a review. J Plant Nutr 43(2):297–315

Serfling A, Kopahnke D, Habekuss A, Novakazi F, Ordon F (2016) Wheat diseases: an overview. In: Langridge P (ed) Achieving sustainable cultivation of wheat, vol 1. Burleigh Dodds Science Publishing, pp 319–326

Shah DA, De Wolf ED, Paul PA, Madden LV (2014) Predicting Fusarium head blight epidemics with boosted regression trees. Phytopathology 104(7):702–714

Sharma M, Tarafdar A, Ghosh R, Gopalakrishanan S (2017) Biological control as a tool for eco-friendly management of plant pathogens. In: Advances in soil microbiology: recent trends and future prospects. Springer, Singapore, pp 153–188

Shi AN, Leath S, Murphy JP (1998) A major gene for powdery mildew resistance transferred to common wheat from wild einkorn wheat. Phytopathology 88(2):144–147

Shiferaw B, Smale M, Braun HJ, Duveiller E, Reynolds M, Muricho G (2013) Crops that feed the world 10. Past successes and future challenges to the role played by wheat in global food security. Food Secur 5(3):291–317

Silo-Suh LA, Lethbridge BJ, Raffel SJ, He H, Clardy J, Handelsman J (1994) Biological activities of two fungistatic antibiotics produced by *Bacillus cereus* UW85. Appl Environ Microbiol 60:2023–2030

Singh G, Rajaram S, Montoya J, Fuentes-Davila G (1995) Genetic analysis of karnal bunt resistance in 14 Mexican bread-wheat genotypes. Plant Breed 114(5):439–441

Singh JS, Koushal S, Kumar A, Vimal SR, Gupta VK (2016) Book review: microbial inoculants in sustainable agricultural productivity-Vol. II: functional application. Front Microbiol 7:2105

Spanu PD, Panstruga R (2017) Biotrophic plant-microbe interactions. Front Plant Sci 8:192

Sztejnberg A, Galper S, Mazar S, Lisker N (1989) *Ampelomyces quisqualis* for biological and integrated control of powdery mildews in Israel. J Phytopathol 124(4):285–295

Teli B, Chattopadhyay A, Meena SC, Gangwar GP, Pandey SK (2016) Present status of Fusarium head blight of wheat and barley in India. In: Diseases of wheat and their management. Astral International, NewDelhi, pp 79–92

Thomashow L, Weller DM, Bonsall RF, Pierson LS (1990) Production of the antibiotic phenazine-1-carboxylic acid by fluorescent Pseudomonas species in the rhizosphere of wheat. Appl Environ Microbiol 56:908–912. https://doi.org/10.1128/AEM.56.4.908-912.1990

Torarbi M, Mardoukhi V, Jaliani N (1996) First report on the occurrence of partial bunt on wheat in the Southern parts of Iran. Seed Plant 12:8–9 (English), 59–60 (Persian)

Ullah HMZ, Haque MI, Rauf CA, Akhtar LH, Munir M (2012) Comparative virulence in isolates of *Tilletia indica* and host resistance against karnal bunt of wheat. J Anim Plant Sci 22(2):467–472

Urashima AS, Igarashi S, Kato H (1993) Host range, mating type, and fertility of *Pyricularia grisea* from wheat in Brazil. Plant Dis 77:1211–1216

Valencia-Botin AJ, Cisneros-Lopez ME (2012) A review of the studies and interactions of Pseudomonas syringae pathovars on wheat. Int J Agron 2012:692350, 5p. https://doi.org/10.1155/2012/692350

Van Ginkel M, Rajaram S (1998) Breeding for resistance to spot blotch in wheat: global perspective. In: Helminthosporium blights of wheat: spot blotch and tan spot. CIMMYT, Mexico, pp 162–170

Venkataramana M, Navya K, Chandranayaka S, Priyanka SR, Murali HS, Batra HV (2014) Development and validation of an immunochromatographic assay for rapid detection of fumonisin B_1 from cereal samples. J Food Sci Technol 51:1920–1928

Viani A, Sinha P, Sharma T, Bhar LM (2017) A model for forecasting spot blotch disease in wheat. Australas Plant Pathol 46(6):601–609

Wang H, Hwang SF, Chang KF, Turnbull GD, Howard RJ (2003) Suppression of important pea diseases by bacterial antagonists. BioControl 48(4):447–460

Wang J, Fang R, Jiang QY, Xiao M (2012) Effects of carrier and protective agent on the biological activities of *Pseudomonas aurantiaca* JD37 strain microbial fertilizer. J Shanghai Norm Univ (Nat Sci) 41(2):179–185

Weller DM, Landa BB, Mavrodi OV, Schroeder KL, De La Fuente L, Blouin Bankhead S, Thomashow LS (2007) Role of 2,4-diacetylphloroglucinol-producing fluorescent *Pseudomonas spp.* in the defense of plant roots. Plant Biol 9(1):4–20

Whipps JM (2001) Microbial interactions and biocontrol in the rhizosphere. J Exp Bot 52:487–511

Yi X, Cheng J, Jiang Z, Hu W, Bie T, Gao D, Cheng X (2018) Genetic analysis of Fusarium head blight resistance in CIMMYT bread wheat line C615 using traditional and conditional QTL mapping. Front Plant Sci 9:573

Zayan SA (2019) Impact of climate change on plant diseases and IPM strategies. In: Plant pathology and management of plant diseases. IntechOpen, London

Chapter 9
Phytohormones as Fundamental Regulators of Plant–Microbe Associations Under Stress Conditions

Khushboo Choudhary, V. Vivekanand, and Nidhi Pareek

Abstract Stress conditions, be they abiotic or biotic, have detrimental impacts on agricultural yields. They also slow down bioremediation and lead to changes in ecosystems. These effects are primarily caused by rapid climate change due to various different factors and activities. To adapt to climate change conditions, plants have developed complex physiological and molecular mechanisms to prevent disaster. Phytohormones produced by root-associated microbes are essential for plant growth and also contribute to stimulation of plant tolerance of various stresses. Hormones act either by activating secondary messengers or via phosphorylation cascades involved in gene regulation. The roles of microbes under various types of environmental stress can be appreciated with a particular focus on production of phytohormones and their associations with host plants. Moreover, they also contribute to tolerance of biotic stresses such as pathogenic organisms via activation of induced systemic resistance and systemic acquired resistance mechanisms in plants. The combination of plants, plant growth–promoting microbes and phytohormones represents a tripartite consortium to provide a suitable environment for the spread of beneficial microbes, which, in turn, enhance plant growth. However, the association of such microbes with plants for management of stresses in agricultural systems still needs to be explored in greater depth.

Keywords Microorganisms · Phytohormones · Abiotic stress · Biotic stress

---

K. Choudhary · N. Pareek (✉)
Department of Microbiology, School of Life Sciences, Central University of Rajasthan, Ajmer, Rajasthan, India
e-mail: nidhipareek@curaj.ac.in

V. Vivekanand
Centre for Energy and Environment, Malaviya National Institute of Technology, Jaipur, Rajasthan, India

S. K. Dubey, S. K. Verma (eds.), *Plant, Soil and Microbes in Tropical Ecosystems*, Rhizosphere Biology, https://doi.org/10.1007/978-981-16-3364-5_9

9.1 Introduction

Plant species throughout the world are greatly affected by biological stresses (both biotic and abiotic) and by anthropogenic activities, preventing plants from reaching their full capacity for growth and production (Ogbe et al. 2020). An important step in plant defence is timely perception of stress conditions so they can be responded to quickly and efficiently. After detection, the constitutive basal defence mechanisms of plants lead to activation of complex signalling cascades that protect the plants from different stresses (Pandey et al. 2017).

Plants have developed very complex immune systems that enable them, as individual organisms, to tolerate not only individual stresses but also combinations of stresses. In plants, biotic and abiotic stresses prompt a broad range of defence responses at the molecular and cellular levels (Nejat and Mantri 2017). Better understanding of various tolerance strategies to maintain agriculture productivity by manipulation of environmental conditions can be helpful for exploiting the maximum genetic potential of crops (Egamberdieva et al. 2017).

Phytohormones are important growth regulators in specific plant organs. They have major effects on plant metabolism and play important roles in stress mitigation (Kazan 2013; Egamberdieva et al. 2017). Under conditions of biotic and abiotic stress, phytohormones control the allocation of resources to combat the most severe stress and activate several signalling pathways to control the balance of plant growth and defence responses (Yang et al. 2019). It is important to understand the similarities and differences in phytohormone signalling in agriculture production.

Phytohormones are a group of small quantities of growth regulators and signalling molecules, including abscisic acid (ABA), gibberellins (GAs), ethylene (ET), cytokinins (CKs), jasmonic acid (JA), auxins (AUXs), brassinosteroids (BRs), strigolactones (SLs) and salicylic acid (SA) (Kazan 2015). Some phytohormones—such as salicylic acid, ethylene, abscisic acid and jasmonates—are known for their positive roles in providing stress tolerance in plants (Pieterse et al. 2012). Salicylic acid, ethylene, abscisic acid and jasmonates are involved in crosstalk of auxins, gibberellin and cytokinin for regulation of plant defence response mechanisms (Nishiyama et al. 2013). It is essential to understand the complex communication of crosstalk between phytohormones (Khan et al. 2020).

Plant growth–promoting microbes can play beneficial roles, protecting plants from potential pathogens and providing adaptive benefits to plants, along with improving growth, health and production. Microbiomes are composed of many different types of microorganisms, viz. fungi, bacteria, archaea, protozoa and viruses (Mueller and Sachs 2015). Microbes modulate hormones level in plant tissues, and they have been found to have effects similar to those of exogenous phytohormone applications (Shahzad et al. 2016).

This chapter, based on the available literature on the effects of phytohormones on plant tolerance, seeks to improve understanding of microbial phytohormones and the impacts of their interactions with plants by defining their effects on plant morphological and physiological properties. The focus here is on plant-associated microbes, their physiology, their diversity and their involvement in plant tolerance of biotic and abiotic stresses.

9.2 Roles of Various Phytohormones in Plant Tolerance of Stresses

Various phytohormones are involved in plant tolerance of different types of stress (Tables 9.1 and 9.2; Figs. 9.1 and 9.2).

9.2.1 Cytokinins

Cytokinins are a very important group of phytohormones and are involved in many activities in plant growth and development, such as shoot and root meristem activity, regulation of organ size and development, shoot and root branching, and control of leaf senescence (Cortleven et al. 2019). Under conditions of water stress, especially in the grain-filling phase, it was observed that the 'stay-green' genotype has the potential to exhibit increased tolerance (Egamberdieva et al. 2017). It was shown that cytokinins enhanced tolerance of drought in transgenic cassava compared with that in wild type plants. Genes involved in biosynthesis of cytokinins are overexpressed, and their role in stress tolerance has been decoded (Zhang et al. 2010).

Exogenous application of cytokinins has been used to optimize internal cytokinin concentrations. It has also been documented that heavy metals, such as zinc and lead, severely inhibit seedling growth in chickpea through inhibition of gibberellic acid (GA_3) concentrations (Mohapatra et al. 2011). In one study, application of kinetin to chickpea stimulated plant growth and development under salt stress, and in another study, kinetin alleviated cadmium stress in eggplant by enhancing its antioxidant potential (Egamberdieva et al. 2017).

High cytokinin levels in plants increase resistance to pathogens, including fungi, bacteria and pest insects; the same is true of plant susceptibility to disease (Akhtar et al. 2019). The role of cytokinins in interactions with insects has been known for decades, and the discovery of cytokinin-mediated resistance to microbial pathogens in *Arabidopsis* and tobacco has been extended to other species (Dowd et al. 2017; Akhtar et al. 2019). There is experimental support for a possible dual role of fungi in modulating host immunity and optimizing nutrient supply (Akhtar et al. 2019). Similarly, bacteria cause cytokinin-induced resistance to bacterial pathogens in *Arabidopsis* (Großkinsky et al. 2016).

Table 9.1 Phytohormone-producing microbes and their actions against stress conditions in plants

Microbes	Stresses	Plants	Phytohormones	References
Bacillus licheniformis	Salinity stress	*Triticum aestivum*	Indole-3-acetic acid	Singh and Jha (2016)
Staphylococcus arlettae	Chromium (heavy metal) stress	*Helianthus annuus*	Indole-3-acetic acid, gibberellic acid, salicylic acid	Qadir et al. (2020)
Bacillus cereus, Bacillus megaterium, Trichoderma longibrachiatum, Trichoderma simmonsii	Drought and salt stress	*Glycine max*	Indole-3-acetic acid	Bakhshandeh et al. (2020)
Bacillus strains	Salinity stress	*Pennisetum glaucum*	Indole-3-acetic acid	Kushwaha et al. (2020)
Porostereum spadiceum	Salinity stress	*Glycine max*	Gibberellic acid	Hamayun et al. (2017)
Pseudomonas fluorescens	Water stress	*Vitis vinifera*	Abscisic acid	Salomon et al. (2014)
Arthrobacter woluwensis	Salinity stress	*Glycine max*	Abscisic acid, gibberellic acid, indole-3-acetic acid, jasmonic acid	Khan et al. (2019)
Micrococcus luteus	Drought stress	*Zea mays*	Cytokinin	Raza and Faisal (2013)
Sinorhizobium meliloti	Salinity stress	*Medicago sativa*	Indole-3-acetic acid, cytokinin	Provorov et al. (2016)
Serratia marcescens	Salinity stress	*Zea mays*	Salicylic acid	Lavania and Nautiyal (2013)
Gluconacetobacter diazotrophicus	Drought stress	*Oryza sativa*	Indole-3-acetic acid	Silva et al. (2020)
Bacillus aryabhattai	Heat stress	*Glycine max*	Indole-3-acetic acid, abscisic acid, gibberellic acid	Park et al. (2017)
Enterobacter sp.	Metal stress	*Hibiscus cannabinus*	Indole-3-acetic acid	Chen et al. (2017)
Rhizophagus irregularis	Drought, cold and salinity stress	*Digitaria eriantha*	Jasmonic acid	Pedranzani et al. (2016)

9.2.2 *Auxins*

Auxins are crucial phytohormones. They promote multiple growth and development events, such as elongation, cell division and differentiation (Asgher et al. 2015). Ljung (2013) described various modulations in synthesis transport metabolism and activity of auxins after plant exposure to stresses.

Table 9.2 Genetic delude of plant hormones from various transgenic plant origins and their roles in stress tolerance by plants

Phytohormones	Associated genes	Function	Function in plants	References
Abscisic acid	*LOS5*	Regulation of abscisic acid biosynthesis	Increased abscisic acid levels in transgenic *Zea mays*	Wani et al. (2016)
Cytokinin	*CKX*	Cytokinin inactivation	Drought resistance in *Arabidopsis thaliana*	Werner et al. (2010)
Ethylene	ACC synthase gene	Catalysis of the rate-limiting step in ethylene biosynthesis	Reduced ethylene levels with good drought resistance in *Zea mays*	Habben et al. (2014)
Indole-3-acetic acid (auxin)	*YUCCA6*	Indole-3-acetic acid biosynthesis	Drought resistance	Ke et al. (2015)
Brassinosteroids	*AtHSD1*	Brassinosteroid biosynthesis	Salinity resistance, enhanced growth and development	Tiwari et al. (2020)
Abscisic acid	*NCED*	Abscisic acid biosynthesis for feedback control	Stomatal conductance, enhanced drought resistance	Wani et al. (2016), Estrada-Melo et al. (2015)

ACC 1-aminocyclopropane-1-carboxylase

Auxins play important roles, directly or indirectly, in promoting heavy metal tolerance. Heavy metals have a negative effect on biosynthesis of auxins (Hu et al. 2013). The toxic effect of lead on sunflower plant growth was minimized by addition of a low concentration of auxin, which stimulated an increase in root volume, surface area and diameter (Fässler et al. 2010).

Plants are exposed to many different microbes around them, with disease being the exception rather than the rule for plants. The occurrence of disease is relatively infrequent because plants are able to detect potential pathogens in their vicinity and induce a basal host defence that prevents most environmental microbes from colonizing them and causing disease (Kunkel and Harper 2018).

Auxins plays important roles in numerous plant–microbe associations. Several plant-associated microbes—nitrogen-fixing symbionts, plant growth–promoting rhizobacteria (PGPRs), pathogens etc.—produce auxin hormones (Yin et al. 2014). When grown in a culture medium, some plant pathogenic bacteria (such as *Pseudomonas savastanoi*, *Pantoea agglomerans*, *Dickeya* sp. and *Xanthomonas campestris*) produce auxins (McClerklin et al. 2018; Kunkel and Harper 2018). Enhancement of the auxin stratum in contagious host tissue prompts a number of different processes associated with pathogenesis, such as inhibition of host protection, epiphytic colonization, stimulation of host cell division and pathogen development in plant tissue (Kazan and Lyons 2014). In many cases, the pathogen itself produces auxin, and auxin can be seen as a virulence factor in this interaction. However, in other interactions, the pathogen stimulates auxin accumulation or auxin

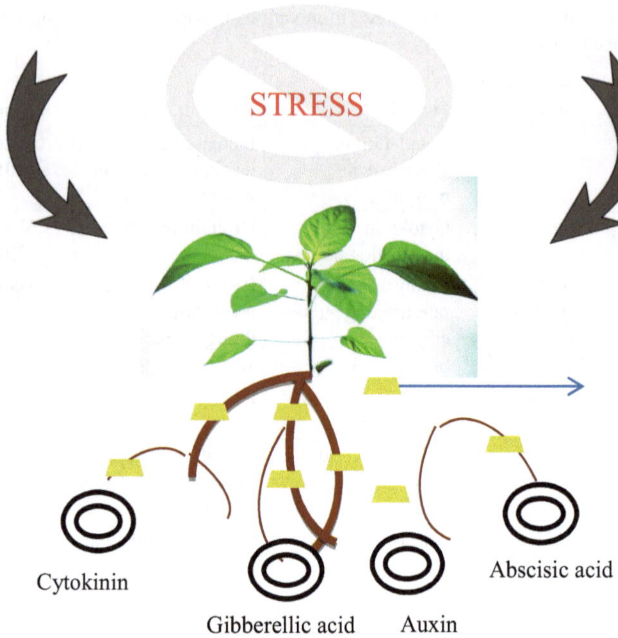

Fig. 9.1 Mechanisms of microbial phytohormone–mediated plant stress tolerance. Various root-associated microbes produce several phytohormones, which help plants to withstand stress by enhancing their antioxidant potential

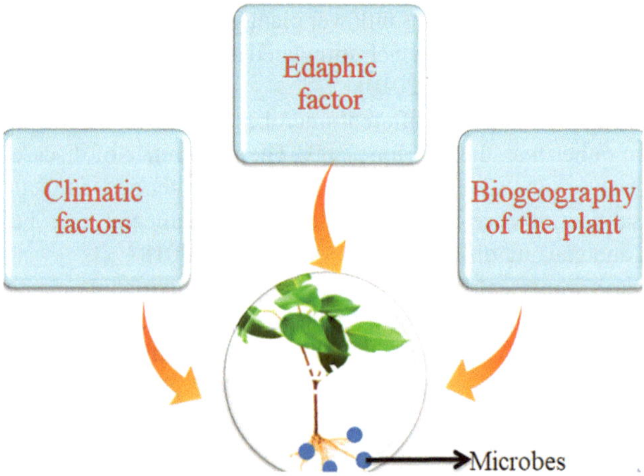

Fig. 9.2 Factors affecting plant-associated microorganisms

signalling in the host, which has evolved to modulate host auxin biology through the action of a viral factor (Kunkel and Harper 2018).

9.2.3 *Abscisic Acid*

Like other phytohormones, abscisic acid is known to play a crucial role in plants by improving stress impedance and adaptation. It is a naturally occurring member of the sesquiterpenoids, a group of major phytohormones involved in regulation of development. Several reports have described the role of abscisic acid in integrating signalling during stress exposure with subsequent control of downstream responses (Wilkinson et al. 2012). Stress response gene regulation through abscisic acid promotes and regulates signalling under abiotic stresses (Sah et al. 2016).

Abscisic acid has been described as controlling root development and water content under drought stress conditions (Cutler et al. 2010). However, during stress, a sudden increase in abscisic acid concentrations can cause growth retardation and modulate tolerance responses to stress. Even so, there is information indicating a useful effect of abscisic acid in countering the side effects of stresses, including cold stress, chilling, salinity and drought stress (Egamberdieva et al. 2017).

Exogenous utilization of abscisic acid under drought stress conditions to promote the activities of antioxidants to ameliorate stress passivity has been proposed as an effective tool for stress mitigation (Bano et al. 2012). Exogenous application of abscisic acid under drought stress conditions to improve carbon metabolism, stress tolerance and protein transport was found to significantly affect the proteome of tea plants (Zhou et al. 2014).

The major roles of abscisic acid in plant protection against pathogenic microbes are multifaceted. Abscisic acid–induced stomatal closing by regulation of guard cell ion flux in response to pathogenic attacks is important in preventing penetration of bacterial pathogens through the foramen (Lu and Yao 2018). The main components of abscisic acid–mediated stomatal function (immunity) are the serine protein kinase Open Stomata 1 (OST1), the regulatory component of the abscisic acid receptor and 2C-type protein phosphatase (Lim et al. 2015). A flagellin peptide from *Pseudomonas syringae*, a member of the pathogen-associated molecular patterns (PAMPs), induced stomatal closure through stimulation of SLAC1/SLAH3 in guard cells in an OST1-dependent manner (Chen et al. 2020). Su et al. (2017) reported that MKK4/5-MPK3 is an interdependent function in the organic acid metabolism cascade that mediates stomatal function (immunity) with abscisic acid.

9.2.4 *Gibberellic Acid*

Gibberellins are important plant development regulators and part of a large family of tetracyclic diterpenoids, which play vital roles in aspects such as lateral shoot growth, seed dormancy and establishment of floral organs (Olszewski et al. 2002). Khan et al. (2004) observed increased fruit production, leaflet development, and potassium, nitrogen and phosphorus levels in tomatoes as a result of exogenous application of gibberellic acid.

Gibberellic acid was found to trigger plant development under several different types of abiotic stress, such as salinity, drought and cold (Ahmad 2010). Increased plant water levels and reduced stomatal resistance were observed in gibberellic acid–treated tomato plants grown in saline stress conditions. Gibberellic acid influenced uptake and partitioning of ions in roots and shoots, promoting growth and maintaining plant metabolism under ordinary and stress conditions (Maggio et al. 2010; Iqbal and Ashraf 2013). An increase in osmatic components was observed in plants exposed to salt stress, and their content was further increased through gibberellin acid treatment. Endogenous use of gibberellin influenced osmatic stress in plants and preservation of tissue water content (Egamberdieva et al. 2017).

In *Arabidopsis thaliana*, gibberellic acid enhanced resistance to the bacteria *Pseudomonas syringae* and conferred disease immunity to the fungus *Alternaria brassicola* (Yimer et al. 2018). Softening during storage and development of *Alternaria* black spot disease, caused by *Alternaria alternata*, are the main post-harvest factors that reduce the storability and quality of *Diospyros* fruit. Pre-harvest application of gibberellic acid significantly enhanced fruit storage, as evaluated through fruit preservation and levels of *Alternaria* black spot (Maurer et al. 2019).

9.3 Plant-Associated Microbes

In the environment, vigorous and healthy plants live in association with various plant microbes consisting of all types of microorganisms—including fungi, archaebacteria, bacteria and protists—which create complex microbial consortia and influence plant development, health and productivity (Hassani et al. 2018). These microbes are present on the surfaces of leaves, sprouted seeds, roots and fruits, or they live inside the plants (Hardoim et al. 2015). Plants have developed their own adjustments to mitigate most stresses (abiotic and biotic) in their environments. They also depend on their associated microbes to help them survive and protect themselves against microbial attacks (Turner et al. 2013).

The relationships between plants and their associated microbial communities are not unidirectional; the host plants also provide novel metabolic capabilities for their associated microbes, leading to adaptations to specialized niches that have either positive, neutral, or variable impacts on plant health (Thrall et al. 2007). The microorganisms that promote plant development are plant growth–promoting bacteria (PGPBs), ectomycorrhizal fungi, arbuscular mycorrhizal fungi and vesicular arbuscular mycorrhizae, which live in association with plants and moderate levels of phytohormones.

9.3.1 Plant Growth–Promoting Bacteria

Plants are close allies with a numerous variety of bacteria, which play important roles in their development, disease prevention and stress tolerance. A number of beneficial bacterial strains, defined by Kloepper and Schroth (1981) as plant growth–promoting bacteria, have been isolated from the phyllospheres, rhizospheres and endospheres of a wide variety of plant types (Rilling et al. 2019). Some bacteria have become intracellular endophytes that assist in plant–microbe co-development (Bulgarelli et al. 2013). Among these bacterial taxa are PGPRs, which exert beneficial effects on plants via indirect and direct mechanisms. Beneficial rhizobacteria are used by plants to increase their water and nutrient uptake, and their biotic and abiotic stress tolerance. Although many soil bacteria species have been studied to encourage plant growth, the systems of processes by which bacteria perform their beneficial activities are usually not easy to elucidate. The molecular bases of the plant–bacteria interaction mechanisms accountable for physiological changes is now starting to be identified, mainly through new 'omics' approaches (Backer et al. 2018).

9.3.1.1 Phytohormones Produced by Plant Growth–Promoting Rhizobacteria

Phytohormones produced by PGPRs are key performers in regulating plant development. They also act as molecular signals in response to environmental factors that limit plant development or become lethal if otherwise uncontrolled (Fahad et al. 2015). Many rhizosphere bacteria species are able to secrete hormones for root uptake or maintenance of hormone balance in plants to enhance growth and biotic and abiotic stress responses (Backer et al. 2018).

PGPRs that produce auxins have been characterized through transcriptional changes in hormones and have been found to enhance root biomass, confer protection, stimulate root lengthening and cell wall modification, reduce stomatal size and induce expression of auxin-inhibiting genes that improve plant growth (Spaepen et al. 2014; Ruzzi and Aroca 2015; Llorente et al. 2016). Rhizobacteria can produce relatively large amounts of gibberellic acid, leading to improved plant shoot development (Jha and Saraf 2015). Production of cytokinins by rhizobacteria can also lead to increased root exudate production by plants, potentially increasing the numbers of rhizobacteria associated with the plants (Backer et al. 2018). The hormone ethylene plays a crucial role in plant stress tolerance (Nadeem et al. 2014). PGPRs produce 1-aminocyclopropane-1-carboxylase (ACC) deaminase, which decreases ethylene output in plants (Vejan et al. 2016). Several studies have demonstrated increased stress (biotic and abiotic) tolerance in plants inoculated with rhizobacteria that produce ACC deaminase. This appears to occur when the rhizobacteria are able to raise the level of ethylene to a sufficient level to reduce plant development, as has been shown with *Camelina sativa* (Ahemad and Kibret 2014; Pérez-Montaño et al. 2014; Ruzzi and Aroca 2015; Heydarian et al. 2016).

9.3.1.2 Enhancement of Plant Development by Plant Growth–Promoting Rhizobacteria Under Stress Conditions

The mechanisms regulating stress tolerance in plants are convoluted and complex, as plants are sessile organisms, which have no choice as to where they live (Wani et al. 2016). Development of biotic and abiotic stress tolerance in crop plants through lineal breeding is a time-consuming and expensive procedure, and genetic engineering raises issues related to moral and social ethics.

The roles of beneficial microorganisms are now being exploited for stress management and development of climate change–tolerant agriculture (Backer et al. 2018). *Bacillus amyloliquefaciens* is a biological control agent, used against *Rhizoctonia solani*, which enhances tolerance through increased defence mechanisms in plants. Modulation of phytohormone signalling in colonized plants has revealed sustained maintenance of elicitors, production of secondary metabolites and moderation of the balance between reactive oxygen species (ROS) and ROS scavengers (Srivastava et al. 2016). *Enterobacter asburiae* enhances resistance to viral disease (tomato yellow leaf curl virus) by enhancing expression of defence-related genes and antioxidant enzymes such as lyase, catalase, peroxidase and superoxide dismutase (Li et al. 2016). Thus, by performing biocontrol functions, rhizobacteria defend plants against pathogens by prompting biochemical and molecular defence responses inside the plants (Lugtenberg and Kamilova 2009).

PGPRs can promote induced systemic resistance (ISR) in their host plants by triggering expression of pathogenesis-related genes, mediated via phytohormone signalling pathways and defence regulatory proteins, to arm plants against future pathogen attacks (Pieterse et al. 2014). *Pseudomonas putida* MTCC5279 was shown to ameliorate drought stress in *Cicer arietinum* (chickpea) plants by regulating ROS scavenging efficiency, membrane integrity and osmolyte (betaine, proline and glycine) accumulation.

Stress tolerance is positively regulated by bacteria through differential expression of genes involved in ethylene biosynthesis (*ACO* and *ACS*), stress response (*LEA* and *DHN* (dehydrin)), ROS scavenging by antioxidant enzymes (*CAT*, *APX*, *SOD* and *GST*), transcription activation (*DREB1A* (dehydration responsive element binding) and *NAC1*), salicylic acid (*PR1*) and jasmonate signalling (*MYC2*) (Tiwari et al. 2016).

Application of thuricin-17, produced by *Bacillus thuringiensis* NEB17, to *Glycine max* (soybean) under drought conditions resulted in root modifications such as greater root length and increased total N_2 content, nodule biomass and root abscisic acid content (Prudent et al. 2015).

9.3.2 *Arbuscular Mycorrhizal Fungi Associated with Plants*

In natural ecosystems, growth of numerous plants in nutrient-poor soils is viable because they form symbiotic associations with microorganisms for their mutual benefit (Liao et al. 2018). Associations between arbuscular mycorrhizal fungi (which belong to the Glomeromycotina subphylum) and more than 70% of land plants, including the most economically important crops—such as potato, rice and soybean—are considered to be some of the most prevalent and significant symbiotic associations in nature (Brundrett and Tedersoo 2018).

Formation of intracellular fungal structures and the degree of fungal dispersal inside plant roots are tuned dynamically by the plants, and this may prevent excessive colonization and loss of carbon, thereby ensuring that both the plants and the fungi continue to benefit from this association. To accomplish this regulation, extensive transcriptional programming and cellular rearrangements are needed in the plants, along with continuous signalling and exchange between the plants and the fungi (Maclean et al. 2017). Later phases of arbuscular mycorrhizal interactions are controlled by a variety of factors, together with nutrient exchange and phytohormone activity (Gutjahr 2014; Lanfranco et al. 2018). Analysis of arbuscular mycorrhizal symbiont regulation by phytohormones has revealed a complex pattern of modifications in hormonal content or altered responses to hormones in mycorrhizal plants and reciprocal effects of hormones on the symbiotic interaction (Pons et al. 2020). Phytohormones are known to be important signalling regulators, which participate in all physiological processes in plants, including interactions between the plants and microorganisms (Liao et al. 2018). There is growing evidence of the important roles played by various phytohormones—such as strigolactones, gibberellic acid, auxins, abscisic acid and brassinosteroids—which have been identified as positive controls of arbuscular mycorrhiza symbionts.

As phytohormone signalling in arbuscular mycorrhizal growth is a new research area, many novel findings related to phytohormone regulation and potential interactions during establishment of arbuscular mycorrhizal symbiosis have been published in recent years (Liao et al. 2018). DELLA proteins are a small cluster of GRAS transcriptional controls, which have been found to act as a central node in numerous signalling pathways, including hormonal crosstalk during nodulation and arbuscular mycorrhizal colonization.

9.3.2.1 Phytohormones Produced by Arbuscular Mycorrhizal Fungi

This chapter mainly discusses the following key aspects of the contributions of phytohormones to arbuscular mycorrhizal symbiosis: investigation of plant mutants affected by phytohormone synthesis or perception, and exogenous hormone treatment of mycorrhizal plants (Pons et al. 2020). Studies of phytohormone perception mutants have focused on the effects of phytohormones on plants. Both exogenous treatment and phytohormone deficiency lead to modified hormonal content in colonized roots, which can affect either or both of the symbiosis partners. Despite this, and because phytohormones are commonly perceived as plant signals, the reported outcomes of these studies have usually concentrated only on the effects on plants (Liao et al. 2018). Similarly, hormonal content changes measured in mycorrhizal plants are usually attributed to hormonal metabolism changes in plant cells. This interpretation overlooks the potential influence of the fungi on the hormonal pool. However, many microbes can produce phytohormones, and this could be the case with arbuscular mycorrhizal (AM) fungi (Kudoyarova et al. 2019). Among the soil microbes associated with plants, fungi and PGPRs have been found to produce several phytohormones (such as abscisic acid, auxin, gibberellic acid and cytokinins) that can have growth-promoting effects (Hamayun et al. 2010; Kang et al. 2012; Spaepen et al. 2014; Kudoyarova et al. 2019). In the fungal kingdom, phytohormone production has been documented in both symbionts (such as mycorrhizal fungi) and pathogens (Chanclud and Morel 2016).

Ethylene is commonly produced by fungal species and in certain cases, the biosynthesis pathways have been described (Splivallo et al. 2009). Ethylene-forming enzyme (EFE), characterized in *Penicillium digitatum* and *Fusarium oxysporum*, produces ethylene via two simultaneous reactions using L-arginine and 2-oxoglutarate as co-substrates (Pons et al. 2020). Both pathways differ from the major one used for ethylene production in plants, which is a methionine- and light-independent pathway involving ACC synthase and amino-cyclopropane-carboxylate oxidase (ACO).

Considering that numerous plant-associated microbes produce phytohormones, it is possible that arbuscular mycorrhizal fungi do so too, given that they have evolved together with their host plants for more than 400 million years (Pons et al. 2020). This possibility is not easy to study experimentally, because these fungi are obligate biotrophs that can be isolated and cultured only for short periods, limiting the availability of biological material for such study (Liao et al. 2018). There is indirect evidence from previous studies that phytohormones may be present in some mycorrhizal fungi. Enzyme-linked immunosorbent assay (ELISA) tests have shown that the spores and hyphal sheaths of *Rhizophagus* species may contain aglycone and glycosylated abscisic acid, and indirect bioassays have indicated the presence of gibberellin and cytokinin-like molecules (Pons et al. 2020). Genes encoding CLAVATA3/Embryo Surrounding Region-Related (CLE) peptide hormone, which positively modulates the symbiosis process, have been identified in arbuscular mycorrhizal fungal genomes (Le Marquer et al. 2019).

9.3.2.2 Enhancement of Plant Development by Arbuscular Mycorrhizal Fungi Under Stress Conditions

As microbial symbionts, arbuscular mycorrhizal fungi play important roles in the plant micro-ecosystem. They are found on plant organs and inhabit internal plant tissues in natural and managed ecosystems (Card et al. 2016). Arbuscular mycorrhizal fungi help their host plants to thrive in stressful conditions via complex processes in both the plants and the fungal species, increasing photosynthesis, other gas exchange–related processes and water uptake. Numerous reports have described how fungal symbiosis improves plant resistance to a variety of stresses, such as extreme temperatures, disease, drought, salinity and metal contamination (Begum et al. 2019).

Drought stress has various impacts on plant health, such as deficiency of water supply to the roots, a reduction in the transpiration rate and stimulation of oxidative stress (Impa et al. 2012; Hasanuzzaman et al. 2013). It also has deleterious impacts on plant development and growth by affecting enzyme activity, nutrient assimilation and ion uptake (Ahanger and Agarwal 2017; Ahanger et al. 2017). However, there is strong evidence that arbuscular mycorrhizal fungi reduce drought stress in various crops, including wheat, soybean, barley, strawberry, maize and onion (Mena-Violante et al. 2006; Ruiz-Lozano et al. 2016; Yooyongwech et al. 2016; Moradtalab et al. 2019). Plant tolerance of drought may be mainly due to the large volume of soil that is accessible to the roots via the extended hyphae of the fungi (Gianinazzi et al. 2010; Orfanoudakis et al. 2010; Zhang et al. 2017). This symbiotic consortium is known to modulate diverse physio-biochemical processes in plants, such as enhanced osmotic adjustment, stomatal management through control of abscisic acid metabolism and increases in proline and glutathione levels (Kubikova et al. 2001; Ruiz-Sánchez et al. 2010; Yooyongwech et al. 2013; Rani 2016). Onion (*Allium sativum*) plants inoculated with arbuscular mycorrhizal fungi demonstrated better development and growth traits, including a higher leaf area index and greater fresh and dry biomass, under salinity stress conditions (Borde et al. 2010).

The strong impacts of arbuscular mycorrhizal fungi on plant development under intensely stressful conditions are most likely due to the efficiency of these fungi in optimizing morphological and physiological processes, thereby increasing the plant biomass and uptake of vital nutrients such as P, Zn and Cu, and decreasing the toxic effects of metals on the host plants (Kanwal et al. 2015; Miransari 2017).

Root colonization with arbuscular mycorrhizal fungi increases plant resistance to soilborne pathogenic fungi (Wang et al. 2018). Arbuscular mycorrhizal fungi provide resistance to blackleg disease in *Solanum tuberosum* (potato), which is caused by the pathogenic bacterial strain *Pectobacterium carotovora* subsp. *atrosepticum* (Bagy et al. 2019), and bioprotective effects that help plants withstand both viral diseases and soilborne fungal pathogens that cause wilting or root rot. Arbuscular mycorrhizal symbiosis also stimulates host plant resistance to chewing insects, shoot pathogens and nematodes. Various mechanisms such as regulation of plant tolerance, manipulation of induced systemic resistance and altered vector pressure are involved in these interactions (Hao et al. 2019).

9.3.3 Ectomycorrhizal Fungi Associated with Plants

Ectomycorrhizal fungi belonging to the Basidiomycota and Ascomycota are the major symbionts of many plants in numerous ecosystems worldwide (Smith and

Read 2008; Tedersoo 2017). They clearly affect mineral nutrient uptake in their host plants (angiosperms, shrubs and gymnosperms) and play roles in essential forest ecosystem processes such as nutrient cycling, carbon sequestration and breakdown of organic substances. They also help their host plants to tolerate abiotic stresses (Read and Perez-Moreno 2003; Clemmensen et al. 2015; Shah et al. 2016; Mello and Balestrini 2018).

Most ectomycorrhizal plants are completely dependent on their mycorrhizal symbiosis and cannot complete their life cycle without this root association (Vlk et al. 2020). Stimulation of root growth and development during ectomycorrhizal fungus formation depends partially on changes in plant metabolism or susceptibility to phytohormones, which are the chief regulators of plant responses to growth, development and environmental factors (Garcia et al. 2015). Various ectomycorrhizal fungi, including basidiomycetes and ascomycetes, can produce phytohormones such as auxins, ethylene, jasmonate and gibberellic acid, thereby improving the entire nutritional condition of the plants in response to numerous different factors (Guerrero-Galán et al. 2019).

The expansion of the nutrient exchange surface provided through the mycelia of ectomycorrhizal fungi is a crucial factor in increased absorption of mineral nutrients and water by the host plants because the hyphae are potentially able to penetrate nearby soil pores (Bogeat-Triboulot et al. 2004; Lehto and Zwiazek 2011). An additional beneficial influence is improvement of the soil texture by the mycelia, facilitating plant root formation (Rillig and Mummey 2006). All of these influences boost growth, development and biomass accumulation by mycorrhizal plants, making them stronger and better adapted to challenging environments than nonmycorrhizal plants (Smith and Read 2008).

The evolutionary diversity of ectomycorrhizal fungi suggests that they may perform various different functional roles in the physiology of the host. Little is known about the precise mechanisms by which ectomycorrhizal fungi reduce the impact of salinity on their host plants (Guerrero-Galán et al. 2019).

9.3.3.1 Phytohormones Produced by Ectomycorrhizal Fungi

Auxins are phytohormones that facilitate root colonization in ectomycorrhizal plants (Vayssières et al. 2015). In addition, ectomycorrhizal fungi can induce plant ethylene and auxin signalling to encourage lateral root growth and root hair elongation (Ditengou et al. 2000; Reboutier et al. 2002; Felten et al. 2009; Splivallo et al. 2009; Vayssières et al. 2015).

Salicylic acid signalling plays a crucial role in plant defence mechanisms, acting as an antagonist of ethylene and jasmonate signalling (Glazebrook 2005; Spoel and Dong 2008; Pieterse et al. 2012). In addition, exogenous salicylic acid treatment does not influence fungal colonization.

Ultimately, the crosstalk between gibberellic acid and jasmonate signalling regulates plant responses (Hou et al. 2010; Wild et al. 2012; Yang et al. 2012; Song et al. 2014). Initial reports have suggested that exogenous gibberellic acid prevents hyphal development in various ectomycorrhizal species (Basso et al. 2020).

9.3.3.2 Enhancement of Plant Development by Ectomycorrhizal Fungi Under Stress Conditions

Ectomycorrhizal fungi form symbiotic associations with plant roots and help to promote growth and protect the plant from various biotic and abiotic stresses. The association with ectomycorrhizal fungus symbionts has been suggested to be a major factor in improved tolerance of woody plant species to salinity stress, decreasing sodium uptake by photosynthetic organs (Guerrero-Galán et al. 2019).

Plants are more sensitive to increased concentrations of heavy metals in the rhizosphere than microbes, but this may be at least partially due to evolutionary selection of tolerant fungi (Gadd 2007; Amir et al. 2014). Ectomycorrhizal fungi are able to alleviate stress caused by the presence of phytotoxic substances (Joner and Leyval 2003; Amir et al. 2014). The efficiency of ectomycorrhizal fungi in defending their host plants may be due to development of the hyphal sheath, which reduces direct contact between the roots and the elements stored in the soil.

Mycorrhizal fungi exhibit mechanisms that preserve the host's health under drought stress. Ectomycorrhizal fungi induces expression of plant aquaporins in drought conditions, which improve the host plants' drought tolerance via regulation of stomatal, root and shoot conductance, and thereby regulate transpiration in the host plants (Lehto and Zwiazek 2011). Because of their extensive mycelial biomass and development of rhizomorphs, ectomycorrhizal fungi are able to transport soil water more proficiently and access moisture in the substratum (Egerton-Warburton et al. 2003). Mycorrhizal plant seedlings tolerate drought stress better than nonmycorrhizal seedlings (Augé 2001; Lehto and Zwiazek 2011).

The extent to which ectomycorrhizal trees control their photosynthesis depends on the type of ectomycorrhiza they have. Waterlogging reduces the oxygen content of the soil. Numerous wetland trees have developed mechanisms for transporting oxygen to feeder roots.

9.4 Conclusion

Agricultural crops suffer various environmental stresses (biotic as well as abiotic ones), which adversely affect their productivity. Scientific methods and high-throughput technologies have made substantial contributions in addressing these concerns but have met with limited success. There is substantial evidence that application of exogenous phytohormones from microbial sources could be a crucial tool for enhancing plant tolerance of both biotic and abiotic stresses, furnish potential practical usages under realignment or highest environmental conditions. The beneficial impacts that microorganisms have on plants—such as plant growth stimulation, resistance of biotic stresses (pathogens) and tolerance of abiotic stresses—are due to the efficiency of the microorganisms in producing various phytohormones (including auxins, abscisic acid, cytokinins, gibberellic acid and salicylic acid) in plant tissues. Moreover, plant-associated microorganisms have the ability to regulate phytohormone levels and changes in plant tissues through biochemical processes that limit the damaging impacts of abiotic stresses, such as nutrient deficiency, drought, heavy metal contamination and salinity. The symbiotic alliance of host plants with microorganisms (particularly fungi), including ectomycorrhizal and arbuscular mycorrhizal fungi, provide distinct benefits for plant species. Genetic interplay between plant hormones for enhanced tolerance towards stress conditions presents substantial opportunities to help agricultural systems adapt to climate change and enhance agricultural production.

References

Ahanger MA, Agarwal RM (2017) Potassium up-regulates antioxidant metabolism and alleviates growth inhibition under water and osmotic stress in wheat (*Triticum aestivum* L.). Protoplasma 254:1471–1486

Ahanger MA, Tittal M, Mir RA, Agarwal RM (2017) Alleviation of water and osmotic stress-induced changes in nitrogen metabolizing enzymes in *Triticum aestivum* L. cultivars by potassium. Protoplasma 254:1953–1963

Ahemad M, Kibret M (2014) Mechanisms and applications of plant growth promoting rhizobacteria: current perspective. J King Saud Univ Sci 26:1–20. https://doi.org/10.1016/j.jksus.2013.05.001

Ahmad P (2010) Growth and antioxidant responses in mustard (*Brassica juncea* L.) plants subjected to combined effect of gibberellic acid and salinity. Arch Agron Soil Sci 56:575–588. https://doi.org/10.1080/03650340903164231

Akhtar SS, Mekureyaw MF, Pandey C, Roitsch T (2019) Role of cytokinins for interactions of plants with microbial pathogens and pest insects. Front Plant Sci 10:1777. https://doi.org/10.3389/fpls.2019.01777

Amir H, Jourand P, Cavaloc Y, Ducousso M (2014) Role of mycorrhizal fungi in the alleviation of heavy metal toxicity in plants. In: Solaiman ZM, Abbott LK, Varma A (eds) Mycorrhizal fungi: use in sustainable agriculture and land restoration. Springer, Berlin, pp 241–258. https://doi.org/10.1007/978-3-662-45370-4_15

Asgher M, Khan MI, Anjum NA, Khan NA (2015) Minimising toxicity of cadmium in plants—role of plant growth regulators. Protoplasma 252:399–413

Augé RM (2001) Water relations, drought and vesicular-arbuscular mycorrhizal symbiosis. Mycorrhiza 11:3–42. https://doi.org/10.1007/s005720100097

Backer R, Rokem JS, Ilangumaran G, Lamont J, Praslickova D, Ricci E, Subramanian S, Smith DL (2018) Plant growth-promoting rhizobacteria: context, mechanisms of action, and roadmap to commercialization of biostimulants for sustainable agriculture. Front Plant Sci 9:1473. https://doi.org/10.3389/fpls.2018.01473

Bagy HM, Hassan EA, Nafady NA, Dawood MF (2019) Efficacy of arbuscular mycorrhizal fungi and endophytic strain *Epicoccum nigrum* ASU11 as biocontrol agents against blackleg disease of potato caused by bacterial strain *Pectobacterium carotovora* subsp. *atrosepticum* PHY7. Biol Control 134:103–113

Bakhshandeh E, Gholamhosseini M, Yaghoubian Y, Pirdashti H (2020) Plant growth promoting microorganisms can improve germination, seedling growth and potassium uptake of soybean under drought and salt stress. Plant Growth Regul 90:123–136

Bano A, Ullah F, Nosheen A (2012) Role of abscisic acid and drought stress on the activities of antioxidant enzymes in wheat. Plant Soil Environ 58:181–185

Basso V, Kohler A, Miyauchi S, Singan V, Guinet F, Šimura J, Novák O, Barry KW, Amirebrahimi M, Block J, Daguerre Y (2020) An ectomycorrhizal fungus alters sensitivity to jasmonate, salicylate, gibberellin, and ethylene in host roots. Plant Cell Environ 43:1047–1068

Begum N, Qin C, Ahanger MA, Raza S, Khan MI, Ahmed N, Ashraf M, Zhang L (2019) Role of arbuscular mycorrhizal fungi in plant growth regulation: implications in abiotic stress tolerance. Front Plant Sci 10:1068

Bogeat-Triboulot MB, Bartoli F, Garbaye J, Marmeisse R, Tagu D (2004) Fungal ectomycorrhizal community and drought affect root hydraulic properties and soil adherence to roots of *Pinus pinaster* seedlings. Plant Soil 267:213–223

Borde M, Dudhane M, Jite PK (2010) AM fungi influences the photosynthetic activity, growth and antioxidant enzymes in *Allium sativum* L. under salinity condition. Not Sci Biol 2:64–71

Brundrett MC, Tedersoo L (2018) Evolutionary history of mycorrhizal symbioses and global host plant diversity. New Phytol 220:1108–1115

Bulgarelli D, Schlaeppi K, Spaepen S, Van Themaat EV, Schulze-Lefert P (2013) Structure and functions of the bacterial microbiota of plants. Annu Rev Plant Biol 64:807–838

Card S, Johnson L, Teasdale S, Caradus J (2016) Deciphering endophyte behaviour: the link between endophyte biology and efficacious biological control agents. FEMS Microbiol Ecol 92(8):fiw11492. https://doi.org/10.1093/femsec/fiw114

Chanclud E, Morel JB (2016) Plant hormones: a fungal point of view. Mol Plant Pathol 17:1289–1297

Chen Y, Yang W, Chao Y, Wang S, Tang YT, Qiu RL (2017) Metal-tolerant *Enterobacter* sp. strain EG16 enhanced phytoremediation using *Hibiscus cannabinus* via siderophore-mediated plant growth promotion under metal contamination. Plant Soil 413:203–216

Chen K, Li GJ, Bressan RA, Song CP, Zhu JK, Zhao Y (2020) Abscisic acid dynamics, signaling, and functions in plants. J Integr Plant Biol 62:25–54

Clemmensen KE, Finlay RD, Dahlberg A, Stenlid J, Wardle DA, Lindahl BD (2015) Carbon sequestration is related to mycorrhizal fungal community shifts during long-term succession in boreal forests. New Phytol 205:1525–1536

Cortleven A, Leuendorf JE, Frank M, Pezzetta D, Bolt S, Schmülling T (2019) Cytokinin action in response to abiotic and biotic stresses in plants. Plant Cell Environ 42:998–1018

Cutler SR, Rodriguez PL, Finkelstein RR, Abrams SR (2010) Abscisic acid: emergence of a core signaling network. Annu Rev Plant Biol 61:651–679

Ditengou FA, Béguiristain T, Lapeyrie F (2000) Root hair elongation is inhibited by hypaphorine, the indole alkaloid from the ectomycorrhizal fungus Pisolithus tinctorius, and restored by indole-3-acetic acid. Planta 211:722–728. https://doi.org/10.1007/s004250000342

Dowd CD, Chronis D, Radakovic ZS, Siddique S, Schmülling T, Werner T, Kakimoto T, Grundler FM, Mitchum MG (2017) Divergent expression of cytokinin biosynthesis, signaling and catabolism genes underlying differences in feeding sites induced by cyst and root-knot nematodes. Plant J 92:211–228

Egamberdieva D, Wirth SJ, Alqarawi AA, Abd Allah EF, Hashem A (2017) Phytohormones and beneficial microbes: essential components for plants to balance stress and fitness. Front Microbiol 8:2104. https://doi.org/10.3389/fmicb.2017.02104

Egerton-Warburton LM, Graham RC, Hubbert KR (2003) Spatial variability in mycorrhizal hyphae and nutrient and water availability in a soil-weathered bedrock profile. Plant Soil 249:331–342

Estrada-Melo AC, Reid MS, Jiang CZ (2015) Overexpression of an ABA biosynthesis gene using a stress-inducible promoter enhances drought resistance in petunia. Horti Res 2:1–9

Fahad S, Hussain S, Bano A, Saud S, Hassan S, Shan D, Khan FA, Khan F, Chen Y, Wu C, Tabassum MA (2015) Potential role of phytohormones and plant growth-promoting rhizobacteria in abiotic stresses: consequences for changing environment. Environ Sci Pollut Res 22:4907–4921

Fässler E, Evangelou MW, Robinson BH, Schulin R (2010) Effects of indole-3-acetic acid (IAA) on sunflower growth and heavy metal uptake in combination with ethylene diamine disuccinic acid (EDDS). Chemosphere 80:901–907

Felten J, Kohler A, Morin E, Bhalerao RP, Palme K, Martin F, Ditengou FA, Legué V (2009) The ectomycorrhizal fungus *Laccaria bicolor* stimulates lateral root formation in poplar and *Arabidopsis* through auxin transport and signaling. Plant Physiol 151:1991–2005. https://doi.org/10.1104/pp.109.147231

Gadd GM (2007) Geomycology: biogeochemical transformations of rocks, minerals, metals and radionuclides by fungi, bioweathering and bioremediation. Mycol Res 111:3–49. https://doi.org/10.1016/j.mycres.2006.12.001

Garcia K, Delaux PM, Cope KR, Ané JM (2015) Molecular signals required for the establishment and maintenance of ectomycorrhizal symbioses. New Phytol 208:79–87. https://doi.org/10.1111/nph.13423

Gianinazzi S, Gollotte A, Binet MN, van Tuinen D, Redecker D, Wipf D (2010) Agroecology: the key role of arbuscular mycorrhizas in ecosystem services. Mycorrhiza 20:519–530

Glazebrook J (2005) Contrasting mechanisms of defense against biotrophic and necrotrophic pathogens. Annu Rev Phytopathol 43:205–227

Großkinsky DK, Tafner R, Moreno MV, Stenglein SA, De Salamone IE, Nelson LM, Novák O, Strnad M, Van Der Graaff E, Roitsch T (2016) Cytokinin production by *Pseudomonas fluorescens* G20-18 determines biocontrol activity against *Pseudomonas syringae* in *Arabidopsis*. Sci Rep 6:23310

Guerrero-Galán C, Calvo-Polanco M, Zimmermann SD (2019) Ectomycorrhizal symbiosis helps plants to challenge salt stress conditions. Mycorrhiza 29:291–301. https://doi.org/10.1007/s00572-019-00894-2

Gutjahr C (2014) Phytohormone signaling in arbuscular mycorrhiza development. Curr Opin Plant Biol 20:26–34

Habben JE, Bao X, Bate NJ, DeBruin JL, Dolan D, Hasegawa D, Helentjaris TG, Lafitte RH, Lovan N, Mo H, Reimann K (2014) Transgenic alteration of ethylene biosynthesis increases grain yield in maize under field drought-stress conditions. Plant Biotechnol J 12:685–693

Hamayun M, Khan SA, Khan AL, Rehman G, Kim YH, Iqbal I, Hussain J, Sohn EY, Lee IJ (2010) Gibberellin production and plant growth promotion from pure cultures of *Cladosporium* sp. MH-6 isolated from cucumber (*Cucumis sativus* L.). Mycologia 102:989–995

Hamayun M, Hussain A, Khan SA, Kim HY, Khan AL, Waqas M, Irshad M, Iqbal A, Rehman G, Jan S, Lee IJ (2017) Gibberellins producing endophytic fungus *Porostereum spadiceum* AGH786 rescues growth of salt affected soybean. Front Microbiol 8:686

Hao Z, Xie W, Chen B (2019) Arbuscular mycorrhizal symbiosis affects plant immunity to viral infection and accumulation. Viruses 11:534. https://doi.org/10.3390/v11060534

Hardoim PR, Van Overbeek LS, Berg G, Pirttilä AM, Compant S, Campisano A, Döring M, Sessitsch A (2015) The hidden world within plants: ecological and evolutionary considerations for defining functioning of microbial endophytes. Microbiol Mol Biol Rev 79:293–320

Hasanuzzaman M, Gill SS, Fujita M (2013) Physiological role of nitric oxide in plants grown under adverse environmental conditions. In: Tuteja N, Singh GS (eds) Plant acclimation to environmental stress. Springer, New York, pp 269–322. https://doi.org/10.1007/978-1-4614-5001-6_11

Hassani MA, Durán P, Hacquard S (2018) Microbial interactions within the plant holobiont. Microbiome 6:58

Heydarian Z, Yu M, Gruber M, Glick BR, Zhou R, Hegedus DD (2016) Inoculation of soil with plant growth promoting bacteria producing 1-aminocyclopropane-1-carboxylate deaminase or expression of the corresponding *acdS* gene in transgenic plants increases salinity tolerance in *Camelina sativa*. Front Microbiol 7:1966. https://doi.org/10.3389/fmicb.2016.01966

Hou X, Lee LY, Xia K, Yan Y, Yu H (2010) DELLAs modulate jasmonate signaling via competitive binding to JAZs. Dev Cell 19:884–894

Hu YF, Zhou G, Na XF, Yang L, Nan WB, Liu X, Zhang YQ, Li JL, Bi YR (2013) Cadmium interferes with maintenance of auxin homeostasis in *Arabidopsis* seedlings. J Plant Physiol 170:965–975

Impa SM, Nadaradjan S, Jagadish SV (2012) Drought stress induced reactive oxygen species and anti-oxidants in plants. In: Ahmad P, Prasad MNV (eds) Abiotic stress responses in plants. Springer, New York, pp 131–147

Iqbal M, Ashraf M (2013) Gibberellic acid mediated induction of salt tolerance in wheat plants: growth, ionic partitioning, photosynthesis, yield and hormonal homeostasis. Environ Exp Bot 86:76–85. https://doi.org/10.1016/j.envexpbot.2010.06.002

Jha CK, Saraf M (2015) Plant growth promoting rhizobacteria (PGPR): a review. J Agric Res Dev 5:108–119

Joner E, Leyval C (2003) Phytoremediation of organic pollutants using mycorrhizal plants: a new aspect of rhizosphere interactions. In: Lichtfouse E, Navarrete M, Debaeke P, Véronique S, Alberola C (eds) Sustainable agriculture. Springer, Dordrecht, pp 885–894. https://doi.org/10.1007/978-90-481-2666-8_54

Kang SM, Khan AL, Hamayun M, Hussain J, Joo GJ, You YH, Kim JG, Lee IJ (2012) Gibberellin-producing *Promicromonospora* sp. SE188 improves *Solanum lycopersicum* plant growth and influences endogenous plant hormones. J Microbiol 50:902–909

Kanwal S, Bano A, Malik RN (2015) Effects of arbuscular mycorrhizal fungi on metals uptake, physiological and biochemical response of *Medicago sativa* L. with increasing Zn and Cd concentrations in soil. Am J Plant Sci 6:2906

Kazan K (2013) Auxin and the integration of environmental signals into plant root development. Ann Bot 112:1655–1665

Kazan K (2015) Diverse roles of jasmonates and ethylene in abiotic stress tolerance. Trends Plant Sci 20:219–229

Kazan K, Lyons R (2014) Intervention of phytohormone pathways by pathogen effectors. Plant Cell 26:2285–2309

Ke Q, Wang Z, Ji CY, Jeong JC, Lee HS, Li H, Xu B, Deng X, Kwak SS (2015) Transgenic poplar expressing *Arabidopsis* YUCCA6 exhibits auxin-overproduction phenotypes and increased tolerance to abiotic stress. Plant Physiol Biochem 94:19–27

Khan MA, Gul B, Weber DJ (2004) Action of plant growth regulators and salinity on seed germination of *Ceratoides lanata*. Can J Bot 82:37–42. https://doi.org/10.1139/b03-140

Khan MA, Ullah I, Waqas M, Hamayun M, Khan AL, Asaf S, Kang SM, Kim KM, Jan R, Lee IJ (2019) Halo-tolerant rhizospheric *Arthrobacter woluwensis* AK1 mitigates salt stress and induces physio-hormonal changes and expression of *GmST1* and *GmLAX3* in soybean. Symbiosis 77:9–21. https://doi.org/10.1007/s13199-018-0562-3

Khan N, Bano A, Ali S, Babar MA (2020) Crosstalk amongst phytohormones from planta and PGPR under biotic and abiotic stresses. Plant Growth Regul 90:189–203. https://doi.org/10.1007/s10725-020-00571-x

Kloepper JW, Schroth MN (1981) Relationship of in vitro antibiosis of plant growth-promoting rhizobacteria to plant growth and the displacement of root microflora. Phytopathology 71:1020–1024

Kubikova E, Jennifer LM, Bonnie HO, Michael DM, Augé MR (2001) Mycorrhizal impact on osmotic adjustment in *Ocimum basilicum* during a lethal drying episode. J Plant Physiol 158:1227–1230

Kudoyarova G, Arkhipova TN, Korshunova T, Bakaeva M, Loginov O, Dodd IC (2019) Phytohormone mediation of interactions between plants and non-symbiotic growth promoting bacteria under edaphic stresses. Front Plant Sci 10:1368

Kunkel BN, Harper CP (2018) The roles of auxin during interactions between bacterial plant pathogens and their hosts. J Exp Bot 69:245–254

Kushwaha P, Kashyap PL, Kuppusamy P, Srivastava AK, Tiwari RK (2020) Functional characterization of endophytic bacilli from pearl millet (*Pennisetum glaucum*) and their possible role in multiple stress tolerance. Plant Biosyst 154:503–514. https://doi.org/10.1080/11263504.2019.1651773

Lanfranco L, Fiorilli V, Gutjahr C (2018) Partner communication and role of nutrients in the arbuscular mycorrhizal symbiosis. New Phytol 220:1031–1046

Lavania M, Nautiyal CS (2013) Solubilization of tricalcium phosphate by temperature and salt tolerant *Serratia marcescens* NBRI1213 isolated from alkaline soils. Afr J Microbiol Res 7:4403–4413

Le Marquer M, Bécard G, Frei Dit Frey N (2019) Arbuscular mycorrhizal fungi possess a CLAVATA3/embryo surrounding region-related gene that positively regulates symbiosis. New Phytol 222:1030–1042. https://doi.org/10.1111/nph.15643

Lehto T, Zwiazek JJ (2011) Ectomycorrhizas and water relations of trees: a review. Mycorrhiza 21:71–90

Li H, Ding X, Wang C, Ke H, Wu Z, Wang Y, Liu H, Guo J (2016) Control of tomato yellow leaf curl virus disease by *Enterobacter asburiae* BQ9 as a result of priming plant resistance in tomatoes. Turk J Biol 40:150–159. https://doi.org/10.3906/biy-1502-12

Liao D, Wang S, Cui M, Liu J, Chen A, Xu G (2018) Phytohormones regulate the development of arbuscular mycorrhizal symbiosis. Int J Mol Sci 19:3146. https://doi.org/10.3390/ijms19103146

Lim CW, Baek W, Jung J, Kim JH, Lee SC (2015) Function of ABA in stomatal defense against biotic and drought stresses. Int J Mol Sci 16:15251–15270. https://doi.org/10.3390/ijms160715251

Ljung K (2013) Auxin metabolism and homeostasis during plant development. Development 140:943–950. https://doi.org/10.1242/dev.086363

Llorente BE, Alasia MA, Larraburu EE (2016) Biofertilization with *Azospirillum brasilense* improves in vitro culture of *Handroanthus ochraceus*, a forestry, ornamental and medicinal plant. New Biotechnol 33:32–40. https://doi.org/10.1016/j.nbt.2015.07.006

Lu Y, Yao J (2018) Chloroplasts at the crossroad of photosynthesis, pathogen infection and plant defense. Int J Mol Sci 19:3900

Lugtenberg B, Kamilova F (2009) Plant-growth-promoting rhizobacteria. Annu Rev Microbiol 63:541–556

MacLean AM, Bravo A, Harrison MJ (2017) Plant signaling and metabolic pathways enabling arbuscular mycorrhizal symbiosis. Plant Cell 29:2319–2335

Maggio A, Barbieri G, Raimondi G, De Pascale S (2010) Contrasting effects of GA $_3$ treatments on tomato plants exposed to increasing salinity. J Plant Growth Regul 29:63–72. https://doi.org/10.1007/s00344-009-9114-7

Maurer D, Feygenberg O, Tzoor A, Atzmon G, Glidai S, Prusky D (2019) Postharvest dips of persimmon fruit in gibberellic acid: an efficient treatment to improve storability and reduce *Alternaria* black spot. Horticulturae 5:23. https://doi.org/10.3390/horticulturae5010023

McClerklin SA, Lee SG, Harper CP, Nwumeh R, Jez JM, Kunkel BN (2018) Indole-3-acetaldehyde dehydrogenase–dependent auxin synthesis contributes to virulence of *Pseudomonas syringae* strain DC3000. PLoS Pathog 14:1006811

Mello A, Balestrini R (2018) Recent insights on biological and ecological aspects of ectomycorrhizal fungi and their interactions. Front Microbiol 9:216

Mena-Violante HG, Ocampo-Jiménez O, Dendooven L, Martínez-Soto G, González-Castañeda J, Davies FT, Olalde-Portugal V (2006) Arbuscular mycorrhizal fungi enhance fruit growth and quality of chile ancho (*Capsicum annuum* L. cv San Luis) plants exposed to drought. Mycorrhiza 16:261–267. https://doi.org/10.1007/s00572-006-0043-z

Miransari M (2017) Arbuscular mycorrhizal fungi and heavy metal tolerance in plants. In: Wu QS (ed) Arbuscular mycorrhizas and stress tolerance of plants. Springer, Singapore, pp 147–161. https://doi.org/10.1007/978-981-10-4115-0_7

Mohapatra PK, Panigrahi R, Turner NC (2011) Physiology of spikelet development on the rice panicle: is manipulation of apical dominance crucial for grain yield improvement? Adv Agron 110:333–359. https://doi.org/10.1016/B978-0-12-385531-2.00005-0

Moradtalab N, Hajiboland R, Aliasgharzad N, Hartmann TE, Neumann G (2019) Silicon and the association with an arbuscular-mycorrhizal fungus (*Rhizophagus clarus*) mitigate the adverse effects of drought stress on strawberry. Agronomy 9:41

Mueller UG, Sachs JL (2015) Engineering microbiomes to improve plant and animal health. Trends Microbiol 23:606–617

Nadeem SM, Ahmad M, Zahir ZA, Javaid A, Ashraf M (2014) The role of mycorrhizae and plant growth promoting rhizobacteria (PGPR) in improving crop productivity under stressful environments. Biotechnol Adv 32:429–448. https://doi.org/10.1016/j.biotechadv.2013.12.005

Nejat N, Mantri N (2017) Plant immune system: crosstalk between responses to biotic and abiotic stresses the missing link in understanding plant defence. Curr Issues Mol Biol 23:1–16. https://doi.org/10.21775/cimb.023.001

Nishiyama R, Watanabe Y, Leyva-Gonzalez MA, Van Ha C, Fujita Y, Tanaka M, Seki M, Yamaguchi-Shinozaki K, Shinozaki K, Herrera-Estrella L, Tran LS (2013) *Arabidopsis* AHP2, AHP3, and AHP5 histidine phosphotransfer proteins function as redundant negative regulators of drought stress response. Proc Natl Acad Sci 110:4840–4845

Ogbe AA, Finnie JF, Van Staden J (2020) The role of endophytes in secondary metabolites accumulation in medicinal plants under abiotic stress. S Afr J Bot 134:126–134. https://doi.org/10.1016/j.sajb.2020.06.023

Olszewski N, Sun TP, Gubler F (2002) Gibberellin signaling: biosynthesis, catabolism, and response pathways. Plant Cell 14:61–80. https://doi.org/10.1105/tpc.010476

Orfanoudakis M, Wheeler CT, Hooker JE (2010) Both the arbuscular mycorrhizal fungus *Gigaspora rosea* and *Frankia* increase root system branching and reduce root hair frequency in *Alnus glutinosa*. Mycorrhiza 20:117–126

Pandey P, Irulappan V, Bagavathiannan MV, Senthil-Kumar M (2017) Impact of combined abiotic and biotic stresses on plant growth and avenues for crop improvement by exploiting physio-morphological traits. Front Plant Sci 8:537

Park YG, Mun BG, Kang SM, Hussain A, Shahzad R, Seo CW, Kim AY, Lee SU, Oh KY, Lee DY, Lee IJ (2017) *Bacillus aryabhattai* SRB02 tolerates oxidative and nitrosative stress and promotes the growth of soybean by modulating the production of phytohormones. PLoS One 12:0173203

Pedranzani H, Rodríguez-Rivera M, Gutiérrez M, Porcel R, Hause B, Ruiz-Lozano JM (2016) Arbuscular mycorrhizal symbiosis regulates physiology and performance of *Digitaria eriantha* plants subjected to abiotic stresses by modulating antioxidant and jasmonate levels. Mycorrhiza 26:141–152

Pérez-Montaño F, Alías-Villegas C, Bellogín RA, Del Cerro P, Espuny MR, Jiménez-Guerrero I, López-Baena FJ, Ollero FJ, Cubo T (2014) Plant growth promotion in cereal and leguminous agricultural important plants: from microorganism capacities to crop production. Microbiol Res 169:325–336

Pieterse CM, Van der Does D, Zamioudis C, Leon-Reyes A, Van Wees SC (2012) Hormonal modulation of plant immunity. Annu Rev Cell Dev Biol 28:489–521. https://doi.org/10.1146/annurev-cellbio-092910-154055

Pieterse CM, Zamioudis C, Berendsen RL, Weller DM, Van Wees SC, Bakker PA (2014) Induced systemic resistance by beneficial microbes. Annu Rev Phytopathol 52:347–375. https://doi.org/10.1146/annurev-phyto-082712-102340

Pons S, Fournier S, Chervin C, Bécard G, Rochange S, Frei Dit Frey N, Puech Pagès V (2020) Phytohormone production by the arbuscular mycorrhizal fungus *Rhizophagus irregularis*. PLoS One 15(10):e0240886240886. https://doi.org/10.1371/journal.pone.0240886

Provorov NA, Onishchuk OP, Kurchak ON (2016) Impacts of inoculation with *Sinorhizobium meliloti* strains differing in salt tolerance on the productivity and habitus of alfalfa (*Medicago sativa* L.). Agric Biol 51:343–350

Prudent M, Salon C, Souleimanov A, Emery RN, Smith DL (2015) Soybean is less impacted by water stress using *Bradyrhizobium japonicum* and thuricin-17 from *Bacillus thuringiensis*. Agron Sustain Dev 35:749–757. https://doi.org/10.1007/s13593-014-0256-z

Qadir M, Hussain A, Hamayun M, Shah M, Iqbal A, Murad W (2020) Phytohormones producing rhizobacterium alleviates chromium toxicity in *Helianthus annuus* L. by reducing chromate uptake and strengthening antioxidant system. Chemosphere 258:127386. https://doi.org/10.1016/j.chemosphere.2020.127386

Rani B (2016) Effect of arbuscular mycorrhiza fungi on biochemical parameters in wheat (*Triticum aestivum* L.) under drought conditions. Dissertation, Chaudhary Charan Singh Haryana Agricultural University, Hisar (Doctoral dissertation, CCSHAU)

Raza FA, Faisal M (2013) Growth promotion of maize by desiccation tolerant *Micrococcus luteus*-chp37 isolated from Cholistan desert, Pakistan. Aust J Crop Sci 7:1693

Read DJ, Perez-Moreno J (2003) Mycorrhizas and nutrient cycling in ecosystems—a journey towards relevance? New Phytol 157:475–492

Reboutier D, Bianchi M, Brault M, Roux C, Dauphin A, Rona JP, Legué V, Lapeyrie F, Bouteau F (2002) The indolic compound hypaphorine produced by ectomycorrhizal fungus interferes with auxin action and evokes early responses in nonhost *Arabidopsis thaliana*. Mol Plant-Microbe Interact 15:932–938. https://doi.org/10.1094/MPMI.2002.15.9.932

Rillig MC, Mummey DL (2006) Mycorrhizas and soil structure. New Phytol 171:41–53

Rilling JI, Acuña JJ, Nannipieri P, Cassan F, Maruyama F, Jorquera MA (2019) Current opinion and perspectives on the methods for tracking and monitoring plant growth-promoting bacteria. Soil Biol Biochem 130:205–219

Ruiz-Lozano JM, Aroca R, Zamarreño ÁM, Molina S, Andreo-Jiménez B, Porcel R, García-Mina JM, Ruyter-Spira C, López-Ráez JA (2016) Arbuscular mycorrhizal symbiosis induces strigolactone biosynthesis under drought and improves drought tolerance in lettuce and tomato. Plant Cell Environ 39:441–452

Ruiz-Sánchez M, Aroca R, Muñoz Y, Polón R, Ruiz-Lozano JM (2010) The arbuscular mycorrhizal symbiosis enhances the photosynthetic efficiency and the antioxidative response of rice plants subjected to drought stress. J Plant Physiol 167:862–869

Ruzzi M, Aroca R (2015) Plant growth-promoting rhizobacteria act as biostimulants in horticulture. Sci Hortic 196:124–134. https://doi.org/10.1016/j.scienta.2015.08.042

Sah SK, Reddy KR, Li J (2016) Abscisic acid and abiotic stress tolerance in crop plants. Front Plant Sci 7:571. https://doi.org/10.3389/fpls.2016.00571

Salomon MV, Bottini R, de Souza Filho GA, Cohen AC, Moreno D, Gil M, Piccoli P (2014) Bacteria isolated from roots and rhizosphere of *Vitis vinifera* retard water losses, induce abscisic acid accumulation and synthesis of defense-related terpenes in in vitro cultured grapevine. Physiol Plant 151:359–374

Shah F, Nicolás C, Bentzer J, Ellström M, Smits M, Rineau F, Canbäck B, Floudas D, Carleer R, Lackner G, Braesel J (2016) Ectomycorrhizal fungi decompose soil organic matter using oxidative mechanisms adapted from saprotrophic ancestors. New Phytol 209:1705–1719

Shahzad R, Waqas M, Khan AL, Asaf S, Khan MA, Kang SM, Yun BW, Lee IJ (2016) Seed-borne endophytic *Bacillus amyloliquefaciens* RWL-1 produces gibberellins and regulates endogenous phytohormones of *Oryza sativa*. Plant Physiol Biochem 106:236–243

Silva R, Filgueiras L, Santos B, Coelho M, Silva M, Estrada-Bonilla G, Vidal M, Baldani JI, Meneses C (2020) *Gluconacetobacter diazotrophicus* changes the molecular mechanisms of root development in *Oryza sativa* L. growing under water stress. Int J Mol Sci 21:333

Singh RP, Jha PN (2016) A halotolerant bacterium *Bacillus licheniformis* HSW-16 augments induced systemic tolerance to salt stress in wheat plant (*Triticum aestivum*). Front Plant Sci 7:1890

Smith SE, Read DJ (2008) The symbionts forming arbuscular mycorrhizas. In: Mycorrhizal symbiosis, vol 2. Academic Press, New York, pp 13–41. https://doi.org/10.1016/B978-012370526-6.50003-9

Song S, Qi T, Wasternack C, Xie D (2014) Jasmonate signaling and crosstalk with gibberellin and ethylene. Curr Opin Plant Biol 21:112–119

Spaepen S, Bossuyt S, Engelen K, Marchal K, Vanderleyden J (2014) Phenotypical and molecular responses of *Arabidopsis thaliana* roots as a result of inoculation with the auxin-producing bacterium *Azospirillum brasilense*. New Phytol 201(3):850–861. https://doi.org/10.1111/nph.12590

Splivallo R, Fischer U, Göbel C, Feussner I, Karlovsky P (2009) Truffles regulate plant root morphogenesis via the production of auxin and ethylene. Plant Physiol 150:2018–2029

Spoel SH, Dong X (2008) Making sense of hormone crosstalk during plant immune responses. Cell Host Microbe 3:348–351

Srivastava S, Bist V, Srivastava S, Singh PC, Trivedi PK, Asif MH, Chauhan PS, Nautiyal CS (2016) Unraveling aspects of *Bacillus amyloliquefaciens* mediated enhanced production of rice under biotic stress of *Rhizoctonia solani*. Front Plant Sci 7:587. https://doi.org/10.3389/fpls.2016.00587

Tedersoo L (2017) Global biogeography and invasions of ectomycorrhizal plants: past, present and future. In: Tedersoo L (ed) Biogeography of mycorrhizal symbiosis. Springer, Cham, pp 469–531. https://doi.org/10.1007/978-3-319-56363-3_20

Thrall PH, Hochberg ME, Burdon JJ, Bever JD (2007) Coevolution of symbiotic mutualists and parasites in a community context. Trends Ecol Evol 22:120–126. https://doi.org/10.1016/j.tree.2006.11.007

Tiwari S, Lata C, Chauhan PS, Nautiyal CS (2016) *Pseudomonas putida* attunes morphophysiological, biochemical and molecular responses in *Cicer arietinum* L. during drought stress and recovery. Plant Physiol Biochem 99:108–117. https://doi.org/10.1016/j.plaphy.2015.11.001

Tiwari P, Bajpai M, Singh LK, Mishra S, Yadav AN (2020) Phytohormones producing fungal communities: metabolic engineering for abiotic stress tolerance in crops. In: Yadav A, Mishra S, Kour D, Yadav N, Kumar A (eds) Agriculturally important fungi for sustainable agriculture. Springer, Cham, pp 171–197. https://doi.org/10.1007/978-3-030-45971-0_8

Turner TR, James EK, Poole PS (2013) The plant microbiome. Genome Biol 14:1–10

Vayssières A, Pěnčík A, Felten J, Kohler A, Ljung K, Martin F, Legué V (2015) Development of the poplar-*Laccaria bicolor* ectomycorrhiza modifies root auxin metabolism, signaling, and response. Plant Physiol 169:890–902

Vejan P, Abdullah R, Khadiran T, Ismail S, Nasrulhaq Boyce A (2016) Role of plant growth promoting rhizobacteria in agricultural sustainability—a review. Molecules 21:573

Vlk L, Tedersoo L, Antl T, Větrovský T, Abarenkov K, Pergl J, Albrechtová J, Vosátka M, Baldrian P, Pyšek P, Kohout P (2020) Early successional ectomycorrhizal fungi are more likely to naturalize outside their native range than other ectomycorrhizal fungi. New Phytol 227:1289–1293. https://doi.org/10.1111/nph.16557

Wang Y, Wang M, Li Y, Wu A, Huang J (2018) Effects of arbuscular mycorrhizal fungi on growth and nitrogen uptake of *Chrysanthemum morifolium* under salt stress. PLoS One 13:0196408

Wani SH, Kumar V, Shriram V, Sah SK (2016) Phytohormones and their metabolic engineering for abiotic stress tolerance in crop plants. Crop J 4:162–176

Werner T, Nehnevajova E, Köllmer I, Novák O, Strnad M, Krämer U, Schmülling T (2010) Root-specific reduction of cytokinin causes enhanced root growth, drought tolerance, and leaf mineral enrichment in *Arabidopsis* and tobacco. Plant Cell 22:3905–3920. https://doi.org/10.1105/tpc.109.072694

Wild M, Davière JM, Cheminant S, Regnault T, Baumberger N, Heintz D, Baltz R, Genschik P, Achard P (2012) The *Arabidopsis* DELLA RGA-LIKE$_3$ is a direct target of MYC2 and modulates jasmonate signaling responses. Plant Cell 24:3307–3319. https://doi.org/10.1105/tpc.112.101428

Wilkinson S, Kudoyarova GR, Veselov DS, Arkhipova TN, Davies WJ (2012) Plant hormone interactions: innovative targets for crop breeding and management. J Exp Bot 63:3499–3509

Yang DL, Yao J, Mei CS, Tong XH, Zeng LJ, Li Q, Xiao LT, Sun TP, Li J, Deng XW, Lee CM (2012) Plant hormone jasmonate prioritizes defense over growth by interfering with gibberellin signaling cascade. Proc Natl Acad Sci 109:1192–1200

Yang J, Duan G, Li C, Liu L, Han G, Zhang Y, Wang C (2019) The crosstalks between jasmonic acid and other plant hormone signaling highlight the involvement of jasmonic acid as a core component in plant response to biotic and abiotic stresses. Front Plant Sci 10:1349. https://doi.org/10.3389/fpls.2019.01349

Yimer HZ, Nahar K, Kyndt T, Haeck A, Van Meulebroek L, Vanhaecke L, Demeestere K, Höfte M, Gheysen G (2018) Gibberellin antagonizes jasmonate-induced defense against *Meloidogyne graminicola* in rice. New Phytol 218:646–660

Yin C, Park JJ, Gang DR, Hulbert SH (2014) Characterization of a tryptophan 2-monooxygenase gene from *Puccinia graminis* f. sp. *tritici* involved in auxin biosynthesis and rust pathogenicity. Mol Plant-Microbe Interact 27:227–235. https://doi.org/10.1094/MPMI-09-13-0289-FI

Yooyongwech S, Phaukinsang N, Cha-um S, Supaibulwatana K (2013) Arbuscular mycorrhiza improved growth performance in *Macadamia tetraphylla* L. grown under water deficit stress involves soluble sugar and proline accumulation. Plant Growth Regul 69:285–293. https://doi.org/10.1007/s10725-012-9771-6

Yooyongwech S, Samphumphuang T, Tisarum R, Theerawitaya C, Cha-um S (2016) Arbuscular mycorrhizal fungi (AMF) improved water deficit tolerance in two different sweet potato genotypes involves osmotic adjustments via soluble sugar and free proline. Sci Hortic 198:107–117. https://doi.org/10.1016/j.scienta.2015.11.002

Zhang P, Wang WQ, Zhang GL, Kaminek M, Dobrev P, Xu J, Gruissem W (2010) Senescence-inducible expression of isopentenyl transferase extends leaf life, increases drought stress resistance and alters cytokinin metabolism in cassava. J Integr Plant Biol 52:653–669. https://doi.org/10.1111/j.1744-7909.2010.00956.x

Zhang X, Wang L, Ma F, Yang J, Su M (2017) Effects of arbuscular mycorrhizal fungi inoculation on carbon and nitrogen distribution and grain yield and nutritional quality in rice (*Oryza sativa* L.). J Sci Food Agric 97:2919–2925

Zhou L, Xu H, Mischke S, Meinhardt LW, Zhang D, Zhu X, Li X, Fang W (2014) Exogenous abscisic acid significantly affects proteome in tea plant (*Camellia sinensis*) exposed to drought stress. Hortic Res 1:14029. https://doi.org/10.1038/hortres.2014.29

Chapter 10
Use of PGPR to Optimize Soil and Crop Productivity Under Abiotic Stress

Md. Mahtab Rashid, Surabhi Chaturvedi, Anukool Vaishnav, and Devendra Kumar Choudhary

Abstract Based on natural and anthropogenic activities, the soil quality has depleted gradually which is also mediated through unpredictably changed environmental conditions. As a consequence, a challenge has been raised before farmers and nations to compensate the degraded soil quality. To replenish the soil quality, farmers deploy synthetic fertilizer which is an unsustainable practice and further lead conditions worse by the diminished biological activity, increased level of toxicity, decreased fertility, etc. As a comparatively safe and sustainable alternative, PGPRs have been characterized as they could always assist plants against various challenges like nutrient unavailability, abiotic stresses, and pathogens. Many of those PGPRs which have been studied for their beneficial impacts are now used on a commercial scale for alleviating abiotic and biotic stresses of crop plants. Some of the PGPRs with wonderful ex situ and in situ performances are *Azospirillum brasilense*, *Azospirillum lipoferum*, *Bacillus*, *Pseudomonas*, *Acinetobacter*, *Alcaligenes faecalis*, *Stenotrophomonas*, *Pseudomonas*, *Rahnella*, *Pseudomonas fluorescens*, *Bacillus megaterium*, *Bacillus licheniformis*, *Proteus mirabilis*, *Achromobacter xylosoxidans*, *Gluconoacetobacter diazotrophicus*, *Azoarcus*, *Pseudomonas migulae*, *Brachybacterium saurashtrense*, *Brevibacterium casei*, *Haererohalobacter*, and many more which have surely assisted plant including a

M. M. Rashid
Department of Mycology and Plant Pathology, Institute of Agricultural Sciences, Banaras Hindu University, Varanasi, Uttar Pradesh, India

S. Chaturvedi
Lab No. 10, Department of Botany, University of Rajasthan, Jaipur, Rajasthan, India

A. Vaishnav (✉)
Department of Mycology and Plant Pathology, Institute of Agricultural Sciences, Banaras Hindu University, Varanasi, Uttar Pradesh, India

Department of Biotechnology, Institute of Applied Sciences & Humanities, GLA University, Mathura, Uttar Pradesh, India

D. K. Choudhary
Amity Institute of Microbial Technology (AIMT), Amity University Uttar Pradesh, Noida, Uttar Pradesh, India
e-mail: dkchoudhary1@amity.edu

S. K. Dubey, S. K. Verma (eds.), *Plant, Soil and Microbes in Tropical Ecosystems*, Rhizosphere Biology, https://doi.org/10.1007/978-981-16-3364-5_10

list of *Arabidopsis*, maize, wheat, potato, tomato, capsicum, etc. This chapter unfolds important mechanisms and strategies used by PGPRs to help crop plant cope up the various biotic and abiotic stresses and increase soil and plant health.

Keywords PGPRs · Mechanism · Biotic and abiotic stress · Crop plants

10.1 Introduction

Around the globe, the plant biologists involve in research with key point "global food security," to fulfill the needs of the human population which is growing day by day at a very alarming rate. As per the estimates, the present global population is approximately 7.7 billion which is going to increase by 2 billion in another 30 years. By 2100, the total number of residents around the world would be 11 billion (UN 2019). One way of meeting the food demands is to increase the area of crop production, but the overgrowing population also has the requirement of dwelling lands, and most of the arable lands are being changed into the urban and industrial area. This is coupled with the environmental challenges which also add to the magnitude of burden that the agricultural system has to bear. This has led crop scientists to come up with peculiar innovations in order to meet the food demands with such diminishing land resources. Since the prospect of increasing the area of crop production is out of the alternatives, increasing the productivity of crops on the available agricultural land is an appropriate preference. The crop productivity depends upon four major factors viz.: the genetic make-up of the crop, properties of the soil on which it is grown, environmental conditions of the area/region in which the crop is grown, and the microorganisms which are associated a particular crop (Chaudhry et al. 2012).

Better crop productivity is directly proportional to the productivity of soil on which it is grown as the soil quality is a major factor driving the growth and development of a crop. However, in present times the soil quality is also depleting due to the anthropogenic activities conjoined with the erratic environmental conditions. In order to compensate the degraded soil quality, farmers are impelled to use synthetic fertilizer which is not recognized as a sustainable practice. The injudicious use of synthetic fertilizers by the farmers to obtain improved yield and income has instead made the land to pay for the consequences. The diminished biological activity, increased level of toxicity, decreased fertility, etc. are some of those consequences. In addition to this, the changing of environmental conditions into adverse forms has only added to the problem. The soil factors and the environmental factors which adversely affect the production and productivity of crop are altogether known as abiotic stresses. They have now become a severe threat to the agricultural productivity and are accountable for principal crop yield losses (Wang et al. 2003; Padgham 2009; Wani et al. 2016). As per an estimate by the Food and Agricultural Organization, a mere 3.5% of the total global land is free from any of the environmental coercions (Van Velthuizen 2007). The most prominent abiotic stresses include salinity, drought, high/low temperature, soil acidity, nutrient unavailability, nutrient toxicity, anaerobiosis, submergence, ultraviolet irradiations, etc. (Wang et al. 2003; Chaves and Oliveira 2004; Agarwal and Grover 2006; Hirel et al. 2007; Meena et al. 2017). These abiotic stresses account for 50–82% yield losses worldwide and are extremely detrimental when occurring in combination (Mittler 2006; Gao et al. 2007).

In the drought-prone areas of arid and semiarid conditions, drought stress induced due to water deficiency is the predominant factor which hampers the crop yield and productivity. About 64% of the total world lands are affected by this stress and essentially requires an integrated approach for amelioration (Mittler 2006; Cramer et al. 2011). Soil salinity is the most persistent stress which has only aggravated over the time all over the world and has an adverse effect on about 20–50% of the land used for growing irrigated crops (Flowers and Flowers 2005; Munns and Tester 2008; Yuan et al. 2015). Extremes of temperature both high (heat stress) and low (cold stress) are harmful to the crop. Heat stress causes have both the direct effect like increase in the internal temperature of the plants and indirect effects such as water deficit by excessive transpiration and low water potential in the soil surrounding the root (Hall 2000). Cold stress on the other hand causes reduced rate of seed germination, stunted growth of seedlings, chlorosis, reduced expansion of leaves, hampered reproductive development, etc. (Yadav 2010). Since the plants are sessile organisms, they have to affront these stresses and develop potential tactics in order to withstand and/or to avoid them and survive. Plants generally endure these stresses through their intrinsic abilities to survive in unfavorable conditions (Simontacchi et al. 2015). In the past decades or so, plant breeders have tried using the breeding approaches to incorporate these abilities to crop plants so that abiotic stress tolerance or resistance can be built in them. This is a very time taking process and a very tedious task as screening and selection in many generations of the breeding population is essentially required. The use of genetic engineering methods to impart these abilities to crop plants on others is dealt with high criticism and negativity owing to the environmental hazards and ethical issues. Therefore, there is an obligation to come with a more suitable and adoptable method or approach which can impart tolerance or resistance to the crop plants against these behemoth abiotic stresses.

Plant growth-promoting bacteria (PGPR) are beneficial bacterial species that colonize the rhizospheric plane of the soil, are free-living, and promote plant growth (Schroth and Hancock 1982; Kloepper et al. 1989; Dimkpa et al. 2009; Beneduzi et al. 2012). The soil rhizosphere is an exclusive region of the soil which is explicitly influenced by the roots of growing plant (Dobbelaere et al. 2003). This zone or area of the soil has an abundance of nutrients in comparison to the bulk soil because of the accumulated varieties of exudates from roots of the plants like sugars and amino acids that are required for many bacterial species for their growth and colony development (Gray and Smith 2005). Hence, there is almost 10–100 times more bacterial count in the rhizospheric soil in comparison to bulk soil (Weller and Thomashow 1994). These PGPRs usually colonize the rhizosphere or rhizoplane and sometimes the roots of growing plants as per their nature and are the most abundant as well as elaborate of all those microbes which are associated with higher plants (Gray and Smith 2005). The microbial community which is associated to terrestrial plants has been present all along the evolution period and has assisted them against various challenges like nutrient unavailability, abiotic stresses, and pathogens (Smith et al. 2015). The green revolution came at many environmental costs owing to the ill effects of various chemical inputs such as pesticides, herbicides, and chemical fertilizers. There is a need to bring about a novel revolution which can sustain the demands of the growing population as well as the changing climatic conditions. PGPRs can be very crucial in bringing one such revolution as it comes under the approach of using biological inputs to improve the crop growth and at the same time reduce and/or avoid environmental degradation. The utilization of beneficial microbes as agricultural inputs is an age-old technique which began with the inoculation of compatible rhizobial bacteria to legumes in the early 20s (Desbrosses and Stougaard 2011). Since then, there have been numerous studies on the rhizospheric microbes especially PGPRs to find out their impact on the crop plants under various biotic as well as abiotic stress. Many of those PGPRs which have been studied for their beneficial impacts are now used on a commercial scale for alleviating abiotic and biotic stresses of crop plants. In this chapter, we will discuss different PGPRs which help crop plants to cope up the abiotic stress condition and progress the crop as well as soil productivity and the mechanisms involved in doing so.

10.2 PGPR Mechanisms in Relation to Crop Productivity Under Abiotic Stresses

The crop plants essentially require water, light, and mineral nutrients for their normal development and growth followed by reproduction in order to continue their generation cycle. Any fluctuations from these optimal conditions of environment and nutrients alter their growth and developmental process. Plants, in general, can sense any abiotic stresses, viz. drought, salinity, thermal stress, cold stress, etc., and react accordingly to sustain themselves in these unfavorable conditions (Crane et al. 2011; Ahmad et al. 2015; Jiang et al. 2016). The life cycle of a plant consists of two modes: the growth mode, where they employ all their resources for growth and developmental processes, and the defense mode, where they deploy all their resources to combat the stresses (Huot et al. 2014; Karasov et al. 2017). When in growth mode, plants grow lavishly without any hesitation, reach reproductive phase, and complete its life cycle. However, in defense mode, the growth of plant becomes stagnant, and there is a delay in reaching to their reproductive phase. This ultimately hampers their reproductive potential which in the case of agriculture is, in general, the production potential. Various PGPRs help plants to continue in their growth mode, and the former combat the stresses on behalf of the latter. There are various direct and indirect mechanisms of PGPR species which are involved in the maintenance and/or enhancement of crop productivity under abiotic stresses which are as described further.

10.2.1 PGPRs Under Drought Stress

As described earlier, drought is regarded as the chief reason for yield losses in crop plants by adversely affecting their physiological processes (Lambers et al. 2008). In response to drought stress, plant generates increased production of abscisic acid (cause increased water uptake, stomatal closure, and reduced leaf expansion), osmolytes, proline level, etc. (Barnabás et al. 2008; Farooq et al. 2009; Ashraf 2010; Farooq et al. 2012). PGPRs have evolved with different survival mechanism to tolerate drought stress. Majorly, they accumulate various osmolytes to increase their osmotic potential, develop thick cell wall to avoid loss of water, enter the dormant stage to escape the drought period, and produce exo-polysaccharides (Kumar and Verma 2018). These microbial species have various mechanisms to negate the adverse impacts of drought on crop plants and soil. The probable mechanisms include (1) production of phytohormones like indole-3-acetic acid (IAA), abscisic acid (ABA), and cytokinin, (2) synthesis of exopolysaccharides (EPSs), (3) production of 1-aminocyclopropane-1-carboxylate (ACC) deaminase, and (4) induced systemic tolerance.

Phytohormones are responsible for relieving the plants from stress due to abiotic factors and increase their chance of existence (Skirycz and Inze 2010; Fahad et al. 2015; Vurukonda et al. 2016). IAA, a major auxin that regulates various physiological processes in plants, is the molecule whose production is one of the most directly used mechanism of PGPR on crops. It is known to change the architectural structure of the roots under drought along with the increase in the number of root tips and surface which ultimately leads to embellished water and nutrient uptake by the plants (Vacheron et al. 2013). Application of *Enterobacter lignolyticus* strain TG1 under greenhouse conditions recorded a significant amount of IAA in treated tea clones with a 4.3-fold increased root biomass and 2.2-fold increased root length in comparison to control (Dutta et al. 2015). *Bacillus amyloliquefaciens* and *Paenibacillus polymyxa* BFKC01 have the genes which are essential for the biosynthesis of IAA which enables them to produce the hormone in sufficient amounts that could be utilized for growth enhancement of plants (Zhou et al. 2016; Rosier et al. 2018). ABA also has a major role during drought stress as controls in the regulation of water loss in plants by controlling the opening and closing of stomatal pore. It also ameliorates the drought stress by regulating the expression of drought-related genes and hydraulic conductivity of the root). Additionally, ABA also orchestrates the root architecture under salinity and drought stress in order to enhance branching and increase in water absorption (Tardieu et al. 2010; Etesami and Maheshwari 2018). For example, *Azospirillum brasilense* and *Azospirillum lipoferum* ameliorated drought stress in *Arabidopsis* and maize, respectively, by increasing the level of ABA. Exopolysaccharides (EPSs) are produced by many microbial species under unfavorable conditions which can either be slime or capsular, and it protects them by stabilizing the cell membrane (Donot et al. 2012; Mishra and Jha 2013; Ojuederie and Babalola 2017; Fukami et al. 2018). The EPS-producing genera such as *Bacillus*, *Pseudomonas*, *Acinetobacter*, and many other colonize in the rhizosphere and rhizoplane of the plants and make a stable soil aggregate for better absorption of water (Sandhya et al. 2009; Cheng et al. 2019). *Pseudomonas putida* GAP-P45 inoculation alleviated sunflower plants from drought stress through a high level of EPS (Sandhya et al. 2009) and three bacterial strains: *Proteus penneri* Pp1, *Pseudomonas aeruginosa* Pa2, and *Alcaligenes faecalis* did the same in maize plants (Naseem and Bano 2014). ACC is an immediate precursor of ethylene which is a phytohormone controlling senescence, aging, fruit ripening, abscission, etc. in plants whose high concentration is deleterious. ACC deaminase produced by PGPRs hydrolyses ACC into ammonia and α-ketobutyrate (Bal et al. 2013). This ACC deaminase is required by plants to degrade the ACC molecules that are produced under drought stress and consequently reduce the level of ethylene production for normal development. *Achromobacter piechaudii* ARV8 alleviated the drought stress effects on tomato and pepper through the production of ACC deaminase (Mayak et al. 2004a, b). In a similar fashion, *Pseudomonas fluorescens* DR7 and *Pseudomonas fluorescens* D11 which have an elevated ACC deaminase activity increased the germination rate in foxtail millets under very low water potentials (Niu et al. 2018). Some of the important microbe-assisting plants coping up draught stress are given in Table 10.1.

Another possible mechanism of PGPRs to impart drought tolerance in crop plants is the production of osmolytes. Out of the many compounds that are exudated by root zone inhabiting bacterial species, some are osmolyte in nature. Glycine betaine is one such compound that is produced by osmo-tolerant bacterial species which can probably act in synergism with plant-produced one to embellish the drought stress response. In consistence to the statement, osmolyte-producing rhizobacteria had a significant beneficial effect on rice under more severe drought stress. The PGPR is also known to decrease the antioxidant activity in plants and increase the production of free amino acids, proline, and sugars (Vardharajula et al. 2011). The accumulation of proline in root and shoot of plants associated with *Pseudomonas putida* and *Bacillus thuringiensis* decreases the stomatal conductance and leakage of electrolytes (Ortiz et al. 2015). Additionally, compatible solutes like trehalose, proline, glycine, and betaine are produced by many bacterial species which acts as osmoprotectants for the plants under drought stress (Tiwari et al. 2016; Cura et al. 2017).

10.2.2 PGPRs Under Salinity Stress

Salinity brings alteration to flowering and fruiting pattern and aberrated reproductive physiological process which causes loss of crop yield and biomass accumulation (Ghanem et al. 2009; Khan et al. 2012; Farooq et al. 2017). Soil salinity is also a

Table 10.1 Microbe-assisted drought stress tolerance in plant

S. no.	Microbe inoculation	Test plant	Tolerance strategy generated	Reference
1.	*Rhizobium tropici* with *Paenibacillus polymyxa*	*Phaseolus vulgaris*	Upregulation of genes involved in stress tolerance	Figueiredo et al. (2008)
2.	*Burkholderia phytofirmans*	*Zea mays*	Increased photosynthesis, root and shoot biomass under drought conditions	Naveed et al. (2014a, b)
	Enterobacter sp. FD17			
3.	*Bacillus thuringiensis* AZP2	*Triticum aestivum*	Production of volatile organic compounds	Timmusk et al. (2014)
4.	*Pseudomonas chlororaphis* O6	*Arabidopsis thaliana*	Production of 2*R*,3*R* butanediol—a volatile compound	Cho et al. (2008)
5.	*Pseudomonas putida* strain GAP-P45	*Helianthus annuus*	Epoxypolysaccharide production	Sandhya et al. (2009)
6.	*Bacillus cereus* AR156, *B. subtilis* SM21 and *Serratia* sp. XY21	*Cucumis sativa*	Production of monodehydroascorbate, proline, and antioxidant enzyme, expression of genes	Wang et al. (2012)
7.	*Bacillus* spp. strains KB122, KB129, KB133, and KB14	Sorghum (*Sorghum bicolor*)	Production of siderophore IAA and solubilization of phosphate	Naveed et al. (2014a, b)
8.	*Ochrobactrum pseudogrignonense* RJ12, *Pseudomonas* sp. RJ15 and *Bacillus subtilis* RJ46	Black gram (*Vigna mungo* L.) and Garden pea (*Pisum sativum* L.)	Synthesis of siderophore, ACC deaminase activity, indole-3-acetic acid production, and phosphate solubilization	Saikia et al. (2018)
9.	*Pseudomonas aeruginosa* (JHA6) and *Bacillus amyloliquefaciens* (ROH14)	Pepper (*Capsicum annum* L.; Solanaceae)	Synthesis of siderophore, ACC deaminase activity, and indole-3-acetic acid production	Gupta et al. (2019)
10.	*Achromobacter piechaudii*	Tomato (*L. esculentum*), pepper (*Capsicum annuum*)	Reduced ethylene	Mayak et al. (2004a, b)
11.	*A. brasilense*	Maize (*Z. mays*	Proline accumulation in leaves and roots, increase in phosphatidylcholine content	Casanovas et al. (2002)
12.	*A. brasilense*	Common bean (*P. vulgaris*)	Increased IAA production	German et al. (2000)

cause of poor microbial activity because of the toxic effects of ion concentrations and osmotic stress. The higher water potential due to soil salinity also makes water and nutrient uptake difficult for the crop plants. The major direct mechanisms of PGPRs which are involved to negate the effect of soil salinity on crop productivity are formation of biofilms, production of phytohormones, nutrient mobilization, antioxidant enzymes production, osmoprotectant production, siderophore production, and nitrogen fixation (Hayat et al. 2010; Mishra et al. 2018; Egamberdieva et al. 2019). Most of these mechanisms cause changes in pattern of root growth leading to an increase in root growth and number and thereby uptake of water and nutrients (Egamberdieva and Kucharova 2009). The biofilm-forming bacterial species alleviates deleterious effects caused by soil salinity (Kasim et al. 2016).

PGPRs are well recognized to produce phytohormones of which IAA is required by the plants for their cells to divide and elongate in order to cope with soil salinity. Few of the most known PGPRs to produce IAA under soil salinity are *Arthrobacter*, *Azotobacter*, *Azospirillum*, *Stenotrophomonas*, *Pseudomonas*, and *Rahnella* (Egamberdieva et al. 2008, 2018; Piccoli et al. 2011; Abd Allah et al. 2018). *Pseudomonas putida* application to cotton led to the modulated synthesis of IAA in plant tissue and upgraded growth parameters (Yao et al. 2010). Salt-tolerant *Streptomyces* isolates were also seen to improve the plant's ability of IAA production in wheat under soil salinity (Sadeghi et al. 2012). Cytokinins produced by many salt-tolerant bacterium such as *Arthrobacter*, *Azospirillum*, *Bacillus*, *Halomonas*, and *Pseudomonas* are very much important for the processes of cell proliferation and cell differentiation in plants under soil salinity (García de Salamone et al. 2001; Karadeniz et al. 2006; Naz et al. 2009; Parray et al. 2016). ABA, which is also a major phytohormone which influences plant processes under soil salinity, is synthesized by many PGPRs like *Pseudomonas fluorescens*, *Bacillus megaterium*, *Bacillus licheniformis*, *Proteus mirabilis*, and *Achromobacter xylosoxidans* (Karadeniz et al. 2006; Forchetti et al. 2007; Salomon et al. 2014). Gibberellins are also produced by many salt-tolerant bacteria which is helpful in alleviating crop productivity under soil salinity (Bottini et al. 2004). In a recent study, salicylic acid has also been reported to ameliorate salt stress and enhance growth in sunflower under soil salinity (Tewari and Arora 2018).

ACC deaminase has a very key role in combating salt stress also. In many studies, ACC deaminase producing PGPRs like *Stenotrophomonas rhizophila*, *Pseudomonas fluorescens*, *Pseudomonas migulae*, *Brachybacterium saurashtrense*, *Brevibacterium casei*, and *Haererohalobacter* improved the tolerance level to soil salinity in many crop plants (Egamberdieva et al. 2011; Shukla et al. 2012; Ali et al. 2014). In a similar trend, salt-tolerant species of bacterial genera *Arthrobacter*, *Brevibacterium*, *Bacillus*, *Gracilibacillus*, *Salinicoccus*, *Virgibacillus*, *Pseudomonas*, and *Exiguobacterium*, which produced ACC deaminase, stimulated maize growth under soil salinity (Aslam and Ali 2018). Many of the PGPRs produce osmoprotective products such as proline, glycine, betaine, trehalose, polyamines, sugars, etc. which aids plant to overcome salt stress (Rajendrakumar et al. 1997; Saum and Muller 2007; Bremer and Kramer 2019; Kushwaha et al. 2019; Shim et al. 2019). EPS produced from the PGPRs form a physical barrier around the plant's roots which support plant progression under higher soil salinity (Vaishnav et al. 2016). Antioxidative enzyme-producing PGPRs minimize the negative effects of oxidative stress in plants upon inoculation (Manaf and Zayed 2015; Islam et al. 2016). Many antioxidative enzymes like POD, CAT, SOD, NR, and GR are produced by PGPRs in high concentration under soil salinity and influence the crop productivity (Kohler et al. 2009; Jha and Subramanian, 2013; Patel and Saraf 2013; Sen and Chandrasekhar 2015; Hidri et al. 2016; Ansari et al. 2019; El-Esawi et al. 2019). Some PGPRs like *Pseudomonas fluorescens* OKC also are known to influence the stress-responsive transcription factors under salinity stress (Kumar et al. 2019) as summarized in Table 10.2.

10.2.3 PGPRs Under Thermic Stress

The changing global climatic scenario has led to an increase in the occurrence along with strength of temperature stress, and both the heat and cold stresses are becoming a compelling abiotic stress factor hampering the crop production. Many plant processes are regulated by temperatures, such as root elongation, transpiration, photosynthesis, enzymatic action, and cell division. The plant uses various mechanisms to negate the temperature stress which constitutes of production and accumulation of enzymes and osmolytes (Kotak et al. 2007; Qu et al. 2013). Although not all plants possess such kind of mechanisms, hence it becomes important to find out other ways for their survival under heat and cold stress. There are many bacterial species which are proved to be present in extreme temperature conditions and have adaptations to high and low temperatures. Production of heat shock proteins by PGPRs living under high-temperature conditions is one such adaptation which helps in their survival. Trehalose accumulation is another adaptation which helps PGPRs to sustain heat and cold shock injury along with oxidative stress (Kumar and Verma 2018).

Research on PGPR interactions with crop plants under temperature is relatively scarce, and the mechanisms are also not well defined. However, in a study, it was seen that the interactions of various bacterial strains with soybean growth and physiology were temperature-dependent under suboptimal root-zone temperature (Zhang et al. 1997). Inoculation of *Burkholderia phtofirmans* PsJN on 18 clones of potato imparted better tuberization under high temperature (Bensalim et al. 1998). Inoculation of a same bacterial strain to grapevine was able to lower the rate of biomass destruction and escape of electrolytes during cold stress as well as enhanced the recovery process post cold injury (Barka et al. 2006). Another mechanism of PGPRs which aids in protecting plants from temperature stress is embellished accumulation of sugars, proline, and anthocyanin, since there was also a significantly higher amount of carbohydrates, proline, and phenols in *Burkholderia phtofirmans* inoculated grapevine after cold stress (Barka et al. 2006). Similarly, *Pseudomonas cedrina*, *Brevundimonas terrae*, and *Arthrobacter nicotianae* which are adapted to low temperatures also show plant growth-promoting abilities (Yadav et al. 2014) as

Table 10.2 Microbe-assisted salinity stress tolerance in plants

S. no.	Microbe inoculation	Test plant	Tolerance strategy generated	Reference
1.	*Bacillus subtilis* GB03	*Arabidopsis thaliana*	Tissue-specific regulation of sodium transporter HKT1	Zhang et al. (2008)
2.	*Pseudomonas simiae* 4-	*Glycine max*	Nitroguaiacol and quinoline promote soybean seed germination	Vaishnav et al. (2016)
3.	*Pseudomonas syringae* DC3000, *Bacillus* sp. strain L81, *Arthrobacter oxidans*	*Arabidopsis thaliana*	SA-dependent pathway	Barriuso et al. (2008)
5.	Cyanobacteria and cyanobacterial extracts	*Oryza sativa*, *Triticum aestivum*, *Zea mays*, *Gossypium hirsutum*	Phytohormones as elicitor molecule	Singh (2014)
6.	*Pseudomonas koreensis* strain AK-1	*Glycine max* L.	Merrill reduction in NaC level and increase in KC level	Kasotia et al. (2015)
7.	*Glomus etunicatum*	*Glycine max*	Increased root but decreased shoot proline concentrations	Sharifi et al. (2007)
8.	*Burkholderia*, *Arthrobacter*, and *Bacillus Vitis vinifera*	*Capsicum annuum*	Increased accumu-lation of proline	Barka et al. (2006)
9.	*Azospirillum brasilense*	Pea (*Phaseolus vulgaris*)	Induced flavonoid content	Dardanelli et al. (2008)
10.	*Pseudomonas syringae*, *Pseudomonas fluorescens*, *Enterobacter aerogenes*	Maize (*Zea mays*)	ACC deaminase activation	Nadeem et al. (2007)
11.	*P. fluorescens*	Groundnut (*Arachis hypogaea*)	Decreased ethylene production	Saravanakumar and Samiyappan (2007)
12.	*Achromobacter piechaudii*	Tomato (*Lycopersicon esculentum*)	Reduced ethylene production	Mayak et al. (2004a, b)

Table 10.3 Microbe-assisted thermic stress tolerance in plants

S. no.	Microbe inoculation	Test plant	Tolerance strategy generated	Reference
1.	*Bacillus amyloliquefaciens*	*Triticum aestivum*	Reduced regeneration of reactive oxygen species	El-Daim et al. (2014)
2.	*Burkholderia phytofirmans*	Grapevine (*Vitis vinifera*)	Plant growth promotion and disease resistance	Barka et al. (2006)
3.	*Azospirillum brasilense*	*Triticum aestivum*	Preactivation of heat shock transcription factors, changes in metabolome	El-Daim et al. (2014)
4.	*B. phytofirmans*	Potato (*Solanum tuberosum*)		Bensalim et al. (1998)
5.	*Aeromonas hydrophila*, *Serratia liquefaciens*, *Serratia proteamaculans*	Soy bean (*Glycine max*)	Plant growth promotion	Zhang et al. (1997)
6.	*Pseudomonas putida* strain AKMP7	*Triticum* spp.	Reduced membrane injury and the activity of several antioxidant enzymes such as SOD, APX, and CAT	Zulfikar Ali et al. (2011)

summarized in Table 10.3. In a study, bacteria isolated from wheat grown under cool conditions proved to be an efficient colonizer of rhizosphere and improved plant's resistance towards cold climate stress. The same trend was found for high temperature when wheat was inoculated with bacterial strains isolated from warmer environments (Egamberdiyeva and Hoflich 2003). There was an increase in the rate of survival and development under heat treatment of sorghum seedlings inoculated with *Pseudomonas aeruginosa* strain isolated from hot semiarid conditions (Ali et al. 2009). PGPR isolated from root nodules of pea growing in low temperature has efficient bio fertilizer ability at colder temperatures (Meena et al. 2015)

10.3 PGPRs in Soil Productivity Under Abiotic Stresses

Soil is the natural dwelling place for all the microbes which may be either beneficial or harmful to the crop plants. PGPRs also live in this phytomicrobial community and are known to carry out various important processes which enhance the soil health and productivity. Soil productivity is often defined as the aptitude of soil to yield a certain amount of agricultural crops or other plants by use of a distinct set of practices. Soil fertility along with intrinsic and other management-related factors affecting plant growth and development is included in it (Karlen 2005). The major soil functions which influence the productivity of soil are soil structure, air, plant-available water, and essential nutrients. Injudicious use of synthetic insecticides, fungicides, fertilizers, etc. has posed great environmental hazard along with depletion of soil productivity. Additionally, the aberrant climatic condition has only added to the magnitude of the problem. Accumulation of toxic metals, nutrient immobilization, loss of soil structure, reduction in the population level of beneficial microbes, reduced water holding capacity, less soil organic matter, and many more constitute the ill-effects brought in soil productivity by the indiscriminate use of synthetic fertilizers and chemicals along with faulty agronomic practices. Application of PGPRs can neutralize these harmful effects and can restore the soil productivity and reduce the magnitude of abiotic stresses posed by them onto crop plants.

PGPRs are very well known for their beneficial activities within soil that includes crop residue decomposition, soil organic mineralization, soil nutrient mobilization, phosphorous solubilization, soil organic matter synthesis, nitrification, fixation of atmospheric nitrogen, acquisition of nutrients, phytohormone production, and suppression of phytopathogenic microbes (Prasad et al. 2015). Fixation of atmospheric nitrogen is the most important activity which is performed by PGPRs that help to enhance soil productivity. Rhizobia include various Proteobacteria well known to colonize and fix atmospheric nitrogen in plant roots, and a part of this fixed nitrogen is provided to plants as an exchange material of photosynthates (Long 1989). These include *Allorhizobium*, *Azorhizobium*, *Bradyrhizobium*, *Mesorhizobium*, *Ensifer/ Sinorhizobium*, and *Rhizobium* (Reddy 2014). These bacterium species grow symbiotically with leguminous crops enhancing both crop and soil productivity. There are also certain PGPRs which fix atmospheric nitrogen but are free-living and/or associative such as *Azospirillum*, *Azotobacter*, *Gluconoacetobacter diazotrophicus*, *Azocarus*, *Nostoc*, etc. (Bhattacharyya and Jha 2012). The second most important nutrient is phosphorous which limits the growth of crop plants. It is present abundantly as organic and inorganic forms in the soil, but the plant-available forms are relatively lower which decreases the soil productivity (Khan et al. 2009). There are certain phosphorus-solubilizing bacteria (PSB) which convert the fixed phosphorous into plant-available forms that come under genera like *Azotobacter*, *Bacillus*, *Beijerinckia*, *Burkholderia*, *Enterobacter*, *Erwinia*, *Flavobacterium*, *Pseudomonas*, *Rhizobium*, *Microbacterium*, and *Serratia* (Bhattacharyya and Jha 2012). In addition of making phosphorus availability to plants, PSBs also stimulate the efficiency of biological nitrogen fixation and embellish the availability of other trace elements (Suman et al. 2001; Ahmad et al. 2008; Zaidi et al. 2009).

For almost all the life forms, iron has been identified as a vital nutrient; hence, the level of plant absorbable iron forms also constitutes the soil productivity. The bacterial species obtain iron from their surrounding by secretion of a low-molecular-weight iron chelator known as siderophores that have a high binding affinity to the complexing iron. The siderophore is a molecule which is mostly water-soluble and is found in two different forms, viz. extracellular siderophores and intracellular siderophores. Their function is to solubilize iron from its organic as well as inorganic form, seven under limited iron conditions (Indiragandhi et al. 2008). Plants are able to assimilate iron from these siderophores through many ways of mechanisms like chelation, through direct uptake of siderophore-iron complex, or through ligand exchange (Schmidt 1999). The exhaustive agricultural practices, industrialization, and anthropogenic activities have led to heavy metal contamination in the soil which has also caused decrease in soil productivity. Microbes have the ability to remove heavy metal contamination by the process of bio-accumulation. Major mechanisms through which PGPRs alleviate the heavy metal contamination in the soil are extracellular and intracellular accumulation, sequestration, and bio-transformation of highly toxic compounds to less toxic ones (Qian et al. 2012; Babu et al. 2013). The bacterial groups from Proteobacteria, Firmicutes, and Actinobacteria have the potential to remove manganese, lead, and arsenic metals from the soil when in higher concentrations (Zhang et al. 2015). *Viciafaba* growth with inoculation of PGPR had less effect of copper toxicity in comparison to the control (Fatnassi et al. 2015). Heavy metal-resistant bacteria like *Bacillus*, *Lysinibacillus*, and *Pseudomonas* produce chelating agents which improve the phytoremediation of heavy metals and hence improve soil productivity (Vigliotta et al. 2016).

In a study, it was seen that barley produced 120% greater grain yield with twofold reduced cadmium content in grains when inoculated with *Klebsiella mobilis* CIAM 880 and grown under cadmium-contaminated soil (Pishchik et al. 2009). In another study, *Brassica juncea* which when grown with inoculation of IAA and bacteria could produce siderophores showed tolerance to chromium contamination without any alteration in nutrient uptake (Rajkumar et al. 2005). Rice production under wetland conditions is greatly hampered by toxicity due to metal iron. This constraint is indicated to be mitigated to some extent by several *Bacillus* strains (Asch and Padham 2005; Terre et al. 2007). PGPRs also improve the decomposition rate of many organic residues in soil and thereby help in increasing the soil organic matter content. PGPRs produce lytic enzymes such as chitinases, cellulase, β-glucanases, lipases, dehydrogenase, phosphatases, proteases, etc. (Lanteigne et al. 2012; Joshi et al. 2012). These lytic enzymes can be very much helpful in the fast decomposition of organic residues in soil.

10.4 PGPRs as a Remedial Agent Against Abiotic Stresses

The broad-spectrum application of PGPRs to the crop plants in any form as seed inoculants, seedling inoculants, and/or soil inoculants would reduce the use of chemical pesticides and synthetic fertilizers that pollute the environment. The increasing monetary cost of pesticides and the growing consumer demand have made it essential to find a substitute for these chemical inputs, and there is a global market for pesticide- and fertilizer-free food. Additionally, there are certain crop diseases and pest for which there are very few or ineffective chemical solutions which add to the scope of PGPR application. An approach which utilizes biological control to all these problems is considered appropriate, but the only lacuna in their widespread application is the inconsistent performance in the field. With the increasing progress in the study of PGPR application under field conditions, there is also an increase in chances of its success under field conditions. The potential of abiotic stress-tolerant PGPRs can be harnessed for improving crop and soil productivity in the increasing magnitude of abiotic stresses. The elucidation of mechanisms behind all these abiotic stress tolerance would be very helpful in the long-term goal of improving crop yield under stress conditions. The understanding of PGPR mechanisms for providing stress tolerance in crop plants would serve as a favorable measure for rebutting the ill-effects of abiotic stress and improving the global production of food. Although a more illustrated investigation of microbial responses to the abiotic stresses is required for their more efficient utilization against unfavorable conditions. The crop-specific field trials of PGPRs have given strong suggestions about their use to mitigate abiotic stresses in various crops. The replication of these findings at different geographic locations and different crop plants with the same extent of success has been a drawback. This is the reason for the erratic results of PGPR application at field conditions (Souza et al. 2015; Ambrosini et al. 2016).

Commercialization of PGPRs has been at a slow pace due to their inconsistent behavior under field conditions. It is therefore very necessary to find out the outcome of the release of such bacterial species under stressful environments before moving forward to their commercialization. The commercialization of Pseudomonad strains failed because of their inability to long-term survival owing to their as sporogenous behavior. Instead, the commercialization of *Bacillus* species of PGPRs has however been carried forward due to their long-term viability because of endospore formation. Survival of introduced PGPR species in any given community of microbe is also a major factor in determining their commercial application. The survival depends upon the competing microbial organisms of any given community for the limited resources, and furthermore, PGPRs should be able to survive in poor soils where the development of an indigenous microbial community is inhibited (Strigul and Kravchenko 2006; Solano et al. 2007). Altogether, PGPRs perform well and promote the growth of crop plants under normal conditions but are sensitive to environmental fluctuations and have inconsistent performance. The successful application of PGPRs under field conditions is for bio-fertilization, phytoremediation, and bio-protection. *Azospirillum*, *Herbaspirillum*, *Acetobacter*, *Azotobacter*, and *Azoarocus* are successfully used as biofertilizers under field conditions. There is a long way forward for the large-scale application and commercialization of PGPRs for the alleviation of abiotic stress and increasing crop as well as soil productivity. There is a need to fill the research gaps which is essentially required to go forward with PGPRs, a potential driver of increasing soil as well as crop productivity under increasing abiotic stress conditions.

10.5 Conclusion and Future Implication

In the times of changing weather pattern and increasing degradation of soil, it has essential to bring changes to the system of crop production and cultivation of the soil. PGPRs have been proved then and again to induce resistance and/or tolerance to crop plants against biotic as well as abiotic stress. The scientific community should go on with the research to find out specific PGPRs that could provide defense against multiple abiotic stresses and simultaneously increase the crop yield. Successful utilization of PGPRs as a biofertilizer, phytoremediation, or bio-protector would depend on the establishment of the desired bacterial strains into the already present soil microbial community. In addition to this, a proper and efficient inoculation method of their PGPRs to the crop plants is also needed to be studied in order to find out the right time and method of application of these beneficial microbes under the field conditions so that they can establish themselves and survive before providing beneficial effects to crop and soil. Various other parameters such as type of soil, chemical properties of soil, as well the management practices should be in coherence to the establishment of preselected PGPR population under the given set of conditions. There also should be a defined source of energy for PGPRs before they could establish them and reach the threshold population and could live on their own or in symbiotic association with crop plants.

In addition to the isolation methods, biotechnological methods should be applied for the transfer of gene of interest to already studied and characterized PGPR in order to increase their scope of application. The crop plants should also be bred in such a way that the produced crop varieties should interact with the beneficial PGPRs upon inoculation and help in the establishment in order to promote a specific beneficial plant-bacteria combination as in case of legumes and rhizobial groups. There should also be research on finding the compatible consortium of PGPRs which would work synergistically upon inoculation with crop plants under abiotic stress. There are many defined PGPRs which work very well under certain abiotic stress condition. There should be future studies to find out ways through which these stress-specific PGPRs could be used together as a consortium in crop plants so that they could alleviate multiple abiotic stresses. A better formulation product of PGPRs would help to increase the shelf-life of its commercial product. So there should also be future research in order to make better formulations for proper and efficient application of PGPRs. In conclusion, it is evident that the climate is changing with that the level of abiotic stresses is increasing as well. In order to sustain the human demands, the farming community has to come up with better approaches to maintain the crop and soil productivity to meet those demands. Application of PGPRs to crop plants and to the soil is a very good option as it is of biological origin and also does not degrade the environment further. There are certain lacunas in their application, but it can be getting fulfilled with further research and development in scientific advancements. The relationship between plants and the rhizospheric microbial community which has been set all along the process of evolution is the key factor in mitigation of the abiotic stresses, and application of PGPRs would represent a strong step ahead in taking the beat out of the interaction between the two.

References

Abd Allah EF, Alqarawi AA, Hashem A, Radhakrishnan R, Al-Huqail AA, Al-Otibi FON et al (2018) Endophytic bacterium Bacillus subtilis (BERA 71) improves salt tolerance in chickpea plants by regulating the plant defense mechanisms. J Plant Interact 13(1):37–44

Agarwal S, Grover A (2006) Molecular biology, biotechnology and genomics of flooding-associated low O2 stress response in plants. Crit Rev Plant Sci 25(1):1–21

Ahmad F, Ahmad I, Khan MS (2008) Screening of free-living rhizospheric bacteria for their multiple plant growth promoting activities. Microbiol Res 163(2):173–181

Ahmad P, Hashem A, Abd-Allah EF, Alqarawi AA, John R, Egamberdieva D, Gucel S (2015) Role of Trichoderma harzianum in mitigating NaCl stress in Indian mustard (Brassica juncea L) through antioxidative defense system. Front Plant Sci 6:868

Ali SZ, Sandhya V, Grover M, Kishore N, Rao LV, Venkateswarlu B (2009) *Pseudomonas* sp. strain AKM-P6 enhances tolerance of sorghum seedlings to elevated temperatures. Biol Fertil Soils 46(1):45–55

Ali SZ, Sandhya V, Grover M, Linga VR, Bandi V (2011) Effect of inoculation with a thermotolerant plant growth promoting Pseudomonas putida strain AKMP7 on growth of wheat (Triticum spp.) under heat stress. J Plant Interact 6(4):239–246. https://doi.org/10.1080/17429145.2010.545147

Ali S, Charles TC, Glick BR (2014) Amelioration of high salinity stress damage by plant growth-promoting bacterial endophytes that contain ACC deaminase. Plant Physiol Biochem 80:160–167

Ambrosini A, de Souza R, Passaglia LM (2016) Ecological role of bacterial inoculants and their potential impact on soil microbial diversity. Plant Soil 400(1–2):193–207

Ansari FA, Ahmad I, Pichtel J (2019) Growth stimulation and alleviation of salinity stress to wheat by the biofilm forming *Bacillus pumilus* strain FAB10. Appl Soil Ecol 143:45–54

Asch F, Padham JL (2005) Root associated bacteria suppress symptoms of iron toxicity in lowland rice. In: Tielkes E, Hülsebusch C, Häuser I, Deininger A, Becker K (eds) The global food & product chain-dynamics, innovations, conflicts, strategies. MDD GmbH, Stuttgart, p 276

Ashraf M (2010) Inducing drought tolerance in plants: recent advances. Biotechnol Adv 28 (1):169–183

Aslam F, Ali B (2018) Halotolerant bacterial diversity associated with Suaeda fruticosa (L.) Forssk. improved growth of maize under salinity stress. Agronomy 8(8):131

Babu AG, Kim JD, Oh BT (2013) Enhancement of heavy metal phytoremediation by Alnus firma with endophytic Bacillus thuringiensis GDB-1. J Hazard Mater 250:477–483

Bal HB, Nayak L, Das S, Adhya TK (2013) Isolation of ACC deaminase producing PGPR from rice rhizosphere and evaluating their plant growth promoting activity under salt stress. Plant Soil 366 (1–2):93–105

Barka EA, Nowak J, Clément C (2006) Enhancement of chilling resistance of inoculated grapevine plantlets with a plant growth-promoting rhizobacterium, *Burkholderia phytofirmans* strain PsJN. Appl Environ Microbiol 72(11):7246–7252

Barnabás B, Jäger K, Fehér A (2008) The effect of drought and heat stress on reproductive processes in cereals. Plant Cell Environ 31(1):11–38

Barriuso J, Solano BR, Mañero FJG (2008) Protection against pathogen and salt stress by four plant growth-promoting rhizobacteria isolated from Pinus sp. on Arabidopsis thaliana. Phytopathology 98:666–672. https://doi.org/10.1094/PHYTO-98-6-0666

Beneduzi A, Ambrosini A, Passaglia LM (2012) Plant growth-promoting rhizobacteria (PGPR): their potential as antagonists and biocontrol agents. Genet Mol Biol 35(4):1044–1051

Bensalim S, Nowak J, Asiedu SK (1998) A plant growth promoting rhizobacterium and temperature effects on performance of 18 clones of potato. Am J Potato Res 75:145–152

Bhattacharyya PN, Jha DK (2012) Plant growth-promoting rhizobacteria (PGPR): emergence in agriculture. World J Microbiol Biotechnol 28(4):1327–1350

Bottini R, Cassán F, Piccoli P (2004) Gibberellin production by bacteria and its involvement in plant growth promotion and yield increase. Appl Microbiol Biotechnol 65(5):497–503

Bremer E, Kramer R (2019) Responses of microorganisms to osmotic stress. Annu Rev Microbiol 73:313–334

Casanovas EM, Barassi CA, Sueldo RJ (2002) Azospirillum inoculation mitigates water stress effects in maize seedlings. Cereal Res Commun 30:343–350

Chaudhry V, Srivastava S, Chauhan PS, Singh PC, Mishra A, Nautiyal CS (2012) Plant growth promoting rhizobacteria-mediated amelioration of abiotic and biotic stresses for increasing crop productivity. Improving crop productivity in sustainable agriculture. Wiley-VCH, Weinheim, pp 133–153

Chaves MM, Oliveira MM (2004) Mechanisms underlying plant resilience to water deficits: prospects for water-saving agriculture. J Exp Bot 55(407):2365–2384

Cheng X, Huang L, Li KT (2019) Antioxidant activity changes of exopolysaccharides with different carbon sources from Lactobacillus plantarum LPC-1 and its metabolomic analysis. World J Microbiol Biotechnol 35(5):68

Cho SM, Kang BR, Han SH, Anderson AJ, Park JY, Lee YH, Cho BH, Yang KY, Ryu CM, Kim YC (2008) 2R,3R-butanediol, a bacterial volatile produced by Pseudomonas chlororaphis O6, is involved in induction of systemic tolerance to drought in Arabidopsis thaliana. Mol Plant-Microbe Interact 21:1067–1075. https://doi.org/10.1094/MPMI-21-8-1067

Cramer GR, Urano K, Delrot S, Pezzotti M, Shinozaki K (2011) Effects of abiotic stress on plants: a systems biology perspective. BMC Plant Biol 11(1):163

Crane TA, Roncoli C, Hoogenboom G (2011) Adaptation to climate change and climate variability: the importance of understanding agriculture as performance. NJAS-Wageningen J Life Sci 57 (3–4):179–185

Cura JA, Franz DR, Filosofía JE, Balestrasse KB, Burgueno LE (2017) Inoculation with Azospirillum sp. and Herbaspirillum sp. bacteria increases the tolerance of maize to drought stress. Microorganisms 5(3):41

Dardanelli MS, Fernandez de Cordoba FJ, Rosario Espuny M, Rodríguez Carvajal MA, SoriaDiaz ME, Gil Serrano AM, Okon Y, Megias M (2008) Effect of Azospirillum brasilense coinoculated with rhizobium on Phaseolus vulgaris flavonoids and nod factor production under salt stress. Soil Biol Biochem 40:2713–2721

Desbrosses GJ, Stougaard J (2011) Root nodulation: a paradigm for how plant-microbe symbiosis influences host developmental pathways. Cell Host Microbe 10(4):348–358

Dimkpa C, Weinand T, Asch F (2009) Plant–rhizobacteria interactions alleviate abiotic stress conditions. Plant Cell Environ 32(12):1682–1694

Dobbelaere S, Vanderleyden J, Okon Y (2003) Plant growth-promoting effects of diazotrophs in the rhizosphere. Crit Rev Plant Sci 22(2):107–149

Donot F, Fontana A, Baccou JC, Schorr-Galindo S (2012) Microbial exopolysaccharides: main examples of synthesis, excretion, genetics and extraction. Carbohydr Polym 87(2):951–962

Dutta J, Handique PJ, Thakur D (2015) Assessment of culturable tea rhizobacteria isolated from tea estates of Assam, India for growth promotion in commercial tea cultivars. Front Microbiol 6:1252

Egamberdieva D, Kucharova Z (2009) Selection for root colonising bacteria stimulating wheat growth in saline soils. Biol Fertil Soils 45(6):563–571

Egamberdieva D, Kamilova F, Validov S, Gafurova L, Kucharova Z, Lugtenberg B (2008) High incidence of plant growth-stimulating bacteria associated with the rhizosphere of wheat grown on salinated soil in Uzbekistan. Environ Microbiol 10(1):1–9

Egamberdieva D, Kucharova Z, Davranov K, Berg G, Makarova N, Azarova T, Lugtenberg B (2011) Bacteria able to control foot and root rot and to promote growth of cucumber in salinated soils. Biol Fertil Soils 47(2):197–205

Egamberdieva D, Jabborova D, Wirth SJ, Alam P, Alyemeni MN, Ahmad P (2018) Interactive effects of nutrients and Bradyrhizobium japonicum on the growth and root architecture of soybean (Glycine max L.). Front Microbiol 9:1000

Egamberdieva D, Wirth S, Bellingrath-Kimura SD, Mishra J, Arora NK (2019) Salt-tolerant plant growth promoting rhizobacteria for enhancing crop productivity of saline soils. Front Microbiol 10:2791

Egamberdiyeva D, Hoflich G (2003) Influence of growth-promoting bacteria on the growth of wheat in different soils and temperatures. Soil Biol Biochem 35(7):973–978

El-Daim IAA, Bejai S, Meijer J (2014) Improved heat stress tolerance of wheat seedlings by bacterial seed treatment. Plant Soil 379:337–350. https://doi.org/10.1007/s11104-014-2063-3

El-Esawi MA, Al-Ghamdi AA, Ali HM, Alayafi AA (2019) Azospirillum lipoferum FK1 confers improved salt tolerance in chickpea (Cicer arietinum L.) by modulating osmolytes, antioxidant machinery and stress-related genes expression. Environ Exp Bot 159:55–65

Etesami H, Maheshwari DK (2018) Use of plant growth promoting rhizobacteria (PGPRs) with multiple plant growth promoting traits in stress agriculture: action mechanisms and future prospects. Ecotoxicol Environ Saf 156:225–246

Fahad S, Hussain S, Bano A, Saud S, Hassan S, Shan D, Tabassum MA (2015) Potential role of phytohormones and plant growth-promoting rhizobacteria in abiotic stresses: consequences for changing environment. Environ Sci Pollut Res 22(7):4907–4921

Farooq M, Wahid A, Kobayashi N, Fujita DBSMA, Basra SMA (2009) Plant drought stress: effects, mechanisms and management. In: Sustainable agriculture. Springer, Dordrecht, pp 153–188

Farooq M, Hussain M, Wahid A, Siddique KHM (2012) Drought stress in plants: an overview. In: Plant responses to drought stress. Springer, Berlin, pp 1–33

Farooq M, Gogoi N, Hussain M, Barthakur S, Paul S, Bharadwaj N, Siddique KH (2017) Effects, tolerance mechanisms and management of salt stress in grain legumes. Plant Physiol Biochem 118:199–217

Fatnassi IC, Chiboub M, Saadani O, Jebara M, Jebara SH (2015) Impact of dual inoculation with rhizobium and PGPR on growth and antioxidant status of Viciafaba L. under copper stress. C R Biol 338(4):241–254

Figueiredo MVB, Burity HA, Martinez CR, Chanway CP (2008) Alleviation of drought stress in common bean (Phaseolus vulgaris L.) by coinoculation with Paenibacillus polymyxa and Rhizobium tropici. Appl Soil Ecol 40:182–188. https://doi.org/10.1016/j.apsoil.2008.04.005

Flowers TJ, Flowers SA (2005) Why does salinity pose such a difficult problem for plant breeders. Agric Water Manag 78(1–2):15–24

Forchetti G, Masciarelli O, Alemano S, Alvarez D, Abdala G (2007) Endophytic bacteria in sunflower (Helianthus annuus L.): isolation, characterization, and production of jasmonates and abscisic acid in culture medium. Appl Microbiol Biotechnol 76(5):1145–1152

Fukami J, Cerezini P, Hungria M (2018) Azospirillum: benefits that go far beyond biological nitrogen fixation. AMB Express 8:73

Gao JP, Chao DY, Lin HX (2007) Understanding abiotic stress tolerance mechanisms: recent studies on stress response in rice. J Integr Plant Biol 49(6):742–750

García de Salamone IE, Hynes RK, Nelson LM (2001) Cytokinin production by plant growth promoting rhizobacteria and selected mutants. Can J Microbiol 47(5):404–411

German MA, Burdman S, Okon Y, Kigel J (2000) Effects of Azospirillum brasilense on root morphology of common bean (Phaseolus vulgaris L.) under different water regimes. Biol Fertil Soils 32:259–264

Ghanem ME, van Elteren J, Albacete A, Quinet M, Martínez-Andújar C, Kinet JM, Lutts S (2009) Impact of salinity on early reproductive physiology of tomato (Solanum lycopersicum) in relation to a heterogeneous distribution of toxic ions in flower organs. Funct Plant Biol 36 (2):125–136

Gray EJ, Smith DL (2005) Intracellular and extracellular PGPR: commonalities and distinctions in the plant–bacterium signaling processes. Soil Biol Biochem 37(3):395–412

Gupta S, Kaushal R, Sood G, Dipta B, Kirti S, Spehia RS (2019) Water stress amelioration and plant growth promotion in capsicum plants by osmotic stress tolerant bacteria. Int J Plant Soil Sci 29:1–12

Hall AE (2000) Crop responses to environment. CRC Press, Boca Raton

Hayat R, Ali S, Amara U, Khalid R, Ahmed I (2010) Soil beneficial bacteria and their role in plant growth promotion: a review. Ann Microbiol 60(4):579–598

Hidri R, Barea JM, Mahmoud OMB, Abdelly C, Azcon R (2016) Impact of microbial inoculation on biomass accumulation by Sulla carnosa provenances, and in regulating nutrition, physiological and antioxidant activities of this species under non-saline and saline conditions. J Plant Physiol 201:28–41

Hirel B, Le Gouis J, Ney B, Gallais A (2007) The challenge of improving nitrogen use efficiency in crop plants: towards a more central role for genetic variability and quantitative genetics within integrated approaches. J Exp Bot 58(9):2369–2387

Huot B, Yao J, Montgomery BL, He SY (2014) Growth-defense tradeoffs in plants: a balancing act to optimize fitness. Mol Plant 7(8):1267–1287

Indiragandhi P, Anandham R, Madhaiyan M, Sa TM (2008) Characterization of plant growth-promoting traits of bacteria isolated from larval guts of diamondback moth Plutella xylostella (Lepidoptera: Plutellidae). Curr Microbiol 56(4):327–333

Islam F, Yasmeen T, Arif MS, Ali S, Ali B, Hameed S, Zhou W (2016) Plant growth promoting bacteria confer salt tolerance in Vigna radiata by up-regulating antioxidant defense and biological soil fertility. Plant Growth Regul 80(1):23–36

Jha Y, Subramanian RB (2013) Paddy plants inoculated with PGPR show better growth physiology and nutrient content under saline condition. Chilean J Agric Res 73(3):213–219

Jiang QY, Zhuo F, Long SH, Zhao HD, Yang DJ, Ye ZH, Jing YX (2016) Can arbuscular mycorrhizal fungi reduce Cd uptake and alleviate Cd toxicity of Lonicera japonica grown in Cd-added soils? Sci Rep 6(1):1–9

Joshi M, Srivastava R, Sharma AK, Prakash A (2012) Screening of resistant varieties and antagonistic Fusarium oxysporum for biocontrol of Fusarium wilt of chilli. J Plant Pathol Microb 3 (134):2

Karadeniz A, Topcuoglu SF, Inan S (2006) Auxin, gibberellin, cytokinin and abscisic acid production in some bacteria. World J Microbiol Biotechnol 22(10):1061–1064

Karasov TL, Chae E, Herman JJ, Bergelson J (2017) Mechanisms to mitigate the trade-off between growth and defense. Plant Cell 29(4):666–680

Karlen DL (2005) Tilth. Encyclopedia of soils in the environment. Elsevier Academic Press, San Diego, pp 168–174

Kasim WA, Gaafar RM, Abou-Ali RM, Omar MN, Hewait HM (2016) Effect of biofilm forming plant growth promoting rhizobacteria on salinity tolerance in barley. Ann Agric Sci 61 (2):217–227

Kasotia A, Varma A, Choudhary DK (2015) Pseudomonas-mediated mitigation of salt stress and growth promotion in Glycine max. Agric Res 4:31–41. https://doi.org/10.1007/s40003-014-0139-1

Khan MS, Zaidi A, Wani PA, Oves M (2009) Role of plant growth promoting rhizobacteria in the remediation of metal contaminated soils. Environ Chem Lett 7(1):1–19

Khan MA, Shaukat SS, Shahzad A, Arif H (2012) Growth and yield responses of pearl millet (Pennisetum glaucum [L.] R. Br.) irrigated with treated effluent from waste stabilization ponds. Pak J Bot 44(3):905–910

Kloepper JW, Lifshitz R, Zablotowicz RM (1989) Free-living bacterial inocula for enhancing crop productivity. Trends Biotechnol 7(2):39–44

Kohler J, Hernandez JA, Caravaca F, Roldan A (2009) Induction of antioxidant enzymes is involved in the greater effectiveness of a PGPR versus AM fungi with respect to increasing the tolerance of lettuce to severe salt stress. Environ Exp Bot 65(2–3):245–252

Kotak S, Larkindale J, Lee U, von Koskull-Doring P, Vierling E, Scharf KD (2007) Complexity of the heat stress response in plants. Curr Opin Plant Biol 10(3):310–316

Kumar A, Verma JP (2018) Does plant–microbe interaction confer stress tolerance in plants: a review. Microbiol Res 207:41–52

Kumar G, Bajpai R, Sarkar A, Mishra RK, Gupta VK, Singh HB, Sarma BK (2019) Identification, characterization and expression profiles of Fusarium udum stress-responsive WRKY transcription factors in Cajanus cajan under the influence of NaCl stress and Pseudomonas fluorescens OKC. Sci Rep 9(1):1–9

Kushwaha B, Jadhav I, Verma HN, Geethadevi A, Parashar D, Jadhav K (2019) Betaine accumulation suppresses the de-novo synthesis of ectoine at a low osmotic concentration in Halomonas sp. SBS 10, a bacterium with broad salinity tolerance. Mol Biol Rep 46(5):4779–4786

Lambers H, Chapin FS III, Pons TL (2008) Plant physiological ecology. Springer Science & Business Media, Berlin

Lanteigne C, Gadkar VJ, Wallon T, Novinscak A, Filion M (2012) Production of DAPG and HCN by Pseudomonas sp. LBUM300 contributes to the biological control of bacterial canker of tomato. Phytopathology 102(10):967–973

Long SR (1989) Rhizobium-legume nodulation: life together in the underground. Cell 56 (2):203–214

Manaf HH, Zayed MS (2015) Productivity of cowpea as affected by salt stress in presence of endomycorrhizae and Pseudomonas fluorescens. Ann Agric Sci 60(2):219–226

Mayak S, Tirosh T, Glick BR (2004a) Plant growth-promoting bacteria confer resistance in tomato plants to salt stress. Plant Physiol Biochem 42:565–572

Mayak S, Tirosh T, Glick BR (2004b) Plant growth-promoting bacteria that confer resistance to water stress in tomatoes and peppers. Plant Sci 166(2):525–530

Meena RK, Singh RK, Singh NP, Meena SK, Meena VS (2015) Isolation of low temperature surviving plant growth-promoting rhizobacteria (PGPR) from pea (Pisum sativum L.) and documentation of their plant growth promoting traits. Biocatal Agric Biotechnol 4(4):806–811

Meena KK, Sorty AM, Bitla UM, Choudhary K, Gupta P, Pareek A, Singh HB (2017) Abiotic stress responses and microbe-mediated mitigation in plants: the omics strategies. Front Plant Sci 8:172

Mishra A, Jha B (2013) Microbial exopolysaccharides. In: Rosenberg E, De Long EF, Lory S, Stackebrandt E, Thompson F (eds) The prokaryotes. Springer, Berlin, pp 179–192

Mishra J, Fatima T, Arora NK (2018) Role of secondary metabolites from plant growth-promoting rhizobacteria in combating salinity stress. In: Plant microbiome: stress response. Springer, Singapore, pp 127–163

Mittler R (2006) Abiotic stress, the field environment and stress combination. Trends Plant Sci 11 (1):15–19

Munns R, Tester M (2008) Mechanisms of salinity tolerance. Annu Rev Plant Biol 59:651–681

Nadeem SM, Zahir ZA, Naveed M, Arshad M (2007) Preliminary investigations on inducing salt tolerance in maize through inoculation with rhizobacteria containing ACC deaminase activity. Can J Microbiol 53:1141–1149

Naseem H, Bano A (2014) Role of plant growth-promoting rhizobacteria and their exopolysaccharide in drought tolerance of maize. J Plant Interact 9(1):689–701

Naveed M, Hussain MB, Zahir ZA, Mitter B, Sessitsch A (2014a) Drought stress amelioration in wheat through inoculation with Burkholderia phytofirmans strain PsJN. Plant Growth Regul 73:121–131

Naveed M, Mitter B, Reichenauer TG, Wieczorek K, Sessitsch A (2014b) Increased drought stress resilience of maize through endophytic colonization by Burkholderia phytofirmans PsJN and Enterobacter sp. FD17. Environ Exp Bot 97:30–39. https://doi.org/10.1016/j.envexpbot.2013.09.014

Naz I, Bano A, Ul-Hassan T (2009) Isolation of phytohormones producing plant growth promoting rhizobacteria from weeds growing in Khewra salt range, Pakistan and their implication in providing salt tolerance to Glycine max L. Afr J Biotechnol 8(21):5762–5766

Niu X, Song L, Xiao Y, Ge W (2018) Drought-tolerant plant growth-promoting rhizobacteria associated with foxtail millet in a semi-arid agroecosystem and their potential in alleviating drought stress. Front Microbiol 8:2580

Ojuederie OB, Babalola OO (2017) Microbial and plant-assisted bioremediation of heavy metal polluted environments: a review. Int J Environ Res Public Health 14(12):1504

Ortiz N, Armada E, Duque E, Roldan A, Azcon R (2015) Contribution of arbuscular mycorrhizal fungi and/or bacteria to enhancing plant drought tolerance under natural soil conditions: effectiveness of autochthonous or allochthonous strains. J Plant Physiol 174:87–96

Padgham J (2009) Agricultural development under a changing climate: opportunities and challenges for adaptation. World Bank, Washington, DC

Parray JA, Jan S, Kamili AN, Qadri RA, Egamberdieva D, Ahmad P (2016) Current perspectives on plant growth-promoting rhizobacteria. J Plant Growth Regul 35(3):877–902

Patel D, Saraf M (2013) Influence of soil ameliorants and microflora on induction of antioxidant enzymes and growth promotion of Jatropha curcas L. under saline condition. Eur J Soil Biol 55:47–54

Piccoli P, Travaglia C, Cohen A, Sosa L, Cornejo P, Masuelli R, Bottini R (2011) An endophytic bacterium isolated from roots of the halophyte Prosopis strombulifera produces ABA, IAA, gibberellins A1 and A3 and jasmonic acid in chemically-defined culture medium. Plant Growth Regul 64(2):207–210

Pishchik VN, Provorov NA, Vorobyov NI, Chizevskaya EP, Safronova VI, Tuev AN, Kozhemyakov AP (2009) Interactions between plants and associated bacteria in soils contaminated with heavy metals. Microbiology 78(6):785–793

Prasad R, Kumar M, Varma A (2015) Role of PGPR in soil fertility and plant health. In: Plant-growth-promoting rhizobacteria (PGPR) and medicinal plants. Springer, Cham, pp 247–260

Qian J, Li D, Zhan G, Zhang L, Su W, Gao P (2012) Simultaneous biodegradation of Ni–citrate complexes and removal of nickel from solutions by Pseudomonas alcaliphila. Bioresour Technol 116:66–73

Qu AL, Ding YF, Jiang Q, Zhu C (2013) Molecular mechanisms of the plant heat stress response. Biochem Biophys Res Commun 432(2):203–207

Rajendrakumar CS, Suryanarayana T, Reddy AR (1997) DNA helix destabilization by proline and betaine: possible role in the salinity tolerance process. FEBS Lett 410(2–3):201–205

Rajkumar M, Lee KJ, Lee WH, Banu JR (2005) Growth of Brassica juncea under chromium stress: influence of siderophores and indole 3 acetic acid producing rhizosphere bacteria. J Environ Biol 26(4):693–699

Reddy PP (2014) Plant growth promoting rhizobacteria for horticultural crop protection. Springer India, New Delhi

Rosier A, Medeiros FH, Bais HP (2018) Defining plant growth promoting rhizobacteria molecular and biochemical networks in beneficial plant-microbe interactions. Plant Soil 428:35–55

Sadeghi A, Karimi E, Dahaji PA, Javid MG, Dalvand Y, Askari H (2012) Plant growth promoting activity of an auxin and siderophore producing isolate of Streptomyces under saline soil conditions. World J Microbiol Biotechnol 28(4):1503–1509

Saikia J, Sarma RK, Dhandia R, Yadav A, Bharali R, Gupta VK, Saikia R (2018) Alleviation of drought stress in pulse crops with ACC deaminase producing rhizobacteria isolated from acidic soil of Northeast India. Sci Rep 8:3560

Salomon MV, Bottini R, de Souza Filho GA, Cohen AC, Moreno D, Gil M, Piccoli P (2014) Bacteria isolated from roots and rhizosphere of Vitisvinifera retard water losses, induce abscisic acid accumulation and synthesis of defense-related terpenes in in vitro cultured grapevine. Physiol Plant 151(4):359–374

Sandhya V, Ali SKZ, Grover M, Reddy G, Venkateswarlu B (2009) Alleviation of drought stress effects in sunflower seedlings by the exopolysaccharides producing Pseudomonas putida strain GAP-P45. Biol Fertil Soils 46:17–26. https://doi.org/10.1007/s00374-009-0401-z

Saravanakumar D, Samiyappan R (2007) ACC deaminase from Pseudomonas fluorescens mediated saline resistance in groundnut (Arachis hypogaea) plants. J Appl Microbiol 102:1283–1292

Saum SH, Muller V (2007) Salinity-dependent switching of osmolyte strategies in a moderately halophilic bacterium: glutamate induces proline biosynthesis in Halobacillus halophilus. J Bacteriol 189(19):6968–6975

Schmidt W (1999) Mechanisms and regulation of reduction-based iron uptake in plants. New Phytol 141(1):1–26

Schroth MN, Hancock JG (1982) Disease-suppressive soil and root-colonizing bacteria. Science 216(4553):1376–1381

Sen S, Chandrasekhar CN (2015) Effect of PGPR on enzymatic activities of rice (Oryza sativa L.) under salt stress. Asian J Plant Sci Res 5:44–48

Sharifi M, Ghorbanli M, Ebrahimzadeh H (2007) Improved growth of salinity-stressed soybean after inoculation with salt pre-treated mycorrhizal fungi. J Plant Physiol 164:1144–1151. https://doi.org/10.1016/j.jplph.2006.06.016

Shim JS, Seo JS, Seo JS, Kim Y, Koo Y, Do Choi Y, Jung C (2019) Heterologous expression of bacterial trehalose biosynthetic genes enhances trehalose accumulation in potato plants without adverse growth effects. Plant Biotechnol Rep 13(4):409–418

Shukla PS, Agarwal PK, Jha B (2012) Improved salinity tolerance of Arachishypogaea (L.) by the interaction of halotolerant plant-growth-promoting rhizobacteria. J Plant Growth Regul 31 (2):195–206

Simontacchi M, Galatro A, Ramos-Artuso F, Santa-Maria GE (2015) Plant survival in a changing environment: the role of nitric oxide in plant responses to abiotic stress. Front Plant Sci 6:977

Singh S (2014) A review on possible elicitor molecules of cyanobacteria: their role in improving plant growth and providing tolerance against biotic or abiotic stress. J Appl Microbiol 117:1221–1244. https://doi.org/10.1111/jam.12612

Skirycz A, Inze D (2010) More from less: plant growth under limited water. Curr Opin Biotechnol 21(2):197–203

Smith DL, Praslickova D, Ilangumaran G (2015) Inter-organismal signaling and management of the phytomicrobiome. Front Plant Sci 6:722

Solano BR, de la Iglesia MP, Probanza A, García JL, Megías M, Manero FG (2007) Screening for PGPR to improve growth of Cistus ladanifer seedlings for reforestation of degraded Mediterranean ecosystems. In: First international meeting on microbial phosphate solubilization. Springer, Dordrecht, pp 59–68

Souza RD, Ambrosini A, Passaglia LM (2015) Plant growth-promoting bacteria as inoculants in agricultural soils. Genet Mol Biol 38(4):401–419

Strigul NS, Kravchenko LV (2006) Mathematical modeling of PGPR inoculation into the rhizosphere. Environ Model Softw 21(8):1158–1171

Suman A, Shasany AK, Singh M, Shahi HN, Gaur A, Khanuja SPS (2001) Molecular assessment of diversity among endophytic diazotrophs isolated from subtropical Indian sugarcane. World J Microbiol Biotechnol 17(1):39–45

Tardieu F, Parent B, Simonneau TJP (2010) Control of leaf growth by abscisic acid: hydraulic or non-hydraulic processes? Plant Cell Environ 33:636–647

Terre S, Asch F, Padham J, Sikora RA, Becker M (2007) Influence of root zone bacteria on root iron plaque formation in rice subjected to iron toxicity. Utilisation of diversity in land use systems: sustainable and organic approaches to meet human needs. In: Tropentag, Witzenhausen, Germany

Tewari S, Arora NK (2018) Role of salicylic acid from Pseudomonas aeruginosa PF23 EPS+ in growth promotion of sunflower in saline soils infested with phytopathogen Macrophomina phaseolina. Environ Sustain 1(1):49–59

Timmusk S, El-Daim IAA, Copolovici L, Tanilas T, Kannaste A, Behers L, Nevo E, Seisenbaeva G, Stenstrom E, Niinemets U (2014) Drought-tolerance of wheat improved by rhizosphere bacteria from harsh environments: enhanced biomass production and reduced emissions of stress volatiles. PLoS One 9:96086. https://doi.org/10.1371/journal.pone.0096086

Tiwari S, Lata C, Chauhan PS, Nautiyal CS (2016) Pseudomonas putida attunes morphophysiological, biochemical and molecular responses in Cicer arietinum L. during drought stress and recovery. Plant Physiol Biochem 99:108–117

United Nations, Department of Economic and Social Affairs, Population Division (2019) Population facts no. 2019/5, December 2019: potential impact of later childbearing on future population

Vacheron J, Desbrosses G, Bouffaud ML, Touraine B, Moenne-Loccoz Y, Muller D, Prigent-Combaret C (2013) Plant growth-promoting rhizobacteria and root system functioning. Front Plant Sci 4:356

Vaishnav A, Kumari S, Jain S, Verma A, Tuteja N, Choudhary DK (2016) PGPR-mediated expression of salt tolerance gene in soybean through volatiles under sodium nitroprusside. J Basic Microbiol 56:1274–1288. https://doi.org/10.1002/jobm.201600188

Van Velthuizen H (2007) Mapping biophysical factors that influence agricultural production and rural vulnerability (no. 11). Food & Agriculture Organization, Rome

Vardharajula S, Zulfikar Ali S, Grover M, Reddy G, Bandi V (2011) Drought-tolerant plant growth promoting Bacillus spp.: effect on growth, osmolytes, and antioxidant status of maize under drought stress. J Plant Interact 6(1):1–14

Vigliotta G, Matrella S, Cicatelli A, Guarino F, Castiglione S (2016) Effects of heavy metals and chelants on phytoremediation capacity and on rhizobacterial communities of maize. J Environ Manag 179:93–102

Vurukonda SSKP, Vardharajula S, Shrivastava M, Skz A (2016) Enhancement of drought stress tolerance in crops by plant growth promoting rhizobacteria. Microbiol Res 184:13–24

Wang W, Vinocur B, Altman A (2003) Plant responses to drought, salinity and extreme temperatures: towards genetic engineering for stress tolerance. Planta 218(1):1–14

Wang C, Yang W, Wang C, Gu C, Niu D, Liu H (2012) Induction of drought tolerance in cucumber plants by a consortium of three plant growth-promoting rhizobacterium strains. PLoS One 7:0052565. https://doi.org/10.1371/journal.pone.0052565

Wani SH, Kumar V, Shriram V, Sah SK (2016) Phytohormones and their metabolic engineering for abiotic stress tolerance in crop plants. Crop J 4(3):162–176

Weller DM, Thomashow LS (1994) Current challenges in introducing beneficial microorganisms into the rhizosphere. In: Molecular ecology of rhizosphere microorganisms: biotechnology and the release of GMOs. VCH Verlagsgesellschaft, Weinheim, pp 1–18

Yadav SK (2010) Cold stress tolerance mechanisms in plants. A review. Agron Sustain Dev 30 (3):515–527

Yadav J, Verma JP, Jaiswal DK, Kumar A (2014) Evaluation of PGPR and different concentration of phosphorus level on plant growth, yield and nutrient content of rice (Oryza sativa). Ecol Eng 62:123–128

Yao L, Wu Z, Zheng Y, Kaleem I, Li C (2010) Growth promotion and protection against salt stress by Pseudomonas putida Rs-198 on cotton. Eur J Soil Biol 46(1):49–54

Yuan F, Lyu MJA, Leng BY, Zheng GY, Feng ZT, Li PH, Wang BS (2015) Comparative transcriptome analysis of developmental stages of the Limonium bicolor leaf generates insights into salt gland differentiation. Plant Cell Environ 38(8):1637–1657

Zaidi A, Khan M, Ahemad M, Oves M (2009) Plant growth promotion by phosphate solubilizing bacteria. Acta Microbiol Immunol Hung 56(3):263–284

Zhang F, Dashti N, Hynes RK, Smith DL (1997) Plant growth-promoting rhizobacteria and soybean [Glycine max (L.) Merr.] growth and physiology at suboptimal root zone temperatures. Ann Bot 79(3):243–249

Zhang H, Kim MS, Sun Y, Dowd SE, Shi H, Pare PW (2008) Soil bacteria confer plant salt tolerance by tissue-specific regulation of the sodium transporter HKT1. Mol Plant-Microbe Interact 21:737–744. https://doi.org/10.1094/MPMI-21-6-0737

Zhang J, Wang LH, Yang JC, Liu H, Dai JL (2015) Health risk to residents and stimulation to inherent bacteria of various heavy metals in soil. Sci Total Environ 508:29–36

Zhou C, Guo J, Zhu L, Xiao X, Xie Y, Zhu J, Ma Z, Wang J (2016) Paenibacillus polymyxa BFKC01 enhances plant iron absorption via improved root systems and activated iron acquisition mechanisms. Plant Physiol Biochem 105:162–173

Chapter 11
Framework for Studying Rhizospheric Microflora Under the Effect of Improved Crop Variety

Shipra Shahi, Suresh Kumar Dubey, and Pranjali Vishwakarma

Abstract Recent advances in microbial ecology have incited research on rhizosphere microflora under the effect of improved crop variety. Effects have often been assessed with the use of culture-dependent and culture-independent methods, but these usually delude the conclusions of evaluation due to improper hypothesis and redundant use of DNA fingerprinting methods. In view of our incomplete knowledge of the microbial communities and processes in plant-soil systems, recent technological and conceptual improvements do offer a way forward. We propose a framework that encompasses both, a sensible selection of process and general parameters for assessing the impact of improved crop variety on rhizosphere microorganisms.

Keywords Rhizosphere · Modification · Root exudates · Improved crop variety

11.1 Introduction

The biodiversity of soil bacterial communities contributes significantly to plant nutrition and health or soil fertility (Gyaneshwar et al. 2002; Jeffries et al. 2003). The steady increase in management interventions like tillage, crop species composition, and soil amendments (Vishwakarma et al. 2009; Liu et al. 2019; Jaramillo et al. 2017) influences the rhizosphere community composition. The concept of rhizosphere, i.e., a narrow region of soil surrounding the plant roots, was first coined by Hiltner in 1904. The rhizosphere forms the site for hot-spot of microbial abundance and activity owing to the presence of plant exudates as well as rhizodeposits (Pathan et al. 2020; Korenbluma et al. 2020; Kuzyakov and Blagodatskaya 2015). For decades, the importance of the rhizo-microbiome for plant functioning has been

---

S. Shahi · P. Vishwakarma (✉)
Biosafety Support Unit, Regional Centre of Biotechnology, Faridabad, India

S. K. Dubey
Department of Botany, Institute of Science, Banaras Hindu University, Varanasi, India

S. K. Dubey, S. K. Verma (eds.), *Plant, Soil and Microbes in Tropical Ecosystems*, Rhizosphere Biology, https://doi.org/10.1007/978-981-16-3364-5_11

recognized; however, the tools for investigating several interactions were not available.

Plants differ greatly in their interaction with soil-borne organisms via properties such as root structure, duration of growing season, and exudate patterns. Cultivars of the same crop can also be significantly different from each other. With the advent of new technologies in plant improvement, there has been considerable renewed interest in studying the interaction between plants and specific soil microorganisms inhabiting their rhizosphere (Lundberg et al. 2012). Crops with improved characteristics are produced by both conventional plant breeding and genetic modification. In conventional breeding, genetic makeup is changed by crossing together plants and selecting the offspring with the desired characteristics, whereas in a genetically modified plant, a new gene or genes are added to the genome of the crop plant. Hence, plant variety developed via both the techniques has altered character. A large number of studies have been carried out to see the effect of the conventionally improved crop (Bulgarelli et al. 2015; Hussain et al. 2011) and also the effect of the genetically modified plant on soil micro-flora in rhizosphere (Kowalchuk et al. 2013; Singh and Dubey 2016; Mandal et al. 2020). However, a major lacuna was that most of these studies were limited to identifying microbial populations and properties expected to respond to the effect of cultivars on soil microbes in the rhizosphere.

Therefore, we hereby propose a framework for experimental design for studies on the impact of improved crops on the surrounding rhizosphere and the elucidation of the related mechanisms, covering a range from simple model experiments up to the field scale. The frame of this review does not allow detailed method descriptions, the advantages, and limitations of techniques. Nonetheless, herein a perspective for designing an experiment and formulating the hypothesis of study has been presented in four steps: (1) formulation of hypotheses, (2) source of stressor, (3) experimental design for studying the effect of improved plant variety on rhizosphere microbes, and (4) evaluation of result.

11.2 Formulation of Hypothesis

Effect of plant variety on rhizosphere microbe in agriculture ecosystem has been studied for almost three decades now (Luigi et al. 1998; Özgül et al. 2012; Chen et al. 2019); however, a number of gaps in our knowledge impede our understanding of plant rhizosphere system. Such gap may lead to conclusion that may delude scientific opinion to an extent. Therefore, sufficient inference and detailed assessment is a requisite. Use of new and cost-effective techniques can contribute when a scientific experiment is integrated with a well-designed hypothesis. An example of risk hypothesis could be *cultivation of improved variety of crop may effect rhizosphere soil microbial diversity and its function.*

The above-mentioned hypothesis provides a foundation of selecting basic information for the study. Approach and methods to evaluate effect of biotic or abiotic stressor on any environmental ecosystem or factors are dependent on three important aspects: (1) source of stressor, (2) path of stressor to reach the ecological entity, and (3) effect of stressor on ecological entity. Following the route of these three aspects, the steps involved for studying the effect of plant variety on rhizosphere microbe can be easily planned. Herein, the source of stress, i.e., improved/newly developed plant variety and effect on rhizosphere microbe, is constant; however, the paths in between these two constants are variable. To check the existence of true variable, the correct hypothesis should be drawn with set of questions. These questions include crop biology including its root structure, root–microbes interaction, root exudate pattern, persistency in soil, as well as effect on microorganism, likely soil function to be affected and persistence of impact after plant removal. Table 11.1 provides baseline and summary of the prerequisite information to set the perspective and relationships needed for evaluating and thereby predicting the studies. Information required includes crop biology, existing microbial community in receiving environments, modification made in improved plant variety including introduced gene in case of transgenic crop, possible effect on soil properties, and effect on soil microflora.

11.3 Source of Stress to Rhizosphere Soil Microflora

A source is point of origin of stressor; herein it is the plant variety. The cause of origin of stress could be the modification in the plant. Modification can be owed to domestication of wild variety, conventional hybrid breeding, marker-assisted selection of new crop variety, or genetic modification. Modification methods and alteration in plant trait can be the possible origin of source of stressor. As mentioned earlier, source is the first aspect of the cause of stressor. Further, within this source, the exact point of origin and point of contact with rhizosphere microbe need to be examined. For the same, the study shall encompass the site, timing, and extent. Rhizosphere is the primary site where soil microbes come in direct contact of plant. The point of contact with stressor will occur only if alteration in either root architecture or root exudate occurs. The rhizosphere is a densely populated area of intense biological and chemical activity with large array of exudates including carbohydrates, lipids, proteins, organic acids, and secondary metabolites providing nourishment to microorganisms in rhizosphere.

With modification in crop variety, changing agricultural management practices can be second stressor. These sources of variation must be considered for establishing a "baseline of normal variation" against which effects induced in crop varieties can be gauged.

Table. 11.1 Baseline information of crops required to predict and plan the study of microbes in rhizosphere

	Conventional breeding			Genetic modification
	Domestication	Hybridisation	MARS*	Genetic engineering
Relevant control	Wild progenitor	Parents	Parents	Non-modified parent
Information about improved crop	• Plant growth character • Root architecture • Plant litter quality • C:N ratio and NO_3 availability of soil	• Characters and genotype of parent • Pedigree of crop • What characters are modified • Composition of hybrids and inbred and evidence for heterotic patterns for multiple components of the microbiome • Traits of hybrids in comparison to parents • Mutagens used for crop improvement	• Characters and genotype of parent • Pedigree of crop • What characters are modified • Composition of hybrids and inbred and evidence for heterotic patterns for multiple components of the microbiome • Traits of hybrids in comparison to parents • Information about pyramiding • Marker used	• Source of introduced gene • Gene cassette introduced, site of integration • Route of exposure • Nature of expression (either DNA/protein) • Trait of modified plant
Field management for improved crop	Habitat expansion changes in crop management practices	Application of tillage, fertilizer use, pesticide application, herbicide treatment, and crop rotation	Application of tillage, fertilizer use, pesticide application, herbicide treatment, and crop rotation	Application of pesticide application, herbicide treatment
Ecosystem parameters for receiving environment	Difference between natural and agriculture environment	Agroecosystem suitable for parents vs. hybrid	Agroecosystem suitable for parents vs. MAS hybrid	Soil type, season, agroecosystem suitable for new introduced trait vs. non-modified plant, etc.
Literature information	• Similar crop • Native environment of crop before domestication • Process of domestication	• Any effect reported on agroecosystem • Genetic changes across generations • Genotype of parents as per pedigree of crop • Overlapping of many genes and environment interactions	• Any effect reported on agroecosystem • Genetic changes across generations • Genotype of parents as per pedigree of crop • Overlapping of many genes and environment interactions	• History of safe use of introduced protein • Route of exposure of introduced gene

Question for building hypothesis	• Which soil function is affected • Which extent the impact will persist after removal of the plant • Activity of the compounds (e.g., proteins), quantities released, persistency in the soil • Which type of microorganism is most susceptible • Which crop, root structure, duration of growing season, exudate pattern

MARS Marker-assisted recurrent selection

11.4 Experimental Design

While assessing the influence of plants on soil rhizosphere microbes, between the two factors, i.e., parameters contributing to route of exposure to stressor and parameters indicating effect on rhizosphere microbes, lies measures of ecosystem and receptor characteristic. The receptor of stressor in present case is rhizosphere microflora. Figure 11.1 demonstrate a framework for studying effect of improved plant variety on rhizosphere microbes.

The first parameter, i.e., route of exposure, initiates from plant and ends at soil rhizosphere. Root architecture and root exudation are responsible for point of contact to rhizosphere microbes. Therefore, as soon as source of stressor is established, route of exposure needs to be examined. As mentioned in Table 11.1, improved variety may exhibit alteration in root architecture, plant biomass, or root exudate depending upon type of modification. A domesticated plant will show altered root architecture, plant biomass, and root exudate for adapting in agriculture ecosystem. Therefore, while studying the effect of domesticated variety, alteration in physicochemical parameters of natural and agroecosystem needs to be considered. In case of hybrid

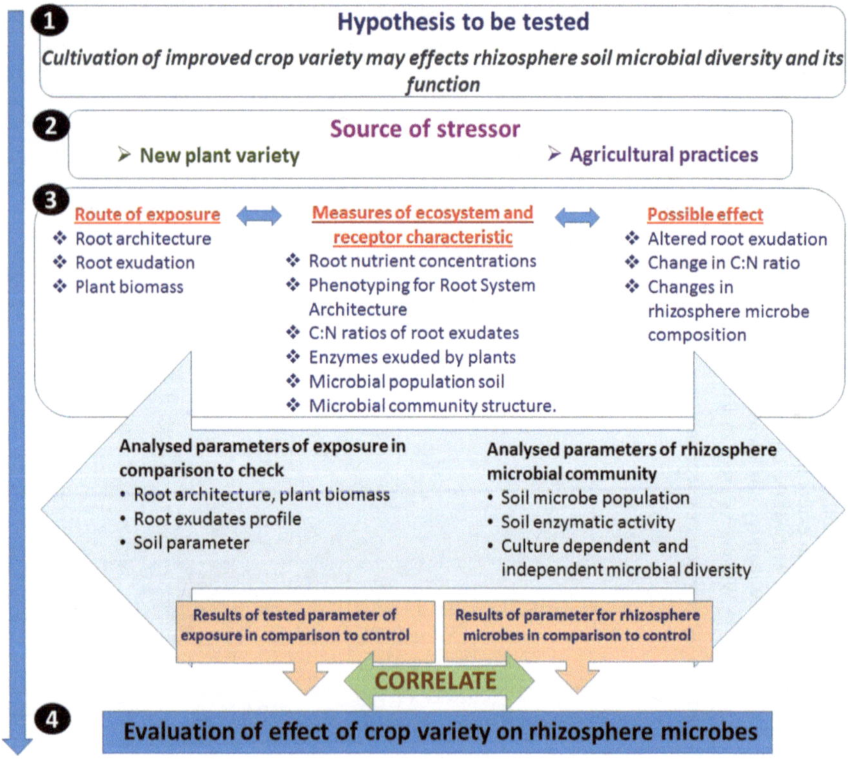

Fig. 11.1 Schematic representation of framework showing four steps to study effect of improved crop variety on rhizosphere microflora

plants (conventional or marker assisted), hybridization between genetically distinct lineages may result into novel characters that may change traits, phenotypic responses in root, and root morphology thereby effecting rhizosphere microbes. Therefore, comparative assessment of the root architecture, plant biomass, and root exudate shall be conducted between hybrid and its parent genotypes. Similarly, for genetically modified crop, also comparative assessment needs to be considered. However, the genetically modified plant may exhibit phenotypic and morphological alteration depending on the introduced trait. Therefore, studies on root architecture, plant biomass, and root exudate need to be considered on a case per case basis and can be exempted in certain cases.

Once the route of stressor is well characterized, the second parameter involves analyzing how this stressor influences characteristic of rhizosphere microbes. It has been well documented that depending upon trait of plant, variation has been reported in the relative demand for nitrogen and phosphorus in single agro-ecosystem (spatial variation), carbon exudate (Phillips et al. 2011; Drake et al. 2013), microbial biomass (Treseder et al. 2010), as well as turnover of soil enzymes (Bhavya et al. 2018). In view of such variations within rhizosphere, the parameter to be analyzed could be C:N ratios of root exudates, enzymes exuded by plants, microbial biomass, microbial population soil, and microbial community structure (Martiny et al. 2006). Examining microbial enzyme within rhizosphere includes dehydrogenase, peroxidase, phosphatase, protease and urease, β-glucosidase, catalase, cellulase, and invertase. In addition to crop variety composition and quantity of rhizodeposits could be also influenced by spatiotemporal variation in physicochemical parameter of receiving environment (Jones et al. 2009; Darwent et al. 2003). Therefore, physicochemical parameter of receiving environment should be compared between test crop and comparator. Further, identification and characterization of the exudates in the rhizosphere of improved variety of crop in comparison to its comparator is an important parameter. For this, soil-based exudate sampling approaches can be studied by hydroponic exudate sampling, soil growth–hydroponic sampling, soil growth, and sampling Rhizobox/REC. However, there are limited studies and experimental setups for exudate sampling techniques to evaluate the effect, quantity, and quality of rhizodepositions. To date, rhizodeposition studies focusing carbon flow in rhizosphere have been carried out based on sterile solution, wherein the real impact of receiving environment in conjugation with crop variety could not be studied. Hydroponic media promotes growth of root system because nutrient is readily available to plant, thereby promoting fast growth without any environmental stresses; contrary to agricultural fields wherein roots of plant struggles to extract nutrients from soil for its growth and development. In such stress, full condition types of exudates could be different than that of hydroponic solution. Additionally, hydroponic solution-based study cannot be performed for full life cycle of plants; thus most studies have focused on young roots (Canarini et al. 2019). Therefore, the turnover and profile of root exudates in rhizosphere needs better characterization. Without information on variation in root exudates, other parameter of rhizosphere and soil physicochemical parameter cannot be correlated between test crop and comparator. Therefore, future studies must include root exudates analysis while studying rhizosphere microbes.

After any difference in the aforementioned parameters is established between improved variety and its comparator, the next step shall involve examining the effect of crop variety on soil microflora. It is to reiterate here that if crops under examination show similarity to comparator with respect to root architecture plant biomass and root exudation, then an anticipation could be drawn that microbes in rhizosphere will not be affected. However, wherever differences are found, study of rhizosphere microbe becomes necessary. During the last few decades, a wide variety of culture-dependent and culture-independent techniques have been developed and are in use for studying structure and function of rhizosphere microbe (Franck et al. 2015; Yang and Crowley 2000; Smalla et al. 2001; Wieland et al. 2001). These techniques are costly but however effective. With advancement in metagenomics, proteomics, and met transcriptomics, high-throughput precise and accurate evaluation is possible. However, it depends on the rationale of researcher to use these techniques in the study of bacteria in the rhizosphere. Herein, our aim is to focus on steps to study rhizosphere microbes; therefore, details about techniques could be seen elsewhere (Rincon-Florez et al. 2013; Shany et al. 2017; Quince et al. 2017; Sergaki et al. 2018).

11.5 Evaluation of Result

As depicted in Fig. 11.1, as soon as source of stressor and hypothesis is defined, the experiments shall be focused on two aspects. The first aspect involves characterizing the route of exposure, and the second aspect involves characterizing ecological entities. Study of measures of receptor provided scientific basis for route exposure as well as its effect on ecological entity. All tested parameters of exposure and of rhizosphere microflora in comparison to relevant comparator shall be assessed and correlated to establish that improved crop variety is responsible for effecting rhizosphere microbes. If the tested parameter and data collected during experiment provides enough evidences for potential effect on rhizosphere microflora, then parameters of route of exposure of stressor shall be correlated to determine whether and to what degree the rhizosphere microbes have been effected. In certain cases, there may be similarity with respect to parameter for route of exposure between tested crop and comparator. In such condition, new crop variety may not be responsible for changes; however, the second stressor, i.e., agricultural practices (agronomic parameter and/or physicochemical parameter and/or field management), may be responsible. With presented framework in this study, if a researcher moves in stepwise manner, then many experimental biases could be neglected.

11.6 Conclusion

With this framework, we suggest that studying the function and diversity of rhizosphere microbes' route of exposure is equally important to other parameters. We also suggest that while designing studies, type of crop and type of modification in the crop shall always be kept into mind. All crops could not be assessed with similar objectives. For the same, importance of baseline information has been also proposed. This is important because in order to design more effective rhizosphere management strategies, it is necessary to determine where microbial communities are associated with different processes in the rhizosphere. A better understanding of processes in the rhizosphere may facilitate the development of crop variety to enhance plant-microbe interaction. This proposed framework has steps for a better understanding of the rhizosphere ecosystem functioning. Indeed, the rhizospheric community has been studied in response to a crop variety. Yet, designing of experiment and stepwise assessment have remained elusive. This study will allow young researchers to design experiments and will help in answering the function of the rhizosphere.

Based on the unique nature of the crop introduced, impact assessment must be considered on a case-by-case basis. For the same, proper experimental design is crucial. Further, ample sampling procedures should be included to control natural variation within the plant-soil systems.

References

Bhavya VP, Anil Kumar S, Kiran SK, Manjunath S (2018) Effect of different cropping system on important soil enzyme activity, organic carbon and microbial activity with different depth. Int J Curr Microbiol Appl Sci 7:315–322

Bulgarelli D, Oter RG, Münch PC, Weiman A, Dröge J, Pan Y, McHardy AC, Paul SL (2015) Structure and function of the bacterial root microbiota in wild and domesticated barley. Cell Host Microbe 17:392–403

Canarini A, Kaiser C, Merchant A, Richter A, Wanek W (2019) Root exudation of primary metabolites: mechanisms and their roles in plant responses to environmental stimuli. Front Plant Sci 10:157. https://doi.org/10.3389/fpls.2019.00157

Chen L, Xin X, Zhang J, Redmile GM, Nie G, Wang Q (2019) Soil characteristics overwhelm cultivar effects on the structure and assembly of root-associated microbiomes of modern maize. Pedosphere 3:360–373

Darwent MJ, Paterson E, McDonald AJS, Tomos AD (2003) Biosensor reporting of root exudation from *Hordeum vulgare* in relation to shoot nitrate concentration. J Exp Bot 54:325–334

Drake JE, Darby BA, Giasson MA, Kramer MA, Phillips RP, Finzi AC (2013) Stoichiometry constrains microbial response to root exudation-insights from a model and a field experiment in a temperate forest. Biogeosciences 10:821–838

Franck OP, Stefani TH, Bell CM, Ivan E, de la Providencia IV, Yassimi AE, St-Arnaud M, Hijri M (2015) Culture-dependent and -independent methods capture different microbial community fractions in hydrocarbon-contaminated soils. PLoS One 10:1–16

Gyaneshwar P, Naresh Kumar G, Parekh LJ, Poole PS (2002) Role of soil microorganisms in improving P nutrition of plants. Plant Soil 245:83–93

Hiltner L (1904) Uber neuere Erfahrungen und Probleme auf demGebiete der Bodenbakteriologie unter bes onderden beruck sichtigung und Brache. Arb Dtsch Landwirtsch Gesellschaft 98:59–78

Hussain Q, Liu Y, Zhang A, Pan G, Li L, Zhang X, Song X, Cui L, Jin Z (2011) Variation of bacterial and fungal community structures in the rhizosphere of hybrid and standard rice cultivars and linkage to CO2 flux. FEMS Microbiol Ecol 78:116–128

Jaramillo JEP, Carrión VJ, Luiz MB, Ferrão FV, de Hollander M, Garcia AAF, Ramírez CA, Mendes R, Raaijmakers JM (2017) Linking rhizosphere microbiome composition of wild and domesticated Phaseolus vulgaris to genotypic and root phenotypic traits. ISME 11:2244–2257

Jeffries P, Gianinazzi S, Perotto S, Turnau K, Barea JM (2003) Interaction between selected bacterial strains and the contribution of arbuscular mycorrhizal fungi in sustainable maintenance of plant health and soil fertility. Biol Fertil Soils 37:1–16

Jones DL, Nguyen C, Finlay RD (2009) Carbon flow in the rhizosphere: carbon trading at the soil-root interface. Plant Soil 321:5–33

Korenbluma E, Donga Y, Szymanskib J, Pandaa S, Jozwiaka A, Massalhaa H, Meira S, Rogacheva I, Aharonia A (2020) Rhizosphere microbiome mediates systemic root metabolite exudation by root-to-root signalling. PNAS 117:3874–3883

Kowalchuk GA, Bruinsma M, Johannes AV (2013) Assessing responses of soil microorganisms to GM plants. Trends Ecol Evol 18:403–410

Kuzyakov Y, Blagodatskaya E (2015) Microbial hotspots and hot moments in soil: concept & review. Soil Biol Biochem 83:184–199

Liu F, Hewezi T, Lebeis SL, Pantalone V, Grewal PS, Staton ME (2019) Soil indigenous microbiome and plant genotypes cooperatively modify soybean rhizosphere microbiome assembly. BMC Microbiol 19:201

Luigi C, Annamaria B, Claudia D, Silvia T (1998) Influence of plant development, cultivar and soil type on microbial colonisation of maize root. Appl Soil Ecol 8:11–18

Lundberg DS, Lebeis SL, Paredes SH, Yourstone S, Gehring J, Malfatti S, Tremblay J, Engelbrektson A, Kunin V, Glavina TR, Edgar RC, Eickhorst T, Ley RE, Hugenholt P, Tringe SG, Dangl JL (2012) Defining the core Arabidopsis thaliana root microbiome. Nature 488:86–90

Mandal A, Sarkar B, Owens G, Thakur JK, Manna MC, Khan NN, Jayaraman S, Patra AK (2020) Impact of genetically modified crops on rhizosphere microorganisms and processes: a review focusing on Bt cotton. Appl Soil Ecol 148:103492. https://doi.org/10.1016/j.apsoil.2019.103492

Martiny JBH, Bohannan BJM, Brown JH, Colwell RK, Fuhrman JA, Green JL, Horner Devine MC, Kane M, Krumins JA, Kuske CR et al (2006) Microbial biogeography: putting microorganisms on the map. Nat Rev Microbiol 4:102–112

Özgül İ, Joana FS, Jan Dirk VE (2012) Soil and cultivar type shape the bacterial community in the potato rhizosphere. Microb Ecol 63:460–470

Pathan SI, Teresa M, Sunseri CF, Sunseri F, Lupini A, Lupini A (2020) Rhizosphere as hotspot for plant-soil-microbe interaction: carbon and nitrogen cycling in soil. Springer, Berlin, pp 978–981

Phillips RP, Finzi AC, Bernhardt ES (2011) Enhanced root exudation induces microbial feedbacks to N cycling in a pine forest under long-term CO2 fumigation. Ecol Lett 14:187–194

Quince C, Walker AW, Simpson JT, Loman N, Segata N (2017) Shotgun metagenomics, from sampling to analysis. Nat Biotechnol 35:833–844

Rincon-Florez VA, Carvalhais LC, Schenk PM (2013) Culture-independent molecular tools for soil and rhizosphere microbiology. Diversity 5:581–612

Sergaki C, Lagunas B, Lidbury I, Gifford ML, Schäfer P (2018) Challenges and approaches in microbiome research: from fundamental to applied. Front Plant Sci 9:1205. https://doi.org/10.3389/fpls.2018.01205

Shany O, Maya OL, Noa S, Jiandong J, Yechezke K, Dror M, Shiri F (2017) Analysis of microbial functions in the rhizosphere using a metabolic-network based framework for metagenomics interpretation. Front Microbiol 8:1606. https://doi.org/10.3389/fmicb.2017.01606

Singh AK, Dubey SK (2016) Current trends in Bt crops and their fate on associated microbial community dynamics: a review. Protoplasma 253:663–681

Smalla K, Wieland G, Buchner A, Zock A, Parzy J, Kaiser S, Roskot N, Heuer H, Berg G (2001) Bulk and rhizosphere soil bacterial communities studied by denaturing gradient gel electrophoresis: plant dependent enrichment and seasonal shifts revealed. Appl Environ Microbiol 67:4742–4751

Treseder KK, Schimel JP, Garcia MO, Whiteside MD (2010) Slow turnover and production of fungal hyphae during a Californian dry season. Soil Biol Biochem 42:1657–1660

Vishwakarma P, Dumont MG, Bodrossy L, Stralis-Pavese N, Murrell JC, Dubey SK (2009) Ecological and molecular analyses of the rhizospheric methanotroph community in tropical rice soil: effect of crop phenology and land use history. Curr Sci 96:1082–1089

Wieland G, Neumann R, Backhaus H (2001) Variation of microbial communities in soil, rhizosphere, and rhizoplane in response to crop species, soil type, and crop development. Appl Environ Microbiol 67:5849–5854

Yang C-H, Crowley DE (2000) Rhizosphere microbial community structure in relation to root location and plant iron nutritional status. Appl Environ Microbiol 66:345–351

Chapter 12
Role of Rhizospheric Bacteria in Disease Suppression During Seedling Formation in Millet

Kanchan Kumar, Gaurav Pal, Anand Verma, and Satish Kumar Verma

Abstract Bacteria present in the rhizospheric area of the plant are called rhizospheric bacteria. Rhizospheric bacteria play crucial role in plant development and growth starting from seed germination and also protect the seedlings from fungal phytopathogens. These rhizobacteria are known to produce growth hormones; siderophore; lytic enzymes such as chitinase, lipase, protease, and β-1, 3-glucanase; organic acids; lipopeptides; volatile compounds; and some antibiotics. Some of the common rhizospheric bacteria are *Pseudomonas chlororaphis*, *Bacillus subtilis*, *Bacillus licheniformis*, *Pseudomonas fluorescens*, *Chromobacterium violaceum*, *Bacillus cereus*, and *Bacillus stearothermophilus* which have been found to suppress the growth of fungal pathogens including *Macrophomina phaseolina*, *Magnaporthe grisea*, and *Fusarium oxysporum*. Lytic enzymes such as chitinase, protease, and β-1, 3- glucanase produced by the rhizobacteria degrade the chitin, glucan, and proteins of the fungal cell wall, respectively. Secondary metabolites produced by the rhizobacteria inhibit the growth of pathogenic fungi by reducing the spore germination, swelling in fungal mycelia, making pore formation in hyphae, cytoplasmic leakages from fungal cells, and finally lysis of hyphae. *Pseudomonas* and *Bacillus* are known to induce the induced systemic resistance (ISR) in plants and make them disease resistant against phytopathogens. Millets are group of very important small grain crop which seedling establishment is affected by many soil pathogens. The present chapter is focused on the role of beneficial rhizospheric bacteria in disease suppression in millet crop.

Keywords Rhizospheric bacteria · Pathogenic fungi · Disease suppression · Siderophore · Lytic enzymes

K. Kumar · G. Pal · A. Verma · S. K. Verma (✉)
Centre of Advanced Study in Botany, Banaras Hindu University, Varanasi, Uttar Pradesh, India

S. K. Dubey, S. K. Verma (eds.), *Plant, Soil and Microbes in Tropical Ecosystems*, Rhizosphere Biology, https://doi.org/10.1007/978-981-16-3364-5_12

12.1 Introduction

Narrow region around every plant is colonised by different types of microbes including bacteria and fungi called rhizoshpere, and these microorganism are called rhizospheric microorganism (Hirsch and Mauchline 2012; Darrah 1993; Philippot et al. 2013). Rhizospheric areas directly influence the chemical secretion from roots of the plant (called root exudates) and also determine the growth and distribution of microbes around it. Root exudates contain different types of organic acids, sugars, vitamins, and phenolic compounds, which act as signalling molecules playing important role in the recruitment of microbes; it is also used by microbes as food material (Lugtenberg et al. 2001; Dini-Andreote and van Elsas 2013; Philippot et al. 2013). Microbes present in the rhizospheric zone struggle for nutrition and space; some of the microbes make mutualistic or symbiotic relationships with the roots of the plants (N-fixing bacteria) and improve their growth (Bazin et al. 1990; Lugtenberg et al. 2001). Some of the pathogenic microbes (acting as parasites) harm the plant's health by causing several diseases that affect the growth of the plant (Ahmad et al. 2008). Rhizospheric bacteria play several roles in the growth and development of plants including seed germination and establishment, development of root shoot length, increasing biomass of the plant through plant growth-promoting activities such as the production of auxins, solubilising inorganic phosphate, and increasing nutrient uptake by nitrogen fixation. Such PGPRs (plant growth-promoting rhizobacteria) also control the growth of the plant pathogens in the rhizospheric region by secreting hydrolytic enzymes, hydrogen cyanide, siderophore production, and secreting several types of antimicrobial compounds and also induce systemic resistance inside the plant; rhizospheric bacteria also improve survival of the plant in abiotic stress conditions. *Azotobacter chroococcum*, *Azospirillum lipoferum*, *Bacillus cereus*, *Serratia marcescens*, *Chryseomonas luteola*, *Bacillus subtilis*, *Acinetobacter*, *Pseudomonas fluorescens*, *Acetobacter and Azospirillum*, *Bacillus amyloliquefaciens*, *Burkholderia phytofirmans*, and *Pseudomonas* sp. are reported as rhizospheric bacteria which play an important role in plant growth-promoting activities of the pearl millet, sorghum, foxtail millet, and finger millet (Jogaiah et al. 2007; Idris et al. 2007; Rokhbakhsh-Zamin et al. 2011; Saxena et al. 2013; Khatri et al. 2016; Mounde et al. 2015; Sekar et al. 2018). *Azospirillum* and *Acetobacter* protected the pearl millet against downy mildew (Jogaiah et al. 2007); *Pseudomonas* sp. protects finger millet from the blast disease caused by *Pyricularia grisea* (Sekar et al. 2018). Rhizobacteria including *Pseudomonas migulae*, *Pseudomonas fluorescens*, and *Enterobacter hormaechei* are reported as drought-tolerant bacteria; this might be a reason why some of the millets crop including foxtail millet grow in drought conditions (Niu et al. 2018).

12.2 Millets Crop

Millets are small seeded, annual, cereal crops which are grown mostly in developing countries of Asia and Africa (FAO 1972; Sarita and Singh 2016). Millets are favoured due to their productivity and high nutritional values and because they easily grow in dry, low fertility soil condition. Millets include pearl millet, foxtail millet, finger millet, kodo millet, barnyard millet, browntop millet, proso millet, little millet, teff millet, and fonio millet (FAO 1972; Rao 1989; Dendy 1995; Hulse et al. 1980; Doggett 1989). Sorghum, due to their large size grain, is known as great millet (Adeyeye 2008). Millets have high mineral contents including iron, fibre, protein, and calcium. Finger millet has highest calcium content as compared to other millets and other main grains like rice and wheat (Sarita and Singh 2016; Chauhan et al. 2018; Kumar et al. 2018; Ambati and Sucharitha 2019).

12.3 Millets Seedling Disease

Large numbers of pathogenic fungi have been reported which are responsible for causing several diseases in different types of millets during germination of seeds and seedling development. Some seedling diseases are seed rot, seed decay, seedling blight, pre-emergence and post-emergence damping-off, seedling root rot, downy mildew, and blast disease. (Das 2017; Leukel and Martin 1943; Little and Perumal 2019; Nagaraja and Das 2016; Wilson 2000; Raghunathan 1968). Common seedling disease caused by fungal pathogens is listed in Table 12.1.

12.4 Role of Rhizospheric Bacteria in Disease Suppression

Several fungi are known to cause disease in sorghum and other millets at different stages of their life. Production of millets crop has been severely affected due to fungal infection during seedling formation (Das 2017; Wilson 2000). Many rhizospheric bacteria have been reported to play important roles in disease suppression in millet plants. *Macrophomina phaseolina* is responsible for seedling blight and charcoal rot disease in sorghum, and one rhizospheric bacterium *Pseudomonas chlororaphis* SRB127 has shown to suppress the growth of the pathogenic fungus and minimise the severity of charcoal rot disease in sorghum under field conditions (Das et al. 2008). Rhizospheric bacteria including *Bacillus subtilis*, *Bacillus licheniformis*, *Chromobacterium violaceum*, *Bacillus cereus*, and *Bacillus stearothermophilus* have been known to suppress the mycelial growth of *Fusarium oxysporum* and control the root and crown rot disease in sorghum (Idris et al. 2007; Al-Jedabi 2009). *Bacillus subtilis* and *Bacillus cereus* were isolated from rhizospheric regions of sorghum; they significantly suppressed the growth of

Table 12.1 List of diseases caused by pathogenic fungi during seed germination and seedling development

Disease symptoms	Plant host	Pathogens	Reference
Seed rot, damping-off, seedling blight	Sorghum	*Colletotrichum sublineolum* *Bipolaris turcica* *Rhizoctonia bataticola* *Pythium* spp. *Fusarium* spp.	Das (2017) Leukel and Martin (1943) Little and Perumal (2019)
Seedling blight	Sorghum	*Macrophomina phaseolina*	Das (2017)
Reduced germination and seedling death	Sorghum	*Alternaria alternata*	Little and Perumal (2019)
Seedling root rot	Sorghum	*Pythium arrhenomanes*	Leukel and Martin (1943)
Downy mildew	Sorghum	*Peronosclerospora sorghi*	Das (2017)
Downy mildew	Pearl millet	*Sclerospora graminicola*	Das (2017)
Blast	Pearl millet	*Pyricularia grisea*	Nagaraja and Das (2016)
Seed decay, damping-off, stem lesions on seedlings	Pearl millet	*Rhizoctonia solani, Sclerotium rolfsii*	Wilson (2000)
Reduced germination, seedling blight	Pearl millet	*Curvularia penniseti* *Drechslera setariae* *Exserohilum rostratum* *Fusarium moniliforme* *Fusarium solani* *Fusarium equiseti* *Fusarium fusarioides* *Phyllosticta penicillariae*	Wilson (2000)
Blast	Finger millet	*Pyricularia grisea*	Das (2017)
Seedling and leaf blight	Finger millet	*Drechslera nodulosum*	Das (2017)
Damping-off	Finger millet	*Pythium aphanidermatum*	Raghunathan (1968)
Blast	Foxtail millet	*Pyricularia setariae*	Das (2017)
Chlorosis of the seedling leaves	Foxtail millet	*Sclerospora graminicola*	Nagaraja and Das (2016)
Damping-off	Teff millet	*Helminthosporium poae*	Nagaraja and Das (2016)
Seed rotting, coleoptile spot, seedling blight	Proso millet	*Bipolaris panici-miliacei*	Das (2017)
Blast	Barnyard millet	*Pyricularia grisea*	Das (2017)

Pythium ultimum and also controlled the root rot in sorghum (Idris et al. 2008). *Pseudomonas fluorescens* having chitinase activity was shown to suppress the growth of *Magnaporthe grisea* and control the blast disease in finger millet/ragi (Negi et al. 2017). Downy mildew is a very damaging disease caused by *Sclerospora graminicola* in pearl millet. Treatment of seeds with *Pseudomonas fluorescens, Acetobacter*, *Azospirillum* strain, *Bacillus subtilis*, and *Bacillus pumilus* reduced the downy mildew disease in pearl millets (Raj et al. 2003; Jogaiah et al. 2007). Smut disease is caused by *Ustilago crameri* in foxtail millet, and a research found that rhizospheric bacterial community plays significant role in minimising disease occurrence and loss of productivity in fox tail millet (Han et al. 2017).

12.5 Mechanism of Disease Suppression

Rhizobacteria control the growth of fungal phytopathogens directly by producing antifungal antibiotics, siderophores, volatile compounds, antifungal lipopeptides, and lytic enzymes and indirectly by induced systemic resistance in the crop plants and protect the crops from fungal infections (Duffy and Défago 1999; Bhattacharyya and Jha 2012; Lugtenberg and Kamilova 2009; Glick 2012; Negi et al. 2017). These rhizospheric bacteria are considered as better prospect for eco-friendly cultivation.

12.6 Lytic Enzymes

The cell wall of fungi is made up mainly of chitin, glucans, and glycoproteins. Chitin provides rigidity and structural support to the cell wall. Several hydrolytic enzymes including chitinase, glucanase, protease, lipases, and cellulase production have been reported from plant growth-promoting bacteria including *Serratia marcescens, Paenibacillus, Streptomyces* spp., *Bacillus cepacia, Lysobacter antibioticus, Bacillus licheniformis, Bacillus cereus, Bacillus circulans, Bacillus thuringiensis, Enterobacter agglomerans, Stenotrophomonas maltophilia, Pseudomonas aeruginosa, Pseudomonas fluorescens, S. plymuthica, Pseudomonas stutzeri, Paenibacillus ehimensis*, etc. which play important roles in several fungal disease control (Dunne et al. 1997; Neiendam Nielsen and Sørensen 1999; Sadfi et al. 2001; Xiao-Jing et al. 2005; Compant et al. 2005; Radjacommare et al. 2010; Sekar and Prabavathy 2014; Negi et al. 2017; Tariq et al. 2017). Chitinase enzymes degrade the chitin by breaking the β-1, 4 glycosidic bonds in between the two *N*-acetyl-D-glycosamines (Fleuri et al. 2009; Kim et al. 2003; Webster and Weber 2007; Jadhav et al. 2017). The β-1,3-glucanase enzymes break the β-1,3 glucosidic bonds in β-1,3-glucans (Gupta et al. 2013; Jadhav et al. 2017). Protease plays important role in the breakdown of the membrane integrity in the cell wall of fungi by hydrolysing the proteins into small peptide chains through the breaking of the peptide bond (Jadhav et al. 2017). During interaction with fungi, plant growth-promoting bacteria release these hydrolytic enzymes in the interaction zone and inhibit the growth of pathogenic fungi around that. Due to lytic enzymes, several changes are observed in fungal structure like swelling in a fungal hyphae or lysis of hyphae, deformed mycelia, pore formation in the tips of hyphae, and leakage of cytoplasmic material. Lytic enzymes are also known to retard the growth of fungal pathogens by reducing the spore germination and suppressing the elongation of germ tubes (Budi et al. 2000; Kim et al. 2003; Negi et al. 2017). Chitinases produced by fluorescent *Pseudomonas* showed antifungal activity against *Colletotrichum falcatum* which is responsible for red rot disease in sugarcane (Viswanathan and Samiyappan 2000). Rhizobacteria that belong to *Serratia* genus produce hydrolytic enzymes such as chitinases and β-1,3-glucanases which showed greater antagonism against *Verticillium dahliae*, *Rhizoctonia solani*, and *Sclerotinia sclerotiorum* causing diseases of oilseed rape (Kalbe et al. 1996). *Paenibacillus ehimensis* IB-X-b secretes both glucanase and chitinase, which are responsible for the cell wall degradation of fungi (Aktuganov et al. 2008). Rhizobacteria *Stenotrophomonas maltophilia* strain W81 suppresses the growth of *Pythium ultimum* by producing protease enzyme and protects the sugar beet from damping-off (Dunne et al. 1997). *Bacillus subtilis* and *Pseudomonas fluorescens* produce chitinase and inhibit the growth of root rot causal fungal pathogens such as *Rhizoctonia solani* and *Fusarium solani* (El-Mougy et al. 2011). Chitinolytic enzymes produced by *Serratia plymuthica* HRO-C48 was shown to retard the growth of phytopathogens *Botrytis cinerea* by inhibiting spore germination and elongation of germ tube (Frankowski et al. 2001). Rhizobacteria *Serratia marcescens* suppressed the mycelial growth of *Sclerotium rolfsii* by producing chitinase (Ordentlich et al. 1988). Lytic enzymes including chitinase, β-1,3-glucanase, lipase, and protease are produced by *Lysobacter antibioticus* HS124 and are known to inhibit the growth of *Phytophthora capsici* by partial swelling or lysis of fungal hyphae (Ko et al. 2009).

12.7 Antibiotics

Antibiotics produced by several PGPRs including *Pseudomonas* spp., *Bacillus subtilis*, and *Bacillus cereus* are known to suppress the growth of pathogenic fungi and finally protect the crop from fungal infections (Idris et al. 2008; Das et al. 2008; Sekar et al. 2018). Antibiotics produced by rhizobacteria reduce the spore formation, lyse the fungal hyphae, make pore formation at the tips of hyphae, and increase vacuolisation in the fungal cells (Das et al. 2008; Sekar et al. 2018). Antibiotics produced by *Bacillus subtilis* and *Bacillus cereus* played a major role in disease suppression caused by *Pythium ultimum* in sorghum root (Idris et al. 2008). Volatile compounds, siderophore, and antibiotics produced by *Pseudomonas chlororaphis* SRB127 inhibited the growth of *Macrophomina phaseolina* by inhibiting the growth of mycelia and reduced microsclerotia and spore germination and control the charcoal rot disease in sorghum (Das et al. 2008). 2, 4-DAPG, chitinase, and protease produced by *Pseudomonas* spp. inhibited the growth of *Erwinia persicina*,

Pyricularia grisea, *Xanthomonas campestris*, *Gaeumannomyces graminis*, and *Fusarium oxysporum* (Sekar and Prabavathy 2014). *Pseudomonas* sp. MSSRFD41 isolated from the rhizospheric region of the finger millet has been reported to produce several antifungal compounds such as derivatives of 2,4-DAPG, pyrrolo [1, 2-a]pyrazine-1, 4-dione, octasiloxane, 2, 5-piperazinedione, 1, 2 benzenedicarboxylic acid, pyran, 2-propenoic acid and dasycarpidan-1-methanol, n-hexadecanoic acid, 1, 2-benzenedicarboxylic acid, and 9-octadecenoic acid and also produce lytic enzymes such as chitinase, protease, and lipase; these above activities are responsible for suppressing the growth of *Pyricularia grisea*. Antifungal compounds and lytic enzymes make the changes in structures of *Pyricularia grisea* like abnormal mycelia, loss of smoothness, and unusual bulges in the fungal hyphae then suppress the growth of fungi (Sekar et al. 2018).

12.8 Volatile Organic Compounds (VOCs)

Bacillus subtilis by producing several VOCs such as acetophenone, aniline, benzothiazole, 5-methyl-2-hexanone, 6-methyl-2-heptanone, m-tolunitrile, and 2-ethylhexanol inhibited the growth of *Alternaria solani*. Similarly, volatile organic compounds produced by *Bacillus subtilis* is shown to reduce the conidia germination, penetrate the fungal hyphae, and decompose the cell wall, resulting in inhibition of the fungal growth (Zhang et al. 2020). VOCs such as acetic acid and 2-nonanone were produced by *Pseudomonas* spp. which caused changes in the structure of fungal mycelia and partial lysis of fungal hyphae and degraded the cell wall, and finally leakages of cytoplasm material reduced the growth of *Sclerotinia sclerotium* (Giorgio et al. 2015).

12.9 Siderophore

Siderophore is a low-molecular-weight iron-chelating agent that plays important role in antagonistic activity against fungi by reducing iron contents in the rhizospheric region. Generally, iron is present in the soil as insoluble ferric ion; some of the bacteria chelate the ferric ions from soils by secreting siderophore. Siderophore has a high affinity toward the ferric form of iron so siderophore makes a complex with ferric ion called ferric-siderophore complex, and this complex is taken up by the cell membrane. After reaching in cell cytoplasm, ferric ion is reduced into ferrous ion and siderophore dissociates from the complex due to its low affinity toward ferrous ion. As a result, the availability of iron in the soil is reduced due to which fungal spore germination is inhibited (Beneduzi et al. 2012; Ali and Vidhale 2013; Patil et al. 2014; Dimkpa 2016). In a study by producing siderophore, *Pseudomonas aeruginosa* FP6 suppressed the growth of *Rhizoctonia solani* by 72.25% in the absence of ferric chloride and by 12% in the presence of ferric chloride indicating that siderophores play important role in antagonistic activity against fungal pathogens (Sasirekha and Srividya 2016).

12.10 Lipopeptides

By producing lipopeptides such as surfactin, fengycin, and iturin, rhizobacteria play important role in the suppression of growth of fungal phytopathogens (Ongena and Jacques 2008). Rhizobacteria *Bacillus velezensis* produces lipopeptides that retard the growth of *Fusarium oxysporum* by inhibiting the spore germination (Cao et al. 2018). WH1fungin, a new surfactin produced by *Bacillus amyloliquefaciens*, inhibited the growth of fungal pathogens. Low level of WH1fungin induces apoptosis process in fungi, but when treated with a high dose of WH1fungin, it creates pores in the cell membrane. WH1fungin also stops the synthesis of glucan part of the cell wall by inhibiting the activity of glucan synthase. When *Rhizoctonia solani* was treated with WH1fungin, then pores in the cell membrane and leakage of the cytoplasm from the pores were observed which ultimately caused to cell death in fungi (Qi et al. 2010).

12.11 Induced Systemic Resistance

Beneficial rhizobacteria play important role in inducing disease resistance in plants toward pathogens called induced systemic resistance (ISR) (Van Loon et al. 1998; Ramamoorthy et al. 2001). *P. fluorescens* strain WCS417 protected the *Dianthus caryophyllus* plant by induced systemic resistance against *F. oxysporum* (Van Peer et al. 1991). Many species of rhizobacteria belonging to the genus of *Pseudomonas* and *Bacillus* are known to induce systemic resistance in plants by inducing defence gene expressions (Van Peer et al. 1991; Kloepper et al. 2004; Van Wees et al. 2008). Both jasmonic acid and ethylene signalling pathways play major role in enhancement of the induced systemic resistance in plants (Pieterse et al. 1998; Beneduzi et al. 2012).

12.12 Conclusion

Numerous reports on rhizospheric microbes of millets have suggested the potential of rhizospheric bacteria in biocontrol of phytopathogenic fungi through the production of lytic enzymes, VOCs, siderophores, antibiotics, etc. However, majority of the studies have been confined to controlled experiments in sterilised soil and in pots. Field trials must be conducted regularly in order to justify the true biocontrol potential of PGPRs. Development of an effective microbial consortia against a wide range of phytopathogens can do wonders in the field of biofertilisers and pesticides. Further, lack of interest of commercial players in the biocontrol of phytopathogens by rhizospheric microorganisms is also a limiting factor.

Acknowledgement The authors thank the Head and Coordinator CAS, FIST of Botany, B.H.U., Varanasi, India for providing necessary facilities. Authors thank the UGC for financial help in the form of project M-14-26. KK and GP acknowledge the CSIR for financial help as SRF.

References

Adeyeye EI (2008) The intercorrelation of the amino acid quality between raw, steeped and germinated Guinea corn (*Sorghum bicolor*) grains. Bull Chem Soc Ethiop 22(1). https://doi.org/10.4314/bcse.v22i1.61320

Ahmad F, Ahmad I, Khan MS (2008) Screening of free-living rhizospheric bacteria for their multiple plant growth promoting activities. Microbiol Res 163(2):173–181

Aktuganov G, Melentjev A, Galimzianova N, Khalikova E, Korpela T, Susi P (2008) Wide-range antifungal antagonism of *Paenibacillus ehimensis* IB-Xb and its dependence on chitinase and β-1, 3-glucanase production. Can J Microbiol 54(7):577–587

Ali SS, Vidhale NN (2013) Bacterial siderophore and their application: a review. Int J Curr Microbiol App Sci 2(12):303–312

Al-Jedabi AA (2009) Biological control of *Fusarium* root-rot of sorghum. Res J Agric Biol Sci 5 (4):465–473

Ambati K, Sucharitha KV (2019) Millets-review on nutritional profiles and health benefits. Int J Recent Sci Res 10(7):33943–33948

Bazin MJ, Markham P, Scott EM, Lynch JM (1990) Population dynamics and rhizosphere interactions. In: Lynch JM (ed) The rhizosphere. Wiley, Chichester, pp 99–127

Beneduzi A, Ambrosini A, Passaglia LM (2012) Plant growth-promoting rhizobacteria (PGPR): their potential as antagonists and biocontrol agents. Genet Mol Biol 35(4):1044–1051

Bhattacharyya PN, Jha DK (2012) Plant growth-promoting rhizobacteria (PGPR): emergence in agriculture. World J Microbiol Biotechnol 28(4):1327–1350

Budi SW, van Tuinen D, Arnould C, Dumas-Gaudot E, Gianinazzi-Pearson V, Gianinazzi S (2000) Hydrolytic enzyme activity of *Paenibacillus* sp. strain B2 and effects of the antagonistic bacterium on cell integrity of two soil-borne pathogenic fungi. Appl Soil Ecol 15(2):191–199

Cao Y, Pi H, Chandrangsu P, Li Y, Wang Y, Zhou H, Cai Y (2018) Antagonism of two plant-growth promoting *Bacillus velezensis* isolates against *Ralstonia solanacearum* and *Fusarium oxysporum*. Sci Rep 8:1–14

Chauhan M, Sonawane SK, Arya SS (2018) Nutritional and nutraceutical properties of millets: a review. Clin J Nutrit Dietet 1(1):1–10

Compant S, Duffy B, Nowak J, Clément C, Barka EA (2005) Use of plant growth-promoting bacteria for biocontrol of plant diseases: principles, mechanisms of action, and future prospects. Appl Environ Microbiol 71(9):4951–4959

Darrah PR (1993) The rhizosphere and plant nutrition: a quantitative approach. Plant Soil 155 (1):1–20

Das I (2017) Millet diseases: current status and their management. In: Patil JV (ed) Millets and sorghum: biology and genetic improvement. Wiley, New Jersey, pp 291–322

Das IK, Indira S, Annapurna A, Seetharama N (2008) Biocontrol of charcoal rot in sorghum by fluorescent pseudomonads associated with the rhizosphere. Crop Protect 27(11):1407–1414

Dendy DA (1995) Sorghum and millets: production and importance. In: Denby DAV (ed) Sorghum and millets: chemistry and technology. Am. Assoc. Cereal Chemists, Inc., St. Paul, MN, pp 11–26

Dimkpa C (2016) Microbial siderophores: production, detection and application in agriculture and environment. Endocytob Cell Res 27(2):7–16

Dini-Andreote F, van Elsas JD (2013) Back to the basics: the need for ecophysiological insights to enhance our understanding of microbial behavior in the rhizosphere. Plant Soil 373(1–2):1–15

Doggett H (1989) Small millets—a selective overview. In: Seetharam A, Riley KW, Harinarayana G (eds) Small millets in global agriculture. Oxford & IBH, New Delhi, pp 59–70

Duffy BK, Défago G (1999) Environmental factors modulating antibiotic and siderophore biosynthesis by *Pseudomonas fluorescens* biocontrol strains. Appl Environ Microbiol 65 (6):2429–2438

Dunne C, Crowley JJ, Moënne-Loccoz Y, Dowling DN, O'Gara F (1997) Biological control of *Pythium ultimum* by *Stenotrophomonas maltophilia* W81 is mediated by an extracellular proteolytic activity. Microbiology 143(12):3921–3931

El-Mougy NS, Abdel-Kader MM, Alhabeb RS (2011) In vitro antifungal activity of chitinolytic enzymes produced by bio-agents against root rot pathogenic fungi. Arch Phytopathol Plant Protect 44(7):613–622

FAO (1972) Food and nutrition, sorghum and millets in human nutrition, Series, No 27

Fleuri LF, Kawaguti HY, Sato HH (2009) Production, purification and application of extracellular chitinase from *Cellulosimicrobium cellulans* 191. Braz J Microbiol 40(3):623–630

Frankowski J, Lorito M, Scala F, Schmid R, Berg G, Bahl H (2001) Purification and properties of two chitinolytic enzymes of *Serratia plymuthica* HRO-C48. Arch Microbiol 176(6):421–426

Giorgio A, De Stradis A, Lo Cantore P, Iacobellis NS (2015) Biocide effects of volatile organic compounds produced by potential biocontrol rhizobacteria on *Sclerotinia sclerotiorum*. Front Microbiol 6:1056

Glick BR (2012) Plant growth-promoting bacteria: mechanisms and applications. Scientifica

Gupta P, Ravi I, Sharma V (2013) Induction of β-1, 3-glucanase and chitinase activity in the defense response of *Eruca sativa* plants against the fungal pathogen *Alternaria brassicicola*. J Plant Interact 8(2):155–161

Han Y, Xu L, Liu L, Yi M, Guo E, Zhang A, Yi H (2017) Illumina sequencing reveals a rhizosphere bacterial community associated with foxtail millet smut disease suppression. Plant Soil 410 (1–2):411–421

Hirsch PR, Mauchline TH (2012) Who's who in the plant root microbiome? Nat Biotechnol 30 (10):961

Hulse JH, Laing EM, Pearson OE (1980) Sorghum and millets their composition and nutritive value. Academic Press, New York

Idris HA, Labuschagne N, Korsten L (2007) Screening rhizobacteria for biological control of Fusarium root and crown rot of sorghum in Ethiopia. Biol Control 40(1):97–106

Idris HA, Labuschagne N, Korsten L (2008) Suppression of *Pythium ultimum* root rot of sorghum by rhizobacterial isolates from Ethiopia and South Africa. Biol Control 45(1):72–84

Jadhav HP, Shaikh SS, Sayyed RZ (2017) Role of hydrolytic enzymes of rhizoflora in biocontrol of fungal phytopathogens: an overview. In: Rhizotrophs: plant growth promotion to bioremediation. Springer, Singapore, pp 183–203

Jogaiah S, Mitani S, Kestur Nagaraj A, Huntrike Shekar S (2007) Activity of cyazofamid against *Sclerospora graminicola*, a downy mildew disease of pearl millet. Pest Manag Sci 63 (7):722–727

Kalbe C, Marten P, Berg G (1996) Strains of the genus *Serratia* as beneficial rhizobacteria of oilseed rape with antifungal properties. Microbiol Res 151(4):433–439

Khatri D, Durgapal A, Joshi PK (2016) Biofertilization enhances productivity and nutrient uptake of foxtail millet plants. J Crop Improv 30(1):32–46

Kim KJ, Yang YJ, Kim JG (2003) Purification and characterization of chitinase from *Streptomyces* sp. M-20. J Biochem Mol Boil 36(2):185–189

Kloepper JW, Ryu CM, Zhang S (2004) Induced systemic resistance and promotion of plant growth by *Bacillus* spp. Phytopathology 94(11):1259–1266

Ko HS, Jin RD, Krishnan HB, Lee SB, Kim KY (2009) Biocontrol ability of *Lysobacter antibioticus* HS124 against *Phytophthora* blight is mediated by the production of 4-hydroxyphenylacetic acid and several lytic enzymes. Curr Microbiol 59(6):608–615

Kumar A, Tomer V, Kaur A, Kumar V, Gupta K (2018) Millets: a solution to agrarian and nutritional challenges. Agricult Food Secur 7(1):31

Leukel RW, Martin JH (1943) Seed rot and seedling blight of sorghum (No 1488-2016-124410)

Little CR, Perumal R (2019) The biology and control of sorghum diseases. In: Ciampitti IA, Vara Prasad PV (eds) Sorghum: a state of the art and future perspectives, vol 58, pp 297–346

Lugtenberg B, Kamilova F (2009) Plant-growth-promoting rhizobacteria. Annu Rev Microbiol 63:541–556

Lugtenberg BJ, Dekkers L, Bloemberg GV (2001) Molecular determinants of rhizosphere colonization by *Pseudomonas*. Annu Rev Phytopathol 39(1):461–490

Mounde LG, Boh MY, Cotter M, Rasche F (2015) Potential of rhizobacteria for promoting sorghum growth and suppressing *Striga hermonthica* development. J Plant Dis Protect 122(2):100–106

Nagaraja A, Das IK (2016) Disease resistance in pearl millet and small millets. In: Biotic stress resistance in millets. Academic Press, pp 69–104

Negi YK, Prabha D, Garg SK, Kumar J (2017) Biological control of ragi blast disease by chitinase producing fluorescent *Pseudomonas* isolates. Org Agric 7(1):63–71

Neiendam Nielsen M, Sørensen J (1999) Chitinolytic activity of *Pseudomonas fluorescens* isolates from barley and sugar beet rhizosphere. FEMS Microbiol Ecol 30(3):217–227

Niu X, Song L, Xiao Y, Ge W (2018) Drought-tolerant plant growth-promoting rhizobacteria associated with foxtail millet in a semi-arid agroecosystem and their potential in alleviating drought stress. Front Microbiol 8:2580

Ongena M, Jacques P (2008) *Bacillus* lipopeptides: versatile weapons for plant disease biocontrol. Trends Microbiol 16(3):115–125

Ordentlich A, Elad Y, Chet I (1988) The role of chitinase of *Serratia marcescens* in biocontrol of *Sclerotium rolfsii*. Phytopathology 78(1):84–88

Patil S, Bheemaraddi MC, Shivannavar CT, Gaddad SM (2014) Biocontrol activity of siderophore producing Bacillus subtilis CTS-G24 against wilt and dry root rot causing fungi in chickpea. IOSR J Agricult Veterin Sci 7(9):63–68

Philippot L, Raaijmakers JM, Lemanceau P, Van Der Putten WH (2013) Going back to the roots: the microbial ecology of the rhizosphere. Nat Rev Microbiol 11(11):789–799

Pieterse CM, Van Wees SC, Van Pelt JA, Knoester M, Laan R, Gerrits H, Van Loon LC (1998) A novel signaling pathway controlling induced systemic resistance in *Arabidopsis*. Plant Cell 10 (9):1571–1580

Qi G, Zhu F, Du P, Yang X, Qiu D, Yu Z, Zhao X (2010) Lipopeptide induces apoptosis in fungal cells by a mitochondria-dependent pathway. Peptides 31(11):1978–1986

Radjacommare R, Venkatesan S, Samiyappan R (2010) Biological control of phytopathogenic fungi of vanilla through lytic action of *Trichoderma* species and *Pseudomonas fluorescens*. Arch Phytopathol Plant Protect 43(1):1–17

Raghunathan V (1968) Damping-off of green gram, cauliflower, daincha, ragi and cluster beans. Indian Phytopathol 21:456–457

Raj SN, Deepak SA, Basavaraju P, Shetty HS, Reddy MS, Kloepper JW (2003) Comparative performance of formulations of plant growth promoting rhizobacteria in growth promotion and suppression of downy mildew in pearl millet. Crop Prot 22(4):579–588

Ramamoorthy V, Viswanathan R, Raguchander T, Prakasam V, Samiyappan R (2001) Induction of systemic resistance by plant growth promoting rhizobacteria in crop plants against pests and diseases. Crop Prot 20(1):1–11

Rao M (1989) Inaugural address. In: Seetharam A, Riley K, Harinaryana G (eds) Small millets in global agriculture. Oxford IBH, New Delhi, pp 9–11

Rokhbakhsh-Zamin F, Sachdev D, Kazemi-Pour N, Engineer A, Pardesi KR, Zinjarde S, Chopade BA (2011) Characterization of plant-growth-promoting traits of *Acinetobacter* species isolated from rhizosphere of *Pennisetum glaucum*. J Microbiol Biotechnol 21(6):556–566

Sadfi N, Cherif M, Fliss I, Boudabbous A, Antoun H (2001) Evaluation of bacterial isolates from salty soils and *Bacillus thuringiensis* strains for the biocontrol of *Fusarium* dry rot of potato tubers. J Plant Pathol 83:101–117

Sarita ES, Singh E (2016) Potential of millets: nutrients composition and health benefits. J Sci Innov Res 5(2):46–50

Sasirekha B, Srividya S (2016) Siderophore production by *Pseudomonas aeruginosa* FP6, a biocontrol strain for *Rhizoctonia solani* and *Colletotrichum gloeosporioides* causing diseases in chilli. Agric Nat Resour 50(4):250–256

Saxena A, Raj N, Sarosh BR, Kini R, Shetty HS (2013) Rhizobacteria mediated growth enhancement in pearl millet. Indian J Sci Res 4(2):41–44

Sekar J, Prabavathy VR (2014) Novel Phl-producing genotypes of finger millet rhizosphere associated pseudomonads and assessment of their functional and genetic diversity. FEMS Microbiol Ecol 89(1):32–46

Sekar J, Raju K, Duraisamy P, Ramalingam Vaiyapuri P (2018) Potential of finger millet indigenous rhizobacterium *Pseudomonas* sp. mssrfd41 in blast disease management—growth promotion and compatibility with the resident rhizomicrobiome. Front Microbiol 9:1029

Tariq M, Noman M, Ahmed T, Hameed A, Manzoor N, Zafar M (2017) Antagonistic features displayed by plant growth promoting rhizobacteria (PGPR): a review. J Plant Sci Phytopathol 1:38–43

Van Loon LC, Bakker PAHM, Pieterse CMJ (1998) Systemic resistance induced by rhizosphere bacteria. Annu Rev Phytopathol 36(1):453–483

Van Peer R, Niemann GJ, Schippers B (1991) Induced resistance and phytoalexin accumulation in biological control of *Fusarium* wilt of carnation by *Pseudomonas* sp. strain WCS 417 r. Phytopathology 81(7):728–734

Van Wees SC, Van der Ent S, Pieterse CM (2008) Plant immune responses triggered by beneficial microbes. Curr Opin Plant Biol 11(4):443–448

Viswanathan R, Samiyappan R (2000) Efficacy of *Pseudomonas* spp. strains against soil borne and sett borne inoculum of *Colletotrichum falcatum* causing red rot disease in sugarcane. Sugar Tech 2(3):26–29

Webster J, Weber R (2007) Introduction to fungi. Cambridge University Press

Wilson JP (2000) Pearl Millet Diseases: a compilation of information on the known pathogens of pearl millet: *Pennisetum glaucum* (L.) R Br (No 716). US Department of Agriculture, Agricultural Research Service

Xiao-Jing X, Li-Qun Z, You-Yong Z, Wen-Hua T (2005) Improving biocontrol effect of *Pseudomonas fluorescens* P5 on plant diseases by genetic modification with chitinase gene. Chin J Agr Biotechnol 2(1):23–27

Zhang H, Kaushal R, Singh SK, Paré PW (2020) Bacterial volatile-mediated plant abiotic stress tolerance. In: Bacterial volatile compounds as mediators of airborne interactions. Springer, Singapore, pp 187–200

Chapter 13
Metagenomics of Plant Rhizosphere and Endophytic Association: Concepts and Applications

Arpita Ghosh, Ankita Rathore, and Balakuntla Jayanth

Abstract Microbes in the rhizosphere influence plant growth, productivity, susceptibility, and resistance to biotic and abiotic stresses. Various studies have been reported to show diversity and activity of microbes are as high in plants as in endophytes and rhizosphere. The roots harbor more diverse microbes than any other part of the plant. The soil type and its management also influence the microbial diversity. The microbial communities can enhance and facilitate pathogen defense and their role in environmental remediation through different mechanisms. Metagenomics is a growing field that helps understand the genomes in the microbial communities. The high resolution of uncultured microbes and the correlation of the function with the environment can be achieved using functional metagenomics. New emerging subdisciplines of metagenomics are Metatranscriptomics and Metaproteomics, which provide further functional analysis of microbial communities. Integrative metagen"omics" approach results in comprehensive information for the community from genes to RNA to proteins and metabolites. In this chapter, we discuss the plant rhizosphere; types of metagenomics analysis such as 16S (for bacteria), whole metagenomics, and 18S/ITS (for fungus); and application of metagenome associated with rhizosphere and endophytes.

Keywords Metagenome · 16S · ITS · Rhizosphere · Endophytes

13.1 Introduction

Rhizosphere plays an important role in microbial-mediated processes like plant growth promotion, plant protection, and pathogenesis. Rhizosphere is the soil neighboring the roots which is most exposed to the influence of plant's root exudates

A. Ghosh (✉) · A. Rathore
Eurofins Genomics India Pvt. Ltd., Bengaluru, Karnataka, India

B. Jayanth
Elanco IAC, Bengaluru, Karnataka, India

S. K. Dubey, S. K. Verma (eds.), *Plant, Soil and Microbes in Tropical Ecosystems*, Rhizosphere Biology, https://doi.org/10.1007/978-981-16-3364-5_13

(Soni et al. 2017). The rhizosphere microbiology has received significant amount of attention, as it influences the plant both directly and indirectly improving their fitness and health (Sapkota et al. 2015). Rhizosphere microbiota interaction helps plants to deal with abiotic stress and diseases and improves the exchange of substances such as nitrogen fixation, phosphate solubilization, and production of plant growth hormones or by acting as a biocontrol agent to help against pathogens and tolerance to various stresses (Tsurumaru et al. 2015; Elias et al. 2016; Majeed et al. 2015; Massart et al. 2015; Reinhold-Hurek et al. 2015; Vega-Avila et al. 2015; Gallart et al. 2018). It differs from the normal soil because of the biological and physicochemical processes happening due to the plant and microbial association such as root growth, water and nutrient uptake, respiration, and rhizodeposition (López et al. 2012).

Approximately, 10^{10}–10^{11} bacterial cells are present in 1 gram soil (Claire Horner-Devine et al. 2003) belonging to 10^3–10^4 species (Curtis et al. 2002), but approximately 1 gram of plant tissue estimates 10^9 bacterial cells (Chi et al. 2005) which shows the vast diversity of microbes in the rhizosphere. The microbiome includes various functional gene pool from prokaryotic to eukaryotic associated with various habitats of a plant-like rhizosphere and rhizoplane and plays a crucial role in plant protection (Abd-Elsalam et al. 2010; Mendes et al. 2011; Lakshmanan et al. 2014). The structure of microbial communities in the rhizosphere is largely influenced by ambient condition, soil properties, plant genotype, cultivars, and developmental stages of the plant (Broeckling et al. 2008; Qiao et al. 2017). Different plant species host specific microbial communities when grown in the same soil, i.e., plants are able to shape their rhizosphere microbiome (Aira et al. 2010; Berendsen et al. 2012; Bazghaleh et al. 2015; Berlanas et al. 2019).

Different approaches of metagenomics help to provide insight on many of the important aspects such as taxonomic diversity, which organisms are present, and functional metagenomics, what are their roles (Vieites et al. 2009) which in turn allows to characterize microbes in the given environmental sample. It detects the species and also helps understand the metabolic activities and functional roles of the microbes in a given sample (Langille et al. 2013). As some of the microorganisms are culturable under laboratory practices and some are not, still they all are life forms based on DNA as a genetic information can be studied by Metagenomics; this makes this approach very important and extensive.

13.2 Study of Microbial Community in Plant Rhizosphere and Endophytic Association

13.2.1 Sampling, DNA Extraction, and Sequencing

The rhizosphere of a plant is collected along with its adhering soils, refined, and made free from root hairs before processing for metagenomic DNA extraction. To study the microbial community associated with the various crop cycle, the soil can be collected at specific growth stages. For example, the sampling of the rhizosphere soil can be prior to the onset of blooming stage in order to analyze the microbial community structure and function before the most critical stage of the crop cycle. The sample after collection should be stored in −80 °C until the metagenomic DNA extraction is performed (Prabha et al. 2019). The collected rhizosphere is subjected to isolation of the DNA using 2–5 gm of the rhizosphere soil sample by any specialized DNA isolation kit or manual isolation method.

To study the endophytic microbial community roots and leaves, samples are collected, and surface sterilization is performed by repeated immersion in 70% (v/v) ethanol for couple of mins and then 2.5% (v/v) sodium hypochlorite (NaOCl) for 5 min (Barra et al. 2016). Sterile distilled water is used to rinse the roots. The roots and leaves are cut in small pieces, frozen in liquid nitrogen, macerated and homogenized with a mortar and pestle, and followed by storage in −80 °C until DNA extraction. DNA isolation can be performed using kit or manual isolation method (Zhang et al. 2019).

The quality of the extracted DNA is determined using NanoDrop and Qubit. The extracted DNA should be subjected to agarose-gel electrophoresis for quality check. There are two main methods for studying microbial community, namely, amplicon-based and shotgun metagenomics. For amplicon-based sequencing, gene-specified (16S/ITS/18S) primer is designed with Illumina adapters. PCR amplification is performed using the forward and reverse universal primers. The PCR products are purified, and the purified products are used for sequencing on Illumina Sequencer. For shotgun metagenome sequencing, the isolated high-quality DNA is used for metagenomic library preparation with respect to the selected metagenomics approach. This library is used for high-throughput sequencing through NGS platforms (Prabha et al. 2019).

13.2.2 Methods of Metagenomics Analysis

There are two main methods for microbiome analysis using high-throughput omic techniques amplicon-based and shotgun metagenomics as shown in Fig. 13.1. In amplicon-based method, primers are designed to amplify a specific gene such as 16S rRNA for bacteria/archaea, 18S for Eukaryotes, and ITS for fungi, from the genomes present in a given sample. The sequences are then clustered into operational taxonomic units (OTUs), and further taxonomic abundance and diversities are compared across samples. Shotgun metagenomics refer to the study of entire genomic material in the microbiome of a sample. It can shed light on the structure and organization of genomes, gene function, and their evolutionary relationships (Roumpeka et al. 2017).

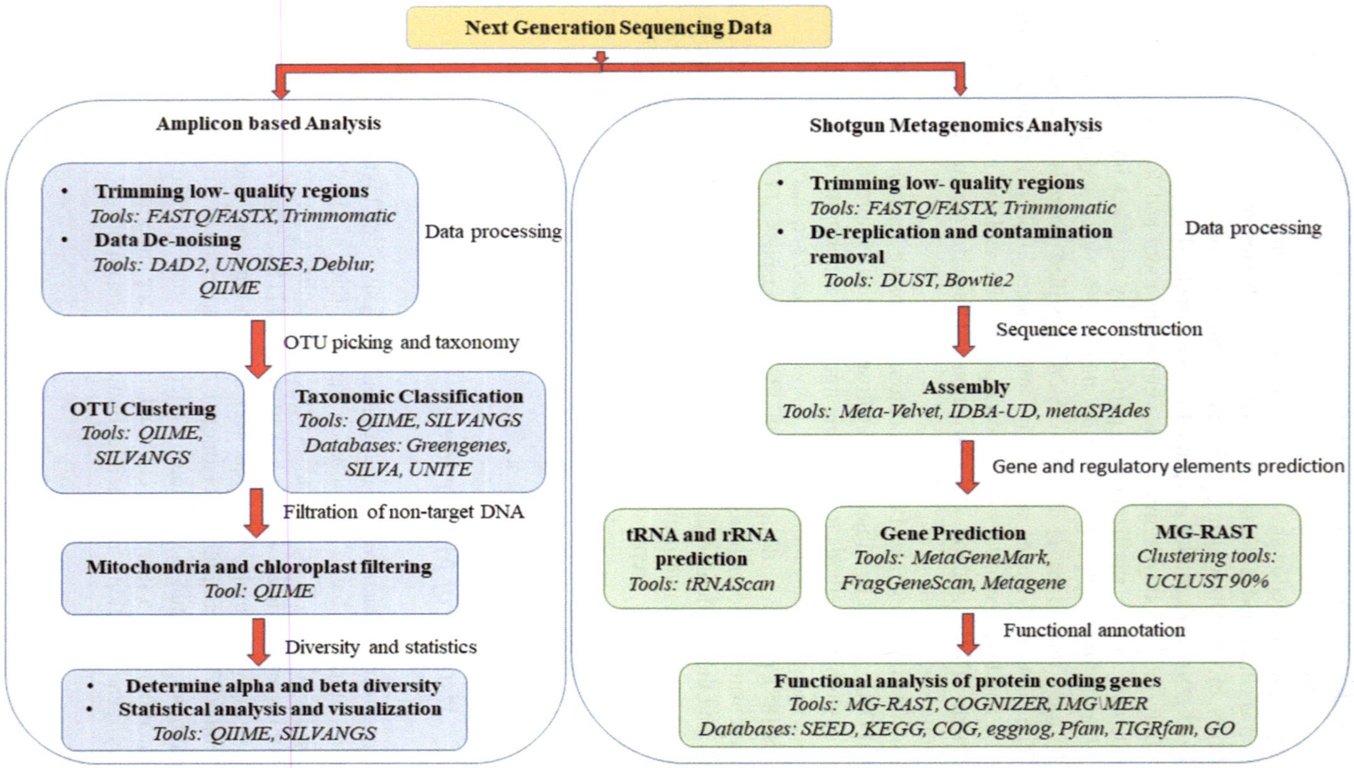

Fig. 13.1 Application of NGS for analyzing rhizospheric and endophytic microbial community

13.2.2.1 Amplicon-Based Metagenomics (16S/18S/ITS)

The prokaryotic 16S ribosomal RNA gene (16S rRNA) is approximately 1500 bp long and contains 9 hypervariable regions (V1–V9) flanked by conserved regions (Chakravorty et al. 2007). These variable regions of 16S rRNA are frequently used in taxonomic classifications in diverse microbial communities. Internal transcribed spacer (ITS) is a highly variable sequence that lies between the 16S and 23S rRNA genes and is of great importance in distinguishing fungal species (Bromberg et al. 2015). The length of ITS regions may vary from 50 bases to several kbs. ITS1 and ITS2 genes were observed to be the most appropriate marker for fungal phylogenetic analysis because of their variable regions, conserved primers, and multicopy nature of the genome (Cuadros-Orellana et al. 2013). The fungal taxonomical studies are based on the nuclear ribosomal gene cluster, which includes 18S or small subunit (SSU), 5.8S subunit, and 28S or large subunit (LSU) genes.

There are six most popular pipelines which are widely used for amplicon-based analysis (Table 13.1): three *OTU based*, QIIME (Kuczynski et al. 2012), MOTHUR (Schloss et al. 2009), and USEARCH-UPARSE (Edgar 2010, 2013), and three *ASV level based*, DADA2 (Callahan et al. 2016), QIIME2-Deblur (Amir et al. 2017), and USEARCH-UNOISE3. The OTU-based three pipelines cluster sequences at 97% identity into operational taxonomic units (OTUs). The latter three pipelines attempt to reconstruct exact biological sequences called amplicon sequence variants (ASVs) present in the sample (Marizzoni et al. 2020; Prodan et al. 2020).

Preprocessing of the sequenced reads: The raw reads are subjected to demultiplexing and quality assessment followed by removal of poor-quality reads prior to analysis (Plummer et al. 2015). Most commonly used tool is Trimmomatic (Bolger et al. 2014). The high-quality PE reads are merged into a unique sequence prior to data analysis. FLASII (Fast Length Adjustment of Short Reads) (Magoč and Salzberg 2011) is used to stitch overlapping paired end reads into single end long reads in 16S analysis. In QIIME-uclust and QIIME 2-Deblur, reads can be filtered and merged externally using USEARCH. DADA2 utilizes a model-based approach for correcting amplicon errors, and reads are merged after denoising of data. After quality filtration and merging of reads, chimeric reads are removed and remaining sequences are clustered into OTUs.

OTU Picking: Clustering of high-throughput 16S sequences into biologically meaningful operational taxonomic units (OTUs) is a challenging task. In OTU picking, 16S sequences are clustered at a certain level of sequence similarity (default 97%). There are three different approaches for OTU picking: de novo, closed-reference, and open-reference. (1) de novo OTU picking method: input sequences are aligned against one another and sequences that align with greater than a user-specified percent identity belongs to the same OTU, without any external reference sequence collection. (2) Closed-reference OTU picking method: sequences are first aligned to a reference sequence collection and any sequences which does not match reference sequence at a user-defined percent identity threshold is excluded from downstream analyses. (3) Open-reference OTU picking method: reads are first

Table 13.1 List of tools and databases

Category	Tools	References
Shotgun	CLARK	Ounit et al. (2015)
	Centrifuge	Kim et al. (2016)
	IDBA-UD	Peng et al. (2012)
	KRAKEN	Wood and Salzberg (2014)
	MetaVelvet	Namiki et al. (2012)
	MetaVelvet-SL	Sato and Sakakibara (2014)
	Ray Meta	Boisvert et al. (2012)
	SOAPdenovo2	Luo et al. (2012)
	metaSPAdes	Nurk et al. (2017)
	MetAMOS	Treangen et al. (2013)
	KAIJU	Menzel et al. (2016)
	Prodigal	Hyatt et al. (2010)
	FragGeneScan	Rho et al. (2010)
	MetaGeneAnnotator	Noguchi et al. (2008)
	MetaGeneMark	Zhu et al. (2010)
	Glimmer-MG	Kelley et al. (2011)
	Kraken2	Wood et al. (2019)
	MetaMaps	Dilthey et al. (2019)
	Megan	Huson and Weber (2013)
	MetaPhlAn	Segata et al. (2012)
	MG-RAST	Wilke et al. (2016)
16S/18S/ITS	QIIME/QIIME2	Caporaso et al. (2010)
	Mothur	Schloss et al. (2009)
	USEARCH	Edgar (2010)
	UPARSE	Edgar (2013)
	UNOISE	Edgar (2016)
	DADA2	Callahan et al. (2016)
	Deblur	Amir et al. (2017)
	PipeCraft	Anslan et al. (2017)
	LotuS	Hildebrand et al. (2014)
	AMPtk	Palmer et al. (2018)
	PIPITS	Gweon et al. (2015)
Functional 16S analysis	PICRUSt	Langille et al. (2013)
Databases	SILVA	Quast et al. (2012)
	Greengenes	DeSantis et al. (2006)
	Ribosomal database (RDP)	Cole et al. (2007)
	KEGG	Ogata et al. (1999)
	GhostKOALA	Kanehisa et al. (2016)
	SEED	Overbeek et al. (2005)
	eggnog	Powell et al. (2014)
	COG/KOG	Tatusov et al. (2000)
	PFAM	Bateman et al. (2004)

(continued)

Table 13.1 (continued)

Category	Tools	References
	TIGRFAM	Haft et al. (2003)
	Reactome	Fabregat et al. (2016)
	MetaCyc	Caspi et al. (2016)
	UNITE	Kõljalg et al. (2013)

aligned to a reference sequence database, and any reads which fail to align are clustered de novo (Rideout et al. 2014). OTU picking method comprises of taxonomic assignment, sequence alignment, and tree-building steps.

Taxonomic Assignment: A crucial step in microbiome amplicon analysis is taxonomic assignment. Taxonomic classification of 16S/18S sequences is accomplished using one of these databases: Greengenes, SILVA, RDP, or NCBI 16S/18S microbial database. The Greengenes database (McDonald et al. 2012) contains Bacteria and Archaea taxonomic information. The SILVA database (Quast et al. 2013; Yilmaz et al. 2014) is designed for Bacteria, Archaea, and Eukarya taxonomic details and is primarily based on phylogenies for small subunit rRNAs (16S for prokaryotes and 18S for Eukarya). The RDP database (Cole et al. 2007, 2014) contains 16S rRNA sequences from Bacteria, Archaea, and 28S rRNA sequences for fungi (Eukarya) available from the International Nucleotide Sequence Database Collaboration (INSDC) (Cochrane et al. 2016) databases (Balvočiūtė and Huson 2017). The most popularly used ITS database for taxonomic assignment is UNITE (Nilsson et al. 2019). In the case of rhizosphere and endophytes, we have an influence of the plant parts. To avoid the non-microbiota such as chloroplast and mitochondria from the data, which are expected due to the presence of the plant part, are removed to obtain only the microbiota using QIIME (Zhang et al. 2019).

Diversity Analysis: Whittaker in 1960 and 1972 described three different types of measures of biodiversity: alpha, beta, and gamma diversity. Alpha diversity is defined as diversity of organisms within a sample or ecosystem and is usually expressed by the number of species (i.e., species richness) in ecosystem. Beta diversity measures difference in diversities across the sample or ecosystem. Gamma diversity measures the diversity of a larger unit such as a region or landscape (Navas-Molina et al. 2013). Alpha diversity measures richness, dominance, and evenness using various diversity metrics such as richness, Chao1, Shannon index, and inverse Simpson index. Beta diversity metrics are namely phylogenetic and non-phylogenetic metrics such as Bray-Curtis distance, Euclidean distance, and unifrac weighted and unweighted that can be calculated using QIIME package and phyloseq R package.

Functional Analysis: The functional composition of 16S microbial communities can be performed using PICRUSt. Ancestral-state reconstruction algorithm is used to predict the gene families and then combines gene families to estimate the composite metagenome. It provides the insight about the metabolic activities and functional roles of the microbes in the sample. The result of the annotation for predicted gene family counts is orthologous groups of the gene families or KOs, COGs, or Pfams (Langille et al. 2013).

13.2.2.2 Shotgun Metagenomics

Metagenomics, also referred to as WGS- or shotgun-metagenomics, allows researchers to comprehensively sequence and study the entire genomic material present in the microbiome sample. Sequencing the genomes of all organisms present in metagenomic sample can furnish detailed information of the structure and organization of genomes, function of predicted genes, evolutionary relationships, and identification of novel genes (Roumpeka et al. 2017). The extensive advantage of metagenomic approach is that it provides high taxonomic and functional resolution. Insight into gene functions and characterization of specific strains of these microbial communities from rhizosphere/endophytes can reveal plant growth promotion predicted coding genes (Romero et al. 2019).

A wide range of bioinformatic tools are available to execute the shotgun metagenomic analysis as shown in Table 13.1. The bioinformatics analysis generally includes the following steps: (a) the assembly of sequenced metagenomic fragments to construct contiguous sequences, (b) gene prediction from assembled sequences, and (c) identification of domains, their functions, and metabolic pathways for the putative proteins (Roumpeka et al. 2017).

Preprocessing of Sequenced Reads: Based on quality assessment of sequenced data, reads are trimmed to retain high-quality pair-end data. Most commonly used trimming tools are Trimmomatic (Bolger et al. 2014) and Cutadapt (Martin 2011) that remove low-quality bases from both terminals of each sequence. Removal of bad quality reads greatly improved the accuracy and contig lengths of resulting assembly.

Metagenomic Assembly: To assemble all of the genomes present within a metagenomic sample, we have many tools based on de novo metagenomic assemblers which uses de Bruijn graph approach for assembly (Pevzner et al. 2001). One of the widely used metagenomic de novo assembler is MetaVelvet (Afiahayati et al. 2015; Namiki et al. 2012). For a given set of metagenomic reads, it first constructs a large de Bruijn graph, and then mixed de Bruijn graph is decomposed into subgraphs which can be used to construct longer contiguous genome sequences. It is reported that MetaVelvet tool surpasses other commonly used assemblers like IDBA-UD (Peng et al. 2011, 2012) and Ray Meta (Boisvert et al. 2012). Another method which metagenomics assembler commonly uses is K-mer-based method: KRAKEN (Wood and Salzberg 2014), CLARK (Ounit et al. 2015), KAIJU (Menzel et al. 2016), and Centrifuge (Kim et al. 2016) are the popular tools which used this method. K-mer based methods extract kmers from each read pair, and heuristic searches were performed against the user-specified database. They are ultrafast, and sensitivity depends on the choice of the database. Another framework which combines available bioinformatics tools into a metagenomic analysis pipeline is MetAMOS (Treangen et al. 2013). This pipeline first assembles the metagenome reads, and scaffolds are created. Finally, in post-assembling stage, assembled scaffolds are annotated and taxonomically classified.

Gene Prediction: Annotating the assembled data and predicting genes and regulatory elements are important steps in a metagenomic analysis pipeline. A metagenomic gene-finding algorithm, MetaGeneAnnotator (Noguchi et al. 2008), can predict genes from uncharacterized metagenomic communities. Glimmer-MG (Kelley et al. 2012), an extension of Glimmer which is a popular bacterial gene prediction tool, clusters metagenomic data which likely belong to the same organism and also considers insertions and deletions during the gene prediction. FragGeneScan (Rho et al. 2010) is another tool based on hidden Markov models (HMMs), specifically designed to predict fragmented genes directly without the need of assembly; however, the software can also run on assembled sequences. MetaGeneMark (Zhu et al. 2010) is an ab-initio gene prediction tool specifically designed for metagenome sample to identify protein coding regions.

Taxonomic classification: Many software has recently been deployed to classify metagenomics data taxonomically and estimate their taxonomic abundance profiles. Certain bioinformatics tools like CosmosID, Inc. (CosmosID, Inc., Rockville, MD, USA), Kraken2 (Wood et al. 2019), MetaMaps (Dilthey et al. 2019), and MetaPhlAn (Segata et al. 2012) are designed to identify taxonomic level till species, subspecies, and strain level using assembled/unassembled metagenomic data. MG-RAST (Glass et al. 2010; Wilke et al. 2016) is a widely used metagenomics analysis web-server which can identify taxonomic information below the genus level.

Functional Annotation: To infer functional annotation from metagenomics data, many reference databases like KEGG (Kanehisa et al. 2012), COG/KOG (Tatusov et al. 1997), eggNOG (Powell et al. 2012), PFAM (Punta et al. 2012), and TIGRFAM (Selengut et al. 2007) are available. MetaCyc (Caspi et al. 2016) is considered as largest comprehensive database of curated metabolic pathways and enzymes from all domains of life. Reactome (Fabregat et al. 2016) is another open-source and curated database of biological pathways. The metabolic pathway analysis can also be done using GhostKOALA (Kanehisa et al. 2016). It correlates taxonomy with their functional annotation, and user can visualize metabolic pathways from different taxa in the same map.

13.2.2.3 Metatranscriptomics and Metaproteomics

Metatranscriptomics and metaproteomics are reasonably recent subtypes of metagenomics, which enables us to look into functional analysis of microbial communities (Ghosh et al. 2019). The study of microbial communities based on RNA sequencing in a complex ecosystem is known as metatranscriptomics (Zhang et al. 2017). The co-expressed gene clusters of the ecologically relevant trends are identified followed by the transcripts abundance, and functional annotation is studied in the environmental samples (Oyserman et al. 2016). To get high-quality RNA from the environment samples is the biggest challenge associated with this method. However, it is an efficient approach to elucidate gene expression and has the capability to discover novel gene in the microbial community (Frias-Lopez et al. 2008; Tartar et al. 2009).

The study of proteome expressed in the microbial community at a particular time is known as metaproteomics. This method allows to discover the microbial activities based on the metabolic pathways in the microbial ecosystem (Zampieri et al. 2016). Metaproteomics is an emerging field along with metagenomics, which allows to characterize the proteins from a microbiota such as human gut (Petriz and Franco 2017). The study of metagenomics, metatranscriptomics, and metaproteomics provides information of the functional dynamics, activities, and production capabilities of microbial community (Simon and Daniel 2011).

13.3 Future Perspective and Applications

Recent studies have highlighted the plant-plant and plant-microbe interactions along with their complexities as an interlinked ecosystem. It is inhabited by diverse microbial communities that are structurally and functionally affected by plant and soil type (Yurgel et al. 2019). Genomics has given rise to metagenomics, an approach that will enable us to explore the as-yet-uncultured microbes which represents the vast majority of organisms in most environments on earth. The high-throughput and "omics" techniques could shed light on the composition and structure of beneficial rhizobiome communities and what role the host may play in the enrollment and control of its microbiome.

Crop production is reliant on pesticides to manage diseases and pests and on chemical fertilizers to provide sufficient nutrients to enhance crop yields. However, the wide use of pesticides and chemically synthesized fertilizers may lead to pesticide resistance pathogens, environmental pollution, contamination of surface along with the groundwater, and detrimental effects on humans, beneficial soil microbes, and other organisms (Liu et al. 2018). One way to address these issues is to utilize rhizosphere engineering which may lessen our dependency on agrochemicals by substituting their functions with beneficial microbes and biodegradable biostimulants and can manipulate plant/microorganism interactions accordingly (Ryan et al. 2009).

There are increasing evidences to suggest that the rhizobiome can enhance plant growth directly, improve drought tolerance, and play important role in environmental remediation through different mechanisms (Jones et al. 2019). The microbe-mediated nutrient uptake, disease resistance, and stress tolerance are some examples of microbial functions crucial to agricultural production systems. Moreover, they are engaged in the secretion of a diverse range of chemicals that can be classified as signaling compounds, and may serve as nutrient solubilizers (Verma et al. 2018). Microbes in the rhizosphere could serve as candidate taxa for biofertilizers and growth supplements and may act as proficient innovative tools for the sustainability of agro-ecosystems. Understanding the hidden mechanisms of the host-based selection of microbiome could further guide insight into microbiome-based breeding programs (Poudel et al. 2019).

Recent metagenomic approaches can help in deciphering these interactions in a comprehensive manner and can enable us to have a reasonable agriculture yield with improved crop management. Besides this, the researcher also suggests that the rhizosphere microflora can benefit plants by increasing tolerance to abiotic stresses like temperature, salinity, and heavy metal stress. It also increases plant-defensive measures by protecting against deadly pathogens through microbial antagonism (Jones et al. 2019).

Rhizosphere has been witnessed as one of the most crucial interfaces for life on earth. The microbial root colonization activates multiple types of physical and chemical interconnections between microbes and plants. Rhizodeposition of discrete exudates acts as an important substrate for the soil microbial community, and there is complex coaction between this community and type of compounds released (Ramakrishnan et al. 2009). The culture-independent rhizosphere and endosphere microbe's analysis will provide insight on plant-microbe interaction, by understanding the variability of beneficial microbes in a various different environment which will, in turn, help crop management practices. Using the metagenomics information from a different niche, we can modulate the composition of root microbiomes to improve crop growth and health (Rascovan et al. 2016).

References

Abd-Elsalam KA, Almohimeed I, Moslem MA, Bahkali AH (2010) M13-microsatellite PCR and rDNA sequence markers for identification of Trichoderma (Hypocreaceae) species in Saudi Arabian soil. Genet Mol Res 9(4):2016–2024

Afiahayati, Sato K, Sakakibara Y (2015) MetaVelvet-SL: an extension of the velvet assembler to a de novo metagenomic assembler utilizing supervised learning. DNA Res 22(1):69–77

Aira M, Gómez-Brandón M, Lazcano C, Bååth E, Domínguez J (2010) Plant genotype strongly modifies the structure and growth of maize rhizosphere microbial communities. Soil Biol Biochem 42(12):2276–2281

Amir A, McDonald D, Navas-Molina JA, Kopylova E, Morton JT, Zech Xu Z, Kightley EP, Thompson LR, Hyde ER, Gonzalez A, Knight R (2017) Deblur rapidly resolves single-nucleotide community sequence patterns. mSystems 2(2):e00191–e00116

Anslan S, Bahram M, Hiiesalu I, Tedersoo L (2017) PipeCraft: flexible open-source toolkit for bioinformatics analysis of custom high-throughput amplicon sequencing data. Mol Ecol Resour 17(6):e234–e240

Balvočiūtė M, Huson DH (2017) SILVA, RDP, greengenes, NCBI and OTT - how do these taxonomies compare? BMC Genomics 18(Suppl 2):114

Barra PJ, Inostroza NG, Acuña JJ, Mora ML, Crowley DE, Jorquera MA (2016) Formulation of bacterial consortia from avocado (Persea americana Mill.) and their effect on growth, biomass and superoxide dismutase activity of wheat seedlings under salt stress. Appl Soil Ecol 102:80–91

Bateman A, Coin L, Durbin R, Finn RD, Hollich V, Griffiths-Jones S, Khanna A, Marshall M, Moxon S, Sonnhammer EL, Studholme DJ, Yeats C, Eddy SR (2004) The Pfam protein families database. Nucleic Acids Res 32:D138–D141

Bazghaleh N, Hamel C, Gan Y, Tar'an B, Knight JD (2015) Genotype-specific variation in the structure of root fungal communities is related to chickpea plant productivity. Appl Environ Microbiol 81(7):2368–2377

Berendsen RL, Pieterse CM, Bakker PA (2012) The rhizosphere microbiome and plant health. Trends Plant Sci 17(8):478–486

Berlanas C, Berbegal M, Elena G, Laidani M, Cibriain JF, Sagües A, Gramaje D (2019) The fungal and bacterial rhizosphere microbiome associated with grapevine rootstock genotypes in mature and young vineyards. Front Microbiol 10:1142

Boisvert S, Raymond F, Godzaridis E, Laviolette F, Corbeil J (2012) Ray Meta: scalable de novo metagenome assembly and profiling. Genome Biol 13(12):R122

Bolger AM, Lohse M, Usadel B (2014) Trimmomatic: a flexible trimmer for Illumina sequence data. Bioinformatics 30(15):2114–2120

Broeckling CD, Broz AK, Bergelson J, Manter DK, Vivanco JM (2008) Root exudates regulate soil fungal community composition and diversity. Appl Environ Microbiol 74(3):738–744

Bromberg JS, Fricke WF, Brinkman CC, Simon T, Mongodin EF (2015) Microbiota—implications for immunity and transplantation. Nat Rev Nephrol 11(6):342–353

Callahan BJ, McMurdie PJ, Rosen MJ, Han AW, Johnson AJ, Holmes SP (2016) DADA2: high-resolution sample inference from Illumina amplicon data. Nat Methods 13(7):581–583

Caporaso JG, Kuczynski J, Stombaugh J, Bittinger K, Bushman FD, Costello EK, Fierer N, Peña AG, Goodrich JK, Gordon JI, Huttley GA, Kelley ST, Knights D, Koenig JE, Ley RE, Lozupone CA, McDonald D, Muegge BD, Pirrung M, Reeder J, Sevinsky JR, Turnbaugh PJ, Walters WA, Widmann J, Yatsunenko T, Zaneveld J, Knight R (2010) QIIME allows analysis of high-throughput community sequencing data. Nat Methods 7(5):335–336

Caspi R, Billington R, Ferrer L, Foerster H, Fulcher CA, Keseler IM, Kothari A, Krummenacker M, Latendresse M, Mueller LA, Ong Q, Paley S, Subhraveti P, Weaver DS, Karp PD (2016) The Metacyc database of metabolic pathways and enzymes and the BioCyc collection of pathway/ genome databases. Nucleic Acids Res 44(D1):D471–D480

Chakravorty S, Helb D, Burday M, Connell N, Alland D (2007) A detailed analysis of 16S ribosomal RNA gene segments for the diagnosis of pathogenic bacteria. J Microbiol Methods 69(2):330–339

Chi F, Shen SH, Cheng HP, Jing YX, Yanni YG, Dazzo FB (2005) Ascending migration of endophytic rhizobia, from roots to leaves, inside rice plants and assessment of benefits to rice growth physiology. Appl Environ Microbiol 71(11):7271–7278

Claire Horner-Devine M, Leibold MA, Smith VH, Bohannan BJ (2003) Bacterial diversity patterns along a gradient of primary productivity. Ecol Lett 6(7):613–622

Cochrane G, Karsch-Mizrachi I, Takagi T (2016) The international nucleotide sequence database collaboration. Nucleic Acids Res 44(D1):D48–D50

Cole JR, Chai B, Farris RJ, Wang Q, Kulam-Syed-Mohideen AS, McGarrell DM, Bandela AM, Cardenas E, Garrity GM, Tiedje JM (2007) The ribosomal database project (RDP-II): introducing myRDP space and quality controlled public data. Nucleic Acids Res 35:D169–D172

Cole JR, Wang Q, Fish JA, Chai B, McGarrell DM, Sun Y, Brown CT, Porras-Alfaro A, Kuske CR, Tiedje JM (2014) Ribosomal database project: data and tools for high throughput rRNA analysis. Nucleic Acids Res 42:D633–D642

Cuadros-Orellana S, Leite LR, Smith A, Medeiros JD, Badotti F, Fonseca PL, Vaz AB, Oliveira G, Góes-Neto A (2013) Assessment of fungal diversity in the environment using metagenomics: a decade in review. Fungal Genom Biol 3(2):1

Curtis TP, Sloan WT, Scannell JW (2002) Estimating prokaryotic diversity and its limits. Proc Natl Acad Sci U S A 99(16):10494–10499

DeSantis TZ, Hugenholtz P, Larsen N, Rojas M, Brodie EL, Keller K, Huber T, Dalevi D, Hu P, Andersen GL (2006) Greengenes, a chimera-checked 16S rRNA gene database and workbench compatible with ARB. Appl Environ Microbiol 72(7):5069–5072

Dilthey AT, Jain C, Koren S, Phillippy AM (2019) Strain-level metagenomic assignment and compositional estimation for long reads with MetaMaps. Nat Commun 10(1):3066

Edgar RC (2010) Search and clustering orders of magnitude faster than BLAST. Bioinformatics 26 (19):2460–2461

Edgar RC (2013) UPARSE: highly accurate OTU sequences from microbial amplicon reads. Nat Methods 10(10):996–998

Edgar RC (2016) UNOISE2: improved error-correction for Illumina 16S and ITS amplicon sequencing. BioRxiv. https://doi.org/10.1101/081257

Elias F, Woyessa D, Muleta D Phosphate solubilization potential of rhizosphere fungi isolated from plants in Jimma zone, Southwest Ethiopia. Int J Microbiol 2016, 2016:5472601

Fabregat A, Sidiropoulos K, Garapati P, Gillespie M, Hausmann K, Haw R, Jassal B, Jupe S, Korninger F, McKay S, Matthews L, May B, Milacic M, Rothfels K, Shamovsky V, Webber M, Weiser J, Williams M, Wu G, Stein L, Hermjakob H, D'Eustachio P (2016) The reactome pathway knowledgebase. Nucleic Acids Res 44(D1):D481–D487

Frias-Lopez J, Shi Y, Tyson GW, Coleman ML, Schuster SC, Chisholm SW, Delong EF (2008) Microbial community gene expression in ocean surface waters. Proc Natl Acad Sci U S A 105 (10):3805–3810

Gallart M, Adair KL, Love J, Meason DF, Clinton PW, Xue J, Turnbull MH (2018) Host genotype and nitrogen form shape the root microbiome of Pinus radiata. Microb Ecol 75(2):419–433

Ghosh A, Mehta A, Khan AM (2019) Metagenomic analysis and its applications. Encyclop Bioinform Comput Biol 3:184–193

Glass EM, Wilkening J, Wilke A, Antonopoulos D, Meyer F (2010) Using the metagenomics RAST server (MG-RAST) for analyzing shotgun metagenomes. Cold Spring Harb Protoc (1): pdb.prot5368

Gweon HS, Oliver A, Taylor J, Booth T, Gibbs M, Read DS, Griffiths RI, Schonrogge K (2015) PIPITS: an automated pipeline for analyses of fungal internal transcribed spacer sequences from the Illumina sequencing platform. Methods Ecol Evol 6(8):973–980

Haft DH, Selengut JD, White O (2003) The TIGRFAMs database of protein families. Nucleic Acids Res 31(1):371–373

Hildebrand F, Tadeo R, Voigt AY, Bork P, Raes J (2014) LotuS: an efficient and user-friendly OTU processing pipeline. Microbiome 2(1):30

Huson DH, Weber N (2013) Microbial community analysis using MEGAN. Methods Enzymol 531:465–485

Hyatt D, Chen GL, Locascio PF, Land ML, Larimer FW, Hauser LJ (2010) Prodigal: prokaryotic gene recognition and translation initiation site identification. BMC Bioinform 11:119

Jones P, Garcia BJ, Furches A, Tuskan GA, Jacobson D (2019) Plant host-associated mechanisms for microbial selection. Front Plant Sci 10:862

Kanehisa M, Goto S, Sato Y, Furumichi M, Tanabe M (2012) KEGG for integration and interpretation of large-scale molecular data sets. Nucleic Acids Res 40:D109–D114

Kanehisa M, Sato Y, Morishima K (2016) BlastKOALA and GhostKOALA: KEGG tools for functional characterization of genome and metagenome sequences. J Mol Biol 428(4):726–731

Kelley DR, Liu B, Delcher AL, Pop M, Salzberg SL (2011) Gene prediction with Glimmer for metagenomic sequences augmented by classification and clustering. Nucl Acids Res 40(1):e9

Kelley DR, Liu B, Delcher AL, Pop M, Salzberg SL (2012) Gene prediction with glimmer for metagenomic sequences augmented by classification and clustering. Nucleic Acids Res 40(1):e9

Kim D, Song L, Breitwieser FP, Salzberg SL (2016) Centrifuge: rapid and sensitive classification of metagenomic sequences. Genome Res 26(12):1721–1729

Kõljalg U, Nilsson RH, Abarenkov K, Tedersoo L, Taylor AF, Bahram M, Bates ST, Bruns TD, Bengtsson-Palme J, Callaghan TM, Douglas B, Drenkhan T, Eberhardt U, Dueñas M, Grebenc T, Griffith GW, Hartmann M, Kirk PM, Kohout P, Larsson E, Lindahl BD, Lücking R, Martín MP, Matheny PB, Nguyen NH, Niskanen T, Oja J, Peay KG, Peintner U, Peterson M, Põldmaa K, Saag L, Saar I, Schüßler A, Scott JA, Senés C, Smith ME, Suija A, Taylor DL, Telleria MT, Weiss M, Larsson KH (2013) Towards a unified paradigm for sequence-based identification of fungi. Mol Ecol 22(21):5271–5277

Kuczynski J, Stombaugh J, Walters WA, González A, Caporaso JG, Knight R (2012) Using QIIME to analyze 16S rRNA gene sequences from microbial communities. Curr Protoc Microbiol. Chapter 1:Unit 1E.5

Lakshmanan V, Selvaraj G, Bais HP (2014) Functional soil microbiome: belowground solutions to an aboveground problem. Plant Physiol 166(2):689–700

Langille MG, Zaneveld J, Caporaso JG, McDonald D, Knights D, Reyes JA, Clemente JC, Burkepile DE, Vega Thurber RL, Knight R, Beiko RG, Huttenhower C (2013) Predictive functional profiling of microbial communities using 16S rRNA marker gene sequences. Nat Biotechnol 31(9):814–821

Liu K, McInroy JA, Hu CH, Kloepper JW (2018) Mixtures of plant-growth-promoting rhizobacteria enhance biological control of multiple plant diseases and plant-growth promotion in the presence of pathogens. Plant Dis 102(1):67–72

López MF, Ramirez-Saad HC, Martínez-Abarca F, Aguirre-Garrido JF, Toro N (2012) Rhizosphere metagenomics. In: Nelson KE (ed) Encyclopedia of metagenomics. Springer, New York, pp 1–8

Luo R, Liu B, Xie Y, Li Z, Huang W, Yuan J, He G, Chen Y, Pan Q, Liu Y, Tang J, Wu G, Zhang H, Shi Y, Liu Y, Yu C, Wang B, Lu Y, Han C, Cheung DW, Yiu SM, Peng S, Xiaoqian Z, Liu G, Liao X, Li Y, Yang H, Wang J, Lam TW, Wang J (2012) SOAPdenovo2: an empirically improved memory-efficient short-read de novo assembler [published correction appears in Gigascience. 2015;4:30]. Gigascience 1(1):18

Magoč T, Salzberg SL (2011) FLASH: fast length adjustment of short reads to improve genome assemblies. Bioinformatics 27(21):2957–2963

Majeed A, Abbasi MK, Hameed S, Imran A, Rahim N (2015) Isolation and characterization of plant growth-promoting rhizobacteria from wheat rhi q zosphere and their effect on plant growth promotion. Front Microbiol 6:198

Marizzoni M, Gurry T, Provasi S, Greub G, Lopizzo N, Ribaldi F, Festari C, Mazzelli M, Mombelli E, Salvatore M, Mirabelli P, Franzese M, Soricelli A, Frisoni GB, Cattaneo A (2020) Comparison of bioinformatics pipelines and operating systems for the analyses of 16S rRNA gene amplicon sequences in human fecal samples. Front Microbiol 11:1262

Martin M (2011) Cutadapt removes adapter sequences from high-throughput sequencing reads. EMBnet J 17(1):10–12

Massart S, Martinez-Medina M, Jijakli MH (2015) Biological control in the microbiome era: challenges and opportunities. Biol Control 89:98–108

McDonald D, Price MN, Goodrich J, Nawrocki EP, DeSantis TZ, Probst A, Andersen GL, Knight R, Hugenholtz P (2012) An improved Greengenes taxonomy with explicit ranks for ecological and evolutionary analyses of bacteria and archaea. ISME J 6(3):610–618

Mendes R, Kruijt M, de Bruijn I, Dekkers E, van der Voort M, Schneider JH, Piceno YM, DeSantis TZ, Andersen GL, Bakker PA, Raaijmakers JM (2011) Deciphering the rhizosphere microbiome for disease-suppressive bacteria. Science 332(6033):1097–1100

Menzel P, Ng KL, Krogh A (2016) Fast and sensitive taxonomic classification for metagenomics with Kaiju. Nat Commun 7:11257

Namiki T, Hachiya T, Tanaka H, Sakakibara Y (2012) MetaVelvet: an extension of velvet assembler to de novo metagenome assembly from short sequence reads. Nucleic Acids Res 40(20):e155

Navas-Molina JA, Peralta-Sánchez JM, González A, McMurdie PJ, Vázquez-Baeza Y, Xu Z, Ursell LK, Lauber C, Zhou H, Song SJ, Huntley J, Ackermann GL, Berg-Lyons D, Holmes S, Caporaso JG, Knight R (2013) Advancing our understanding of the human microbiome using QIIME. Methods Enzymol 531:371–444

Nilsson RH, Larsson KH, Taylor AFS, Bengtsson-Palme J, Jeppesen TS, Schigel D, Kennedy P, Picard K, Glöckner FO, Tedersoo L, Saar I, Kõljalg U, Abarenkov K (2019) The UNITE database for molecular identification of fungi: handling dark taxa and parallel taxonomic classifications. Nucleic Acids Res 47(D1):D259–D264

Noguchi H, Taniguchi T, Itoh T (2008) MetaGeneAnnotator: detecting species-specific patterns of ribosomal binding site for precise gene prediction in anonymous prokaryotic and phage genomes. DNA Res 15(6):387–396

Nurk S, Meleshko D, Korobeynikov A, Pevzner PA (2017) metaSPAdes: a new versatile metagenomic assembler. Genome Res 27(5):824–834

Ogata H, Goto S, Sato K, Fujibuchi W, Bono H, Kanehisa M (1999) KEGG: Kyoto encyclopedia of genes and genomes. Nucleic Acids Res 27(1):29–34

Ounit R, Wanamaker S, Close TJ, Lonardi S (2015) CLARK: fast and accurate classification of metagenomic and genomic sequences using discriminative k-mers. BMC Genomics 16(1):236

Overbeek R, Begley T, Butler RM, Choudhuri JV, Chuang HY, Cohoon M, de Crécy-Lagard V, Diaz N, Disz T, Edwards R, Fonstein M, Frank ED, Gerdes S, Glass EM, Goesmann A, Hanson A, Iwata-Reuyl D, Jensen R, Jamshidi N, Krause L, Kubal M, Larsen N, Linke B, McHardy AC, Meyer F, Neuweger H, Olsen G, Olson R, Osterman A, Portnoy V, Pusch GD, Rodionov DA, Rückert C, Steiner J, Stevens R, Thiele I, Vassieva O, Ye Y, Zagnitko O, Vonstein V (2005) The subsystems approach to genome annotation and its use in the project to annotate 1000 genomes. Nucleic Acids Res 33(17):5691–5702

Oyserman BO, Noguera DR, del Rio TG, Tringe SG, McMahon KD (2016) Metatranscriptomic insights on gene expression and regulatory controls in Candidatus Accumulibacter phosphatis. ISME J 10(4):810–822

Palmer JM, Jusino MA, Banik MT, Lindner DL (2018) Non-biological synthetic spike-in controls and the AMPtk software pipeline improve mycobiome data. PeerJ 6:e4925

Peng Y, Leung HC, Yiu SM, Chin FY (2011) Meta-IDBA: a de Novo assembler for metagenomic data. Bioinformatics 27(13):i94–i101

Peng Y, Leung HC, Yiu SM, Chin FY (2012) IDBA-UD: a de novo assembler for single-cell and metagenomic sequencing data with highly uneven depth. Bioinformatics 28(11):1420–1428

Petriz BA, Franco OL (2017) Metaproteomics as a complementary approach to gut microbiota in health and disease. Front Chem 5:4

Pevzner PA, Tang H, Waterman MS (2001) An Eulerian path approach to DNA fragment assembly. Proc Natl Acad Sci U S A 98(17):9748–9753

Plummer E, Twin J, Bulach DM, Garland SM, Tabrizi SN (2015) A comparison of three bioinformatics pipelines for the analysis of preterm gut microbiota using 16S rRNA gene sequencing data. J Proteom Bioinform 8(12):283–291

Poudel R, Jumpponen A, Kennelly MM, Rivard CL, Gomez-Montano L, Garrett KA (2019) Rootstocks shape the rhizobiome: rhizosphere and endosphere bacterial communities in the grafted tomato system. Appl Environ Microbiol 85(2):e01765–e01718

Powell S, Szklarczyk D, Trachana K, Roth A, Kuhn M, Muller J, Arnold R, Rattei T, Letunic I, Doerks T, Jensen LJ, von Mering C, Bork P (2012) eggNOG v3.0: orthologous groups covering 1133 organisms at 41 different taxonomic ranges. Nucleic Acids Res 40(Database issue):D284–D289

Powell S, Forslund K, Szklarczyk D, Trachana K, Roth A, Huerta-Cepas J, Gabaldón T, Rattei T, Creevey C, Kuhn M, Jensen LJ, von Mering C, Bork P (2014) eggNOG v4.0: nested orthology inference across 3686 organisms. Nucleic Acids Res 42(Database issue):D231–D239

Prabha R, Singh DP, Gupta S, Gupta VK, El-Enshasy HA, Verma MK (2019) Rhizosphere metagenomics of Paspalum scrobiculatum L. (Kodo millet) reveals rhizobiome multifunctionalities. Microorganisms 7(12):608

Prodan A, Tremaroli V, Brolin H, Zwinderman AH, Nieuwdorp M, Levin E (2020) Comparing bioinformatic pipelines for microbial 16S rRNA amplicon sequencing. PLoS One 15(1): e0227434

Punta M, Coggill PC, Eberhardt RY, Mistry J, Tate J, Boursnell C, Pang N, Forslund K, Ceric G, Clements J, Heger A, Holm L, Sonnhammer EL, Eddy SR, Bateman A, Finn RD (2012) The Pfam protein families database. Nucleic Acids Res 40:D290–D301

Qiao Q, Wang F, Zhang J, Chen Y, Zhang C, Liu G, Zhang H, Ma C, Zhang J (2017) The variation in the rhizosphere microbiome of cotton with soil type, genotype and developmental stage. Sci Rep 7(1):3940

Quast C, Pruesse E, Yilmaz P, Gerken J, Schweer T, Yarza P, Peplies J, Glöckner FO (2012) The SILVA ribosomal RNA gene database project: improved data processing and web-based tools. Nucl Acids Res 41(D1):D590–D596

Quast C, Pruesse E, Yilmaz P, Gerken J, Schweer T, Yarza P, Peplies J, Glöckner FO (2013) The SILVA ribosomal RNA gene database project: improved data processing and web-based tools. Nucleic Acids Res 41:D590–D596

Ramakrishnan J, Shunmugasundaram M, Narayanan M (2009) Streptomyces sp. SCBT isolated from rhizosphere soil of medicinal plants is antagonistic to pathogenic bacteria. Iran J Biotechnol 7(2):75–81

Rascovan N, Carbonetto B, Perrig D, Díaz M, Canciani W, Abalo M, Alloati J, González-Anta G, Vazquez MP (2016) Integrated analysis of root microbiomes of soybean and wheat from agricultural fields. Sci Rep 6:28084

Reinhold-Hurek B, Bünger W, Burbano CS, Sabale M, Hurek T (2015) Roots shaping their microbiome: global hotspots for microbial activity. Annu Rev Phytopathol 53:403–424

Rho M, Tang H, Ye Y (2010) FragGeneScan: predicting genes in short and error-prone reads. Nucleic Acids Res 38(20):e191

Rideout JR, He Y, Navas-Molina JA, Walters WA, Ursell LK, Gibbons SM, Chase J, McDonald D, Gonzalez A, Robbins-Pianka A, Clemente JC, Gilbert JA, Huse SM, Zhou HW, Knight R, Caporaso JG (2014) Subsampled open-reference clustering creates consistent, comprehensive OTU definitions and scales to billions of sequences. PeerJ 2:e545

Romero M, Gallego D, Blaz J, Lechuga A, Martínez JF, Barajas HR, Hayano-Kanashiro C, Peimbert M, Cruz-Ortega R, Molina-Freaner FE, Alcaraz LD (2019) Rhizosphere metagenomics of mine tailings colonizing plants: assembling and selecting synthetic bacterial communities to enhance in situ bioremediation. https://doi.org/10.1101/664805

Roumpeka DD, Wallace RJ, Escalettes F, Fotheringham I, Watson M (2017) A review of bioinformatics tools for bio-prospecting from metagenomic sequence data. Front Genet 8:23

Ryan PR, Dessaux Y, Thomashow LS, Weller DM (2009) Rhizosphere engineering and management for sustainable agriculture. Plant Soil 321(1–2):363–383

Sapkota R, Knorr K, Jørgensen LN, O'Hanlon KA, Nicolaisen M (2015) Host genotype is an important determinant of the cereal phyllosphere mycobiome. New Phytol 207(4):1134–1144

Sato K, Sakakibara Y (2014) MetaVelvet-SL: an extension of the Velvet assembler to a de novo metagenomic assembler utilizing supervised learning. DNA Res 22(1):69–77

Schloss PD, Westcott SL, Ryabin T, Hall JR, Hartmann M, Hollister EB, Lesniewski RA, Oakley BB, Parks DH, Robinson CJ, Sahl JW, Stres B, Thallinger GG, Van Horn DJ, Weber CF (2009) Introducing mothur: open-source, platform-independent, community-supported software for describing and comparing microbial communities. Appl Environ Microbiol 75(23):7537–7541

Segata N, Waldron L, Ballarini A, Narasimhan V, Jousson O, Huttenhower C (2012) Metagenomic microbial community profiling using unique clade-specific marker genes. Nat Methods 9 (8):811–814

Selengut JD, Haft DH, Davidsen T, Ganapathy A, Gwinn-Giglio M, Nelson WC, Richter AR, White O (2007) TIGRFAMs and Genome Properties: tools for the assignment of molecular function and biological process in prokaryotic genomes. Nucleic Acids Res 35:D260–D264

Simon C, Daniel R (2011) Metagenomic analyses: past and future trends. Appl Environ Microbiol 77(4):1153–1161

Soni R, Kumar V, Suyal DC, Jain L, Goel R (2017) Metagenomics of plant rhizosphere microbiome. In: Understanding host-microbiome interactions-an omics approach. Springer, Singapore, pp 193–205

Tartar A, Wheeler MM, Zhou X, Coy MR, Boucias DG, Scharf ME (2009) Parallel metatranscriptome analyses of host and symbiont gene expression in the gut of the termite Reticulitermes flavipes. Biotechnol Biofuels 2:25

Tatusov RL, Koonin EV, Lipman DJ (1997) A genomic perspective on protein families. Science 278(5338):631–637

Tatusov RL, Galperin MY, Natale DA, Koonin EV (2000) The COG database: a tool for genome-scale analysis of protein functions and evolution. Nucleic Acids Res 28(1):33–36

Treangen TJ, Koren S, Sommer DD, Liu B, Astrovskaya I, Ondov B, Darling AE, Phillippy AM, Pop M (2013) MetAMOS: a modular and open source metagenomic assembly and analysis pipeline. Genome Biol 14(1):R2

Tsurumaru H, Okubo T, Okazaki K, Hashimoto M, Kakizaki K, Hanzawa E, Takahashi H, Asanome N, Tanaka F, Sekiyama Y, Ikeda S, Minamisawa K (2015) Metagenomic analysis of the bacterial community associated with the taproot of sugar beet. Microbes Environ 30 (1):63–69

Vega-Avila AD, Gumiere T, Andrade PA, Lima-Perim JE, Durrer A, Baigori M, Vazquez F, Andreote FD (2015) Bacterial communities in the rhizosphere of Vitis vinifera L. cultivated under distinct agricultural practices in Argentina. Antonie Van Leeuwenhoek 107(2):575–588

Verma A, Kumar S, Kumar G, Saini JK, Agrawal R, Satlewal A, Ansari MW (2018) Rhizosphere metabolite profiling: an opportunity to understand plant-microbe interactions for crop improvement. In: Crop improvement through microbial biotechnology. Elsevier, Singapore, pp 343–361

Vieites JM, Guazzaroni ME, Beloqui A, Golyshin PN, Ferrer M (2009) Metagenomics approaches in systems microbiology. FEMS Microbiol Rev 33(1):236–255

Wilke A, Bischof J, Gerlach W, Glass E, Harrison T, Keegan KP, Paczian T, Trimble WL, Bagchi S, Grama A, Chaterji S, Meyer F (2016) The MG-RAST metagenomics database and portal in 2015. Nucleic Acids Res 44(D1):D590–D594

Wood DE, Salzberg SL (2014) Kraken: ultrafast metagenomic sequence classification using exact alignments. Genome Biol 15(3):R46

Wood DE, Lu J, Langmead B (2019) Improved metagenomic analysis with Kraken 2. Genome Biol 20(1):257

Yilmaz P, Parfrey LW, Yarza P, Gerken J, Pruesse E, Quast C, Schweer T, Peplies J, Ludwig W, Glöckner FO (2014) The SILVA and "All-species Living Tree Project (LTP)" taxonomic frameworks. Nucleic Acids Res 42:D643–D648

Yurgel SN, Nearing JT, Douglas GM, Langille MGI (2019) Metagenomic functional shifts to plant induced environmental changes. Front Microbiol 10:1682

Zampieri E, Chiapello M, Daghino S, Bonfante P, Mello A (2016) Soil metaproteomics reveals an inter-kingdom stress response to the presence of black truffles. Sci Rep 6:25773

Zhang Y, Sun J, Mu H, Lun JCY, Qiu JW (2017) Molecular pathology of skeletal growth anomalies in the brain coral Platygyra carnosa: a meta-transcriptomic analysis. Mar Pollut Bull 124 (2):660–667

Zhang Q, Acuña JJ, Inostroza NG, Mora ML, Radic S, Sadowsky MJ, Jorquera MA (2019) Endophytic bacterial communities associated with roots and leaves of plants growing in chilean extreme environments. Sci Rep 9(1):4950

Zhu W, Lomsadze A, Borodovsky M (2010) Ab initio gene identification in metagenomic sequences. Nucleic Acids Res 38(12):e132

Chapter 14
Methods of Assessments of Microbial Diversity and Their Functional Role in Soil Fertility and Crop Productivity

Bhaskar Reddy, Aundy Kumar, Sahil Mehta, and Kuleshwar Prasad Sahu

Abstract In the last couple of decades, advancement in the genomic sciences coupled with computational framework has robustly accelerated the deeper understanding of the microbial diversity. The arrival of next-generation sequencing (NGS) technologies among researchers around the globe has facilitated the vast growth of public genomes as well as metagenomes. This progressive development in genome sequencing and environmental metagenomics has enabled the researcher to fully characterize the whole microbial community with detailed functional pathway mappings and enzymes discovery. Therefore, as an attempt, in the present chapter, we have described the role of NGS technologies for the assessment of microbial community coupled with bioinformatic analysis tools in soil fertility and their role in improved crop production. Furthermore, this present chapter also entails the fundamental basis and planning strategy for designing experiments as well as an analysis framework for their robust output for mankind applications.

Keywords NGS · Biochemical pathway · Metagenome · Microbial diversity · Targeted amplicon

14.1 Introduction

Since the beginning of domestication, food production for life survival is mainly performed by green plants through various agricultural process. Such productivity has been improved with the help of a biotechnological process with improved food quality and quantity (Béné et al. 2016; Jovel et al. 2016). The rapid and enlarged food feeding requirements of agricultural industries have affected the environments,

---

B. Reddy (✉) · A. Kumar · K. P. Sahu
Division of Plant Pathology, ICAR- Indian Agricultural Research Institute, New Delhi, India

S. Mehta
Crop Improvement Group, International Centre for Genetic Engineering and Biotechnology, New Delhi, India

S. K. Dubey, S. K. Verma (eds.), *Plant, Soil and Microbes in Tropical Ecosystems*, Rhizosphere Biology, https://doi.org/10.1007/978-981-16-3364-5_14

which is reflected by the warming condition for being environment-friendly with harmless yields along with the stability of involved resources (Gebbers and Adamchuk 2010).

Plants are surrounded by various microorganism which efficiently enriches the enhanced cell quantity rather than individually by plants. Among such microorganisms, the majority of them survive on the rhizosphere or nearby to plant root surrounding areas (Sharma et al. 2021). The useful microbiota of the rhizosphere has been found to be mediating plant growth as well as improve mineral availability (Lakshmanan et al. 2014). Although, anthropogenic actions and activities influences the soil-residing microorganisms habitat and changes their abundance, co-occurrence dynamics, biochemical pathways, and other functional contents (Mehta et al. 2021a).

The wide application of high-throughput genomics technologies enables various researchers to find the host-pathogen interaction, effect of pesticides and herbicides, and cellulolytic, xenobiotic degrading enzymes and pathways through genome mapping and their potential through molecular gene expression and correlational investigation (Keegan et al. 2016; Anamika et al. 2019; Mehta et al. 2019a; Reddy and Dubey 2021). Such progresses in molecular biology and high-throughput applications have led to the expansion of advanced automated analytic software (Lu et al. 2014). The massive advancements in genomics technologies have brought rapid developments to our understanding of cellular biology, phylogenetic relationship, microbial environments, and biochemical pathways in microbes as well as their host plants (Sahil et al. 2021; Rajput et al. 2021; Mehta et al. 2021b; Bharti et al. 2021; Mehta et al. 2019b; Reddy et al. 2019; Reddy and Dubey 2021) and are progressively unlocking new understandings and uses toward clinical care and personalized medicine (Loman and Pallen 2015; Pareek et al. 2011). Additionally, the scientific community has developed various novel tools, packages, and algorithms to process and explain the genomic data, datasets management, simple software layout, usage, and most importantly privacy of the tremendous data (Vincent et al. 2017; Anamika et al. 2019; Mehta et al. 2019a; Reddy 2019; Kumar et al. 2021).

14.2 Approaches for Soil Microbial Community Assessment

14.2.1 Overview of Microbial Diversity Methods

Since the onset of the twenty-first century, the research regarding microbial diversity was based on techniques like Denaturing Gradient Gel Electrophoresis (DGGE) and Terminal-Restriction Fragment Length Polymorphism (T-RFLP) (Mohanty et al. 2007; Ramakrishnan et al. 2001). The former technique works by using a gradient of denaturing strength on microbial DNA samples (PCR-amplified) along either the

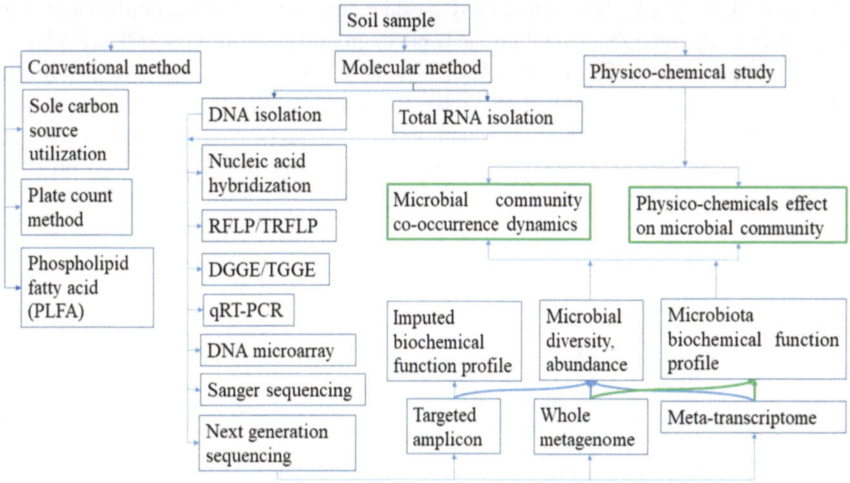

Fig. 14.1 Schematic diagram illustrating the workflow of microbial community structure assessment methodology. Note: The green color box indicates the final output through the various process

horizontal or vertical axis of a polyacrylamide gel followed by electrophoresis (Bo et al. 2020; Leite et al. 2012). While progressing through the gel, the DNA samples at different gel points get separated based on their melting domains, GC clamp, and Tm resulting in a banding pattern of single-stranded branches. In the latter technique, the difference in microbial DNA sequences is detected by a unique blotting pattern generated by using RFLP probes that hybridize specifically with restriction endonucleases-digested different lengths of fragments (Liu et al. 1997). However, with the advent of the sequencing boom, various researchers switched to sequencing-based analysis of microbial diversity. The advantages included low-cost, low rate of errors, high efficiency, high reliability, and time-to-time update. The microbial community structure assessment with various methods is schematically depicted in Fig. 14.1. In the twenty-first century, the majority of researchers are utilizing the molecular methods and depend heavily on next-generation sequencing as compared to the other approaches. NGS provides robust and detailed deep insights of community structure and underlying functional features with significantly reduced labor, time, and cost.

14.2.2 Quantitative Real-Time PCR

Through the literature survey, it has been observed that techniques like DGGE and T-RFLP are only used for qualitative analysis of microbial communities as they reveal the qualitative dynamics diagram among microbial communities (bacteria, archaea, and yeast). However, these techniques cannot be used as a quantitative method (Kanagawa 2003; Neilson et al. 2013). As the name describes, quantitative real-time PCR (qPCR) is a simple PCR-based technique that first amplifies and then quantifies a targeted DNA enabling the users to quantify absolute as well as a relative number of gene copies from a complex DNA sample to reflect the relative abundance of the microbes (Ashajyothi et al. 2020; Bhardwaj et al. 2020). As a result, this technique is extensively applied for quantitative analysis of microbial composition in various ecological habitats such as soil (Franke-Whittle et al. 2015; Ashajyothi et al. 2020), forest soil (Bhardwaj et al. 2020), and rumen (Pitta et al. 2014; Singh et al. 2015a).

14.2.3 Isolation, Library Preparation, and Sequencing

In case of eukaryotic microbes, for example, the isolated fungus is characterized by the internal transcriber region using ITS-4 and ITS-5 markers to confirm the fungus genus, species, and purity of the isolate. Once the fungus is confirmed, genomic DNA (gDNA) is isolated from the pure fungus and further used for library preparation. Generally, each NGS sample processing known as library preparation starts with the shearing/fragmentation/tagmentation of gDNA into desired fragments and followed by end repair. After the end repair, each sample is usually subjected to multiplexing through adapter and barcode/index ligation reaction, referred as sequencing libraries. The prepared library is subjected to the quality and quantity check to make sure prepared libraries are suitable for sequencing. As samples are barcoded, the various samples cab be pooled together through normalization, and then equimolar pooling is carried out. Next, a pooled library is placed for clonal amplification through emulsion PCR (emPCR, in 454 GS FLX and Ion Torrent) and bridge amplification (cluster generation, in Illumina). In 454 GS FLX and Ion Torrent, after emPCR, sample is processed for recovery and enrichment (Fig. 14.2). The finally enriched sample was loaded in a chip and then placed in a machine for sequencing. Whereas in the Illumina platform, the sequencing is followed immediately after the step of cluster generation/cluster amplification. Each machine-generated sequenced sample is stored in the form of nucleotides fastq files, which is a standard output format (Barriuso et al. 2011; Endrullat et al. 2016).

14.2.4 Brief Summary of Sequencing by Reversible Termination

In the year 2006, the instrument Illumina Genome Analyzer (SOLEXA) was launched based on sequencing by reversible termination technology. In this technology, the subjected study material was prepared through random fragmentation, which was followed by the ligation of oligonucleotide adaptors and indexes, referred as prepared libraries which would be subjected to sequencing in the machine. The

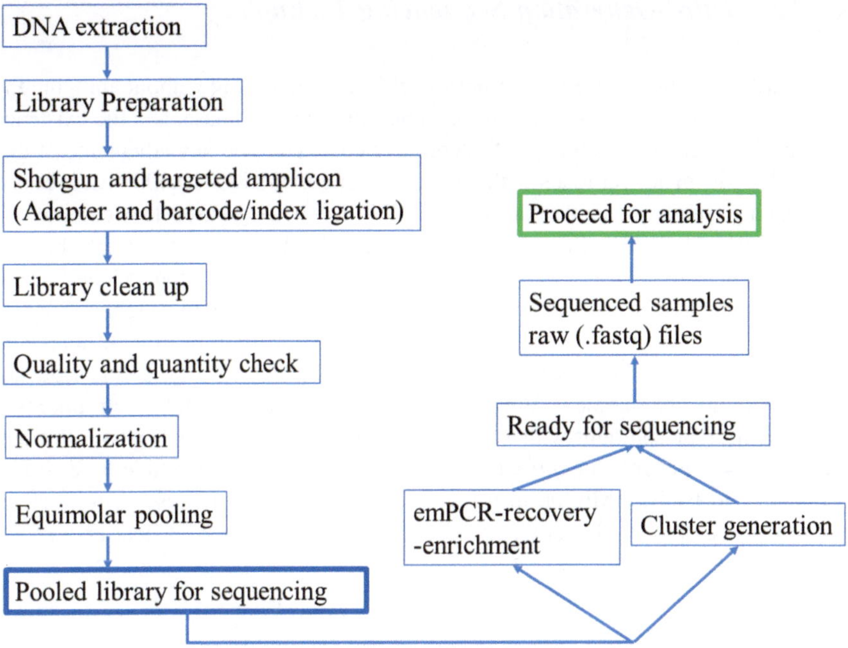

Fig. 14.2 Simple workflow of next-generation sequencing library preparation and sequencing protocol

extensive details of library amplification and sequencing are described further (Adessi et al. 2000; Fedurco et al. 2006; Ju et al. 2006). This technology offers the following two distinct kinds of library preparation, while both kind of libraries are sequenced on compatible Illumina sequencing machine in a default sequencing chemistry.

- Paired-end library preparation:
  - The paired-end (PE) sequencing libraries are prepared using instrument compatible library preparation kit.
  - The insert size of the PE sequencing libraries is usually in range of ~300–550 bp.
  - Each sequencing library will be individually indexed/barcoded for sequencing.
  - Mostly used by researchers for sequencing genomes and metagenomes.
- Mate Pair library preparation:
  - Preparation of Mate Pair library with a jumping distance of 3 and 8 KB average insert size.
  - Each sequencing library will be individually indexed/barcoded for sequencing.
  - Usually used for genome gap finish and polishing.

14.2.5 Third-Generation Sequencing Technology

The third-generation sequencer comprises of DNA sequencing without applying the PCR extension, as extension introduces a bias in sequenced base, and the existence of high GC content influences both depth and coverage. The key advantage of this technology is the longer reads with an average length of 5000–10,000 bases. In this sequencing, single-molecule real-time (SMRT) technology-based first commercial instrument was PacBio Sequel released by Pacific Biosciences and mechanism described here (Eid et al. 2009). The sequenced data (base) output of the PacBio RS II instrument is 0.5–1 billion bases in a single SMRT cell with a higher error rate (10–15%). Another third-generation instrument is the MinIon instrument marketed by Oxford Nanopore Technology in the year of 2014. Specifically, in this sequencing technology, the sample is subjected to a nano-sized pore through electrophoresis, using electrolytic solutions with a fixed electric field. As the template passes through the nanopore, a change in current occurs, and the resultant magnitude is recorded. Compared to PacBio, MinIon instrument is smaller in size and less cost-effective. However, the obtained bases (sequences) display a correctness of near about 88% (Laszlo et al. 2014).

14.3 NGS Reads Processing

Initially in all kind of NGS-based studies, the quality screening and filtration of generated poor bases and reads is a prerequisite. The schematic workflow of NGS reads processing illustrated in Fig. 14.3. The quality passed reads subjected to various kinds of analyses such as whole-genome assembly, metagenome, metatranscriptome, variant calling, and gene expression. In general, for the targeted amplicon sequencing driven taxonomic classification involves the quality passed reads clustering, operational taxonomic unit (OTU) picking, and then OTUs taxonomic classification. On the other hand, targeted amplicon, whole metagenome, and meta-transcriptome approach utilize the reads alignment against the reference database, followed to taxonomic and functional annotation. Whole metagenome and meta-transcriptome classification using de novo assembly provide much more detailed insights of studied samples with significantly increased cost and computation time. Taxonomic classification provides the insights of phylogenetic classification, alpha diversity (number of OTUs, Species richness, Chao1, Shannon index, and Simpson index), beta diversity such as principal coordinate analyses (PCoA), and taxa abundance (number of specific phyla or genera count or percentage). The reads functional classification provides the descriptive insights of underlying metabolic machinery categories obtained against a specific database. A database such as KEGG pathways depicts the classified reads into various biological pathways such as starch and sucrose metabolism, sulfur, propionate, butyrate, and methanogenesis (Anamika et al. 2019). Database CAZymes provide the reads with a property of

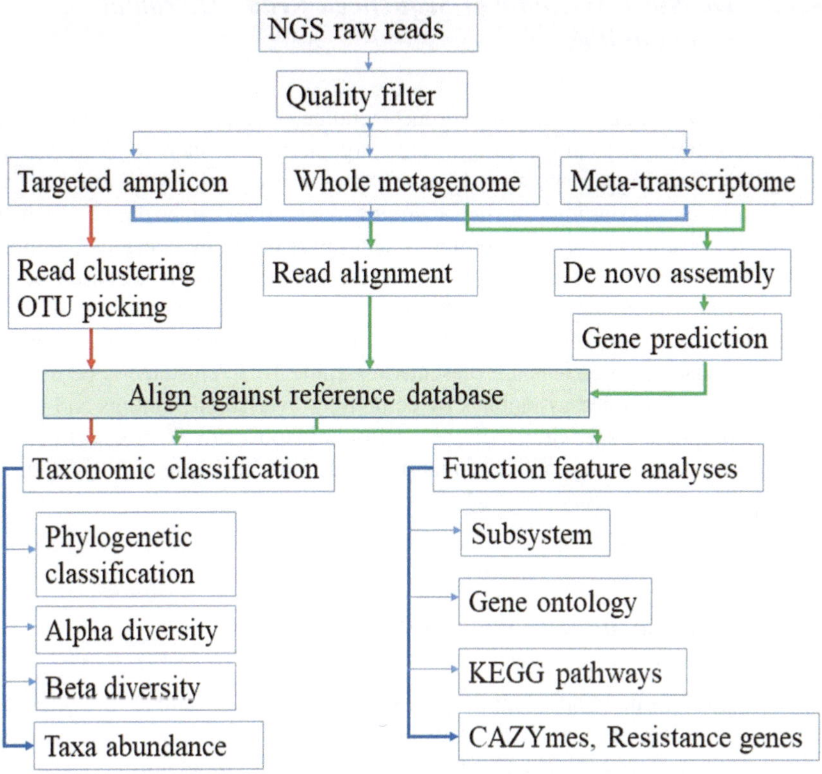

Fig. 14.3 Schematic workflow of NGS raw reads for taxonomic classification and function annotation

cellulose, pectin, and hemicellulose degradation, which offers the metagenomic study could provide the genomic details with great importance for industrial applications.

14.3.1 Generated Read Quality Filtration

Initially, the raw NGS files are processed for the filtering criteria, that is, any read with base quality score $Q < 20$ is filtered, then following to read trimming from 5′ end and 3′ end, if required. The machine-generated raw reads are filtered for the removal of poor bases and reads to obtain high-quality cleaned data. Few quality filtration tools are Trimmomatic, Cutadapt, Trim Galore, PRINSEQ, etc. (Del Fabbro et al. 2013; Pfeifer 2017).

14.3.2 De Novo Assembly of Sequenced Reads Microbial Communities

The quality passed reads are utilized for metagenome assembly, which describes the various steps together as input of fragmented large number of short DNA reads, and placing them back in overlapping fashion generates the original DNA sequence. The word de novo means starting from the beginning. Assemblies can be produced which have fewer gaps, less or no misassemblies, and fewer errors by tweaking the input parameters. The usually used tools for sequenced genome assembly are based on the command-line interface (CLI). Among that, meta-Velvet, Meta-IDBA, MetaSPAdes, and MEGAHIt are widely used. Such assembler algorithm, input data format, and requirements are presented in Table 14.1. This step is performed to optimize the generated assemblies by combining overlapping contigs and introducing appropriate gaps. Some of the scaffolding tools are SSPACE, PBJelly, gapCLoser, etc. More descriptive comparisons are provided here (Vollmers et al. 2017; Ayling et al. 2019).

14.3.3 Analysis of Microbial Diversity

Determination of microbial community in studied ecosystem samples provides the composition of the microbial diversity and composition under the influence of environmental factors and their co-occurrence. To find the community composition, there are various tools available among the scientific community to achieve their objectives. Among that, majority were read alignment against reference database-based annotation such as MG-RAST, MEGAN, EBI-Metagenome, QIIME, and RDP. The further advancement in annotation methodology, approaches such as k-mer, composition, and alignment-free tools, becomes available (Table 14.2). These tools enabled the scientific community to analyze the microbiota associated

Table 14.1 List of some tools available for metagenome assembly currently used by researchers

Assembler	Algorithm	Assembly method	Standard input	Read length	Output format	Availability
MetaMOSS	de Bruijn multiple Kmer	Denovo	fastq, fasta	Arbitrary	fasta	Open source
MetaSPAdes	De Bruijn graphs	Denovo	fastq, fasta	Arbitrary	fasta	Open source
MEGAHIT	de Bruijn graph	Denovo	fastq, fasta	Arbitrary	fasta	Open source
Meta-Velvet	de Bruijn graph	Denovo	fastq	Arbitrary	fasta	Open source
Meta-IBDA	de Bruijn graph	Denovo	fastq	Arbitrary	fasta	Open source
Ray Meta	de Bruijn graph	Denovo	fastq	Arbitrary	fasta	Open source
PRICE	Hybrid	Denovo	fastq	Arbitrary	fasta	Open source

Table 14.2 List of tools employed for 16S rRNA and whole shotgun metagenome data analysis

Available tools	Input	Output	Availability
For 16S rRNA, 18S rRNA, and fungal ITS			
QIIME 1, 2	sff, fasta, fastq	biom, txt	CLI
MOTHUR	sff, fasta, fastq	biom, txt	CLI
RDP	fasta, fastq	txt	CLI, web server
MG-RAST	fasta, fastq	txt	Web server
MEGAN 5, 6	txt, xml, sam	txt	GUI
EBI metagenome	fastq	biom, txt	Web server
MGX	fasta, fastq	txt	GUI
Hybrid_Denovo	fastq	biom, txt	CLI
For shotgun/whole metagenome			
KAAS	fasta	.txt, html	CLI, web server
MG-RAST	fasta, fastq	txt	Web server
MEGAN 5, 6	txt, xml, sam	txt	GUI
InterProScan	fasta	txt	Web server, CLI
dbCAN	fasta	txt	Web server, CLI
RapSearch	fastq	txt	CLI
Diamond	fastq	txt, sam	CLI
BLAST$^+$	fasta	txt, sam	Web server, CLI
EBI metagenome	fasta, fastq	biom, txt	Web server
Kaiju	fasta, fastq	txt	CLI, web server
Kraken	fasta, fastq	txt	CLI
k-Salm	fastq	txt, sam	CLI
CLARK	fasta, fastq	txt	CLI

with dormancy and sporulation, stress response genes, acetogenesis, methanogenesis, carbohydrate, protein metabolism, antibiotic, metal ion resistance genes, and aromatic compound metabolism (Roumpeka et al. 2017; Tamames et al. 2019).

Additionally, tools are also available that automates the matched reads were post-processed to find the community structure such as MG-RAST and EBI-Metagenomics including simple statistical graphical plots (Table 14.2). However, alignment/sequence matching against the reference sequence requires high computation power as the number of reads and length increases, which makes quite challenging and time-consuming tasks such as BLAST$^+$. Meanwhile, methodological advancements, such as *k*-mer and composition-based binning, facilitated the robust way analysis in limited time. In the *k*-mer approach, reads are converted into a small subset of 6 bases, 11 bases, and/or 22 bases called *k*-mers of similar sequences. The generated read *k*-mer composition is then compared to a reference database, and hits are counted to a known organism. For such a task, there are numerous tools available like Kraken, k-SALM, Kaiju, Klark, and the Ray Meta (Table 14.2). Additionally, various web servers are now available for automated whole genome such as RAST, GenSAS, and metagenome annotation for taxonomic and functional annotations such as MG-RAST, EBI-Metagenome, and GALAXY (Roumpeka et al. 2017; Tamames et al. 2019).

14.3.4 Classification of Microbial Diversity with Bioinformatic Tools

14.3.4.1 MG-RAST

Out of all tools, Metagenomics Rapid Annotation using Subsystem Technology (MG-RAST) is the most popular, structured-web server for the analysis of microbial communities abundance at a taxonomic and functional level with graphical result visualization (Keegan et al. 2016). MG-RAST consists of various integrated tools and databases to determine the taxonomic and functional classification of NGS raw datasets. It takes the NGS raw input in form of single-end or pair-end sequencing reads and followed by quality processing. Quality passed reads were then automatically submitted for taxonomic and function analysis. After analysis, the user/researcher can visualize and download the entire result against the various databases. For the functional classification of metagenomes, MG-RAST offers various databases such as subsystem, COG, NOG, and KEGG. These all are hierarchical (up to level 4) type databases that enable the researcher to comprehensively determine the functional roles of sequences obtained from metagenomes. Further KEGG databases extensively provide the mapping of metagenomic sequences to the biochemical pathways such as sulfur metabolism, acetogenesis, methanogenesis, propionic acid metabolism, and starch and sucrose metabolism. The various kinds of visualization are bar, stacked, rarefaction, principal component analysis (PCoA), network, and pathways map.

14.3.4.2 MEGAN

MEtaGenome Analyzer (MEGAN) is a comprehensive locally installation-based stand-alone tool for microbial communities' abundance taxonomic and function analysis. The MEGAN primary requirement is that the sequences should be homology aligned against the database. The aligned sequences are imported/subjected as input to MEGAN and then parsed to taxonomic and functional profiles. In MEGAN, similar to MG-RAST, a researcher can map sequences against subsystem, COG, NOG, and KEGG databases; MEGAN also provides various kinds of visualization and biochemical pathway mappings. MEGAN taxonomic and functional classification can be visualized at various hierarchical levels along with significant statistical values (Huson et al. 2007).

14.3.4.3 QIIME

Quantitative Insights Into Microbial Ecology (QIIME) version 2.0 is a comprehensive tool for the targeted amplicon taxonomic classification and abundance estimation. It is a stand-alone, pipeline nature which consists of various integrated tools such as OTU picking, OTU classification, OTU rarefication, alpha and beta diversity estimation, statistical analysis, and OTU network-based co-occurrence determination. QIIME accepts barcoded, non-barcoded, and single, pair-end raw and quality passed reads. Using this tool, the researcher can classify efficiently amplicon reads such as 16S rRNA and 18S rRNA, fungal ITS, and functional marker-based community classification such as *pmoA*. QIIME also provides integrated rarefaction and statistical graph visualization (Bolyen et al. 2019).

14.3.4.4 MGnify

MGnify is a part of ENA (European Nucleotide Archive) infrastructure and a web server for the analysis of microbial communities' abundance at a taxonomic and functional level with graphical result visualization. For analyzing reads using this tool, the user is required to first deposit the raw read to the ENA database as per the standard of Genome Standard Consortium (GSC). EBI-metagenome enables the researcher to determine the targeted amplicon and whole metagenome taxonomic profile against 16S rRNA and 18S rRNA database, whereas functional classification is performed using gene ontology (GO) approach in a three main broad category, e.g., biological process, molecular function, and cellular component. The EBI graphical visualization includes a bar plot, pie chart, and PCoA plot (Mitchell et al. 2020).

14.3.5 Analysis of Microbial Community Metabolic Potential

The standard metagenome functional annotation pipeline is illustrated in Fig. 14.3, which consists of gene scanning (gene prediction), aligning against the reference sequence, taxonomy, function, and metabolic pathway assignment. The progressive advancement in genomes and metagenomes sequencing has led the development of numerous bioinformatics software for the prediction of genes and gene models. Further, as bioinformatic knowledgebase advanced, it offered to the development of various automated whole genome and metagenome data-based microbial genome binning and functional annotation, while requiring high computation resources (Roumpeka et al. 2017; Vincent et al. 2017). Lately, these developments have even opened up the possibility of "microbiome gene modifications" using CRISPR/Cas technology that will boom the genome editing of higher eukaryotes, especially host plants (Mehta et al. 2020; Dilawari et al. 2021).

Generally, the NGS machines from shotgun metagenome generate the read length from 50 to 600 base. Among that, majority were ranged from 300 to 600 bases, depending on the sequencing platform and chemistry. These short reads are assembled into longer sequences called contigs in a process called assembly. The assembly of short sequences becomes more important when the objective is to find the functional gene and metabolic pathways (Vollmers et al. 2017; Mitchell et al. 2020). Because, earlier, the input DNA is randomly fragmented into short fragments and then sequenced which used to result in a very poor quality of reads which contains a very high number of poor base quality scores. However, using the third generation, the read length is increased to more than 10 K bases, as well as poor base calling. Hence, a combination of both generation sequencers is more reliable for full-length functional gene discovery in genomes and metagenomes. At the current time, numerous tools are available for genomics and metagenomic data analysis. These tools mainly vary from algorithms and code language. Other variations include hardware requirements, user interface, installations, and user-interface (Roumpeka et al. 2017; Vollmers et al. 2017).

For the genome and metagenome functional annotation tools details, algorithm, input data type, and dependencies are given in Table 14.3. In the alignment

Table 14.3 List of software used for gene identification and prediction in genomes and metagenomes

Tools	Input	Single/paired-end	Output format	Availability	Suitability
Reference based					
BLAST+	fasta,fastq	Both	txt, sam, xml	Open source	Genome, metagenome
InterProScan	fasta	Single	txt, xml	Open source	Genome, metagenome
DIAMOND	fasta,fastq	Both	txt, sam, xml	Open source	Genome, metagenome
Usearch	fasta,fastq	Both	standard	Open source	Metagenome
RAPSearch	fasta,fastq	Both	standard	Open source	Genome metagenome
PALADIN	fasta,fastq	Both	standard	Open source	Metagenome
GhostX	fasta	Single	txt, html	Open source	Genome metagenome
Blast2GO	fasta,fastq	Single	txt, xml	License	Genome
Ab-initio gene prediction					
Meta-GeneMark	Fasta,fatsq	Single	txt	Open source	Metagenome
GLIMMER	fasta	Single	txt	Open source	Genome
GLIMMER -MG	Fasta,fatsq	Single	txt	Open source	Metagenome
AUGUSTUS	fasta	Single	txt, gff	Open source	Genome
FragGeneScan	fasta,fastq	Single, paired	txt	Open source	Metagenome
GeneMark	fasta	Single	txt, gff	Open source	Genome
ORF finder	fasta	Single	txt	Open source	Genome
Prodigal	fasta	Single	txt, gff	Open source	Genome

approach, the quality passed reads are matched against reference databases such as NCBI nr and NCBI RefSeq databases using sequence similarity search tools such as DIAMOND, PALADIN, RAPSearch, VSEARCH, and BLAST^{+}. The blast search utilizes the alignment of query sequences against the previously known reference sequence and classifies the sequence to their affiliation to taxonomy and function. InterProScan performs the identification of protein family, conserved domains, and superfamilies in the query sequence (Yadav et al. 2020).

14.4 Application of NGS Technology to Assess Microbial Diversity with Soil Fertility

Earth planet soil is the fundamental site for maintaining the ecological process and equilibrium maintenance. Soil provides the primary site for crop production, vegetation, life survival, biological, various hydrological, and economical processes. Among the biological process, microorganism plays various essential role such as mineralization, nutrient recycle, and maintenance of soil health. Hence, the protection of soil health for prolonged fertility in the agricultural system is highly important. Doran and Zeiss (2000) described health as the potential of soil functionality within an ecosystem and land use borders for sustainable biological productivity, improvement of environmental quality, and enhancement of animal and plant health. In agricultural practice, the microorganism ecosystem is generally balance-altering and dynamics of the microbial community.

It is generally achieved in the agricultural ecosystem through microbes-plant interaction and forms the important phenomenon of soil ecosystems (Bélanger and Avis 2002). In the landscape system, microbes are abundantly distributed in soil, which consists of useful and harmful communities. The plant root-adhered soil bacteria significantly contribute to the enhancement of soil property and release of phosphatase, dehydrogenase, mineralization, and various self-defense molecules such as secondary metabolites (Haas and Keel 2003) and stabilization of soil characteristics (Miller and Jastrow 2000). The microorganism-mediated soil fertility improvement involved (1) nitrogen fixation, (2) phosphate solubilization, (3) siderophore production, and (4) phytohormone production.

14.4.1 Microbial Community Diversity and Composition

The profile and function of soil microbes are connected with variable plants via litter quality, biomass production, root exudates, and root-shoot carbon allocation (Porazinska et al. 2003; Potthoff et al. 2006). Plant-derived alteration in litter inputs affects the microbial diversity and functionality (Habekost et al. 2008; Strecker et al. 2016). Lange et al. (2014) reported that species richness is the fundamental basis of soil microbial community biomass, whereas the ratio of fungi to bacteria was positively affected by active group richness of plants and the existence of legumes. Also, the richness of plant species effect on soil microbial biomass was facilitated through nitrogen inputs and its concentration (Eisenhauer et al. 2010; Bessler et al. 2012).

The descriptive determination of plant microbiota interaction provides not only the remarkable supports for plant biology but additionally the identification and characterization of biochemical machinery for their potential application in biotechnological uses. For example, it can be utilized for improving plant health and growth, development of disease resistance, and various other resistance such as salt, biotic, and abiotic resistance variety development. Further development in the genomic studies facilitated the identification of various biological and biochemical function like virulence (Reddy et al. 2014), resistance against antibiotics and metals (Reddy and Dubey 2019), and energy production through detritus material (Yadav et al. 2020), core microbiome (Kumar et al. 2021) which play a significant role in the agriculture sector (Rialch et al. 2019, Sahu et al. 2020). Thus, detailed information on microbial community and functional ability of soil and rhizospheric microbiota facilitates the manipulation of environmental situations (Alisoltani et al. 2019).

The robust development in high-throughput sequencing technology and the release of vast organism species, strain genomes, and metagenomic studies extensively facilitated the deeper understanding of biochemical pathways (Loman and Pallen 2015; Singh et al. 2015b; Reddy et al. 2019). The technologies available in the twenty-first century have tremendous potential for the illustration/depiction of the taxonomic profile of microbial communities along with the determination of function metabolic pathways. However, the determination of such a taxonomic and functional profile is a tedious process for the microbiologist and hence requires strong computational skills as it consists of pipelines of distinct integrated tools. Although function and metabolic potential determination of microbial communities through metagenome and metatranscriptome are highly suitable for researchers as it provides vast information about the specific function-associated microbial communities (Singh et al. 2015b; Reddy et al. 2019; Reddy 2019).

14.4.2 Application of High-Throughput Sequencing on Soil Fertility

As per the glossary of Soil Science Society of America (SSSA), the soil can be formally defined as complex unconsolidated mixtures of minerals, organic matter, air, water, and countless (non) decayed organisms on the immediate earth's surface (Soil Science Society of America 2020). It forms the "vital skin of the earth" as it supports the earth's life web that consists of plants, animals, humans, and microbes. Since the beginning of civilization, soil fertility seems to sustain the plant's growth and agricultural yield (Sharma et al. 2021). It has been reported to be affected by both genetic (parent material and related-characteristics) and environmental factors (climate, time, landscape, amendments, and macro-, and microorganisms) (Davies et al. 2019; Lisuma et al. 2020). In the present times of modern agriculture, the maintenance of soil fertility is typically required which is achieved by following soil evaluation and conservation practices. One such method is to use the metagenome sequencing for analyzing the soil fertility for various geographical areas. This has been already done significantly by various researchers as sequencing integrated soil fertility management around the globe over the last decade.

One of the very conclusive observations on establishing the role of sequencing in understanding microbial diversity in soil and correlating it with soil fertility was reported by Xue et al. (2011). In their study, they summarized the effect of consecutive years of mono-cropping on microbial populations and diversity. Furthermore, they introduced the advantages of 454 GS-FLX pyrosequencing high-sequencing method for the analysis of microbial populations and diversity. By using pyrosequencing in 146 different soil samples across the globe, Bates et al. (2011) observed consistent correlation among the soil C:N ratio with an abundance of two archaeal members. In the very next year, Hiiesalu et al. (2012) directly compared the multi-time point grassland plant richness below the soil surface by using accurate 454 sequencing of the chloroplast *trn*L(UAA) and related the variations in microbial composition to the fertility of the soil. Gigliotti and group observed the effect of organic addition amendments to the soils results in enhancement of nutrients as well as organic matter, C sequestration, and changes in microbial activity and biodiversity structure (Gigliotti et al. 2013). Furthermore, the use of pyrosequencing revealed that bacterial phyla and fungi species are related to the organic matter turnover in soil. In another study report, the effect of biochar use on re-wiring composition and function of microbes residing in fertile agricultural soils using 16S rRNA tag sequences showed significant differences in the composition of microbial community and the correlation patterns (Nielsen et al. 2014).

By using 454 pyrosequencing, Franke-Whittle and colleagues revealed the significant differences in microbial communities (fungi and bacteria) between replant and fallow soils. Furthermore, they urged to reveal the functional role of associated genera with soil fertility (Franke-Whittle et al. 2015). By employing the pyrosequencing of ITS2 amplicons, Sterkenburg and group observed significant changes in the composition of fungal communities related to plant nutrition and decomposition along a soil fertility gradient in a boreal forest. Through their experiment, they revealed the composition significantly varies at the levels of species, genera, as well as orders. Further, they revealed that ascomycetes fungi were dominant in less fertile forests, while the fungi related to basidiomycetes were highly abundant in more fertile forests, hummus, and litter (Sterkenburg et al. 2015). In a similar manner, the direct impact of fertilization on the composition of below-ground arbuscular mycorrhizal (AM) fungi along the gradient of soil fertility was studied by Liu et al. (2015). They revealed the fertilizer application caused remarkable changes in the genus richness of AM fungi and over-dispersion statistically when fertilizers were applied at higher treatments (Liu et al. 2015).

As per the various experiments, it has been an established fact that soil pH apart from climatic conditions and management practices also regulated the soil fertility as well as impacted the diversity of below-ground communities. This was further supported by the findings by Jeanbille et al. (2016), who characterized the significant differences in bacterial communities enriched with acidic (nutrient-poor) and alkaline soils. Li et al. (2017) highlighted the role of C/N- and C/P-based shifts occurring in succession, composition, and diversity (alpha and beta) of microbial communities along a soil fertility gradient in paddy cultivation (Li et al. 2017). Tu et al. (2018) evaluated the significant effect of fertilizer application on the soil bacteria richness and role related to fertility assessed through 16S rRNA sequencing in dragon tree plantations (Tu et al. 2018). Recently, Burke and group characterized the responses and quantified a high degree of fungal communities in the beech-maple forest. Furthermore, they inferred the fungal taxa strongly associated with P-availability (Burke et al. 2019). More recently, Guo et al. (2020) evidenced the complexity of fungal assemblage in the soil directly correlates with soil fertility gradient by collecting various soil samples from tea plantations and sequencing them further with the Illumina MiSeq platform. In another study, Lisuma et al. (2020) reported work on tobacco plants grown in different Tanzanian landscape soils and cropping patterns linked the changes in rhizospheric bacterial composition with the soil fertility using 16S rRNA sequencing. Furthermore, they inferred the tobacco's rhizospheric bacterial diversity influences the solubilities of various macronutrients such as phosphorous, potassium, sulfur, as well as fix total N in the soil.

14.4.3 Role of High-Throughput Sequencing on Microbial Diversity and Crop Productivity

Ever since their origin millions of years ago, plants have existed in contact with microbes. Among the multitude of host functions that microbes control are nutrient uptake, protection, and phenology (Friesen et al. 2011). The identified microbial composition associated with plant root and their manipulation can be utilized for significantly boosting the quality of crop production by using beneficial microbiomes in agricultural systems (Bakker et al. 2012; Mueller and Sachs 2015). After studies to demonstrate that rhizobium nodules are colonized and the nitrogen fixed for their plant hosts, the Department of Agriculture (USA) advised inoculation of legume crops (Schneider 1892).

The plant microbiome's normal ecological roles leading to plant development, growth, and survival against biotic and abiotic stresses are well recorded (Turner et al. 2013; Müller and Ruppel 2014; Mehta et al. 2021a). Because of their close plant associations, the endophytic microbiome is believed to affect plant growth and production more specifically than epiphytic microbiomes. NGS-based metagenomic analysis is currently widely used to analyze plant endophytic microbiomes, contributing to an increased understanding of the profiles and roles of microbiomes. The endophytic microbiome co-operating plant is now considered a new source of bio-inoculants to improve agricultural productivity. In recent decades, the plants are being inoculated with individual microbes to facilitate growth, nitrogen and phosphorus absorption (Afzal and Bano 2008), drought tolerance (Eke et al. 2019), and resistance to disease (Ashajyothi et al. 2020). However, this initiative was mostly centered almost on an individual strains of microbial species with occurrence of variable performance, which is usually due to the difficulty and habitat settings of experimental site or inoculation place. It is in general requirement for the understanding and administration of the diversified beneficial microbial consortia in cultivation sites to improve soil fertility and enhance support for plant growth. Several initiatives have been taken in the recent past for the above purpose (Reid and Greene 2013; Gilbert et al. 2014; Alivisatos et al. 2015; Stulberg et al. 2016; APS 2016). Because, identification of the "core microbiome" will help to identify plant-associated microbes that should be prioritized for further research and deceptive experiments (Bulgarelli et al. 2012; Lundberg et al. 2012; Sahu et al. 2020). Plant microbiota is highly diverse, yet not all of these microbes play functionally important roles in their host's biology. Defining the core microbiome enables researchers to filter out transient associations and refine the focus on stable taxa with a greater likelihood of influencing host phenotype. In comparison to the very profound sequence of a few plant microbiomes, NGS-based surveys of large numbers of microbiomes of the same plant species from different environments will help in higher progress against that target and follow-up selective cultivation of the candidate core microbiome.

14.5 Conclusion

In the present chapter, we have summarized the various approaches for the characterization of soil microbial community and their function. Furthermore, the involvement of various NGS technology and computational tools for the classification of raw reads has been also covered. The functional classification approach potentially offers the determination of various biochemical pathways and mining of enzymes for uses in industrial applications. The detailed information of the soil community offers the design of policy for manipulating soil microbes, enhancing fertility sustainably, and increasing chances of providing better crop productivity with increased economical values to the farmers, society, and whole mankind. Keeping this point in a long way to the future, the NGS-based assessment will facilitate the development of a sustainable management system for soil fertility and disease prevention.

Acknowledgments All authors are highly thankful to the authority of the departments and institutions for their support in doing this research. All the authors together duly acknowledge the infrastructural support from the Director of ICAR-IARI and Head, Division of Plant Pathology, and also thank-full to World Bank sponsored ICAR-NAHEP-CAAST, New Delhi, for financial support. Furthermore, Director of ICGEB, as well as group leader of Crop Improvement group, ICGEB New Delhi, is also gratefully acknowledged.

References

Adessi C, Matton G, Ayala G et al (2000) Solid phase DNA amplification: characterisation of primer attachment and amplification mechanisms. Nucleic Acids Res 28(20):e87–e87

Afzal A, Bano A (2008) Rhizobium and phosphate solubilizing bacteria improve the yield and phosphorus uptake in wheat (Triticum aestivum). Int J Agric Biol 10(1):85–88

Alisoltani A, Abia ALK, Bester L (2019) Shared microbiome in different ecosystems: a meta-omics perspective. In: Tripathi V, Kumar P, Tripathi P, Kishore A, Kamle M (eds) Microbial genomics in sustainable agroecosystems, vol 2. Springer, Singapore, pp 1–20. https://doi.org/10.1007/978-981-32-9860-6_1

Alivisatos AP, Blaser MJ, Brodie EL et al (2015) A unified initiative to harness earth's microbiomes. Science 350(6260):507–508

American Phytopathological Society (APS) (2016) Phytobiomes: a roadmap for research and translation. Available at http://www.phytobiomes.org/Roadmap/Documents/PhytobiomesRoadmap.pdf

Anamika MS, Singh B et al (2019) Databases: a weapon from the arsenal of bioinformatics for plant abiotic stress research. In: Wani S (ed) Recent approaches in omics for plant resilience to climate change. Springer, Cham, pp 135–169. https://doi.org/10.1007/978-3-030-21687-0_7

Ashajyothi M, Kumar A, Sheoran N et al (2020) Black pepper (Piper nigrum L.) associated endophytic Pseudomonas putida BP25 alters root phenotype and induces defense in rice (Oryza sativa L.) against blast disease incited by Magnaporthe oryzae. Biol Control 143:104181

Ayling M, Clark MD, Leggett RM (2019) New approaches for metagenome assembly with short reads. Brief Bioinform 21(2):584–594. https://doi.org/10.1093/bib/bbz020

Bakker MG, Manter DK, Sheflin AM et al (2012) Harnessing the rhizosphere microbiome through plant breeding and agricultural management. Plant Soil 360(1–2):1–13

Barriuso J, Valverde JR, Mellado RP (2011) Estimation of bacterial diversity using next generation sequencing of 16S rDNA: a comparison of different workflows. BMC Bioinform 12(1):473. https://doi.org/10.1186/1471-2105-12-473

Bates ST, Berg-Lyons D, Caporaso JG et al (2011) Examining the global distribution of dominant archaeal populations in soil. ISME J 5(5):908–917

Bélanger RR, Avis TJ (2002) Ecological processes and interactions occurring in leaf surface fung. In: Lindow SE, Hecht-Poinar EI, Elliott VJ (eds) Phyllosphere microbiology. APS Press, St. Paul, MN, pp 193–207

Béné C, Arthur R, Norbury H et al (2016) Contribution of fisheries and aquaculture to food security and poverty reduction: assessing the current evidence. World Dev 79:177–196. https://doi.org/10.1016/j.worlddev.2015.11.007

Bessler H, Oelmann Y, Roscher C et al (2012) Nitrogen uptake by grassland communities: contribution of N 2 fixation, facilitation, complementarity, and species dominance. Plant Soil 358(1–2):301–322

Bhardwaj Y, Reddy B, Dubey SK (2020) Temporal shift in methanotrophic community and methane oxidation potential in forest soils of dry tropics: high-throughput metagenomics approach. Biol Fertil Soils 56(6):859–867

Bharti J, Sahil MS et al (2021) Mitogen-activated protein kinase, plants, and heat stress. In: Husen A (ed) Harsh environment and plant resilience: molecular and functional aspects. Springer Nature, Cham, pp 323–354. https://doi.org/10.1007/978-3-030-65912-7_13

Bo B, Kim S-A, Han NS (2020) Bacterial and fungal diversity in Laphet, traditional fermented tea leaves in Myanmar, analyzed by culturing, DNA amplicon-based sequencing, and PCR-DGGE methods. Int J Food Microbiol 320:108508

Bolyen E, Rideout JR, Dillon MR et al (2019) Reproducible, interactive, scalable and extensible microbiome data science using QIIME 2. Nat Biotechnol 37(8):852–857. https://doi.org/10.1038/s41587-019-0209-9

Bulgarelli D, Rott M, Schlaeppi K et al (2012) Revealing structure and assembly cues for arabidopsis root-inhabiting bacterial microbiota. Nature 488(7409):91–95

Burke DJ, Carrino-Kyker SR, Burns JH (2019) Is it climate or chemistry? Soil fungal communities respond to soil nutrients in a multi-year high-resolution analysis. Ecosphere 10(10):e02896

Davies G, Kirkpatrick J, Cameron E et al (2019) Ecosystem engineering by digging mammals: effects on soil fertility and condition in Tasmanian temperate woodland. R Soc Open Sci 6 (1):180621

Del Fabbro C, Scalabrin S, Morgante M et al (2013) An extensive evaluation of read trimming effects on illumina ngs data analysis. PLoS One 8(12):e85024. https://doi.org/10.1371/journal.pone.0085024

Dilawari R, Kaur N, Priyadarshi N et al (2021) Genome editing: a tool from the vault of science for engineering climate-resilient cereals. In: Husen A (ed) Harsh environment and plant resilience: molecular and functional aspects. Springer Nature, Cham, pp 45–72. https://doi.org/10.1007/978-3-030-65912-7_3

Doran JW, Zeiss MR (2000) Soil health and sustainability: managing the biotic component of soil quality. Appl Soil Ecol 15(1):3–11. https://doi.org/10.1016/S0929-1393(00)00067-6

Eid J, Fehr A, Gray J et al (2009) Real-time DNA sequencing from single polymerase molecules. Science 323(5910):133–138

Eisenhauer N, Beßler H, Engels C et al (2010) Plant diversity effects on soil microorganisms support the singular hypothesis. Ecology 91(2):485–496

Eke P, Kumar A, Sahu KP et al (2019) Endophytic bacteria of desert cactus (Euphorbia trigonas Mill) confer drought tolerance and induce growth promotion in tomato (Solanum lycopersicum L.). Microbiol Res 228(2019):126302

Endrullat C, Glokler J, Franke P et al (2016) Standardization and quality management in next-generation sequencing. Appl Transl Genom 10:2–9. https://doi.org/10.1016/j.atg.2016.06.001

Fedurco M, Romieu A, Williams S et al (2006) BTA, a novel reagent for DNA attachment on glass and efficient generation of solid-phase amplified DNA colonies. Nucleic Acids Res 34(3):e22–e22

Franke-Whittle IH, Manici LM, Insam H et al (2015) Rhizosphere bacteria and fungi associated with plant growth in soils of three replanted apple orchards. Plant Soil 395(1–2):317–333

Friesen ML, Porter SS, Stark SC et al (2011) Microbially mediated plant functional traits. Annu Rev Ecol Evol Syst 42(1):23–46

Gebbers R, Adamchuk VI (2010) Precision agriculture and food security. Science 327 (5967):828–831. https://doi.org/10.1126/science.1183899

Gigliotti G, Pezzolla D, Zadra C et al (2013) Dynamics of organic matter and microbial populations in amended soil: a multidisciplinary approach. EGUGA:EGU2013-4405

Gilbert JA, Jansson JK, Knight R (2014) The earth microbiome project: successes and aspirations. BMC Biol 12(1):69

Guo J, Ling N, Chen Z et al (2020) Soil fungal assemblage complexity is dependent on soil fertility and dominated by deterministic processes. New Phytol 226(1):232–243

Haas D, Keel C (2003) Regulation of antibiotic production in root-colonizing Peudomonas spp. and relevance for biological control of plant disease. Annu Rev Phytopathol 41:117–153. https://doi.org/10.1146/annurev.phyto.41.052002.095656

Habekost M, Eisenhauer N, Scheu S et al (2008) Seasonal changes in the soil microbial community in a grassland plant diversity gradient four years after establishment. Soil Biol Biochem 40 (10):2588–2595

Hiiesalu I, Oepik M, Metsis M et al (2012) Plant species richness belowground: higher richness and new patterns revealed by next-generation sequencing. Mol Ecol 21(8):2004–2016

Huson DH, Auch AF, Qi J et al (2007) MEGAN analysis of metagenomic data. Genome Res 17 (3):377–386. https://doi.org/10.1101/gr.5969107

Jeanbille M, Buée M, Bach C et al (2016) Soil parameters drive the structure, diversity and metabolic potentials of the bacterial communities across temperate beech forest soil sequences. Microb Ecol 71(2):482–493

Jovel J, Patterson J, Wang W et al (2016) Characterization of the gut microbiome using 16S or shotgun metagenomics. Front Microbiol 7:459. https://doi.org/10.3389/fmicb.2016.00459

Ju J, Kim DH, Bi L et al (2006) Four-color DNA sequencing by synthesis using cleavable fluorescent nucleotide reversible terminators. Proc Natl Acad Sci U S A 103 (52):19635–19640. https://doi.org/10.1073/pnas.0609513103

Kanagawa T (2003) Bias and artifacts in multitemplate polymerase chain reactions (PCR). J Biosci Bioeng 96(4):317–323

Keegan KP, Glass EM, Meyer F (2016) MG-RAST, a metagenomics service for analysis of microbial community structure and function. Methods Mol Biol 1399:207–233. https://doi.org/10.1007/978-1-4939-3369-3_13

Kumar M, Kumar A, Sahu KP et al (2021) Deciphering core-microbiome of rice leaf endosphere: revelation by metagenomic and microbiological analysis of aromatic and non-aromatic genotypes grown in three geographical zones. Microbiol Res 246:126704. https://doi.org/10.1016/j.micres.2021.126704

Lakshmanan V, Selvaraj G, Bais HP (2014) Functional soil microbiome: belowground solutions to an aboveground problem. Plant Physiol 166(2):689–700

Lange M, Habekost M, Eisenhauer N et al (2014) Biotic and abiotic properties mediating plant diversity effects on soil microbial communities in an experimental grassland. PLoS One 9(5): e96182

Laszlo AH, Derrington IM, Ross BC et al (2014) Decoding long nanopore sequencing reads of natural DNA. Nat Biotechnol 32(8):829

Leite AM, Mayo B, Rachid CT et al (2012) Assessment of the microbial diversity of Brazilian kefir grains by PCR-DGGE and pyrosequencing analysis. Food Microbiol 31(2):215–221

Li W, Chen X, Liu M et al (2017) Shifts in microbial communities with increasing soil fertility across a chronosequence of paddy cultivation in subtropical China. Appl Soil Ecol 120:153–159

Lisuma JB, Zuberi Z, Ndakidemi PA et al (2020) Linking rhizosphere bacterial diversity and soil fertility in tobacco plants under different soil types and cropping pattern in Tanzania: a pilot study. Heliyon 6(7):e04278

Liu W-T, Marsh TL, Cheng H et al (1997) Characterization of microbial diversity by determining terminal restriction fragment length polymorphisms of genes encoding 16S rRNA. Appl Environ Microbiol 63(11):4516–4522

Liu Y, Johnson NC, Mao L et al (2015) Phylogenetic structure of arbuscular mycorrhizal community shifts in response to increasing soil fertility. Soil Biol Biochem 89:196–205

Loman NJ, Pallen MJ (2015) Twenty years of bacterial genome sequencing. Nat Rev Microbiol 13 (12):787–794. https://doi.org/10.1038/nrmicro3565

Lu YF, Goldstein DB, Angrist M et al (2014) Personalized medicine and human genetic diversity. Cold Spring Harb Perspect Med 4(9):a008581. https://doi.org/10.1101/cshperspect.a008581

Lundberg DS, Lebeis SL, Paredes SH et al (2012) Defining the core Arabidopsis thaliana root microbiome. Nature 488(7409):86–90

Mehta S, James D, Reddy MK (2019a) Omics technologies for abiotic stress tolerance in plants: current status and prospects. In: Wani S (ed) Recent approaches in omics for plant resilience to climate change. Springer, Cham, pp 1–34. https://doi.org/10.1007/978-3-030-21687-0_1

Mehta S, Singh B, Dhakate P et al (2019b) Rice, marker-assisted breeding, and disease resistance. In: Wani S (ed) Disease resistance in crop plants. Springer, Cham, pp 83–111. https://doi.org/10.1007/978-3-030-20728-1_5

Mehta S, Lal SK, Sahu KP et al (2020) CRISPR/Cas9-edited rice: a new frontier for sustainable agriculture. In: Rakshit A, Singh H, Singh A, Singh U, Fraceto L (eds) New frontiers in stress management for durable agriculture. Springer, Singapore, pp 427–458. https://doi.org/10.1007/978-981-15-1322-0_23

Mehta S, Singh B, Patra A et al (2021a) Maize microbiome: current insights for the sustainable agriculture. In: Solanki MK, Kashyap PL, Ansari RA (eds) Microbiomes and plant health. Academic Press, Massachusetts, pp 267–297. https://doi.org/10.1016/B978-0-12-819715-8.00009-4

Mehta S, Chakraborty A, Roy A et al (2021b) Fight hard or die trying: current status of lipid signaling during plant–pathogen interaction. Plants 10(6):1098

Miller RM, Jastrow JD (2000) Mycorrhizal fungi influence soil structure. In: Kapulnik Y, Douds DD (eds) Arbuscular mycorrhizas: physiology and function. Springer, Dordrecht, pp 3–18. https://doi.org/10.1007/978-94-017-0776-3_1

Mitchell AL, Almeida A, Beracochea M et al (2020) MGnify: the microbiome analysis resource in 2020. Nucleic Acids Res 48(D1):D570–D578. https://doi.org/10.1093/nar/gkz1035

Mohanty SR, Bodelier PL, Conrad R (2007) Effect of temperature on composition of the methanotrophic community in rice field and forest soil. FEMS Microbiol Ecol 62(1):24–31

Mueller UG, Sachs JL (2015) Engineering microbiomes to improve plant and animal health. Trends Microbiol 23(10):606–617

Müller T, Ruppel S (2014) Progress in cultivation-independent phyllosphere microbiology. FEMS Microbiol Ecol 87(1):2–17

Neilson JW, Jordan FL, Maier RM (2013) Analysis of artifacts suggests DGGE should not be used for quantitative diversity analysis. J Microbiol Methods 92(3):256–263

Nielsen S, Minchin T, Kimber S et al (2014) Comparative analysis of the microbial communities in agricultural soil amended with enhanced biochars or traditional fertilisers. Agric Ecosyst Environ 191:73–82

Pareek CS, Smoczynski R, Tretyn A (2011) Sequencing technologies and genome sequencing. J Appl Genet 52(4):413–435. https://doi.org/10.1007/s13353-011-0057-x

Pfeifer SP (2017) From next-generation resequencing reads to a high-quality variant data set. Heredity (Edinb) 118(2):111–124. https://doi.org/10.1038/hdy.2016.102

Pitta DW, Parmar N, Patel AK et al (2014) Bacterial diversity dynamics associated with different diets and different primer pairs in the rumen of Kankrej cattle. PLoS One 9(11):e111710

Porazinska DL, Bardgett RD, Blaauw MB et al (2003) Relationships at the aboveground–belowground interface: plants, soil biota, and soil processes. Ecol Monogr 73(3):377–395

Potthoff M, Steenwerth KL, Jackson LE et al (2006) Soil microbial community composition as affected by restoration practices in California grassland. Soil Biol Biochem 38(7):1851–1860. https://doi.org/10.1016/j.soilbio.2005.12.009

Rajput LS, Aggarwal SK, Mehta S et al (2021) Role of WRKY transcription factor superfamily in plant disease management. In: Giri B, Sharma MP (eds) Plant stress biology. Springer, Singapore, pp 335–361. https://doi.org/10.1007/978-981-15-9380-2_11

Ramakrishnan B, Lueders T, Dunfield PF et al (2001) Archaeal community structures in rice soils from different geographical regions before and after initiation of methane production. FEMS Microbiol Ecol 37(2):175–186

Reddy B (2019) Advancements in microbial genome sequencing and microbial community characterization. In: Tripathi V, Kumar P, Tripathi P, Kishore A, Kamle M (eds) Microbial genomics in sustainable agroecosystems. Springer, Singapore, pp 87–113

Reddy B, Dubey SK (2019) River Ganges water as reservoir of microbes with antibiotic and metal ion resistance genes: High throughput metagenomic approach. Environ Pollut 246:443–451. https://doi.org/10.1016/j.envpol.2018.12.022

Reddy B, Dubey SK (2021) Exploring the allochthonous pollution influence on bacterial community and co-occurrence dynamics of River Ganga water through 16S rRNA-tagged amplicon metagenome. Environ Sci Pollut Res 28:26990–27005

Reddy B, Singh KM, Patel AK et al (2014) Insights into resistome and stress responses genes in Bubalus bubalis rumen through metagenomic analysis. Mol Biol Rep 41(10):6405–6417

Reddy B, Pandey J, Dubey SK (2019) Assessment of environmental gene tags linked with carbohydrate metabolism and chemolithotrophy associated microbial community in River Ganga. Gene 704:31–41. https://doi.org/10.1016/j.gene.2019.04.004

Reid A, Greene SE (2013) How microbes can help feed the world. Issues 105(31):33–37

Rialch I, Singh S, Singh R et al (2019) Plant and microbial genomics in crop improvement. In: Tripathi V, Kumar P, Tripathi P, Kishore A, Kamle M (eds) Microbial genomics in sustainable agroecosystems, vol 2. Springer, Singapore, pp 215–230. https://doi.org/10.1007/978-981-32-9860-6_13

Roumpeka DD, Wallace RJ, Escalettes F et al (2017) A review of bioinformatics tools for bio-prospecting from metagenomic sequence data. Front Genet 8:23. https://doi.org/10.3389/fgene.2017.00023

Sahil KR, Patra A et al (2021) Expression and regulation of stress-responsive genes in plants under harsh environmental conditions. In: Husen A (ed) Harsh environment and plant resilience: molecular and functional aspects. Springer Nature, Cham, pp 25–44. https://doi.org/10.1007/978-3-030-65912-7_2

Sahu KP, Kumar A, Patel A et al (2020) Rice blast lesions: an unexplored phyllosphere microhabitat for novel antagonistic bacterial species against Magnaporthe oryzae. Microb Ecol. https://doi.org/10.1007/s00248-020-01617-3

Schneider A (1892) Observations on some American rhizobia. Bull Torrey Bot Club 19(7):203–218

Sharma P, Sharma MMM, Malik A et al (2021) Rhizosphere, rhizosphere biology, and rhizospheric engineering. In: Mohamed HI, El-Beltagi HEDS, Abd-Elsalam KA (eds) Plant growth-promoting microbes for sustainable biotic and abiotic stress management. Springer, Cham, pp 577–624. https://doi.org/10.1007/978-3-030-66587-6_21

Singh K, Jisha T, Reddy B et al (2015a) Microbial profiles of liquid and solid fraction associated biomaterial in buffalo rumen fed green and dry roughage diets by tagged 16S rRNA gene pyrosequencing. Mol Biol Rep 42(1):95–103

Singh KM, Patel AK, Shah RK et al (2015b) Potential functional gene diversity involved in methanogenesis and methanogenic community structure in Indian buffalo (Bubalus bubalis) rumen. J Appl Genet 56(3):411–426

Soil Science Society of America (2020) Wisconsin, United Nations. https://www.soils.org/. Accessed 6 Oct 2020

Sterkenburg E, Bahr A, Brandström Durling M et al (2015) Changes in fungal communities along a boreal forest soil fertility gradient. New Phytol 207(4):1145–1158

Strecker T, Macé OG, Scheu S et al (2016) Functional composition of plant communities determines the spatial and temporal stability of soil microbial properties in a long-term plant diversity experiment. Oikos 125(12):1743–1754

Stulberg E, Fravel D, Proctor LM et al (2016) An assessment of US microbiome research. Nat Microbiol 1(1):15015

Tamames J, Cobo-Simon M, Puente-Sanchez F (2019) Assessing the performance of different approaches for functional and taxonomic annotation of metagenomes. BMC Genomics 20(1):960. https://doi.org/10.1186/s12864-019-6289-6

Tu J, Qiao J, Zhu Z et al (2018) Soil bacterial community responses to long-term fertilizer treatments in Paulownia plantations in subtropical China. Appl Soil Ecol 124:317–326

Turner TR, James EK, Poole PS (2013) The plant microbiome. Genome Biol 14(6):1–10

Vincent AT, Derome N, Boyle B et al (2017) Next-generation sequencing (NGS) in the microbiological world: how to make the most of your money. J Microbiol Methods 138:60–71. https://doi.org/10.1016/j.mimet.2016.02.016

Vollmers J, Wiegand S, Kaster A-K (2017) Comparing and evaluating metagenome assembly tools from a microbiologist's perspective - not only size matters. PLoS One 12(1):e0169662. https://doi.org/10.1371/journal.pone.0169662

Xue C, Huang Q, Ling N et al (2011) Analysis, regulation and high-throughput sequencing of soil microflora in mono-cropping system. Acta Pedol Sin 48(3):612–618

Yadav S, Reddy B, Dubey SK (2020) De novo genome assembly and comparative annotation reveals metabolic versatility in cellulolytic bacteria from cropland and forest soils. Funct Integr Genomics 20(1):89–101. https://doi.org/10.1007/s10142-019-00704-0

Chapter 15
Development of Biofertilizers and Microbial Consortium an Approach to Sustainable Agriculture Practices

Priyanka Gehlot, Nidhi Pareek, and V. Vivekanand

Abstract Globally, there is excessive use of chemical fertilizer beyond the soil and crop threshold limits which had a deleterious effect on the soil ecosystem. So, now agriculturalists are switching from agrochemical practices to agro-biotechnological practices by using soil microbes as a source of fertilizers. In developed countries, soil microbial communities have been considered as the prime factor for sustainable agricultural practices for the last few decades. The activities and the interaction of these soil microorganisms have been proven to promote plant growth, soil quality, and productivity and maintain the biogeochemical cycle, earth geochemical stability, and climatic conditions of the earth system. Biofertilizers are the formulation of the beneficial microbial strains (bacteria, fungus, and algae) packed on the carrier for mobilization. Biofertilizers can fix the atmospheric nitrogen and mineralize the soil's organic matter. Biofertilizers inoculants may be single species-specific or in the combination of different compatible strains. Microbial consortia are the symbiotic interactions of combinations of two or more compatible microbial strains. A microbial consortium improves the productivity of crop and soil in extreme stress conditions much better than the single-strain inoculants. Therefore, microbial fertilizers and consortium are the best solution to achieve sustainable agricultural practices worldwide.

Keywords Biofertilizers · Microbial consortia · Symbiotic interaction · Compatible microbial strain

---

P. Gehlot · N. Pareek
Department of Microbiology, School of Life Sciences, Central University of Rajasthan, Bandarsindri, Kishangarh, Ajmer, Rajasthan, India

V. Vivekanand (✉)
Centre for Energy and Environment, Malaviya National Institute of Technology, Jaipur, Rajasthan, India
e-mail: vivekanand.cee@mnit.ac.in

S. K. Dubey, S. K. Verma (eds.), *Plant, Soil and Microbes in Tropical Ecosystems*, Rhizosphere Biology, https://doi.org/10.1007/978-981-16-3364-5_15

15.1 Introduction

In sustainable agricultural practices, the soil microbial community has earned a great importance over the past decades. Activities of soil microbes were applied in all the spheres of sustainable ecosystem processes and biotechnological developments (Lladó et al. 2017; Tamayo-Vélez and Osorio 2018). International organizations, policymakers, and practitioners have raised the interest to explore the soil microbiota applications, especially in the field of bioremediations, food and agricultural science, and industrial (Chibuzor et al. 2018; Chuks Kenneth et al. 2019; Company et al. 2010; Madigan et al. 2009; Odoh 2017; Sam et al. 2017; Zabbey et al. 2017; Zuroff and Curtis 2012). Organic farming is a distinctive sustainable agricultural practice which improves the overall crop yield and soil microbiota conditions and lowers the soil deteriorations. A sustainable agricultural practice is an agro-biotechnological method where the existing food requirements are fulfilled without affecting the food security of future generations. The increasing human populations have increased the demand for food thereby pressurizing soil resources to increase the yield per unit area. In 2010, food and agriculture organization reported that the need for agro products would increase by 60% till 2030.

Soil microbial communities are considered most important part of the soil ecosystem. The soil microbes have the ability to increase the food production and help in balancing the earth's climate and biogeochemical cycles (Hansel et al. 2008; Tringe et al. 2005).

The increase in global requirement for food prompted the excessive utilization of chemical fertilizers in agricultural field beyond the threshold limits of crops and soil (Liu et al. 2017; Sun et al. 2015). Therefore, scientists discovered the possible way to replace the agrochemical methods of agricultural practices with the agro-biotechnical approaches which involves the use of soil microorganisms as biofertilizers or microbial consortium. The application of soil microbes in agriculture solves the plant growth problems and fulfils the global needs for sustainable agricultural practices (Hung et al. 2015; Odoh 2017). Biofertilizers and microbial consortium are eco-friendly, affordable, and renewable source of nutrients for the plants, so they have achieved the global acceptance in organic farming. The objective of this chapter is to summarize the development and application of biofertilizers and microbial consortium and their role in sustainable agriculture practices.

15.2 Biofertilizers

Sustainable agriculture practices can be used to reduce the excessive utilization of chemical fertilizers by replacing them with biofertilizers (Mishra and Dash 2014). Usually, the biofertilizers term is translated in different ways like all the things from plant extracts to green manures and through animal manures (El-Ramady et al. 2018).

The advancement in knowledge of interaction between the plants and soil microbes clarified the concept of biofertilizers. In 2003, Vessey defined biofertilizers as a substance composed of beneficial microorganisms which increases the supply of essential nutrients and minerals to the host plant and thus promotes the host plant development (Vessey 2003). Further, the biofertilizers are determined as the substances containing living microorganisms, which improve the growth of host plant different mechanisms. In additional to the above definition, the substances containing beneficial microorganisms that are utilized against plant pathogens are called as biopesticides or biofertilizers (Fuentes-Ramirez and Caballero-Mellado 2006). Similarly, there are phytostimulators and rhizoremediators which improve the plant growth by secreting the plant hormones and biodegrade the organic pollutants respectively, but not every microbial formulation can be considered as biofertilizers directly (Bhattacharyya and Jha 2012; Somers et al. 2004).

The scientific view of biofertilizers is the single microorganism which has the ability to promote the plant growth, but in the agricultural context, biofertilizers are substances composed of different microbial strain(s) which are used for various soil and plant improvement applications. The biofertilizers can also contribute in the improvement of soil microorganism by the addition of useful substances. It was reported that the term "biofertilizer" should not be misinterpreted for biostimulants which are obtained from non-living microbial cell or microbial extract (Malusá and Vassilev 2014; Reddy 2014).

15.2.1 Role of Biofertilizers in Agriculture

The major role of biofertilizers is stimulating the growth of plants without affecting the environment and increasing the crop yield (Mishra et al. 2013) (Fig. 15.1). Studies had reported that with biofertilizer inoculations in field increase the crop yield approximately by 16% compared to non-inoculated field (Schütz et al. 2018). Microbial biofertilizers improve the structure and fertility of soil by maintaining the soil microbial loads (Rashid et al. 2016). Biofertilizers also improve the plant-water relationship, provide strength to the crops to withstand the abiotic and biotic stress conditions, and protect the crops from various pests and soil-borne diseases like disease caused by mycotoxins (Bhattacharjee and Dey 2014; Simarmata et al. 2016; Xiang et al. 2012). Therefore, biofertilizers are considered commercially as the most effective method in sustainable agricultural practices, but there are some limitations like lack of storage, appropriate materials for production, and transportation facilities, highly sensitive towards temperatures, and most importantly having short shelf life (Patil and Solanki 2016). On the other hand, microbial biofertilizers need to be applied in higher concentration to crops for its effective usage, and their results are observed only after their longer usage. The results of biofertilizers are dependent on the soil conditions of the applied zone (Jangid et al. 2012). Scientists are still working to develop new approaches or technologies to defeat the limitations of biofertilizers in the agricultural systems (García-Fraile et al. 2015).

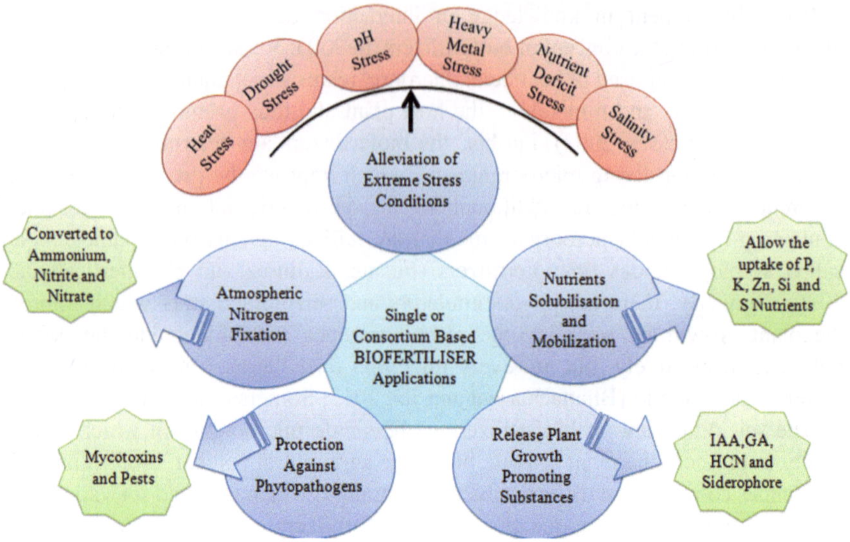

Fig. 15.1 Schematic representation of the role of the single or consortium-based biofertilizer applications

15.2.2 Types of Microbial Fertilizers

There are different types of microbial fertilizers utilized for sustainable agricultural practices. They are grouped according to the microorganism they carry (Itelima et al. 2018). The types of microbial fertilizers are discussed briefly in the following section.

15.2.2.1 Nitrogen Biofertilizers

Nitrogen is considered as the most important nutrient for the crop productions and overall development of plant growth (Thilakarathna et al. 2016). Nitrogen is defined as macronutrient which is the key component of the chlorophyll molecules and also plays a crucial role in most of the enzymatic process in plant cells (Wagner 2011). Nitrogen is most abundantly present in the earth atmosphere, but this atmospheric form of nitrogen is not available for plants and animals due to its triple bond structure which makes its stiff and unbreakable (Figueiredo Mdo et al. 2013).

The most efficient method used by the plants to uptake the atmospheric nitrogen is through the process called biological nitrogen fixation. Microbes involved in biological nitrogen fixation are basically classified as symbiotic and non-symbiotic. In microbial nitrogen fixation process, the atmospheric nitrogen form is converted to the most usable form of nitrogen such as ammonia by the action of nitrogenase enzyme. This ammonia form of atmospheric nitrogen is easily utilized by plants (Galloway et al. 2003; Tairo and Ndakidemi 2013; Vicente and

Dean 2017). The symbiotic nitrogen fixation is basically carried by *Rhizobium* bacteria, which have mutual symbiotic relation with the root nodules of leguminous plants, and the non-symbiotic nitrogen fixation is carried out by the free-living microorganisms like *Cyanobacteria*, *Azotobacter*, and *Azospirillum* species (García-Fraile et al. 2015).

Symbiotic Nitrogen Fixer

The symbiotic nitrogen-fixing bacteria belong mainly to the Rhizobiaceae family and consist of the following genera: *Allorhizobium*, *Rhizobium*, *Bradyrhizobium*, *Azorhizobium*, and *Sinorhizobium* (Patel and Sinha 2011) generally known as *Rhizobia*. The *Rhizobium* develops the mutualistic relationship with the leguminous plants through the formation of the extra structures of root termed as nodules. Inside the root nodules of leguminous plants, the nitrogen fixation process occurs which change the atmospheric nitrogen into the ammonium through the special enzyme called nitrogenase and is further effectively utilized by the plants cells (Shrimant Shridhar 2012). It was reported that the use of rhizobial biofertilizers in the pulse crop field increases the crop yield because of the symbiotic action between host and pulse crop. *Rhizobium* biofertilizers have the ability to fix 15–20 kg N ha^{-1} with 20% increase in crop yields of leguminous plants. The efficiency of these nitrogen biofertilizers depends upon the *rhizobium* strains and the host plant involved; thus in the process of formation of nitrogen biofertilizers, the compatibility of these organisms must be a prime consideration. It was reported that rhizobium fertilizers can fix the 30–643 kg N ha $^{-1}$ in soybean, 25–100 kg N ha $^{-1}$ in green gram, 126–-319 kg N ha $^{-1}$ is groundnut, 125–143 kg N ha $^{-1}$ in black gram, and 77–92 kg N ha $^{-1}$ in pigeon pea (Gopalakrishnan et al. 2015). Similarly, the symbiotic relationship between the vegetable crops and *rhizobium* is also achieved. The most commonly reported vegetables are *Pisum sativum*, *Medicago sativa*, *Trifolium* sp., *Phaseolus vulgaris*, *Lotus corniculatus*, *Cicer arietinum*, and *Glycine max* (Verma et al. 2010). *Rhizobium*, *Mesorhizobium*, and *Bradyrhizobium* have been reported to enhance the growth of legume and supply the nitrogen to the legume plants in the soil populated with metals (Bramhachari et al. 2018). The signature members of the Rhizobiaceae family were reported to secrete the molecules like L-aminocyclopropane-1-carboxylatedeaminase, siderophores, and indoleacetic acid (Wdowiak-Wróbel et al. 2017). It has been observed that the strains of rhizobium which has the ability to secrete L-aminocyclopropane-1-carboxylatedeaminase resulted in the better physiology, growth, and quality of mung bean crops in saline soil conditions.

Further, one more microorganism, *Frankia*, can be used for symbiotic nitrogen biofertilizers. *Frankia* are the gram-positive free-living soil bacteria and have the symbiotic relationship with the actinorhizal plants (Mus et al. 2016). *Frankia* produces root nodules with the actinorhizal plants which are anatomically, morphologically, and functionality different from that of the root nodules of leguminous plants (Hocher et al. 2009). Application of *Frankia*-based nitrogen biofertilizers in the arid soil environments has shown positive impact on actinorhizal tress and also improves the soil fertility of the degraded land (Diagne et al. 2013). In India, South America, China, and Senegal, an agriculturally important tree *Casuarina* was treated with the *Frankia*-based biofertilizers which has shown increase in growth and biomass (Sayed 2011).

Free-Living Non-photosynthetic Nitrogen Fixer

Among the soil bacterial communities, only *Azospirillum* and *Azotobacter* groups are identified as the potent biofertilizers ability in cereals and legume crops (Gupta et al. 2016). *Azotobacter* microbes are aerobic, free-living bacteria which have the ability to fix approximately 20 kg N ha^{-1}/year (Bikash Bag et al. 2017; Mahanty et al. 2017). The most common *Azotobacter* species which are used as biofertilizers to fix atmospheric nitrogen in non-legume crops are *A. beijerinckii*, *A. vinelandii*, *A. chroococcum*, *A. nigricans*, and *A. paspali* (Chandra et al. 2018; Wani et al. 2013). The use of *Azotobacter* sp.-based biofertilizers in maize crops resulted in the improvement in the stem base diameter, plant height, and dry and fresh organic matter content (Iwuagwu et al. 2013). It has been reported that the spraying of *Azotobacter* sp. biofertilizers at oat, clove, and wheat crops increases their dry organic matter by 13–19%, 14–27%, and 10–23%, respectively, compared to control condition (without *Azotobacter* biofertilizer) (Sethi and Adhikary 2012).

In the study conducted by Gothandapani et al. (2017), it was reported that *Azotobacter* species secretes the other useful substances which can improve the growth and development of plants. The beneficial molecules produced by *Azotobacter* species are auxins, cytokines, gibberellins, nicotinic, pantothenic acid, and vitamin B which improve the germination of seeds. Further, it has observed, increase in the seed germination by 20–30%, overall crop yield and provide protection against pathogenic rhizospheric microbes, in the crops inoculated with *Azotobacter sp.* (Mahato and Kafle 2018; Vikhe 2014).

Free-Living Photosynthetic Nitrogen Fixer

Most commonly used free-living photosynthetic microorganism as nitrogen fixer biofertilizers is *Blue green algae* (BGA) or *Cyanobacteria*. They are generally found in lakes, rivers, ponds, and water streams and have the ability to fix the atmospheric nitrogen into the ammonium and nitrogenous compounds (Singh et al. 2016). Among the BGA, the most commonly used genera for the biofertilizer are *Nostoc*, *Cylindrospermum*, *Anabaena*, *Calothrix*, *Stigonema*, *Aulosira*, and *Tolypothrix* which consists of heterocyst, a modified thick-walled nitrogen-fixing cell (Kumar et al. 2010b). Studies have reported that along with heterocyst containing BGA, some non-heterocyst containing unicellular (*Dermocapsa*, *Aphanothece*) and filamentous (*Trichodesmium*, *Oscillatoria*) genera of Cyanobacteria also has the ability to fix the atmospheric nitrogen (Berrendero et al. 2016). According to the study of

Rathod et al. (2018), these cyanobacteria secrete the beneficial substances like antifungal and antibacterial compounds along with some vitamins and amino acids and therefore have the ability to promote the growth of the plants. Likewise, blue green algae have the ability to convert the insoluble phosphate form to soluble phosphate form, thereby increasing the phosphorous availability in the soil for crops (Rai et al. 2019). In India, generally *Aulosira fertilissima* is considered be to the most effective cyanobacterial nitrogen fixer-based biofertilizers for rice crops (Thingujam et al. 2016). *Cyanobacteria* fixes 20–40 kg N ha^{-1} of atmospheric nitrogen; thus they are considered best alternative against the previously used chemical fertilizers (Issa et al. 2014; Singh et al. 2016).

Associative Nitrogen Fixer

Azospirillum sp. is aerobic, free-living, and non-nodulated bacterium which has the potential to fix the atmospheric nitrogen. The *Azospirillum* sp. is reported vital for the growth of crops in greenhouse and trial fields (Vurukonda et al. 2016). In agricultural or wild crops, *Azospirillum* sp. usually grows at the surface and inside the roots, and this type of association is known as rhizosphere association (Gangwar et al. 2017). *Azospirillum* sp.-based biofertilizers are proposed for the non-legume crops like paddy, oilseeds, banana, millets, chilly, coconut, oil palm, sugarcane, and cotton (Pathak et al. 2018) and can fix 20–40 kg N ha^{-1} atmospheric nitrogen into the soil. It has reported that *Azospirillum* sp.-based nitrogen fixer biofertilizers fix approximately 50% of nitrogen for sugarcane crops (Saranraj and Sivasakthivelan 2013). In barley crop, the salt stress was reduced by using *A. brasilense*-based biofertilizers. According to Atta et al. (2018)s studies, these microorganisms seems to secrete various plant hormones which have the ability to modify the physiological and morphological characteristics of applied crops.

15.2.2.2 Phosphorus Biofertilizers

Phosphorous is the second most essential macronutrient, which is readily absorbed by the plants for the overall growth and development of plants. Phosphorous is involved in various plant metabolic pathways (Sharma et al. 2013). The majorly available phosphorous forms in soil are insoluble phosphate and soluble phosphate, determined on the basis of organic and inorganic compounds. Nearly 90% to 98% phosphorous present in soil is not utilized by the plants, while some form of phosphorus is absorbed such as $H_2PO_4^{\ -}$ and HPO $_4$ (Sharma 2011; Vijayabharathi et al. 2016). According to Sharon et al. (2016), the generally used way to tackle with the insufficiency of phosphorous in soil is through the utilization of the phosphate mineralized fertilizers in the form of monopotassium phosphate or monocalcium phosphate, but the use of these chemical fertilizers for long term has the negative effects on the soil ecosystem. In the acidic soil condition, the phosphorus is bonded with aluminium and iron, and similarly in alkaline soil conditions, it is chemically bonded with calcium and magnesium ions, thereby resulting in the unavailability of the phosphorous for plants in soil (Mehrvarz et al. 2008; Ranjan et al. 2013).

So the best approach in sustainable agricultural practices is utilization of the microbial-based biofertilizers which has the ability to convert the insoluble phosphate form to soluble phosphate form and increases the availability of phosphorous in the soil (Barea 2015). Bacterial strains utilized as phosphate biofertilizers are *Agrobacterium* sp., *Pseudomonas* spp., and *Bacillus circulans*, while there are some bacteria which have been reported for phosphorous solubilizing activity such as *Azotobacter*, *Burkholderia*, *Erwinia*, *Bacillus*, *Rhizobium*, *Enterobacter*, *Bradyrhizobium*, *Paenibacillus*, *Serratia*, *Thiobacillus*, *Salmonella*, *Ralstonia*, and *Sinomonas* (Alori et al. 2017; Elias et al. 2016).

Interestingly, even some fungal strains are reported to have phosphorous solubilization and mobilization abilities. The microbial fungal strains detected for phosphorous mobilization activity are *Achrothcium*, *Fusarium*, *Aspergillus*, *Penicillium*, *Cladosporium*, *Alternaria*, *Myrothecium*, *Pichia fermentans*, *Yarrowia*, *Saccharomyces*, *Curvularia*, *Arthrobotrys*, *Rhizopus*, *Cephalosporium*, *Trichoderma*, *Oidiodendron*, *Schwanniomyces*, *Populospora*, *Glomus*, *Phoma*, *Micromonospora*, *Paecilomyces*, *Torula*, and *Mortierella* (Alori et al. 2017; Pal et al. 2015).

15.2.2.3 Plant Growth-Promoting Biofertilizers (PGPB)

Microorganisms of this types of biofertilisers improve the overall growth of plants by secreting various active compounds like siderophores, cyanides, plant hormones (gibberellic acid and indoleacetic acid), antibiotics, chitinase, and volatile organic compounds (Majeed et al. 2015). These agroactive compounds are produced in large amount and have the ability to enhance the morphological features of the host plant (Gouda et al. 2018). The plant growth-promoting biofertilizers are based on the rhizobacteria which belong to the following genera like *Agrobacterium*, *Alcaligenes*, *Azotobacter*, *Rhizobium*, *Achromobacter*, *Pseudomonas* sp., *Flavobacterium*, *Enterobacter*, *Arthrobacter*, *Bradyrhizobium*, *Amorphosporangium*, *Xanthomonas*, *Erwinia*, *Cellulomonas*, and *Bacillus* (Mohammadi and Sohrabi 2012; Vejan et al. 2016). In the study of Anwar et al. (2016), some *actinomycetes* strains produce the agroactive compounds which promote the plant growth and development.

Some plant growth-promoting rhizobacteria (PGPR) show dual functional properties like biofertilizers and biopesticides. For instance, *Burkholderia cepacia* have been detected with biocontrol activities of *Fusarium* sp. which synthesizes the fungal mycotoxins while they also have the ability to secrete the siderophores which improves the growth of maize crops during the iron deficiency conditions (Bhattacharyya and Jha 2012). There are two groups of PGPR based on the affinity with the roots of plants: extracellular PGPR (ePGPR) which is found in rhizospheric region between the cells of cortex or at the rhizoplane and intracellular PGPR (iPGPR) which are found inside the root nodules (Ahemad and Kibret 2014). These PGPR microorganisms enhance the growth of plants either directly or indirectly. In direct method, the PGPR secrete the phytohormones like GA, siderophores, and IAA which improve the soil nitrogen and phosphorus content. While in indirect method, the PGPR secrete the secondary metabolites like antibiotics and lytic enzymes which provide protection to the host plants towards the various phytopathogens and also enhance the induced systemic resistance activity (Beneduzi et al. 2012; Bhattacharyya and Jha 2012). The soyabean crops inoculated with the *Azotobacter chroococcum-* and *Pseudomonas fluorescens*-based phosphorous biofertilizer improve the phosphatase activity around the roots (Rotaru 2015). *Pseudomonas* microbial strains are reported to secrete the toxic secondary antimicrobial compounds like pyoluteorin, viscosinamide, pyrrolnitrin, and phenazines which create negative effects on the various organisms (Flury et al. 2017).

15.2.2.4 Potassium Biofertilizers

Potassium is considered the third most essential macronutrient for developmental and growth process of plant cells. Potassium plays a vital role in enzymatic reactions, degeneration of sugar, photosynthesis reaction, and protein formation (Basak and Biswas 2009). The total percentage of potassium available in soil is estimated to be in the range of 0.04–3%. In the soil, potassium is available in various forms such as exchangeable potassium, non-exchangeable potassium, mineral potassium, and solution potassium, but among these, mineral potassium form is most abundantly present with 90–98% in the soil which is not accessible for the host plants (Etesami et al. 2017). It has reported that microorganisms such as fungi, bacteria, and *actinomycetes* secrete the various beneficial compounds like polysaccharides, organic acids, exchange reactions, acidolysis, chelation, and complexolysis (Etesami et al. 2017; Mishra et al. 2018).

The unavailable potassium ions of soil react with the Si^{4+} ions and form the metal-organic complex thereby releasing the available potassium form into the soil solution. The biofilm has reported to solubilize potassium from anorthite and biotite (Das and Pradhan 2016). Bacteria responsible for the solubilization of potassium are the following: *Bacillus circulans*, *Burkholderia* sp., *Paenibacillus mucilaginosus*, *Cladosporium* sp., *Paenibacillus glucanolyticus*, *Acidithiobacillus ferrooxidans*, *Bacillus edaphicus and Enterobacter hormaechei*, *Arthrobacter* sp., *Sphingomonas* sp., *P. frequentans*, and *Aminobacter* sp. (Meena et al. 2016). The commercially available brands for potassium mobilizing biofertilizers are Biosol-K, K Sol B®, and Symbion-K which are made up of *Frateuria aurantia* and considered to be effective biofertilizers for growth and development of plants (Mishra and Arora 2016). The microbes involved in potassium solubilization method are observed to have positive effects on the development and growth of plants such as cucumber, cotton, tomato, tobacco, rape, sorghum, chili, pepper, sudan grass, and khella (Meena et al. 2016). According to Bashir et al. (2017) studies, inoculation of soil with the potassium solubilizing biofertilizers enhances the potassium uptake by plants and indigenous activities of soil microbes, improves crop and soil qualities, and reduces the utilization of mineralized potassium fertilizers.

15.2.3 Biofertilizer Production

In order to prepare the best and effective quality of biofertilizers, the following each steps as shown in the Fig. 15.2 need to be carried out carefully in the defined environment (Mohod et al. 2015). Total eight steps are involved in the production procedure as follows: searching and isolating the effective microbial strains, discovering the characteristic of the selective microbes at the proper growth conditions and medium, scaling up the production of selective microbial biomass, choosing the appropriate carrier to load microbial culture, formulating the bioinoculant, testing at the field, industrial level of production experiments, and developing the quality control, transportation, and storage systems (Shaikh and Sayyed 2015; Stamenković et al. 2018).

Selected microbes for biofertilizers production must have certain defined characteristics features for their effectiveness and usefulness at agricultural fields. The characteristic features are as follows: should be easy to replicate in bulk, should be compatible with the natural rhizosphere microorganism, must have high rhizosphere competences, should have the ability to enhance the overall development and growth of crops through various mechanisms or by releasing agroactive compounds, and should not cause any negative effects to the ecosystem (Nakkeeran et al. 2006).

The selective media which are used for the mass production of selective microorganism should be easily accessible in the market, inexpensive, and provide all the essential nutrients in the defined amount (Glick 2020). Biofertilizer production involves the fermentation techniques like solid state, liquid, and semisolid for the large-scale productions. It has been reported that chemical-defined media are needed for the maximum growth of selective microbes as they can alter the ratio of the substance affecting the multiplication of microorganisms (Stamenković et al. 2018).

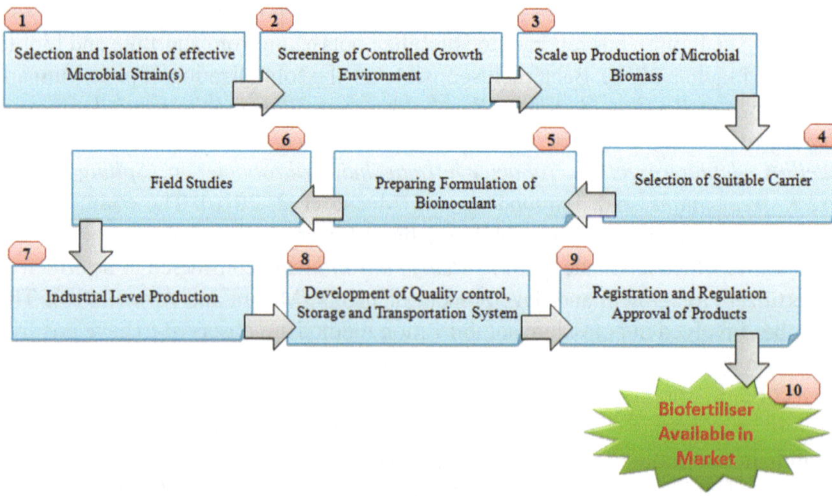

Fig. 15.2 Flow chart of the biofertilizer production procedure

The selection of the suitable carrier is done on the basis of the desired form or quality of end product and microbial strains utilized for biofertilizer production. The next steps is encapsulation of the growth of selective microbial strains or preparation of liquid formulations. Then at last the prepared formulation of biofertilizer is tested at field level and should pass the defined requirements like having positive effect on the growth and yield of crop and having no toxic effect at the ecosystem. After qualifying the minimum defined requirements of biofertilizer, they are applied for registration to grant the approval (Backer et al. 2018; Bashan et al. 2014). The approved biofertilizer formulation is then packed, and the packets should have information mentioning product name, microbial strain (s) composed off, preferred plant name, manufacture and expiry date, producer name and address, and proper instruction and precaution to be followed by the consumers (Bhattacharjee and Dey 2014; García-Fraile et al. 2015).

15.2.4 Quality of Biofertilizer

The parameter of determining biofertilizers quality before commercialization is the most important, so there is need to be performed at each production level carefully (Sethi and Adhikary 2012). In the nations like the USA and European Union, the quality and the production parameters are not clearly defined, but in the nations where sustainable agricultural practices are performed, the rules and regulation of quality control is well defined. The Chinese quality control is defined on the basis of the eight parameters, and among them, the density of the microbial strains used is considered the most important. The eight parameters followed in China are as follows: water and carbon content, size of carrier, amount of microbial load, expiry period, appearance, and contamination. These above parameters are defined for the different microbe groups such as *rhizobium*, phosphorus solubilizing bacteria, silicate solubilizing bacteria, nitrogen fixing bacteria, multistrain consortia, and organic and inorganic phosphorus. For bacterial based liquid formulating biofertilizers, the microbial content must be in the range between $>0.5 \times 10^9$ cfu mL^{-1} and $>1.51 \times 10^9$ cfu mL^{-1}, and in solid product, the microbial content ranges between $>0.1 \times 10^9$ cfu g^{-1} and $> 0.3 \times 10^9$ cfu g^{-1}. According to the parameters approved, the total organic load the biofertilizer should contain is 18–20% irrespective of their phenotypic form and the shelf life at least half a year.

Similarly in India, seven regulatory parameters are defined for the biofertilizers, and the seven parameters are as follows: the phenotypic form, contamination level, size of the carrier, the minimum organic load, pH, water content, and the efficiency parameters. These parameters are defined for the following groups of microorganisms in India like *Rhizobium* sp., *Azospirillum* sp., *Azotobacter* sp., mycorrhizal biofertilizers, and phosphate solubilizing bacteria. For bacteria-based biofertilizer, the minimum amount of organic load in solid carrier system is 5×10^9 cfu g^{-1} and 1×10^8 cfu mL^{-1} for liquid carrier system. In case of mycorrhizal fungal based biofertilisers, 1 g of prepared biofertiliser must compose of at least 100 viable propagules (Sekar et al. 2016).

15.2.5 Application of Biofertilizers

Biofertilizers are applied either directly to the soil or indirectly to seeds, seedling, leaves, etc. (Chen 2006). Each type of approaches has some merits and demerits based on the parameters like type of crop, inoculants used, environmental conditions, and some technical problems from farmers' side (Mahmood et al. 2016). Biofertilizers need to be applied carefully with certain precautions like used biofertilizer solution should not be kept overnight, should be kept in the range 0–35 °C, and should avoid direct contact to sunlight.

Among the approaches used for applying biofertilizers, the seed treatment approach is generally used as it requires small quantity of inoculation product and is very simple to use (Asif et al. 2018). The three ways by which biofertilizers can be applied on the seeds are slurry, seed coating, and dusting (Malusà and Ciesielska 2014). In dusting, the biofertilizers are mixed with the dry seeds, but this technique is not much effective as the interaction between biofertilizer microorganisms and the seed in weak. In slurry approach, biofertilizers are combined with the wet seeds or the seed can be kept in the slurry overnight (Malusà and Ciesielska 2014). It has been reported that the seeds have to be coated with defined number of microbes so the fixative agents like gums, vegetable oils, carboxy methyl cellulose, solution of sucrose, and some harmless marketable products are utilized (Bashan et al. 2014). Alternatively, 1% milk powder or 25% of molasses solution is mixed with the suspension in case biofertilizers do not have any fixative agent. In the third seed coating approach, seeds are added into the slurry suspension of microorganisms and then further coated its outer covering with some inorganic inert substances like charcoal, lime, talc, dolomite, clay, calcium carbonate, and rock phosphate. This outer coating with inert materials protects the seeds from harmful effects of chemical fertilizers and pesticides and from the unfavourable environmental conditions (Malusà and Ciesielska 2014). The bacterial groups involved in seed treatment processes are *Rhizobium*, *Azospirillum*, phosphorous solubilizing microbes, and *Azotobacter* and can also be with the consortium of microbes. According to Brahmaprakash et al. (2017) studies, the seed is first covered with nitrogen fixer microbes, and then further phosphorous solubilizing microbes are coated as outer layer to maintain the viable microbial load. In case large numbers of microbial strains are introduced in the soil field directly, then a soil inoculation technique is required. In this technique of soil inoculation, carrier of granules size 0.5–1.5 mm is favoured, and granular form of soil aggregates, peat, talcum powder, and perlite are mostly utilized in this approach.

Soil treatment process provides protection to the microbial fertilizer strains from the harmful effect of fungicides and pesticides and prevents the destruction of seed coats and the loss of biofertilizers during the seeding machinery activities. The soil inoculation approach improves the chances of interaction between seeds and biofertilizers as compared to seed treatment approach. While there are some technical demerit of this approach like requirement of specialized equipment and high amount of biofertilizers which causes additional transportation and storage problems. In the developed nations, soil inoculation approach with granules is generally employed (Bashan et al. 2014; Deaker et al. 2004).

15.3 Introduction of Microbial Consortia

In nature, microorganisms live in the form of two or more groups called microbial consortium (MC). Therefore, microbial consortium is utilized in the sustainable agricultural practice that has the abilities to perform the activities not possible for individual microorganism. Microbial consortium is formed by the stable symbiotic interaction between the two or more microbial groups for the overall development of crops (Madigan et al. 2009). The microbial consortium has the ability to increase the organic content of soil, make nutrients available for the plants through solubilization and mobilization, and fix the atmospheric nitrogen in the nodules of the leguminous crops (Nuti and Giovannetti 2015). The use and acceptability of microbial consortium in agricultural practices has increased as compared to single strain. Although microbial strains retain their individual characters in the microbial consortium, they still have the ability to respond as a completely different organism in abiotic and biotic stress environment due to their intrinsic beneficial interactions (Nuti and Giovannetti 2015). Microbial consortium has the quorum sensing signalling which helped them to respond and detect the nutrient gradient and microbial density. Quorum sensing mechanism expresses their fascinating biochemical effects that allow their functionality, robustness, stability, and ability to carry out difficult biochemical works.

15.3.1 Microbial Consortium as Biofertilizers

In soil ecosystem, the microbial interactions are complex and dynamic. The biofertilization phenomenon is used to improve the growth and provide the nutrients to the crop (Odoh 2017). Microbial consortium is used as biofertilizers and works in similar way with some advantages over single strain biofertilizers. According to Bradáčová et al. (2019), the comparative analysis of single strain and microbial consortium revealed that the application of microbial consortium has the ability to enhance the crop growth and productivity during the extreme environmental situations. Microbial fertilizers are considered important for the sustainable agricultural practices and maintain the soil fertility for a longer period of time. Microbial fertilizers have the ability to fix the atmospheric nitrogen and solubilize soil phosphorous into the form which can be taken up by the plant roots. Microbial consortium apart from solubilizing the nutrients also has the ability to secrete some bioactive substances like Nod factor and Myc in the signalling pathway (Roberts et al. 2013).

15.3.2 Interaction Between Microbes and Plants

Soil is the topmost layer of the earth crust and is made up of mixtures of minerals, organic matter, gases, and microorganisms that interacts with each other and supports the living system on the earth. Soil system has physical, chemical, and biological properties. The most important and nutritional rich component of soil system is "soil organic matter" (SOM) for plant growth and development. Soil organic matter contains the larger portion of remnants of animals and plant that help in maintaining the soil flora and fauna. It has reported that humic acid substances contribute approximately 60% of SOM, while soil microbes contribute only 8% of SOM, and the remaining is the non-living component of soil system (Htwe et al. 2019; Liste 2003). SOM component of soil is nutrient rich and considered essential portion for sorption of contaminants and cation exchange mechanism, thus promoting the growth of plants and soil microbes (Chibuzor et al. 2018). It has the ability to control soil erosion, and circulation of water and soil also helps in soil aggregation (Guo et al. 2019).

An omics molecular study has disclosed the extent of soil-plant-microbes interactions in the soil system. The advance approach of omics techniques provides the platform to identify, detect, and quantify the diversity of soil microorganism linked with the particular plant. It has reported that the plants are associated with soil microorganism through the various mechanisms that are obligatory for their existences (Schirawski and Perlin 2018). The microorganisms colonizing the rhizosphere of the plants are generally rhizobacteria and mycorrhizal fungi (Hamilton et al. 2016; Nadeem et al. 2014; Yadav et al. 2015a, b). Plant roots not only act as the host for the various soil microorganisms but also secrete the beneficial compounds which provide nutrition to the microbes even after the plant die. These beneficial compounds have the ability to provide the resistance to the plant against the abiotic and biotic stress conditions.

It has documented that high microbial diversity and less nutrient content in rhizoplane part of the soil system generally cause the competition for survival, ability to improve the growth of crops, and development of the adaption mechanism to the stress conditions (Ngumbi and Kloepper 2016).

The beneficial soil microbes interact with the roots of the plants and improve the plant health and growth by the utilizing the biofertilizers, biostimulant, and biocontrol agents (Glick 2014; Nath Yadav et al. 2016; Rashid et al. 2016). Fungal interaction to plant roots aids in the phosphorus solubilizing and mobilizing, protects the plants from many plant pathogens, and provides access to water availability in the drought conditions (Barnawal et al. 2014).

15.3.3 Interaction Among the Bacterial Groups

The interaction between bacteria among the microbial consortium includes the PGPR group like *Pseudomonas*, *Arthrobacter*, *Alcaligenes*, *Burkholderia*, *Klebsiella*, *Bacillus*, *Azospirillum*, *Serratia*, and *Enterobacter*. These PGPR improve the overall growth and development of the crop through various mechanisms (Jambon et al. 2018; Saharan and Nehra 2011). It has reported that the interaction of rhizobacteria with plants improves the ability to segregate the soil pollutants (Chibuzor et al. 2018). The high biochemical and microbial activities have been detected in the rhizospheric environment due to the high availability of nutrients compared to the phylospheric and rhizoplanic components of the soil system (Venturi and Keel 2016).

PGPR also have the ability to improve the nutrient absorption and seed germination, protect the plants from phytopathogens, develop the resistance toward the environmental stress conditions, and increase the shoot and root generations (Odoh 2015). According to Bulgarelli et al. (2012), plant growth-promoting rhizobacteria recruitment process is regulated by the structure of soil microbial communities. It has reported that variation at the genetic level in the plant species is the driving force for the differential recruitment of PGPR communities (Lundberg et al. 2012). These bacterial consortiums are enrolled in the interesting roles like phosphate solubilization, plant development, nitrogen fixation, and the secretion of various plant hormones (Htwe et al. 2019; Odoh 2017).

The bacterial communities of consortium communicate with each other through the chemical signalling process known as quorum sensing (Barriuso 2015). During the quorum sensing, the microbial community's communication and gene expression is regulated by the autoinducers or quorum sensing molecules (QSM). QS signalling is defined as regulatory response to transcribe the particular gene in order to identify the compounds (Venturi and Keel 2016). This QS signalling between the cells are always defined and organized pathogenic activities by adjusting the microbes when the stress conditions are triggered (Jiang et al. 2019). The QSM consists of autoinducing peptide, autoinducer-2, and acyl-homoserine lactone which control the certain biochemical processes like sporulation, biofilm formation, antibiotics productions, releasing out the various virulence factors, and motility (Barriuso 2015; Fleitas Martínez et al. 2018). In this effective cell-to-cell interaction, the high energy-based cost-effective specific tasks are performed only in the presence of the large bacterial population size (Clinton and Rumbaugh 2016).

Additionally, the secretion of nodulation factor (Nod) and volatile organic compounds (VOCs) by *rhizobia* are identified with the ability to assist in the bacterial communications (Hung et al. 2015; Jambon et al. 2018). The VOCs secreted support the long distance communication between the microorganism and plant or between microorganisms and maintain the symbiotic relationship, diffusing the mycorrhizal, harmful microbes and saprophytes (Brilli et al. 2019; Hung et al. 2015; Tyc et al. 2017). VOCs of bacteria improve the plant growth by using acetoin chemical which has the ability to interfere with gene expression of plant and activates the systemic resistances (Bennett et al. 2012). It has been reported that plant roots react to strigolactones and flavonoids as the signalling molecules or host plant symbiosis (Venturi and Keel 2016).

15.3.4 Interaction Between Bacteria and Fungi

The interaction between bacteria and fungi is internally modified through the behavioural characters of the communicating partners (Deveau et al. 2018). There is a close association of biophysical and metabolic activities during the co-occurrences of fungi and bacteria that help in the growth of bacteria and fungi mutuality.

The understanding of microbiomes like *Arabidopsis* root microbiome has been resolved through the characterization of bacteria and fungi interaction (BFI) (Bergelson et al. 2019). This is due to the involvement of molecular techniques which provide account of biomes and environment habitats emphasizing on the microbial diversity (Thompson et al. 2017).

The interaction between the PGPR and arbuscular mycorrhizal fungi (AMF) has been reported to enhance the crop development and growth (Pathak et al. 2017). This interaction also improves the nutrients concentration in the soil and propagates the soil microbiota. It has been reported that PGPR and AMF associations are considered as potent biofertilizers and biocontrol agents for sustainable agricultural practices as they reduced the dependency on the chemical fertilizers (Franco et al. 2011; Pathak et al. 2017).

PGPR are categorized based on the intra- and extracellular PGPB, and in the host plant, they promote plant growth either directly by secreting growth-promoting hormones or indirectly by secreting antimicrobial molecules (Kumar Deshwal and Kumar 2013; Zheng et al. 2018). During the mycorrhization, PGPR and mycorrhizal helping bacteria to interact symbiotically with mycorrhizal roots and fungi in order to uptake the nutrients. Scientists have discovered the PGPR and AMF developed plant resistance by inducing the systemic host immune response (Bramhachari et al. 2018; Zamioudis and Pieterse 2012). The application of PGPR and AMF has proven beneficial for the crops grown in the nutrient-limited soil (Gouda et al. 2018). The usage of PGPR like *Pseudomonas* sp., *Bacillus* sp., and AMF either singly or in combination has reported to produce significant improvement in the growth of crops in various fields (Pathak et al. 2017; Philippot et al. 2013).

15.3.5 Merits and Demerits of Microbial Consortium

In various field of applications, scientists aimed to employ the single pure microbial culture. In spite of the advancement in the microbiology, most of the microbial strains are still not culturable as pure cultures. A co-culture technique has the ability to share the products of the metabolisms and provide strength in stress environmental conditions. Thus, co-culture approach can be utilized to search for the various potentials of unculturable microorganisms. The merits and demerits of utilization of microbial consortium are as follows:

15.3.5.1 Merits

In microbial consortium, different complex carbon sources can be used as mixed microbial culture of microbial consortium producing different enzymes that can degrade the substrates in the different manner (Bhatia et al. 2015). According to Shou et al. (2007), microbial consortium mixed culture cross-feeds the nutrients and regulates the nearby environment to promote each other's development and growth.

Higher productivity is reported in case of mutual interaction as complex multiple step reactions are executed faster than the single strains inoculants. Co-culture technique can utilize the unculturable microorganisms (Stewart 2012). Microbial consortium mixed culture inhibits the growth of unfavourable and toxic microorganisms thereby controlling the contamination.

15.3.5.2 Demerits

The development of the microbial consortium is difficult as the interaction and properties of individual strains of consortia can affect the fermentation process at the industrial level production. During the contamination of microbial consortium, it is difficult to detect the contaminating agent. Unavailability of the prior knowledge of microbial functions, microbial metabolite descriptions, and nutrient demands may restrict the consortium manufacturing process.

Conservation of the microbial consortium through freeze drying is also difficult as the microbial strains have different survival rate at freezing cycles.

15.3.6 Construction of Artificial Microbial Consortium at Industrial Level and Their Interaction

For constructing of non-natural microbial consortium at the industrial level, there are certain parameters that need to be considered such as (a) appropriate inoculums ratio should be taken to avoid the exhaustion of the energy sources, (b) selected microorganism should not have common carbon source for energy, (c) the optimum temperature and pH for microbial growth should be in the physiological range, (d) selected microbe strains should be from the same species as they have the same metabolic behaviour which makes them more compatible, (e) must have prior understanding of the nutrients requirement to design the culture medium suitable for the growth of different strains, and (f) different in silico approaches like flux base analysis (FBA) and constraint-based reconstruction and analysis (COBRA) can be utilized for the better understanding of various complex interactions of microbial groups (Schellenberger et al. 2011).

The efficiency of microbial consortium is based on the mutual interactions of individual strains. Microbial strains have reported different types of phenotypic interactions with one another such as (a) growth of the microbial strains is inhibited at the particular distance from the competitor strains due to the secretion of antibiotics in extracellular environment, known as distance inhibition interaction; (b) there is the formation of the dark precipitation zone when the microbial strain grows larger in size to interact with other strains, known as zone line interactions; (c) in this type of interaction, microbial strains grow enough to contact with other strains with no proof of secondary metabolite release and is known as contact inhibition interactions; and (d) when one microbial colony is taken up by the other colony, it is called overgrowth interaction (Bertrand et al. 2013).

15.3.7 Microbial Consortium in Stress Environment

The climatic factors create obstacle in the agricultural practices as the estimated increase in temperature, drought, and salinity causes abiotic stress condition in the crops, thereby affecting the productivity of crops (Grover et al. 2011; Larson 2013). Plant-associated microbial groups have gained attention as they have the ability to enhance the crop productivity and provide resistance in the stress conditions (Mapelli et al. 2013). Therefore, microbial fertilizers especially consortium application is the best way to mitigate the abiotic stress conditions of plants and improve the growth of the crops in the unfavourable conditions (Jain et al. 2013). Table 15.1 lists the microbial consortiums associated with crops at different extreme environmental conditions. Under 2,4-DNT stress conditions, the microbial consortium degrading 2,4-DNT constitutes of *Pseudomonas*, *Burkholderia*, *Ralstonia*, *Variovorax*, and *Bacillus* spp. has reported to increase the root length of *Arabidopsis* (Thijs et al. 2014).

Application of *R. tropici* and *A. brasilense* co-culture on the bean has no adverse effect of nod gene transcription and salinity situations (Dardanelli et al. 2008). In Jain et al. (2013) studies, the microbial consortium is composed of *Trichoderma* (THU0816), *Rhizobium* (RL091), and *Pseudomonas fluorescens* (PHU094) that has activated the expression of antioxidant enzymes such as peroxidase and superoxide dismutase in stress. In salinity condition, the paddy crops were inoculated with the PGPR-based microbial consortium of *B. pumilus* and *P. pseudoalcaligenes* in has increased the availability of essential nutrients like nitrogen, potassium, phosphorous and reduced the sodium and calcium availability in soil (Jha and Subramanian 2013). Under salt stress, the growth of *P. vulgaris* bacteria is improved with use of *A. brasilense and Rhizobium* consortium (Dardanelli et al. 2008; Smith et al. 2015). Microbial consortium of AMF and *B. thuringiensis* increases the production of

Table 15.1 Microbial consortiums associated with crops at different extreme environmental conditions

Microbial strain(s)	Crop	Extreme environmental conditions	Result	Strategies adapted	References
R. etli	*Phaseolus vulgaris*	Drought	Crop becomes resistant to drought stress and the nitrogen amount, dry weight of nodules and also functionality of nodule reduced	Upregulation of oxidase in Bacteroides	Talbi et al. (2012)
AMF	*Hemarthria altissima, Leymus chinensis*		Increases the photosynthetic rate and rate of passage of CO_2 entering	Improves the activities of antioxidant enzymes	Li et al. (2019)
Azospirillum brasilense	*Glycine max*		Enhanced the plant traits leading the crop resistant to water scarcity stress	Fixation of atmospheric nitrogen and secretion of IAA	Hungria et al. (2015)
	Zea mays	Salinity	Reduces the proline concentration and promotes the plant growth	Alters the selectivity of Na^{+}, Ca^{++}, and K^{+} ions	Fukami et al. (2017, 2018)
Achromobacter piechaudii	*Populus* sp., *Lycopersicon esculentum*	Heavy metal and salt	Promotes the root hair formation and also enhances the root and shoot growth	Synthesises of IAA	Carmen and Roberto (2011), Fahad et al. (2015)
Sinorhizobium arbores	*Acacia Senegal, Cajanus cajan*	Heat	Retain the stability of metabolic actions	Enzyme production like chitinase, esterase, and glucanase	Kumar et al. (2010a), Räsänen et al. (2001)
Brevibacillus brevis	*Cotton crop*		Development and growth of plants	Production of ARA, IAA and ammonia	Nehra et al. (2016)
B. subtilis, B. megaterium, B. thuringiensis	*Triticum aestivum L, Cicer arietinum*	pH, drought, temperature	Increases in the relative water content (RWC), accumulation of biomolecules like sugar, proteins	Phytohormone biosynthesis	Khan et al. (2019)
PGPR and AMF	*Avena sativa*	Hydrocarbon pollution and saline-alkali soil	Improvement in the soil quality and removal of petroleum hydrocarbon	Boost the activities of urease, dehydrogenase, sucrose	Xun et al. (2015)

proline and lowered the risk of oxidative damage to triglyceride in *Zea mays* in drought conditions. In this microbial consortium, *B. thuringiensis* provide the nutrients to the plant, and AMF improves the stress tolerance (Armada et al. 2015). Inoculation of microbial consortium combination of *Anabaena* sp. with *Providencia* sp. and *Anabaena* with *Azotobacter* in maize hybrids has reported to evoked defence response of plant and increases the zinc mobilization (Prasanna et al. 2015).

15.4 Impact on Soil Microorganism

The physiochemical, functional, and structural properties of the soil and soil microorganisms are affected by the uses of biofertilizers (Javoreková et al. 2015). The application of PGPR-based biofertilizer has different effects like some may enhance the growth and others may inhibit the growth, while few of them has neutral or no effect on the microbial growth (Castro-Sowinski et al. 2007).

According to Javoreková et al. (2015) and Rastogi and Sani (2011), the microbial shifts can be evaluated by using the techniques such as terminal restriction fragment length polymorphism (t-RFLP), denaturing gradient gel electrophoresis (DGGE), the community level physiological profiling (CLPP), amplified ribosomal DNA restriction analysis (ARDRA), and single-strand conformation polymorphism (SSCP), with the usage of BIOLOG® plates.

Trabelsi et al. (2011, 2012) have used t-RFLP techniques to demonstrate the application of *rhizobium gallicum8a3*, and *Sinorhizobium meliloti 4H41* has influenced the diversity of *Actinobacteria*, *γ- and α-proteobacteria*, and *Firmicutes*.

Application of the co-culture of *Azospirillum brasilense (40 and 42 M)* strains has altered the community level physiological profiling (CLPP) of microbes related to rice crop (de Salamone et al. 2010). Similarly, the application of *Rhizobium leguminosarum bv. viciae* has also affected the CLPP-associated microorganisms with fababean crop (Siczek and Lipiec 2016). Through SSCP techniques, it has been determined that the application of *Sinorhizobium meliloti L33* strain has increased the number of α-proteobacteria and reduced the number of γ-proteobacteria in the rhizospheric region of *Medicago sativa* plant (Wang et al. 2018). The *Stenotrophomonas acidaminiphila BJ1* is the probiotic strain which has been reported to improve the bacterial growth in the *Vicia faba* rhizosphere polluted with chlorothalonil (Zhang et al. 2017).

15.5 Regulatory Issues of Biofertilizers

Commercialization of the first biofertilizer product was done by the Nobbe and Hiltner in the year 1895 with the Rhizobia-based products under the trade name "Nitragin." In India, the first rhizobium-based biofertilizer was commercialized by

N.V. Joshi for the growth of legume crops (García-Fraile et al. 2015). Through the 95 years of plan, the Agricultural Ministry started promotion and vulgarization of biofertilizer production, designing of standard protocols for different types of biofertilizer, providing hands on training, and applications (Ghosh 2004). The Central and State Government initiated different propagandas to encourage agricultural practitioners to shift from chemical to biological fertilizers and increase the biofertilizer production by giving subsidies and grants at various levels.

The most dominant players in the microbial fertilizers market are Novozymes A/S (Denmark), Camson Bio Technologies Limited (India), Gujarat State Fertilizers and Chemicals Limited (India), Lallemand Inc. (Canada), and Rhizobacter Argentina S.A. (Argentina). In Asian countries, the government subsidies and policies are strongly promoting the biofertilizer market and targeted for green and sustainable agricultural practices. For the development and production of microbial fertilizers and biopesticides, nearly US $1.5 billion has been consumed (García-Fraile et al. 2015). From the last decades then, farmers have shifted from chemical to organic agricultural. In India recently, the demand of production and utilization of biofertilizers has increased, and about 100 Indian private and public firms are established for the production of biofertilizers (García-Fraile et al. 2015; Pindi and Satyanarayana 2012). The average consumption rate of biofertilizers in country is high in comparison to its production rate. The highest production proportion is by Agro Industries Corporations followed by national Biofertilizers Development Centres, State Agricultural Departments, Private Sectors, and State Agricultural Universities (Mazid and Khan 2014).

15.6 Regulatory Issues of Microbial Consortium

The developed nations follow the strict protocols and regulations for the application of the microbial consortium. Before commercializing the product, the foremost step is the successful registration in which the product should meet the specific requirements as mentioned in the guidelines. Prior to the registration, the microbial consortium formulation must have a suitable carrier like biochar or alginate which allows the microbial cells to attach to the seeds during the sowing (Bashan 2016). While in the liquid formulation, microbial consortiums are sprinkled in the seed furrows or can spray on the seeds before sowing.

The microbial viability, biological activity, and survival of the microbial strains can be ensured by the product lifespan and storage. There should be clear knowledge on chronic and the acute applications of the microbial consortium. For example, in acute applications, the microbial consortium is used for a limited time; it can focus on the particular crop development stage and during the abiotic stress conditions. Unlikely in the chronic applications, the microbial consortium is used at the regular interval of spraying or slow release through seed treatment method (Backer et al. 2018). The regulation of microbial consortium is not clear in the American and European countries. Therefore, it is necessary to unified standard protocol, regulatory, and characterization of the microbial consortium across the world and most importantly in Asian and African countries as they have high potential for agricultural practices and large population of uneducated people which are employed for agroactivities.

15.7 Global Biofertilizer Market

The first biofertilizer registered for crop inoculation more than 100 years is *Rhizobium* sp.-based biofertilizer known as Nitragin which is currently available in the market (O'Callaghan 2016). It has been reported that in the available fertilizers in the market, only 5% microbial-based fertilizers are registered for agricultural practices (Verma et al. 2019). Table 15.1 presents the list of some commercial available biofertilizers around the globe. Owen et al. (2015) reported that rhizobial biofertilizers are the most demanded biofertilizer of approximately 80% in the market, while the phosphate-solubilizing biofertilizers and mycorrhizal fungal-based biofertilizers constitute approximately 15% and 7%, respectively, of the total market demand.

It has been expected in the forecast period of 2018–2023 that the global biofertilizers market demand will be reached approximately 2304 million till 2023 through the cumulative annual growth rate of 10%. The biofertilizer commercialization based on the geographical region has divided into Africa, Europe, North America, Middle East, Asia-Pacific, and Latin America. The Asia Pacific biofertilizers are the rapidly growing biofertilizers in the market as countries like India and China have the large area population along with the increasing economics (Table 15.2).

15.8 Global Efforts on Sustainable Agropractices

Across the globe, the increasing environmental issues have led to the loss in agricultural productivity; therefore it demands to develop the global action to overcome these losses. The International Code of Conduct on Pesticides Management was released in 2014 by the combined approval of the World Health Organization and Food and Agricultural Organizations to collect and document the number of deaths from the agrochemical users globally. The death rate due to the usage of chemical fertilizers by the agropracticers is continuously increasing in India despite of following the instructions and protocol strictly and also because of the ill-informed practitioners. The chemical fertilizers available in the market are generally classified as Class 1 chemicals. The government targeted to increase the economic status of the nations by focusing to improve the agricultural practices, but the situation has worsened due to the irregular campaigning in the search to regenerate the agricultural sector. According to the available data, Nigeria produced

Table 15.2 List of the few microbial biofertilizers available in the market across the globe

Product	Organismal consortium	Company	Country
Amnite A100®	*Azotobacter, Bacillus, Rhizobium, Chaetomium, Pseudomonas*	Cleveland biotech	United Kingdom
Bioativo®	*PGPR consortia, organic matter*	Embrafos Ltd.	Brazil
Bactofil A10®	*A. brasilense, B. megaterium, P. fluorescens, A. vinelandii*	Agro bio Hungary kft	Hungary
Biozink®, biomix®, biodine®	*Azotobacter, phosphobacteria, P. fluorescens*	Biomax	India
Life®	*PGPR consortia*		
Calosphere		Camson Bio Technologies Ltd.	
Symbiom-N	*Rhizobium, Acetobacter, Azospirillum, Azotobacter*	T. Stanes and Company Ltd.	
Bio super	*Cellulomonas, Bacillus, Pseudomonas, Rhodococcus*	SKS Bioproducts Pvt Ltd.	
Ceres®	*P. fluorescens*	Biovitis	France
Galtrol®	*Agrobacterium radiobacter strain 84*	AgBioChem	USA
BioYield	*B. amyloliquefaciens, B. subtilis*	Gustafson, Inc., Dallas	
TagTeam®	*Penicillium bilaiae, Rhizobium*	Novozymes	
Hyper coating seeds®	*Legume seed and rhizobium*	Tokachi Federation of Agriculture Cooperatives (TFAC)	Japan
Inomix® biostimulant	*B. subtilis, B. polymyxa*	LAB (Labiotech)	Spain
VitaSoil®	*PGPR consortia*	Symborg	
Nodulator®	*Bradyrhizobium japonicum*	Lallen and plant care BASF Inc.	Canada

about 25% of pesticides with 99% of death due to the use of pesticides in the developing nations (Ojo 2016). This is due to the insufficient education, regarding the use of toxic and cheaper chemicals, careless handling, and unsafe protocol. Similarly, the Taihu Lake in China has been contaminated due to the runoff of fertilizers, pesticides, and herbicides from the agricultural fields. In spite of restriction on DDT and HCH by the Chinese government, still the traces of these toxic chemicals are traced in the sediments (Feng et al. 2003). This is causing harmful effect on the health of the humans and environment and disturbed the biodiversity structures. Thus, it triggers the WHO and FOA to prohibit the use of chemical and hazardous products in the agricultural sectors and also aware the people about the microbial-based fertilizers for the sustainable agricultural practices. The utilization of the microbial fertilizers is considered the best solution to overcome the environmental issues like eutrophication and soil contamination with agrochemical fertilizers.

15.9 Prospects and Challenges of Biofertilizer Application

Microbial fertilizers are considered as the potent source of nutrients to the plants and have achieved global recommendation and acceptances for its usage in the sustainable agricultural productions. Its applications are well noticeable in the European, Asian, and American countries, while its applicability is not completely established in the African countries. This is due to the shortage of proper infrastructure, awareness, skilled manpower, and regulatory protocols. These factors have created restrictions in the sustainable agropractices; therefore the advantages of biofertilizer usage like nitrogen fixation, nutrient uptake, enhancing the crop yield and affordability are not achieved.

15.10 Conclusions

Biofertilizer and microbial consortium are considered potent tools in sustainable agricultural practices as they are a renewable, supplementary, and environmentally friendly nutrients source for plants. They are an essential part in the integrated plant nutrient system as they convert the unusable form of beneficial soil nutrients to become usable without causing harmful effects on the natural ecosystem (Alley and Vanlauwe 2009). Microbial fertilizers are a vital element in improving crop productivity and soil fertility and also increasing the growth of crops during the abiotic stress in extreme environments. Development and application of microbial consortium signify the importance of microbial inoculants in the upcoming years. Despite a large number of PGPR microbes are known for their growth-promoting action, very few are designed to biofertilizers or microbial consortium. Thus, the development of new techniques is required to expand the applications of microbial fertilizers and establish sustainable agriculture practices.

References

Ahemad M, Kibret M (2014) Mechanisms and applications of plant growth promoting rhizobacteria: current perspective. J King Saud Univ Sci 26(1):1–20

Alley MM, Vanlauwe B (2009) The role of fertilizers in integrated plant nutrient management. Paris

Alori ET, Glick BR, Babalola OO (2017) Microbial phosphorus solubilization and its potential for use in sustainable agriculture. Front Microbiol 8:971

Anwar S, Ali B, Sajid I (2016) Screening of rhizospheric actinomycetes for various in-vitro and in-vivo plant growth promoting (PGP) traits and for agroactive compounds. Front Microbiol 7:1334

Armada E, Azcón R, López-Castillo OM, Calvo-Polanco M, Ruiz-Lozano JM (2015) Autochthonous arbuscular mycorrhizal fungi and Bacillus thuringiensis from a degraded Mediterranean area can be used to improve physiological traits and performance of a plant of agronomic interest under drought conditions. Plant Physiol Biochem 90:64–74

Asif M, Mughal AH, Ajaz Malik M et al (2018) Application of different strains of biofertilizers for raising quality forest nursery tree. Int J Curr Microbiol App Sci 7(10):3680–3686

Atta MMM, Abdel-Lattif HM, Hamza M (2018) Soil inoculation by *Azospirillum* affects protein and carbohydrate of maize grain under nitrogen deficiency. J Adv Biol Biotechnol 19(1):1–14

Backer R, Rokem JS, Ilangumaran G, Lamont J, Praslickova D, Ricci E, Subramanian S, Smith DL (2018) Plant growth-promoting rhizobacteria: context, mechanisms of action, and roadmap to commercialization of biostimulants for sustainable agriculture. Front Plant Sci 871:1473

Barea JM (2015) Future challenges and perspectives for applying microbial biotechnology in sustainable agriculture based on a better understanding of plant-microbiome interactions. J Soil Sci Plant Nutr 15(2):261–282

Barnawal D, Bharti N, Maji D, Chanotiya CS, Kalra A (2014) ACC deaminase-containing *Arthrobacter protophormiae* induces NaCl stress tolerance through reduced ACC oxidase activity and ethylene production resulting in improved nodulation and mycorrhization in *Pisum sativum*. J Plant Physiol 171(11):884–894

Barriuso J (2015) Quorum sensing mechanisms in fungi. AIMS Microbiol 1(1):37–47

Basak BB, Biswas DR (2009) Influence of potassium solubilizing microorganism (*Bacillus mucilaginosus*) and waste mica on potassium uptake dynamics by Sudan grass (*Sorghum vulgare Pers.*) grown under two Alfisols. Plant Soil 317(1–2):235–255

Bashan N (2016) Inoculant formulations are essential for successful inoculation with plant growth-promoting bacteria and business opportunities. Indian Phytopathol 69:739–743

Bashan Y, de Bashan LE, Prabhu SR, Hernandez JP (2014) Advances in plant growth-promoting bacterial inoculant technology: formulations and practical perspectives (1998-2013). Plant Soil 378(1–2):1–33

Bashir Z, Zargar MY, Mohiddin FA, Kousar S, Husain M, Rasool F (2017) Phosphorus solubilizing microorganisms: mechanism and diversity. Int J Chem Stud 5:666–673

Beneduzi A, Ambrosini A, Passaglia LMP (2012) Plant growth-promoting rhizobacteria (PGPR): their potential as antagonists and biocontrol agents. Genet Mol Biol 35(4):1044–1051

Bennett JW, Hung R, Lee S, Padhi S (2012) Fungal and bacterial volatile organic compounds: an overview and their role as ecological signaling agents. In: Fungal associations, 2nd edn. Springer, Berlin, Heidelberg, pp 373–393

Bergelson J, Mittelstrass J, Horton MW (2019) Characterizing both bacteria and fungi improves understanding of the *Arabidopsis* root microbiome. Sci Rep 9(1):1–11

Berrendero E, Valiente EF, Perona E, Gómez CL, Loza V, Munõz-Martín MÁ, Mateo P (2016) Nitrogen fixation in a non-heterocystous cyanobacterial mat from a mountain river. Sci Rep 6 (1):30920

Bertrand S, Schumpp O, Bohni N, Bujard A, Azzollini A, Monod M, Gindro K, Wolfender JL (2013) Detection of metabolite induction in fungal co-cultures on solid media by high-throughput differential ultra-high pressure liquid chromatography-time-of-flight mass spectrometry fingerprinting. J Chromatogr A 1292:219–228

Bhatia SK, Yi DH, Kim YH, Kim HJ, Seo HM, Lee JH, Kim JH, Jeon JM, Jang KS, Kim YG, Yang YH (2015) Development of semi-synthetic microbial consortia of *Streptomyces coelicolor* for increased production of biodiesel (fatty acid methyl esters). Fuel 159:189–196

Bhattacharjee R, Dey U (2014) Biofertilizer, a way towards organic agriculture: a review. Afr J Microbiol Res 8(24):2332–2342

Bhattacharyya PN, Jha DK (2012) Plant growth-promoting rhizobacteria (PGPR): emergence in agriculture. World J Microbiol Biotechnol 28(4):1327–1350

Bikash Bag P, Panda P, Paramanik B, Mahato B, Choudhury A, Banga Krishi Viswavidyalaya U, Behar C, Bengal W, Dinajpur D, Vigyan Kendra K, Krishi Vigan Kendra K (2017) Atmospheric nitrogen fixing capacity of *Azotobacter* isolate from Cooch Behar and Jalpaiguri districts soil of West Bengal. Ind Int J Curr Microbiol App Sci 6(3):1775–1788

Bradáčová K, Florea A, Bar-Tal A, Minz D, Yermiyahu U, Shawahna R, Kraut-Cohen J, Zolti A, Erel R, Dietel K, Weinmann M, Zimmermann B, Berger N, Ludewig U, Neumann G, Poşta G

(2019) Microbial consortia versus single-strain inoculants: an advantage in PGPM-assisted tomato production. Agronomy 9(2):105

Brahmaprakash GP, Sahu PK, Lavanya G, Nair SS, Gangaraddi VK, Gupta A (2017) Microbial functions of the rhizosphere. In: Plant-microbe interactions in agro-ecological perspectives. Springer, pp 177–210

Bramhachari PV, Nagaraju GP, Kariali E (2018) Current perspectives on rhizobacterial-EPS interactions in alleviation of stress responses: novel strategies for sustainable agricultural productivity. In: Role of rhizospheric microbes in soil: stress management and agricultural sustainability, vol 1. Springer, Singapore, pp 33–55

Brilli F, Loreto F, Baccelli I (2019) Exploiting plant volatile organic compounds (VOCS) in agriculture to improve sustainable defense strategies and productivity of crops. Front Plant Sci 10:1–8

Bulgarelli D, Rott M, Schlaeppi K, Ver Loren van Themaat E, Ahmadinejad N, Assenza F, Rauf P, Huettel B, Reinhardt R, Schmelzer E, Peplies J, Gloeckner FO, Amann R, Eickhorst T, Schulze-Lefert P (2012) Revealing structure and assembly cues for *Arabidopsis* root-inhabiting bacterial microbiota. Nature 488(7409):91–95

Carmen B, Roberto D (2011) Soil bacteria support and protect plants against abiotic stresses. In: Abiotic stress in plants-mechanisms and daptations, Italy, pp 143–170

Castro-Sowinski S, Herschkovitz Y, Okon Y, Jurkevitch E (2007) Effects of inoculation with plant growth-promoting rhizobacteria on resident rhizosphere microorganisms. FEMS Microbiol Lett 276(1):1–11

Chandra D, Pallavi, Barh A, Sharma IP (2018) Plant growth promoting bacteria: a gateway to sustainable agriculture. In: Bhatt P, Sharma A (eds) Microbial biotechnology in environmental monitoring and cleanup. IGI Global Publication, pp 318–338

Chen JH (2006) The combined use of chemical and organic fertilizers and/or biofertilizer for crop growth and soil fertility. International Workshop on Sustained Management of the Soil-Rhizosphere System for Efficient Crop Production and Fertilizer Use 16(20): 1–11

Chibuzor NE, Chuks KO, Emmanuel AE, Paul IO, Simeon CE, Uchenna JO (2018) Chromium (III) and its effects on soil microbial activities and phytoremediation potentials of *Arachis hypogea* and *Vigna unguiculata*. Afr J Biotechnol 17(38):1207–1214

Chuks Kenneth O, Chibuzor Nwadibe E, Uchenna Kalu A, Victor Unah U (2019) Plant growth promoting rhizobacteria (PGPR): a novel agent for sustainable food production. Am J Agricult Biol Sci 14:35–54

Clinton A, Rumbaugh KP (2016) Interspecies and interkingdom signaling via quorum signals. Israel J Chem 56(5):265–272

Company S, Clément C, Sessitsch A (2010) Plant growth-promoting bacteria in the rhizo- and endosphere of plants: their role, colonization, mechanisms involved and prospects for utilization. Soil Biol Biochem 42(5):669–678

Dardanelli MS, Fernández de Córdoba FJ, Espuny MR, Rodríguez Carvajal MA, Soria Díaz ME, Gil Serrano AM, Okon Y, Megías M (2008) Effect of *Azospirillum brasilense* coinoculated with *Rhizobium* on *Phaseolus vulgaris* flavonoids and Nod factor production under salt stress. Soil Biol Biochem 40(11):2713–2721

Das I, Pradhan M (2016) Potassium-solubilizing microorganisms and their role in enhancing soil fertility and health. In: Potassium solubilizing microorganisms for sustainable agriculture. Springer, New Delhi, pp 281–291

de Salamone IEG, Di Salvo LP, Ortega JSE, Sorte PMFB, Urquiaga S, Teixeira KRS (2010) Field response of rice paddy crop to *Azospirillum* inoculation: physiology of rhizosphere bacterial communities and the genetic diversity of endophytic bacteria in different parts of the plants. Plant Soil 336(1):351–362

Deaker R, Roughley RJ, Kennedy IR (2004) Legume seed inoculation technology - a review. Soil Biol Biochem 36(8):1275–1288

Deveau A, Bonito G, Uehling J, Paoletti M, Becker M, Bindschedler S, Hacquard S, Hervé V, Labbé J, Lastovetsky OA, Mieszkin S, Millet LJ, Vajna B, Junier P, Bonfante P, Krom BP,

Olsson S, van Elsas JD, Wick LY (2018) Bacterial-fungal interactions: ecology, mechanisms and challenges. FEMS Microbiol Rev 42(3):335–352

Diagne N, Arumugam K, Ngom M, Nambiar-Veetil M, Franche C, Narayanan KK, Laplaze L (2013) Use of frankia and actinorhizal plants for degraded lands reclamation. Biomed Res Int 2013:948258

Elias F, Woyessa D, Muleta D (2016) Phosphate solubilization potential of rhizosphere fungi isolated from plants in Jimma zone, Southwest Ethiopia. Int J Microbiol. https://doi.org/10.1155/2016/5472601

El-Ramady H, El-Ghamry A, Mosa A, Alshaal T (2018) Nanofertilizers vs. biofertilizers: new insights. Environ Biodiver Soil Secur 2(1):40–50

Etesami H, Emami S, Alikhani HA (2017) Potassium solubilizing bacteria (KSB): mechanisms, promotion of plant growth, and future prospects - a review. J Soil Sci Plant Nutr 17(4):897–911

Fahad S, Hussain S, Bano A, Saud S, Hassan S, Shan D, Khan FA, Khan F, Chen Y, Wu C, Tabassum MA, Chun MX, Afzal M, Jan A, Jan MT, Huang J (2015) Potential role of phytohormones and plant growth-promoting rhizobacteria in abiotic stresses: consequences for changing environment. Environ Sci Pollut Res 22(7):4907–4921

Feng K, Yu BY, Ge DM, Wong MH, Wang XC, Cao ZH (2003) Organo-chlorine pesticide (DDT and HCH) residues in the Taihu Lake Region and its movement in soil-water system I. Field survey of DDT and HCH residues in ecosystem of the region. Chemosphere 50(6):683–687

Figueiredo Mdo VB, Santo Mergulhão ACdo E, Sobral JK, Junior Mde AL, de Araújo ASF (2013) Biological nitrogen fixation: importance, associated diversity, and estimates. In: Plant microbe symbiosis: fundamentals and advances. Springer, pp 267–289

Fleitas Martínez O, Rigueiras PO, Pires Á d S, Porto WF, Silva ON, de la Fuente-Nunez C, Franco OL (2018) Interference with quorum-sensing signal biosynthesis as a promising therapeutic strategy against multidrug-resistant pathogens. Front Cell Infect Microbiol 8:444

Flury P, Vesga P, Péchy-Tarr M, Aellen N, Dennert F, Hofer N, Kupferschmied KP, Kupferschmied P, Metla Z, Ma Z, Siegfried S, de Weert S, Bloemberg G, Höfte M, Keel CJ, Maurhofer M (2017) Antimicrobial and insecticidal: cyclic lipopeptides and hydrogen cyanide produced by plant-beneficial *Pseudomonas* strains CHA0, CMR12a, and PCL1391 contribute to insect killing. Front Microbiol 8:100

Franco JA, Bañón S, Vicente MJ, Miralles J, Martínez-Sánchez JJ (2011) Root development in horticultural plants grown under abiotic stress conditions - a review. J Hort Sci Biotechnol 86 (6):543–556

Fuentes-Ramirez LE, Caballero-Mellado J (2006) Bacterial biofertilizers. In: PGPR: biocontrol and biofertilization. Springer, Dordrecht, pp 143–172

Fukami J, Ollero FJ, Megías M, Hungria M (2017) Phytohormones and induction of plant-stress tolerance and defense genes by seed and foliar inoculation with *Azospirillum brasilense* cells and metabolites promote maize growth. AMB Express 7(1):1–13

Fukami J, Cerezini P, Hungria M (2018) *Azospirillum*: benefits that go far beyond biological nitrogen fixation. AMB Express 8(1):1–12

Galloway JN, Aber JD, Erisman JW, Seitzinger SP, Howarth RW, Cowling EB, Cosby BJ (2003) The nitrogen cascade. Bio Sci 53(4):341–356

Gangwar M, Saini P, Nikhanj P, Kaur S (2017) Plant growth-promoting microbes (pgpm) as potential microbial bio-agents for eco-friendly agriculture. Springer, Singapore, pp 37–55

García-Fraile P, Menéndez E, Rivas R (2015) Role of bacterial biofertilizers in agriculture and forestry. AIMS Bioeng 2(3):183–205

Ghosh N (2004) Promoting biofertilisers in Indian agriculture. Econ Polit Wkly 39(52):5617–5625

Glick BR (2014) Bacteria with ACC deaminase can promote plant growth and help to feed the world. Microbiol Res 169(1):30–39

Glick BR (2020) Introduction to plant growth-promoting bacteria. In: Beneficial plant-bacterial interactions. Springer, pp 1–37

Gopalakrishnan S, Sathya A, Vijayabharathi R, Varshney RK, Gowda CLL, Krishnamurthy L (2015) Plant growth promoting rhizobia: challenges and opportunities. 3 Biotech 5(4):355–377

Gothandapani S, Sekar S, Padaria JC (2017) *Azotobacter chroococcum*: utilization and potential use for agricultural crop production: an overview. Int J Adv Res Biol Sci 4(3):35–42

Gouda S, Kerry RG, Das G, Paramithiotis S, Shin HS, Patra JK (2018) Revitalization of plant growth promoting rhizobacteria for sustainable development in agriculture. Microbiol Res 206:131–140

Grover M, Ali SZ, Sandhya V, Rasul A, Venkateswarlu B (2011) Role of microorganisms in adaptation of agriculture crops to abiotic stresses. World J Microbiol Biotechnol 27 (5):1231–1240

Guo J, Dong X, Han G, Wang B (2019) Salt-enhanced reproductive development of Suaeda salsa l. coincided with ion transporter gene upregulation in flowers and increased pollen kopenspisupspi+closespisupspi content. Front Plant Sci 10:333

Gupta A, Annapurna K, Jaitley AK (2016) Screening of osmo protectants for liquid formulation of *Azospirillum* bio-inoculant. Int J Serv Technol Manag 5(5):258–267

Hamilton CE, Bever JD, Labbé J, Yang X, Yin H (2016) Mitigating climate change through managing constructed-microbial communities in agriculture. Agric Ecosyst Environ 216:304–308

Hansel CM, Fendorf S, Jardine PM, Francis CA (2008) Changes in bacterial and archaeal community structure and functional diversity along a geochemically variable soil profile. Appl Environ Microbiol 74(5):1620–1633

Hocher V, Auguy F, Bogusz D, Doumas P, Franche C, Gherbi H, Laplaze L, Obertello M, Svistoonoff S (2009) Les symbioses actinorhiziennes fixatrices d'azote: un exemple d'adaptation aux contraintes abiotiques du sol. Cahiers Agricult 18(6):498–505

Htwe AZ, Moh SM, Soe KM, Moe K, Yamakawa T (2019) Effects of biofertilizer produced from *Bradyrhizobium* and *Streptomyces griseoflavus* on plant growth, nodulation, nitrogen fixation, nutrient uptake, and seed yield of mung bean, cowpea, and soybean. Agronomy 9(2):77

Hung R, Lee S, Bennett JW (2015) Fungal volatile organic compounds and their role in ecosystems. Appl Microbiol Biotechnol 99(8):3395–3405

Hungria M, Nogueira MA, Araujo RS (2015) Soybean seed co-inoculation with *Bradyrhizobium spp*. and *Azospirillum brasilense*: a new biotechnological tool to improve yield and sustainability. Embrapa Soja-Artigo Em Periódico Indexado (ALICE). Am J Plant Sci 6:811–817

Issa AA, Abd-Alla MH, Ohyama T (2014) Nitrogen fixing cyanobacteria: future prospect. Adv Biol Ecol Nitrog Fixat 2:24–48

Itelima JU, Bang WJ, Onyimba IA, Oj E (2018) A review: biofertilizer; a key player in enhancing soil fertility and crop productivity. J Microbiol Biotechnol Rep 2(1):22–28

Iwuagwu M, Ks C, Uka U, Amandianeze MC (2013) Effects of biofertilizers on the growth of *Zea mays* L. Asian J Microbiol Biotechnol Environ Sci 15:235–240

Jain A, Singh A, Singh BN, Singh S, Upadhyay RS, Sarma BK, Singh HB (2013) Biotic stress management in agricultural crops using microbial consortium. In: Bacteria in agrobiology: disease management. Springer, Berlin, pp 427–448

Jambon I, Thijs S, Weyens N, Vangronsveld J (2018) Harnessing plant-bacteria-fungi interactions to improve plant growth and degradation of organic pollutants. J Plant Interact 13(1):119–130

Jangid MK, Khan IM, Singh S (2012) Constraints faced by the organic and conventional farmers in adoption of organic farming practices. Ind Res J Exten Educ 2:28–32

Javoreková S, Maková J, Medo J, Kovácsová S, Charousová I, Horák J (2015) Effect of bio-fertilizers application on microbial diversity and physiological profiling of microorganisms in arable soil. Eurasian J Soil Sci 4(1):54

Jha Y, Subramanian RB (2013) Paddy plants inoculated with PGPR show better growth physiology and nutrient content under saline conditions. Chilean J Agric Res 73(3):213–219

Jiang Q, Chen J, Yang C, Yin Y, Yao K, Song D (2019) Quorum sensing: a prospective therapeutic target for bacterial diseases. Biomed Res Int 2019:2015978

Khan N, Bano A, Babar MA (2019) Metabolic and physiological changes induced by plant growth regulators and plant growth promoting rhizobacteria and their impact on drought tolerance in *Cicer arietinum L.* PLoS One 14(3):e0213040

Kumar Deshwal V, Kumar P (2013) Production of plant growth promoting substance by *Pseudomonads*. J Acad Indust Res 2:221–225

Kumar H, Bajpai VK, Dubey RC, Maheshwari DK, Kang SC (2010a) Wilt disease management and enhancement of growth and yield of *Cajanus cajan (L)* var. Manak by bacterial combinations amended with chemical fertilizer. Crop Prot 29(6):591–598

Kumar K, Mella-Herrera RA, Golden JW (2010b) *Cyanobacterial* heterocysts. Cold Spring Harb Perspect Biol 2:a000315

Larson C (2013) Losing arable land, China faces stark choice: adapt or go hungry. Am Assoc Advan Sci 339(6120):644–645

Li J, Meng B, Chai H, Yang X, Song W, Li S, Lu A, Zhang T, Sun W (2019) Arbuscular mycorrhizal fungi alleviate drought stress in C3 (*Leymus chinensis*) and C4 (*Hemarthria altissima*) grasses via altering antioxidant enzyme activities and photosynthesis. Front Plant Sci 10:499

Liste HH (2003) Soil–plant–microbe interactions and their implications for agriculture and environment. Habilitation thesis, Humboldt University, Berlin

Liu Z, Rong Q, Zhou W, Liang G (2017) Effects of inorganic and organic amendment on soil chemical properties, enzyme activities, microbial community and soil quality in yellow clayey soil. PLoS One 12(3):e0172767

Lladó S, López-Mondéjar R, Baldrian P (2017) Forest soil bacteria: diversity, involvement in ecosystem processes, and response to global change. Microbiol Mol Biol Rev 81(2):e00063-16

Lundberg DS, Lebeis SL, Paredes SH, Yourstone S, Gehring J, Malfatti S, Tremblay J, Engelbrektson A, Kunin V, Del Rio TG, Edgar RC, Eickhorst T, Ley RE, Hugenholtz P, Tringe SG, Dangl JL (2012) Defining the core *Arabidopsis thaliana* root microbiome. Nature 488 (7409):86–90

Madigan MT, Martinko JM, Dunlap PV, Clark DP (2009) Brock biology of microorganisms. Edisi 12

Mahanty T, Bhattacharjee S, Goswami M, Bhattacharyya P, Das B, Ghosh A, Tribedi P (2017) Biofertilizers: a potential approach for sustainable agriculture development. Environ Sci Pollut Res 24(4):3315–3335

Mahato S, Kafle A (2018) Comparative study of *Azotobacter* with or without other fertilizers on growth and yield of wheat in Western hills of Nepal. Ann Agrar Sci 16(3):250–256

Mahmood A, Turgay OC, Farooq M, Hayat R (2016) Seed biopriming with plant growth promoting rhizobacteria: a review. FEMS Microbiol Ecol 92(8):112

Majeed A, Kaleem Abbasi M, Hameed S, Imran A, Rahim N (2015) Isolation and characterization of plant growth-promoting rhizobacteria from wheat rhizosphere and their effect on plant growth promotion. Front Microbiol 6:198

Malusà E, Ciesielska J (2014) Biofertilisers: a resource for sustainable plant nutrition. Fertiliser Technol 1(1):282–319

Malusá E, Vassilev N (2014) A contribution to set a legal framework for biofertilisers. Appl Microbiol Biotechnol 98(15):6599–6607

Mapelli F, Marasco R, Rolli E, Barbato M, Cherif H, Guesmi A, Ouzari I, Daffonchio D, Borin S (2013) Potential for plant growth promotion of rhizobacteria associated with Salicornia growing in Tunisian hypersaline soils. Biomed Res Int 2013:248078

Mazid M, Khan TA (2014) Future of bio-fertilizers in Indian agriculture: an overview. Int J Agric Food Res 3(3):10–23

Meena VS, Kumar A, Meena RK (2016) Potassium-solubilizing microorganism in evergreen agriculture: an overview agroforestry and fodder production management view project. Springer, pp 1–20

Mehrvarz S, Chaichi MR, Alikhani HA (2008) Effect of phosphate solubilizing microorganisms and phosphorus chemical fertilizer on forage and grain quality of barely (*Hordeum vulgare L.*). Agric Environ Sci 3(6):822–828

Mishra J, Arora NK (2016) Bioformulations for plant growth promotion and combating phytopathogens: a sustainable approach. In: Bioformulations: for sustainable agriculture. Springer, pp 3–33

Mishra P, Dash D (2014) Rejuvenation of biofertilizer for sustainable agriculture and economic development. Consilience 11:41–61

Mishra D, Rajvir S, Mishra U, Kumar SS (2013) Role of bio-fertilizer in organic agriculture: a review. Res J Recent Sci 2:39–41

Mishra J, Bhimrao B, Arora N, Arora NK (2018) Bioformulations for plant growth promotion and combating phytopathogens: a sustainable approach development of bioformulation for sustainable agriculture view project phyto and rhizoremediation view project bioformulations for plant growth promotion and combating phytopathogens: a sustainable approach 1. Springer, pp 3–33

Mohammadi K, Sohrabi Y (2012) Bacterial biofertilizers for sustainable crop production: a review. ARPN J Agric Biol Sci 7(5):307–316

Mohod S, Lakhawat GP, Deshmukh SK, Ugwekar RP (2015) Production of liquid biofertilizers and its quality control. Int J Emerg Trend Eng Basic Sci 2(2):158–165

Mus F, Crook MB, Garcia K, Costas AG, Geddes BA, Kouri ED, Paramasivan P, Ryu MH, Oldroyd GED, Poole PS (2016) Symbiotic nitrogen fixation and the challenges to its extension to nonlegumes. Appl Environ Microbiol 82(13):3698–3710

Nadeem SM, Ahmad M, Zahir ZA, Javaid A, Ashraf M (2014) The role of mycorrhizae and plant growth promoting rhizobacteria (PGPR) in improving crop productivity under stressful environments. Biotechnol Adv 32(2):429–448

Nakkeeran S, Fernando WGD, Siddiqui ZA (2006) Plant growth promoting rhizobacteria formulations and its scope in commercialization for the management of pests and diseases. In: PGPR: biocontrol and biofertilization. Springer, Dordrecht, pp 257–296

Nath Yadav A, Ghosh Sachan S, Verma P, Kumar Saxena A (2016) Bioprospecting of plant growth promoting psychrotrophic Bacilli from the cold desert of north western Indian Himalayas. Indian J Exp Biol 54:142–150

Nehra V, Saharan BS, Choudhary M (2016) Evaluation of *Brevibacillus brevis* as a potential plant growth promoting rhizobacteria for cotton (*Gossypium hirsutum*) crop. Springerplus 5(1):1–10

Ngumbi E, Kloepper J (2016) Bacterial-mediated drought tolerance: current and future prospects. Appl Soil Ecol 105:109–125

Nuti M, Giovannetti G (2015) Borderline products between bio-fertilizers/bio-effectors and plant protectants: the role of microbial consortia. J Agric Sci Technol A 5:305–315

O'Callaghan M (2016) Microbial inoculation of seed for improved crop performance: issues and opportunities. Appl Microbiol Biotechnol 100(13):5729–5746

Odoh CK (2015) Effects of some heavy metals on soil bacteria, shoot growth and nodulation of cowpea (Vigna unguiculata) and groundnut (Arachis hypogea) grown in sandy loam soil. Research Thesis, Department of Microbiology, University of Nigeria, Nsukka, 1–9

Odoh CK (2017) Plant growth promoting rhizobacteria (PGPR): a bioprotectant bioinoculant for sustainable agrobiology. A review. Int J Adv Res Biol Sci 4(5):123–142

Ojo J (2016) Pesticides use and health in Nigeria. IFE J Sci 18(4):981–991

Owen D, Williams AP, Griffith GW, Withers PJA (2015) Use of commercial bio-inoculants to increase agricultural production through improved phosphrous acquisition. Appl Soil Ecol 86:41–54

Pal S, Singh HB, Farooqui A, Rakshit A (2015) Fungal biofertilizers in Indian agriculture: perception, demand and promotion. J Eco-Friendly Agric 10(2):101–113

Patel U, Sinha S (2011) *Rhizobia* species: a boon for "plant genetic engineering". Indian J Microbiol 51(4):521–527

Pathak D, Lone R, Koul KK (2017) Arbuscular mycorrhizal fungi (AMF) and plant growth-promoting rhizobacteria (PGPR) association in potato (*Solanum tuberosum L.*): a brief review. In: Probiotics and plant health. Springer, Singapore, pp 401–420

Pathak J, Rajneesh Maurya PK, Singh SP, Häder DP, Sinha RP (2018) Cyanobacterial farming for environment friendly sustainable agriculture practices: innovations and perspectives. Front Environ Sci 6:7

Patil HJ, Solanki MK (2016) Microbial inoculant: modern era of fertilizers and pesticides. In: Microbial inoculants in sustainable agricultural productivity: vol. 1: research perspectives. Springer, New Delhi, pp 319–343

Philippot L, Raaijmakers JM, Lemanceau P, Van Der Putten WH (2013) Going back to the roots: the microbial ecology of the rhizosphere. Nat Rev Microbiol 11(11):789–799

Pindi PK, Satyanarayana S (2012) Liquid microbial consortium-a potential tool for sustainable soil health. J Biofertil Biopestici 3:124

Prasanna R, Bidyarani N, Babu S, Hossain F, Shivay YS, Nain L (2015) Cyanobacterial inoculation elicits plant defense response and enhanced Zn mobilization in maize hybrids. Cogent Food Agric 1(1):998507

Rai AN, Singh AK, Syiem MB (2019) Plant growth-promoting abilities in cyanobacteria. In: Cyanobacteria. Elsevier, pp 459–476

Ranjan A, Mahalakshmi MR, Sridevi M (2013) Isolation and characterization of phosphate-solubilizing bacterial species from different crop fields of Salem, Tamil Nadu, India. Int J Nutrit Pharmacol Neurol Dis 3(1):29

Räsänen LA, Elväng AM, Jansson J, Lindström K (2001) Effect of heat stress on cell activity and cell morphology of the tropical *rhizobium*, *Sinorhizobium arboris*. FEMS Microbiol Ecol 34 (3):267–278

Rashid MI, Mujawar LH, Shahzad T, Almeelbi T, Ismail IMI, Oves M (2016) Bacteria and fungi can contribute to nutrients bioavailability and aggregate formation in degraded soils. Microbiol Res 183:26–41

Rastogi G, Sani RK (2011) Molecular techniques to assess microbial community structure, function, and dynamics in the environment. In: Microbes and microbial technology: agricultural and environmental applications. Springer, New York, pp 29–57

Rathod JP, Rathod P, Rathod DR, Gade RM (2018) Study of anabaena ambigua on growth parameters of *Coriandrum sativum* after seed and foliar spray treatment. Int J Curr Microbiol App Sci 7(12):25–32

Reddy PP (2014) Potential role of PGPR in agriculture. In: Plant growth promoting rhizobacteria for horticultural crop protection. Springer, New Delhi, pp 17–34

Roberts NJ, Morieri G, Kalsi G, Rose A, Stiller J, Edwards A, Xie F, Gresshoff PM, Oldroyd GED, Allan Downie J, Etzler ME (2013) Rhizobial and mycorrhizal symbioses in *Lotus japonicus* require Lectin Nucleotide Phosphohydrolase, which acts upstream of calcium signaling. Plant Physiol 161(1):556–567

Rotaru V (2015) Responses of acid phosphatase activity on the root surface and rhizospheric soil of soybean plants to phosphorus fertilization and rhizobacteria application under low water supply. Scient Papers Ser A Agron 58:295

Saharan BS, Nehra V (2011) Plant growth promoting rhizobacteria: a critical review. Life Sci Med Res 21(1):30

Sam K, Coulon F, Prpich G (2017) A multi-attribute methodology for the prioritisation of oil contaminated sites in the Niger Delta. Sci Total Environ 579:1323–1332

Saranraj P, Sivasakthivelan P (2013) *Azospirillum* and its formulations: a review. Int J Microbiol Res 4(3):275–287

Sayed WF (2011) Improving Casuarina growth and symbiosis with *Frankia* under different soil and environmental conditions-review. Folia Microbiol 56(1):1–9

Schellenberger J, Que R, Fleming RMT, Thiele I, Orth JD, Feist AM, Zielinski DC, Bordbar A, Lewis NE, Rahmanian S (2011) Quantitative prediction of cellular metabolism with constraint-based models: the COBRA Toolbox v2. 0. Nat Protoc 6(9):1290

Schirawski J, Perlin M (2018) Plant–microbe interaction 2017—the good, the bad and the diverse. Int J Mol Sci 19(5):1374

Schütz L, Gattinger A, Meier M, Müller A, Boller T, Mäder P, Mathimaran N (2018) Improving crop yield and nutrient use efficiency via biofertilization—a global meta-analysis. Front Plant Sci 8:2204

Sekar J, Raj R, Prabavathy VR (2016) Microbial consortial products for sustainable agriculture: commercialization and regulatory issues in India. In: Agriculturally important microorganisms: commercialization and regulatory requirements in Asia. Springer, Singapore, pp 107–132

Sethi SK, Adhikary SP (2012) Cost effective pilot scale production of biofertilizer using *Rhizobium* and *Azotobacter*. Afr J Biotechnol 11(70):13490–13493

Shaikh SS, Sayyed RZ (2015) Role of plant growth-promoting rhizobacteria and their formulation in biocontrol of plant diseases. In: Plant microbes symbiosis: applied facets. Springer, New Delhi, pp 337–351

Sharma K (2011) Inorganic phosphate solubilization by fungi isolated from agriculture soil. J Phytology 3(4):11–12

Sharma SB, Sayyed RZ, Trivedi MH, Gobi TA (2013) Phosphate solubilizing microbes: sustainable approach for managing phosphorus deficiency in agricultural soils. Springerplus 2(1):1–14

Sharon JA, Hathwaik LT, Glenn GM, Imam SH, Lee CC (2016) Isolation of efficient phosphate solubilizing bacteria capable of enhancing tomato plant growth. J Soil Sci Plant Nutr 16 (2):525–536

Shou W, Ram S, Vilar JMG (2007) Synthetic cooperation in engineered yeast populations. Proc Natl Acad Sci U S A 104(6):1877–1882

Shrimant Shridhar B (2012) Review: nitrogen fixing microorganisms. Int J Microbiol Res 3 (1):46–52

Siczek A, Lipiec J (2016) Impact of faba bean-seed rhizobial inoculation on microbial activity in the rhizosphere soil during growing season. Int J Mol Sci 17(5):784

Simarmata T, Turmuktini T, Fitriatin BN, Setiawati MR (2016) Application of bioameliorant and biofertilizers to increase the soil health and rice productivity. HAYATI J Biosci 23(4):181–184

Singh JS, Kumar A, Rai AN, Singh DP (2016) Cyanobacteria: a precious bio-resource in agriculture, ecosystem, and environmental sustainability. Front Microbiol 7:529

Smith DL, Praslickova D, Ilangumaran G (2015) Inter-organismal signaling and management of the phytomicrobiome. Front Plant Sci 6:722

Somers E, Vanderleyden J, Srinivasan M (2004) Rhizosphere bacterial signalling: a love parade beneath our feet. Crit Rev Microbiol 30(4):205–240

Stamenković S, Beškoski V, Karabegović I, Lazić M, Nikolić N (2018) Microbial fertilizers: a comprehensive review of current findings and future perspectives. Span J Agric Res 16(1):1–18

Stewart EJ (2012) Growing unculturable bacteria. J Bacteriol 194(16):4151–4160

Sun R, Guo X, Wang D, Chu H (2015) Effects of long-term application of chemical and organic fertilizers on the abundance of microbial communities involved in the nitrogen cycle. Appl Soil Ecol 95:171–178

Tairo EV, Ndakidemi PA (2013) Possible benefits of rhizobial inoculation and phosphorus supplementation on nutrition, growth and economic sustainability in grain legumes. Am J Res Commun 1(12):532–556

Talbi C, Sánchez C, Hidalgo-Garcia A, González EM, Arrese-Igor C, Girard L, Bedmar EJ, Delgado MJ (2012) Enhanced expression of *Rhizobium etli* cbb. J Exp Bot 63(14):5035–5043

Tamayo-Vélez Á, Osorio NW (2018) Soil fertility improvement by litter decomposition and inoculation with the fungus *Mortierella* sp. in avocado plantations of Colombia. Commun Soil Sci Plant Anal 49(2):139–147

Thijs S, Weyens N, Sillen W, Gkorezis P, Carleer R, Vangronsveld J (2014) Potential for plant growth promotion by a consortium of stress-tolerant 2,4-dinitrotoluene-degrading bacteria: isolation and characterization of a military soil. Microb Biotechnol 7(4):294–306

Thilakarathna MS, McElroy MS, Chapagain T, Papadopoulos YA, Raizada MN (2016) Belowground nitrogen transfer from legumes to non-legumes under managed herbaceous cropping systems. A review. Agron Sustain Develop 36(4):1–16

Thingujam I, Tiwari ON, Tiwari GL (2016) Screening and characterization of cyanobacterial species isolated from Loktak Lake, Manipur, India with emphasis on biofortification. Int J Adv Res Biol Sci 3(1):88–98

Thompson LR, Sanders JG, McDonald D, Amir A, Ladau J, Locey KJ, Prill RJ, Tripathi A, Gibbons SM, Ackermann G, Navas-Molina JA, Janssen S, Kopylova E, Vázquez-Baeza Y, González A, Morton JT, Mirarab S, Xu ZZ, Jiang L, Zhao H (2017) A communal catalogue reveals Earth's multiscale microbial diversity. Nature 551(7681):457–463

Trabelsi D, Mengoni A, Ben Ammar H, Mhamdi R (2011) Effect of on-field inoculation of *Phaseolus vulgaris* with rhizobia on soil bacterial communities. FEMS Microbiol Ecol 77 (1):211–222

Trabelsi D, Ben Ammar H, Mengoni A, Mhamdi R (2012) Appraisal of the crop-rotation effect of rhizobial inoculation on potato cropping systems in relation to soil bacterial communities. Soil Biol Biochem 54:1–6

Tringe SG, Von Mering C, Kobayashi A, Salamov AA, Chen K, Chang HW, Podar M, Short JM, Mathur EJ, Detter JC, Bork P, Hugenholtz P, Rubin EM (2005) Comparative metagenomics of microbial communities. Science 308(5721):554–557

Tyc O, Song C, Dickschat JS, Vos M, Garbeva P (2017) The ecological role of volatile and soluble secondary metabolites produced by soil bacteria. Trends Microbiol 25(4):280–292

Vejan P, Abdullah R, Khadiran T, Ismail S, Nasrulhaq Boyce A (2016) Role of plant growth promoting rhizobacteria in agricultural sustainability—a review. Molecules 21(5):573

Venturi V, Keel C (2016) Signaling in the rhizosphere. Trends Plant Sci 21(3):187–198

Verma JP, Yadav J, Tiwari KNL, Singh V (2010) Impact of plant growth promoting rhizobacteria on crop production. Int J Agric Res 5(11):954–983

Verma M, Mishra J, Arora NK (2019) Plant growth-promoting rhizobacteria: diversity and applications. In: Environmental biotechnology: for sustainable future. Springer, Singapore, pp 129–173

Vessey JK (2003) Plant growth promoting rhizobacteria as biofertilizers. Plant Soil 255(2):571–586

Vicente EJ, Dean DR (2017) Keeping the nitrogen-fixation dream alive. Proc Natl Acad Sci U S A 114(12):3009–3011

Vijayabharathi R, Sathya A, Gopalakrishnan S, Vijayabharathi R, Sathya A, Gopalakrishnan S (2016) A renaissance in plant growth-promoting and biocontrol agents by endophytes. In: Microbial inoculants in sustainable agricultural productivity. Springer, New Delhi, pp 37–60

Vikhe PS (2014) *Azotobacter* species as a natural plant hormone synthesizer. Res J Recent Sci 3:63–59

Vurukonda SSKP, Vardharajula S, Shrivastava M, SkZ A (2016) Enhancement of drought stress tolerance in crops by plant growth promoting rhizobacteria. Microbiol Res 184:13–24

Wagner SC (2011) Biological nitrogen fixation. Nat Educ Knowled 3:15

Wang J, Li Q, Xu S, Zhao W, Lei Y, Song C, Huang Z (2018) Traits-based integration of multi-species inoculants facilitates shifts of indigenous soil bacterial community. Front Microbiol 9:1692

Wani S, Chand S, Ali T (2013) Potential use of *Azotobacter chroococcum* in crop production: an overview. Curr Agric Res J 1(1):35–38

Wdowiak-Wróbel S, Marek-Kozaczuk M, Kalita M, Karaś M, Wójcik M, Małek W (2017) Diversity and plant growth promoting properties of rhizobia isolated from root nodules of *Ononis arvensis*. Int J Gen Mol Microbiol 110(8):1087–1103

Xiang W, Zhao L, Xu X, Qin Y, Yu G (2012) Mutual information flow between beneficial microorganisms and the roots of host plants determined the bio-functions of biofertilizers. Am J Plant Sci 3(8):1115–1120

Xun F, Xie B, Liu S, Guo C (2015) Effect of plant growth-promoting bacteria (PGPR) and arbuscular mycorrhizal fungi (AMF) inoculation on oats in saline-alkali soil contaminated by petroleum to enhance phytoremediation. Environ Sci Pollut Res 22(1):598–608

Yadav AN, Sachan SG, Verma P, Tyagi SP, Kaushik R, Saxena AK (2015a) Culturable diversity and functional annotation of psychrotrophic bacteria from cold desert of Leh Ladakh (India). World J Microbiol Biotechnol 31(1):95–108

Yadav AN, Sachan SG, Verma P, Saxena AK (2015b) Prospecting cold deserts of north western Himalayas for microbial diversity and plant growth promoting attributes. J Biosci Bioeng 119 (6):683–693

Zabbey N, Sam K, Onyebuchi AT (2017) Remediation of contaminated lands in the Niger Delta, Nigeria: prospects and challenges. Sci Total Environ 586:952–965

Zamioudis C, Pieterse CMJ (2012) Modulation of host immunity by beneficial microbes. Mol Plant-Microbe Interact 25(2):139–150

Zhang Q, Saleem M, Wang C (2017) Probiotic strain *Stenotrophomonas acidaminiphila BJ1* degrades and reduces chlorothalonil toxicity to soil enzymes, microbial communities and plant roots. AMB Exp 7(1):227

Zheng W, Zeng S, Bais H, LaManna JM, Hussey DS, Jacobson DL, Jin Y (2018) Plant growth-promoting rhizobacteria (PGPR) reduce evaporation and increase soil water retention. Water Resour Res 54(5):3673–3687

Zuroff TR, Curtis WR (2012) Developing symbiotic consortia for lignocellulosic biofuel production. Appl Microbiol Biotechnol 93(4):1423–1435

Chapter 16
Biofertilizers as Microbial Consortium for Sustainability in Agriculture

Sudheer Kumar Yadav, Jai Singh Patel, Bansh Narayan Singh, Raina Bajpai, Basavaraj Teli, Mahendra Vikram Singh Rajawat, and Birinchi Kumar Sarma

Abstract In the entire world, the aggregate effect of climate change is ceaselessly expanding deteriorated lands creating pressure on agricultural output and food security. The use of biofertilizers instead of chemical fertilizers can improve crop productivity and food quality under different environmental conditions. The use of bioagents minimizes the deposition of toxic agrochemicals in soil without altering its biological and functional characteristics. Biofertilizers are mainly comprised of a single or combination of microorganisms which can be endophytic or rhizospheric in nature. The communities of plants are directly or indirectly impacted by rhizospheric microorganisms which influence the structure and yield capacity. Considerable data is presently accessible on the composition and different aspects of plants along with microbial population residing in the rhizosphere and their functional capabilities. Hence, belowground microbiota is regarded as a forecaster of variations in plants and overground yield efficiency. Different approaches for microbial population improvement exist, and the use of microbial consortium (MC) as biofertilizer is one of them. Farming practices, environmental factors, and plant genotypes harbor distinct and diverse microbial communities and their functions. Currently, biofertilizer products having individual or combination of microbes exhibit restricted efficiency in specific environmental regimes. MC as a biofertilizer contributes a lot to help the plant to cope up against numerous strains (abiotic and biotic stresses) in different environmental conditions. Therefore, the selection of an appropriate MC for a particular agroecosystem and/or genotype of crops is in the direction of improving interactions between crop and the introduced microbes and is considered to be the way forward for enhancing profitability. However, the benefits of using MC

S. K. Yadav · J. S. Patel · B. N. Singh
Department of Botany, Institute of Science, Banaras Hindu University, Varanasi, Uttar Pradesh, India

R. Bajpai · B. Teli · B. K. Sarma (✉)
Department of Mycology and Plant Pathology, Institute of Agricultural Sciences, Banaras Hindu University, Varanasi, Uttar Pradesh, India

M. V. S. Rajawat
ICAR-National Bureau of Agriculturally Important Microorganism, Mau, Uttar Pradesh, India

S. K. Dubey, S. K. Verma (eds.), *Plant, Soil and Microbes in Tropical Ecosystems*, Rhizosphere Biology, https://doi.org/10.1007/978-981-16-3364-5_16

over the use of individual microbes lie in their multifunctionality unmistakably demonstrated by researches. However, limited attention is being paid by the manufacturers in maintaining quality norms. In the current chapter, we focused on the progress made in the development of biofertilizers comprising MC and their quality, microbiome engineering of biofertilizers, and their impact on plants under various environmental conditions.

Keywords Rhizosphere · Biofertilizer · Microbial consortia · PGPR · Microbial inoculants

16.1 Introduction

Green revolution or the third agricultural revolution had significantly enhanced agricultural food grain production (especially wheat and rice) worldwide, to meet the demand of diet at the beginning of the late 1960s. The rapid increase in agricultural output came from the green revolution because of the enhanced utilization of different chemical inputs (fertilizers, pesticides, and herbicides). This excessive utilization of chemicals caused deleterious impacts on the fertility of the land in addition to the well-being of mankind (Alori and Babalola 2018). The continuous rising of such environmental issues not only enforced researchers to solve the hazardous effect of these chemical fertilizers on the ecosystem but also encouraged farmers for cultivation involving sustainable approaches (Malusá et al. 2012). Globally, the population of mankind at the end of 2050 is anticipated to upsurge nearly 9.6 billion (Yadav et al. 2017). However, on the way to provide food to every individual, two challenges are identified: i) the reduction in arable land due to land acquisition for building residences and ii) availability of quality food to everyone. However, excessive utilization of chemical fertilizers to enhance agricultural production is not feasible in the context of the environment and the wellness of mankind. Under this circumstance, biofertilizer can be implemented as plant growth-promoting agents that would help in reducing the use of chemical fertilizers making the lands more fertile which ultimately will increase yield in addition to reducing different diseases of crop plants (Patel et al. 2016; Singh et al. 2019a). These biofertilizers include live beneficial microorganisms associated with the host and which could provide direct or indirect gain to the host by adapting various mechanisms that lead to enhanced crop production (Fuentes-Ramirez and Caballero-Mellado 2005). Application of these biofertilizer agents either with seed or soil as an inoculant enriches the soil with various important nutrients (micro- and macronutrients) via several ways like nitrogen fixation, nutrients solubilization, and mobilization, secreting compounds, and antibiotics involved in plant development (Singh et al. 2011). Some of these biofertilizers could also help in degrading organic matter and enhancing soil nutrient availability to the plants (Sinha et al. 2010).

In agricultural fields, most of the chemical enrichers applied are leached out (approx. 60–90%), and only the remaining (~10–40%) is available to plants. Yet, these chemicals are not readily available to plants as such due to the formation of complexes with other compounds. The application of biofertilizers enhances agricultural productivity playing a vital part in integrated nutrient management (Bhardwaj et al. 2014; Singh et al. 2019b). Moreover, the implementation of these biofertilizers in soil reduces the chemical input leading to organic farming practices. The demand for organic input in place of a chemical has been recommended for agricultural crop production to improve nutrient supply and maintaining soil fertility. The organic farming system is helpful to ensure food security and enriches soil biodiversity (Yadav et al. 2017). Microbes are found in their natural habitat in communities. Microbial communities in their habitat refer to the formation of microbial consortium (MC) that offers multiple actions like enhancing plant growth and minimizing abiotic and biotic stresses, viz., drought, chilling, temperature variation, pests, and disease infection in plants leading to food safety and security (Sekar et al. 2016). MC is synergistically associated with the host and mimics with the natural condition and plays a diverse role in the rhizospheric zone by solving the most challenging issues raised around the rhizosphere and creates an eco-friendly environment among soil-plant-atmosphere (Jain et al. 2013). In addition to increasing their populations, microbes also provide multiple benefits that support plants for tolerating several abiotic strains (Singh et al. 2013). Subsequently, co-inoculation or soil amendment with consortia based on compatible microbes has very high significance over a single application (Sekar et al. 2016).

The present trend has shown more focus on the application of MC on a small scale to control phytopathogens and improve plant health. Positive outcomes from such studies have attracted more researchers to experiment with MC rather than using single microbial inoculant (Sarma et al. 2015). Furthermore, alternative ways to enhance crop yield is by maintaining soil health and engineering of rhizomicrobiome. The integration of biotechnological approaches over bio-formulations in the cropping system has been adopted globally and would cope with the several challenges raised in plant growth (Odoh 2017). In the current scenario, the colossal application of microbes-based bio-formulations as a biofertilizer in agricultural fields is increasing because of its capacity to preserve the healthiness of soil and lowering down of the environmental concerns. Besides, it can cut down the utilization of inorganic chemicals in agricultural practices. Additionally, the use of biofertilizers is more effective in rainfed agriculture, mainly for the marginal farmers, who cannot afford the high cost of chemical fertilizers (Barman et al. 2017). Biofertilizer application is an ideal, cost-effective, and sustainable approach in farming as it conserves long-term soil fertility (Shelat et al. 2017). Agriculturally, important microorganisms used as biofertilizers include fungal mycorrhiza, cyanobacteria involved in the fixation of nitrogen, and PGPRs (plant growth-promoting rhizobacteria), biocontrol agents, biopesticides, endophytes, and bio-degrading microbes (Singh et al. 2011, 2018). Indeed, microbes are used as supplementary components in the soil which is helpful in promoting cropping practices like crop rotation, crop residue recycling, tillage, and organic manure maintenance. Long-term use of such important microbes in the soil can sustain enhanced yield in many commercial crops (cotton, jute, oilseed, sugarcane, sun hemp, tobacco, tea, coffee, etc.) (Bhardwaj et al. 2014). Therefore, the application of bio-formulations of compatible microbial consortium as biofertilizers around the rhizospheric region could be an efficient tactic for promoting plant growth and development. Similarly, the co-metabolism application of MC might also be a superior approach over single inoculum. This process is manifested during an interaction between microbes, where secreted specific metabolites serve as restricting elements to different communities of microbes within the network (Odoh et al. 2020). This facilitates the availability of limiting nutrients by mineralization of the by-products in addition to augmenting capacities of arable lands.

16.2 Development of Multifunctional MC

Crop productivity can be affected by two major environmental stresses, i.e., abiotic and biotic stresses. Several findings are coming out with tools to minimize these stresses and improve crop productivity. PGPRs are serving an important part against these stresses (Yang et al. 2009). Essential nutrient sources are present in the soil in sufficient amounts, but most of the time they are unavailable to plants. Rhizospheric bacteria present in the soil are majorly known to solubilize these nutrients for plants. Nitrogen availability in the environment is 78%, and plants cannot take it directly from the environment. Here comes the role of microorganisms to fix atmospheric nitrogen for plants and helps to maintain the nitrogen cycle (Rasche and Cadisch 2013). Microbes can solubilize naturally occurring nutrients present in bound form in the environment and maintain the nutrient cycle from source to sink, such as phosphorus (P), sulfur (S), potassium (K), magnesium (Mg), and calcium (Ca). Artificial application of selected microbes could be a strategy to make the soil rich in nutrients and also to minimize the use of chemical fertilizers and thus maintain soil nutrient balance (Ahemad and Kibret 2014).

Microbes may be utilized like biofertilizers or biocontrol agents either individually or in consortia. Microbes should be characterized and well tested scientifically for their specific biofertilizer activity before use. These microbes should fulfill several specific criteria to be a candidate for field use. The use of several microbes having different specific characters could be a strategy to use their potential efficiently. Several reports (Jain et al. 2012; Singh ct al. 2013; Patel et al. 2016, 2017) display the use of compatible MC (either 2–3 bacteria or bacteria and fungi together) for enhancement of plant resistance against stressful factors besides improving the development of the plant. Utilizing different microbes belonging to different rhizosphere and environments can enhance biocontrol efficiency as well as minimize the competition among them. The synergistic effect of these compatible microbes for a plant trait such as crop yield or availability of nutrients resulting in improved plant growth and yield has been reported earlier. The development of potential MC for making the microbes more effective is very important, and it can be a substitute for using harmful pesticides and chemical fertilizers to a great extent. Enhanced uptake of phosphorous and nitrogen along with protection against soil-borne phytopathogens has been reported for the consortium of *Trichoderma*, *Rhizobium*, and PSB, i.e., phosphate-solubilizing bacteria (Rudresh et al. 2005). Sarma et al. (2015) have observed individual microbial components of consortia, and their importance for the protection of plants counters to numerous phytopathogens.

Consortia of fungal mycorrhiza, PGPR, and bacteria living as endophytes have been reported for significantly enhancing plant protection and reducing reliance on chemical fertilizers (Pérez et al. 2007). Utilization of PGPR and bacteria living as endophytes may be considered to use in combination and as it would be a good and highly effective tactic for ensuring sustainability in agriculture and integrated pest management practices. Such a combination can control pests such as fungal pathogens, insects, and weeds effectively. The combined application of PGPR and *Bacillus* sp. has been suggested by Prabhukarthikeyan et al. (2014) for the biocontrol of tomato *Fusarium* wilt as well as tomato fruit borer without using any other chemical pesticides. Rhizobacteria possess the spatial ability to control different plant pathogens. A report for inducing induced systemic resistance (ISR) by *Pseudomonas fluorescens* has been also reported by Bandi and Sivasubramanian (2012) to control damage by an insect pest Thrips (*Thrips tabaci* L.). Researchers are continuously working to understand the roles of microorganisms in the agricultural system, but several puzzles remain unresolved. These microorganisms sustain capacity for improving growth and productivity of plants via different ways such as plant resistance induction and acclimatization to the environment and develop tolerance toward diverse abiotic strains (increased salinity, heavy metal, in addition to high pH) (Ahmad et al. 2018). We should consider the adaptability of MC being inoculated in different kinds of soil other than their natural environment. Earlier studies have not emphasized this point as it will be interesting to note whether the efficiency of the MC is enhanced in non-native soil. Consortia of microbes should be developed by using compatible microbes. It is, therefore, very necessary to test the compatibility of microbes before developing the consortia. If the microbes are not showing a synergistic effect on the targeted traits of plants, the basic concept behind the use of microbial combination will not be fulfilled (Sarma et al. 2015).

The development of microbial combinations that can improve several functions of the plants is a hot topic for current research. The Fig 16.1 highlights the basic methodoligies involved in development of MC for utilizing them as biofertilizers for reducing the chemical use in agriculture practices. Agricultural research is shifting toward the diminishing use of chemical fertilizers and pesticides without compromising production and quality of the produce. Recently, Backer et al. (2018) have shown the potential of PGPR for sustainable agriculture. The strategy of using MC is very old but it has been used for legumes and cereal crops significantly only in the last few years (Sessitsch and Mitter 2015). PGPR can use several mechanisms like deaminase action of ACC, enhanced fixation of N, and solubilization of calcium, besides phosphate solubilization for improvement in wellness of the plants and their yield capacity (Backer et al. 2018). MC activities should be thoroughly studied initially under laboratory conditions to maximize the effect of a consortium to its optimum level (Odoh et al. 2019).

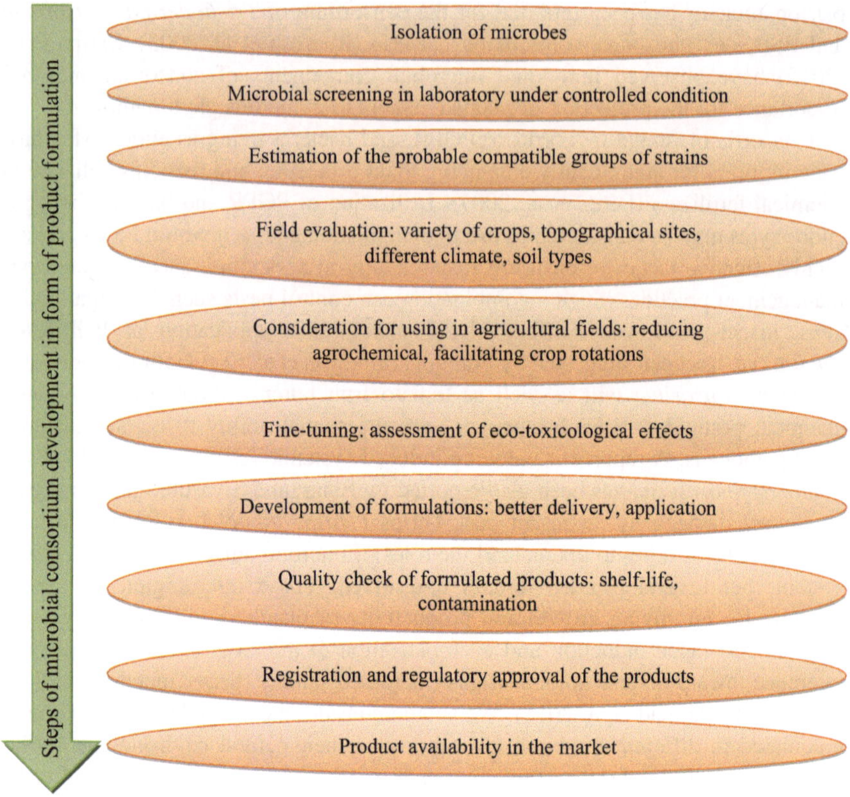

Fig. 16.1 Basic steps for the development of microbial consortia as biofertilizers

Microbial inoculants should be accessed for their shelf life in the particular formulation. Multilocation field trials should be conducted and approved for commercialization. Such testing and approval are important to release any microbe in a particular environment. A recent report (Backer et al. 2018) has shown the necessity to know the microbial load to be inoculated in an agricultural field for efficient colonization in the rhizosphere. The optimal spore dose of *Trichoderma asperellum* varied for different vegetable crops as determined by the growth and germination of the vegetable seeds (Singh et al. 2016a, b). The specific role (such as effect on plant growth, the effect on nutrient uptake, development of host resistance) of each component of MC must be well known including the type of soils suitable for them (Macouzet 2016; Baez-Rogelio et al. 2017). Furthermore, training of staff and farmers is needed for efficient use of these bio-inoculants concerning knowledge about soil specificity, the effect of environmental factors, and complexity of the individual components (Parnell et al. 2016; Bashan 2016; Itelima et al. 2018).

16.3 Impact of MC as Biofertilizer in Different Environmental Conditions

Over the period of evolution, plants are constantly evolving based on their relationship with the associated microorganisms which regulates the well-being and development of plants. These plant-associated microbes, i.e., plant holobiont termed as plant microbiota; plant microbiome which comprises the microbes associated in the different portions of the plants, viz., rhizosphere, phyllosphere, and endospheres; and such microorganisms straight or circuitously have links with the plant's growth in addition to their healthiness (Vorholt 2012; Brader et al. 2017; Lemanceau et al. 2017). For maintaining proper relation, floras actively recruit microbes from various reservoirs, i.e., rhizospheric (soil), phyllospheric (leaf surface and its surrounding environment), the anthosphere (flowers), the spermosphere (seed germination), and the carposphere (fruit area) (Hardoim et al. 2015). Limited information is available related to the structure and different aspects about plant-associated microbes. However, the abundance and species richness information are most commonly revealed by many researchers and tried their best to identify the structural basis of their composition in their community. These microbiomes serve an efficient part to fulfill the requirements of emerging challenges during the production of crops and an emergency prerequisite to constructing innovation in microbial technologies regarding their adaptation to productive agriculture. Plant microbiota has potentiality to reduce farmer's income by utilizing microbes for soil enrichment, nutrient uptake, managing biotic and abiotic stresses, weed management, improving crop nutrient status, and ultimately increasing crop yields (Jangra et al. 2018). Prior to use in agricultural practices, it is necessary to study cultural characteristics and the nature of adaptability of these microbes to know their behavior in different soils (Jiao et al. 2018). However, environmental factors are the principal character in deciding the role of applied microbes and their nature of adaptation concerning soil. Improved approaches for the application of microbes as a group of related strains or their blends could be standardized taking account of soil variability and external parameters. By looking into the crop status of the past years, it is realized that a smart knowledge-based choice of microbes is required to put forward the delivery approach and formulations. Instead, agricultural methods and crop varieties could impact the abundance of plant microbiota and its role in agriculture. Therefore, planning of suitable agricultural techniques beforehand could improve plant microbiome association during and after the cropping season and eventually provide benefit to better adaptability of plant microbiota.

Root microbiota mainly known as rhizobiome harbors a limited group of microbes based on the soil type that surrounds them which can be mostly horizontally transferred, i.e., the difference in soil type and their respective environment. Rhizobiome is extremely complex driven and consists of various microbes. Soil microbes can also target the ecosystem through the biogeochemical cycling of available elements along with the formation of soil surface/sub-surface particles, pollutant degradation, and water quality (Li et al. 2014; Eilers et al. 2012). However, sometimes it can also be vertically transferred via seeds, host plant, available nutrients, and organic matter (Jiao et al. 2018). Seeds also represent a central foundation of microbes as microbes are associated either intrinsically or extrinsically and serve as the initial region for multiplication in the roots in the seedling (Liu et al. 2012). Rhizospheric zone has the ability to provide unique ecological niche and metabolites that help in the attraction of microbiota which consequently provides their effect on the remaining plant parts (Hartmann et al. 2009). However, understanding the rhizobiome with the domesticated plants does not represent the status of native plants as they recruit various microbes during their growth and development (Bulgarelli et al. 2013). Various reports explained the higher richness of bacterial species in root microbiota in the rainforest when compared to other soils. The highest taxonomic ranks for the microbe diversity (bacteria) are given to alpha-proteobacteria <actinobacter<acidobacter in various root-associated studies (Yeoh et al. 2017). Still, the beta-proteobacteria hold better species richness in root association when compared to rhizobiome status suggesting the recruitment process and enrichment of the root environment attract the nearby microbes (Lundberg et al. 2012). Recently, Donn et al. (2015) studied wheat rhizosphere to understand the root-driven bacterial abundance resulting in tenfold increased abundance of actinobacteria, and other microbes, i.e., pseudomonads, oligotrophs, and copiotrophs, in comparison to bulk soil further suggest an alteration of rhizosphere and rhizoplane microbe structure without affecting the bulk soil population.

However, the difference in plant genotypes and relative species can also influence the structure of rhizospheric microbes. The variation in bacterial community is not only affected by rhizosphere or surrounding environment, but the difference in host genetic content can also alter the diversity in microbiodata. Bouffaud et al. (2012) studied the richness of the microbiome in an inbred line of maize landraces using microarray analysis of rhizospheric samples. The dent corn group produced higher discriminating signals targeting the beta-proteobacteria genera, but the flint corn received higher signals from alpha-proteobacteria. Delta-proteobacteria and beta-proteobacteria group bacteria were able to produce high signal intensity in tropical and stiff stalk corn group (Bouffaud et al. 2012), which states the qualitative difference in rhizodeposition and other exudates composition may raise the difference in bacterial community association (Bressan et al. 2009). Zarraonaindia and Gilbert (2015) demonstrated the above-ground microbiome profiling in grapevine and found *Sphingomonas* and *Pseudomonas* were abundant in vine leaves and grapes due to availability of nutrient source and water-limited condition for growth and development of microbes by giving an advantage to crops for disease suppression besides guarding counter to water stress. Flora roots are also colonized internally (root endospheres) by abundant endophytic microbes. The microbe entry takes place through a passive process, root cracks; wounds in roots and emergence of new lateral roots provide the access to a diverse range of microbes (Compant et al. 2005). Correa-Galeote et al. (2018) studied the maize root endophytes showing the predominance of Proteobacteria, Firmicutes, and Bacteroidetes due to soil cultivation practices history. Whereas in rice roots, the microbes belongs to family Rhizobiaceae, Comamonadaceae, Streptomycetaceae, and Bradyrhizobiaceae are occupying the diverse status (Edwards et al. 2015). Similarly, in grapevine, the fullness of Acidobacteria, Proteobacteria, Actinobacteria, Bacteroidetes, and Firmicutes was found in various studies (Burns et al. 2015; Faist et al. 2016; Samad et al. 2017). Apart from the adaptation of microbes to various plant parts and root zone, microbes too serve a vital part against the living and environmental constraints. During this process, the tendency of variation in microbial physiological characters and metabolic pathways occurs due to response from the stress generated. Application of single microbe or consortia is gaining more interest among the researchers and government launching a lot of initiatives to this kind of work by keeping in mind about the reduction in chemical use in agriculture. Similarly, food crops are facing threat due to the climate change scenario created by the downfall of crop production (Odoh et al. 2020). However, the application of microbes in such areas can help in mitigating various biotic and abiotic stresses that result in less crop loss. One such example revealed the meta-transcriptomics analysis resulted in the production of the polyketides, osmotic stress, and cold shock genes in suppressive soils due to the occurrence of *Stenotrophomonas* spp. and *Buttiauxella* spp., whereas oxidative stress genes along with antibiotic synthesis genes were more prevalent in non-suppressive soil in which *Pseudomonas* spp. and *Arthrobacter* spp. were highly present (Hayden et al. 2018). Hence, microbes have the potential to be used as biofertilizers, biopesticides, bioherbicides, and decomposers, and many had already arrived in the market as substitutes of chemicals with wider adaptability in a different environment (Mitter et al. 2016). MC which means mixing of two or more microbes based on the mode of work, i.e., biofertilizer-biopesticides, nodulation growth enhancer, decomposer-growth promotion, nutrient use-crop protection, etc., is new market product strategies to reach more audiences with the same effect when applied in a single form (Yadav et al. 2019). Furthermore, collection of microbes from the extreme habitat and integrating various agri microbial biotechnology tools to transfer the extreme habitat property to the locally adapted microflora will be a sustainable solution for the microbe adaptation without disturbing their community (Timmusk et al. 2017). Different studies proved the significance of microorganisms alleviating abiotic stresses (Kumar et al. 2019; Patel et al. 2017; Srivastava et al. 2015) and biotic stresses (Jain et al. 2012; Singh et al. 2013; Kumar et al. 2017) pretty well. Consortium development integrating microorganisms belonging to varied ecological backgrounds could provide crop protection in different environmental regimes. Thus, proper designing of MC as biofertilizers for particular environmental conditions and crops might prove to be a great move that can help in enhanced crop productivity with reduced chemical uses and environmental damages.

16.4 Microbiome Engineering of Biofertilizers

The historic events during the green revolutions had initiated the cultivation of high yielding varieties. Indeed, an increase in the application of inorganic fertilizers and chemicals created a drastic impact on soil health status leading to depletion of useful microbial diversity. This has led to the extinction or reduction in population dynamics of potential microbes which are working together for sustainable agriculture. However, in the present era, efforts are being made to conserve the potential microbes and engineer them ecologically to meet our requirements and applying those in other fields to meet our needs. However, the microbiome constitutes of a diverse group of microbes which have direct and indirect roles in the ruling soil ecosystem. During the interaction, soil microbes carry out various events in increasing quantitative food production, recycling of biogeochemical cycle, and maintaining soil health status (Hansel et al. 2008). During this process, these biological entities may have positive/negative influence on living and nonliving parameters (Odoh et al. 2019). Advanced studies in the medical field showed the importance of engineered microbes for fast and reliable production of antibiotics (Cycon et al. 2019), and food stains developed from microbes (Sen et al. 2019) by understanding the mechanisms, growth, and development pattern and complexity (Kumar 2016).

Plant-microbe interactions can take place with variation in relationships, i.e., beneficial, neutral, or completely negative. The beneficial interaction between plants and microbes is a matter of interest and is exploited extensively (Farrar et al. 2014). In this category of plant-microbes interaction, the role of arbuscular mycorrhizae (AM) holds a significant position in benefiting the soil and plants grown (Smith and Smith 2011). Similarly, fixation of N in legumes interacts with nodule-forming rhizobacteria (Oldroyd et al. 2011) and pathogenesis (Dodds and Rathjen 2010; Kachroo and Robin 2013; Wirthmueller et al. 2013). This system of symbiosis association between flora and microbes remains well-characterized providing clear information of gene expression, signaling pathway, and many more. However, understanding the plant evolution and adaptation to climate change scenario has made the scientific community think further in such studies (Hirsch 2004). In addition, plants are in interaction with other microbes (bacteria, fungi, algae) in an ecosystem either to get benefits or parasitize them in soil by producing the reciprocal signals during their interactions with other rhizospheric microbes or plants themselves (Badri and Vivanco 2009; Evangelisti et al. 2014). During interactions, microbes instead of acting individually potentially mingle with other microbes as consortia to exhibit the performance (Hirsch 2004). Sometimes the opportunistic microbes integrate with dynamic microbial communities posing threat to plant or humans because of pathogenic behaviors (Berg et al. 2005). MC can be administered by considering the practical parameters rather than selecting the specific microbe species and may undergo tripartite interactions (Bonfante and Anca 2009; Dames and Ridsdale 2012). Progress in procedures is necessary for manipulating the microbiome engineering process through various tools and techniques. Historically, research studies state the information of easily culturable microbial genera/species providing information about growth and medium parameters (Stewart 2012; Vartoukian et al. 2010). Still, some microbes especially endosymbionts are unable to grow in the absence of a living host as mycorrhiza needs a host plant to interact endophytically in the root system (Hildebrandt et al. 2002). Advancements in fluorescent tagging methodologies enable the visualization of the endophytic bacteria community (Elbeltagy et al. 2001) but make the microbe unculturable. This makes the microbiome research more interesting in understanding the microbial need to make it culturable and get benefitted from it. In some cases, microbes stay together in the entire life cycle as "obligate endophyte" which opens a new area of research for resilience in agriculture.

A considerable amount of information about plant microbiome is available now. Similarly, the reports are also available regarding the plant-microbe interactions. Every beneficial microbe does not have all the properties that are linked with structural progress of plant, development, nutrient solubilization besides mobilization, capability for tolerating various abiotic stresses, and biocontrol against various pathogenic microbes. Thus, the information about the microbiome of the plant and the particular genotype of the plant will be helpful to design the MC as a biofertilizer for the various crops in different environmental conditions. Various biotechnological interventions are available to edit the genome of microbes as per the requirement. These tools and techniques will also be helpful to create the compatibility of various incompatible beneficial microbes aimed at the expansion of MC as biofertilizers. By using the knowledge of plant microbiome and biotechnological advancement, microbiome engineering could have great potential for the development of MC as biofertilizers. In this regard, designing of MC must be free from opportunistic pathogenic microbes. Nithya et al. (2014) reported the food poisoning outbreak from lettuce and fresh fruits grown after microbial treatment. However, the engineering in microbiome provides major beneficial properties in the soil for continuous growing crops in rotation (Farrar et al. 2014). Thus, microbiome engineering might be helpful to open a new paradigm shift in sustainable agriculture for better crop production with food safety and security.

16.5 The Standard Norms for Biofertilizers Based on MC

It is very evident that profit due to the application of MC is massive. MC has logged encouraging achievements in different fields like ventures of food production, use in agricultural activities, medical uses, and ecological curative in comparison with a single strain. A challenging and multifaceted network of the prototypical microbial agent is shaped via metabolic modeling and specific strain recreation which is intended for ideal execution and creation of required biochemical agents and biomass (Faust 2019). Due to technical expertise in a few Asian countries like India and China, America, Africa, and Europe have a slight streak between fruitful uses of biofertilizers. The developed countries are focusing on thorough research on biotechnological methods for bioproducts designing, expanding mindfulness on their use although battling for relatively less utilization of chemical fertilizers. Sufficient consideration is yet to be given by developing countries on biofertilizers use as their benefit to the agricultural system is obvious. However, only a few farmers in such countries are using biofertilizers in their cropping practices. This is rather very much non-uniform as in Brazil where nearly the entire harvest protein is produced using biological nitrogen fixation (BNF), but the use of biofertilizer is <1 percent in east and southern Africa. Normally, introduced biofertilizers are formed in accordance with or personalized to its origin nation remembering their regional circumstances like climatic and storing state. Moreover, these constraints assume an immense job in deciding their timeframe of realistic usability and practicality. Through enhanced manpower advancement via training and expanded awareness, research, and innovations, regional impacting circumstances will turn into a significant aspect during the production of indigenous biofertilizers specific to a particular area. This would, in turn, help limit viability loss detected in few biofertilizers available in the marketplace (Jefwa et al. 2014) in which state of storing in addition to management serves as a vital job. However, due to lack of insufficient and comprehensive studies on formulation development which is needed for the spatial crop responses, no country will have the ability to procure profit from the complete capacity of biofertilizers. In nations like India, it would be helpful in the conveyance of better-quality produce where support from the government has improved the production of biofertilizers (Odoh et al. 2020). The quality norm of biofertilizers should incorporate specification which could be recorded in the label or for marketing authorization that would be needed to be given. Known basic features of biofertilizers are considered to be crucial, and they are minimum number of living cells/propagules, nutrient solubilization efficacy and fixation in bacteria, plant inoculation competence in mycorrhizal fungi, time span of usability and/or date of expiry, level of contamination, the pH, the physical structure, and amount of carbon and water. Taking the technical prospects from the producer's part and the facts via researches, a range of standards must be setup for few parameters like minimum number of living cells/propagules and for the efficacy statistics (Malusa and Vassilev 2014). PGPM strains used for commercial purpose should be precisely identified, and it is one of the most essential components. Based on the molecular biology approaches and strain's distinctive features documented in the registration dossier, identification of strains should be done. For conveyance to support commercialization of biofertilizers and conveyance of PGPM in proper physiological state and a reasonable quantity, the choice of the inoculant's carrier is very decisive (Malusá et al. 2012). Thus, other than the regular organic, inorganic, and polymeric complexes utilized as a carrier, presently an innovative way of utilizing biofilm of bacteria or nanocarriers as the carrier is a biological tactic which is under progress (Jayasinghearachchi and Seneviratne 2004; Qureshi et al. 2005; Seneviratne et al. 2007). Efficiency assessment can be very tough for MC biofertilizers. Consortium formed by the combination of a number of PGPM strain stimulates the growth of plant at various developmental phases, and they may also display different mode of activities which may occasionally include a mechanism(s) of plant defense. Improvement in nutrient efficacy by species consortia has been evidenced, for instances, combined inoculation of PSM + AMF, or *Rhizobium* + AMF in single gel preparation could likewise display plant defense characteristics (Vassilev et al. 2006). Moreover, multipurpose goods are more liked by farmers for utilization, and producers favor advertising products having numerous actions as they show higher impact and pull in users.

16.6 Ordinance and Commercialization of MC as Biofertilizer

The authentic definition of a commercialized product like biofertilizer is crucial for the producer's interest to manufacture them. In the United State of America (USA) and the European Union (EU), there is no any authenticate descriptions of biofertilizers or any lawful specifications to explain its features. In the EU, microbes (bacteria, viruses, and fungus) are incorporated as promising contributions in EU Commission Regulation No. 889/2008 on organic manufacture and solitary limited aimed at controlling pests and diseases employing biological approaches (Malusa and Vassilev 2014). A biofertilizer could in this manner be characterized as the developed item containing at least one microorganism that increases the status of plant nutrients (better development and yield) through supplementing soil nutrients additionally leading nutrients further accessible to floras as well as via expanding floras to get nutrients (Malusa and Vassilev 2014).

Worldwide the utilization and interest for biopesticides are ascending because of expanded attention to crops having no or less amount of pesticide residues. The worldwide level estimate for the microbe-based goods in 2014 was US$ 2183 million, and by 2019, it is anticipated twofold by US$ 4556 million with 15.3% of CAGR. Of the few microbial strains, the bacterial portion represented the biggest share (US$ 1.6 billion). Like biopesticides, by 2020 the biofertilizers market worldwide is anticipated to reach US$ 1.88 billion at 14% of CAGR between 2015 and 2020 (Markets and Markets 2015). All around, over 200 active ingredients of biopesticides are enrolled, and roughly 700 such products are accessible in the market. Considering the Indian market, 15 enlisted biopesticides were available during 2008 under the Insecticide Act (IA) (1968), and its market share is only about 4.2% of the total pesticide market. However, the biopesticide market is predicted to grow at a yearly growth rate of about 10% in the coming decade (Suresh 2012). Interestingly, the biopesticides enlisted have grown manifold during the previous years, NAAS (2013), and presently around 400 biopesticides are enlisted, and there are more than 1250 effectively enrolled biopesticide items available in Indian markets. This displays mindfulness among ranchers just like strategy backing of the administration to utilize the biologically sheltered items for bugging the executives. In absence of any particular guidelines, nearly 400 enrolled biopesticides are being marketed independently, and no MC is available (Sekar et al. 2016).

At the worldwide level, the administrative structures contrast broadly among various nations. In the USA, biopesticide creation is standardized beneath a different section as "Biopesticides and Pollution" inside the Environmental Protection Agency (EPA). Succeeding, in 1996 the Japanese Ministry of Agriculture, Forestry, and Fisheries (JMAFF) blended its framework with the rules of EPA. However, in Europe, biopesticides are assessed through the European Pesticide Regulation EC No. 1107/2009 which advances the creation of more safe substances by removing the unsafe products, and it has been advancing the enlistment of generally safe items through (2009/128/EC) basic and straightforward enrollment conventions

(Villaverde et al. 2014). Canada adopts just the security test, and the remainder of the nations need the information of both well-being and viability tests. The EC, JMAFF, and EPA guidelines toward biopesticides are created so that it requires less information when contrasted with synthetic items and decreased an opportunity to process the enrollment applications. In this specific situation, the International Organization for Biological Control of Noxious Animal and Plants (IOBC) completed a worldwide level audit on the utilization of biopesticides and administrative environment friendly measures to control pest and diseases. It focused on the requirement for smoothing out the enrollment procedure through orchestrating information necessities and conventions for hazard appraisals. In India, for any microorganisms utilized for bug and illness, the executives require enrollment for both creation and deal with the Central Insecticides Board (CIB) of the Ministry of Agriculture according to the Insecticides Act (IA), 1968, of the Government of India (GOI) and Insecticides Rules, 1971, which were as of late supplemented by the Pesticides Management Bill 2008. The biopesticides for the most part viewed as protected GRAS under this demonstration, and to advance its creation and use, give the advantage of enrollment the same just as temporary enlistment. In this manner, the makers can enlist the item either for customary enrollment under segment 9 (3) or for temporary enrollment under area 9 (3B) of the IA. While applying for enlistment, the information on item portrayal, well-being, toxicology, adequacy, and marking are fundamental. Notwithstanding the need and temporary enrollment for biopesticides in the Act, the enlistment conventions are made simpler and acknowledge nonexclusive information for many new items containing strains that are as of now enrolled. Such certifiable provisions are inbuilt in the Act which shows the enthusiasm of the administration in advancing the sheltered items for a bug, the board like different nations. So as to manage the business creation of these items, the Government of India has built four unique bodies to control the biopesticide creation. The Central Insecticides Board (CIB) is engaged with creating fitting arrangements, and the enrollment panel registration committee (RC) is capable to enlist the items for creation. While the Central Insecticides Laboratory (CIL) is in control to screen the nature of the items accessible in the market, the State Department of Agriculture (SDA) issues the assembling permit and performs the quality check. On the opposite side, according to the warning dated March 26, 1999, of the Central Insecticides Board, Ministry of Agriculture, biopesticide was put under the Insecticide Schedule Act 1968, and thus, the age of toxicological information turned into an essential for the enlistment of biopesticide.

There are severe guidelines and rules directing the usage and treatment of supplement constructed by the microbe. The main reason behind the call of this item is because definition and effective checking are enlistment in which item should meet explicit administrative necessities. Preceding this, the item must be built up in a transporter, for example, alginate (Bashan 2016) or biochar (Głodowska et al. 2016) via these sticking agents at the time of sowing seeds carrying microbial inoculants. On account of fluid microbe inoculum, these are poured and blend over the seeds preceding to planting or else trickled over seeds wrinkle during sowing time. Capacity besides the item's life expectancy is crucial to guarantee microbial suitability, existence, and/or strain bioactions. There ought to likewise remain lucidity on intense against continuing biomolecule treatment. Much of the time, intense application happens only a few times in a developing period; this could likewise be on an objective phase of crop development, or in light of natural and abiotic situations (dry spell), though in constant treatment, the item can be treated at ordinary period splashes intermission or else a moderate delivery seed treatment (Backer et al. 2018). India stands likely the nation consisting of a maximum comprehensive lawful outline recognized for biofertilizers. Under the request for regulating fertilizers of 1985, the Indian Ministry of Agriculture gave a request in 2006 which was further corrected in 2009, that consists of biofertilizers below the Essential Commodities Act of 1955. Further, the demonstration explains the word biofertilizer as "the product containing transporter based (strong or fluid) living microorganisms which are horticulturally valuable as far as nitrogen fixation, phosphorus solubilization or supplement mobilization, to increment the profitability of the dirt or potentially crop." This word is additionally secured beneath the wide meaning of composts that "means any substance utilized or planned to be utilized as a fertilizer of the dirt and additionally crop" (Malusa and Vassilev 2014).

16.7 Conclusion and Future Prospects

The microbes in the consortium have the capability to provide the best opportunity to enhance crop growth and yield significantly that can be the only way to feed quality food to ever-growing population. The efforts must be taken to identify the compatibility between the microbial strains for the development of a cost-effective product, and this will positively regulate the plant physiology and transcription pattern. Nowadays, there are several bioformulations available in the market as the demand for organic food is increasing to avoid chemical uses in agriculture. The evaluation of microbial count in such formulations is a major problem. Thus, the development of rapid testing kits or methodologies for the evaluation of live microbial count in available products will enhance the quality and doses used during application. The assessment of MC in various field trials for their performance is needed before the preparation of bioformulations. The regulatory and commercialization policies must be strict to develop the microbial formulations in the consortium. Additionally, the metabolites produced from the microbial mixtures in the consortium could be used for enhanced crop production and the modification or upregulation in certain genes that can work in absence of live microbial inoculants. The biotechnological progress in research could be a good way for designing artificially developed MC by editing in the microbial genome.

Acknowledgments SKY is grateful to the Indian Council of Medical Research, New Delhi, India, for financial assistance [Grant 3/1/3/JRF-2012/HRD-66(80689)] as the work is related to Ph.D. BKS is grateful to the Indian Council of Agricultural Sciences, New Delhi, India, for financial assistance (Grant ICAR-NBAIM/AMAAS/2014-15/73).

References

Ahemad M, Kibret M (2014) Mechanisms and applications of plant growth promoting rhizobacteria: current perspective. J King Saud Univ Sci 26:1–20

Ahmad M, Pataczek L, Hilger TH, Zahir ZA, Hussain A, Rasche F, Schafleitner R, Solberg SØ (2018) Perspectives of microbial inoculation for sustainable development and environmental management. Front Microbiol 9:2992

Alori ET, Babalola OO (2018) Microbial inoculants for improving crop quality and human health in Africa. Front Microbiol 9:2213

Backer R, Rokem JS, Ilangumaran G, Lamont J, Praslickova D, Ricci E, Subramanian S, Smith DL (2018) Plant growth-promoting rhizobacteria: context, mechanisms of action, and roadmap to commercialization of biostimulants for sustainable agriculture. Front Plant Sci 9:1473

Badri DV, Vivanco JM (2009) Regulation and function of root exudates. Plant cell Env 6:666–681

Baez-Rogelio A, Morales-García YE, Quintero-Hernández V, Muñoz-Rojas J (2017) Next generation of microbial inoculants for agriculture and bioremediation. Microb Biotechnol 10:19–21

Bandi S, Sivasubramanian P (2012) Management of *Thrips tabaci* Lindeman in onion using *Pseudomonas fluorescens* Migula through induced resistance. J Biopest 5:1–3

Barman M, Paul S, Choudhury AG, Roy P, Sen J (2017) Biofertilizers as prospective input for sustainable agriculture in India. Int J Curr Microbiol Appl Sci 6:1177–1186

Bashan N (2016) Inoculant formulations are essential for successful inoculation with plant growth promoting bacteria and business opportunities. Indian Phytopathol 69:739–743

Berg G, Eberl L, Hartmann A (2005) The rhizosphere as a reservoir for opportunistic human pathogenic bacteria. Environ Microbiol 7:1673–1685

Bhardwaj D, Ansari MW, Sahoo RK, Tuteja N (2014) Biofertilizers function as key player in sustainable agriculture by improving soil fertility, plant tolerance and crop productivity. Microb Cell Fact 13(1):1–10

Bonfante P, Anca IA (2009) Plants, mycorrhizal fungi, and bacteria: a network of interactions. Annu Rev Microbiol 63:363–383

Bouffaud ML, Kyselková M, Gouesnard B, Grundmann G, Muller D, Moënne-Loccoz YVAN (2012) Is diversification history of maize influencing selection of soil bacteria by roots? Mol Ecol 21(1):195–206

Brader G, Compant S, Vescio K, Mitter B, Trognitz F, Ma LJ, Sessitsch A (2017) Ecology and genomic insights into plant-pathogenic and plant-nonpathogenic endophytes. Annu Rev Phytopathol 55:61–83

Bressan M, Roncato M, Bellvert F, Comte G, el Zahar Haichar F, Achouak W, Berge O (2009) Exogenous glucosinolate produced by *Arabidopsis thaliana* has an impact on microbes in the rhizosphere and plant roots. ISME J 3:1243–1257

Bulgarelli D, Schlaeppi K, Spaepen S, van Themaat EVL, Schulze-Lefert P (2013) Structure and functions of the bacterial microbiota of plants. Annu Rev Plant Biol 64:807–838

Burns KN, Kluepfel DA, Strauss SL, Bokulich NA, Cantu D, Steenwerth KL (2015) Vineyard soil bacterial diversity and composition revealed by 16S rRNA genes: differentiation by geographic features. Soil Biol Biochem 91:232–247

Compant S, Reiter B, Sessitsch A, Nowak J, Clément C, Barka EA (2005) Endophytic colonization of *Vitis vinifera* L. by plant growth-promoting bacterium *Burkholderia* sp. strain PsJN. Appl Environ Microbiol 71:1685–1693

Correa-Galeote D, Bedmar EJ, Arone GJ (2018) Maize endophytic bacterial diversity as affected by soil cultivation history. Front Microbiol 9:484

Cycon M, Mrozik A, Piotrowska-Seget Z (2019) Antibiotics in the soil environment degradation and their impact on microbial activity and diversity. Front Microbiol 10:338

Dames JF, Ridsdale CJ (2012) What we know about arbuscular mycorrhizal fungi and associated soil bacteria. Afr J Biotechnol 11:13753–13760

Dodds PN, Rathjen JP (2010) Plant immunity: towards an integrated view of plant–pathogen interactions. Nat Rev Genet 11:539–548

Donn S, Kirkegaard JA, Perera G, Richardson AE, Watt M (2015) Evolution of bacterial communities in the wheat crop rhizosphere. Environ Microbiol 17:610–621

Edwards J, Johnson C, Santos-Medellín C, Lurie E, Podishetty NK, Bhatnagar S, Eisen JA, Sundaresan V (2015) Structure, variation, and assembly of the root-associated microbiomes of rice. Proc Natl Acad Sci 112(8):E911–E920

Eilers KG, Debenport S, Anderson S, Fierer N (2012) Digging deeper to find unique microbial communities: the strong effect of depth on the structure of bacterial and archaeal communities in soil. Soil Biol Biochem 50:58–65

Elbeltagy A, Nishioka K, Sato T, Suzuki H, Ye B, Hamada T, Isawa T, Mitsui H, Minamisawa K (2001) Endophytic colonization and in planta nitrogen fixation by a *Herbaspirillum* sp. isolated from wild rice species. Appl Environ Microbiol 67:5285–5293

Evangelisti E, Rey T, Schornack S (2014) Cross-interference of plant development and plant–microbe interactions. Curr Opin Plant Biol 20:118–126

Faist H, Keller A, Hentschel U, Deeken R (2016) Grapevine (*Vitis vinifera*) crown galls host distinct microbiota. Appl Environ Microbiol 82:5542–5552

Farrar K, Bryant D, Cope-Selby N (2014) Understanding and engineering beneficial plant–microbe interactions: plant growth promotion in energy crops. Plant Biotechnol J 12(9):1193–1206

Faust K (2019) Microbial consortium design benefits from metabolic modeling. Trends Biotechnol 37(2):123–125

Fuentes-Ramirez LE, Caballero-Mellado J (2005) Bacterial biofertilizers. PGPR: biocontrol and biofertilization. Springer, Dordrecht, pp 143–172

Głodowska M, Husk B, Schwinghamer T, Smith D (2016) Biochar is a growth-promoting alternative to peat moss for the inoculation of corn with a pseudomonad. Agron Sustain Dev 36:1–10

Hansel CM, Fendorf S, Jardine PM, Francis CA (2008) Changes in bacterial and archaeal community structure and functional diversity along a geochemically variable soil profile. Appl Environ Microbiol 74(5):1620–1633

Hardoim PR, Van Overbeek LS, Berg G, Pirttilä AM, Compant S, Campisano A, Döring M, Sessitsch A (2015) The hidden world within plants: ecological and evolutionary considerations for defining functioning of microbial endophytes. Microbiol Mol Biol Rev 79(3):293–320

Hartmann A, Schmid M, van Tuinen D, Berg G (2009) Plant-driven selection of microbes. Plant Soil 321:235–275

Hayden HL, Savin KW, Wadeson J, Gupta VV, Mele PM (2018) Comparative metatranscriptomics of wheat rhizosphere microbiomes in disease suppressive and non-suppressive soils for *Rhizoctonia solani* AG8. Front Microbiol 9:859

Hildebrandt U, Janetta K, Bothe H (2002) Towards growth of arbuscular mycorrhizal fungi independent of a plant host. Appl Environ Microbiol 68:1919–1924

Hirsch AM (2004) Plant–microbe symbioses: a continuum from commensalism to parasitism. Symbiosis 37:345–363

Itelima JU, Bang WJ, Onyimba IA, OjE (2018) A review: biofertilizer; a key player in enhancing soil fertility and crop productivity. J Microbiol Biotechnol Rep 2:22–28

Jain A, Singh S, Sarma BK, Singh HB (2012) Microbial consortium-mediated reprogramming of defence network in pea to enhance tolerance against *Sclerotinia sclerotiorum*. J Appl Microbiol 112:537–550

Jain A, Singh A, Singh S, Singh HB (2013) Microbial consortium-induced changes in oxidative stress markers in pea plants challenged with *Sclerotinia sclerotiorum*. J Plant Growth Regul 32 (2):388–398

Jangra M, Jangra S, Nehra K (2018) Role of microbes in agriculture. In: Crop improvement for sustainability. Daya Publishing House, New Delhi, pp 193–222

Jayasinghearachchi HS, Seneviratne G (2004) A bradyrhizobial-*Penicillium* spp. biofilm with nitrogenase activity improves N2 fixing symbiosis of soybean. Biol Fertil Soils 40:432–434

Jefwa JM, Pypers P, Jemo M, Thuita M, Mutegi E, Laditi MA, Faye A, Kavoo A, Munyahali W, Herrmann L, Atieno M (2014) Do commercial biological and chemical products increase crop yields and economic returns under smallholder farmer conditions? In: Challenges and opportunities for agricultural intensification of the humid highland systems of Sub-Saharan Africa. Springer, Cham, pp 81–96

Jiao S, Chen W, Wang J, Du N, Li Q, Wei G (2018) Soil microbiomes with distinct assemblies through vertical soil profiles drive the cycling of multiple nutrients in reforested ecosystems. Microbiome 6(1):1–13

Kachroo A, Robin GP (2013) Systemic signaling during plant defense. Curr Opin Plant Biol 16:527–533

Kumar A (2016) Role of microbes in food and industrial microbiology. J Food Ind Microbiol 2:101

Kumar M, Patel JS, Kumar G, Sarkar A, Singh HB, Sarma BK (2017) Studies on *Pseudomonas* and *Trichoderma*-mediated root exudation pattern in chickpea against *Fusarium oxysporum* f. sp. *ciceris*. J Agric Sci Technol 19(5):969–978

Kumar G, Bajpai R, Sarkar A, Mishra RK, Gupta VK, Singh HB, Sarma BK (2019) Identification, characterization and expression profiles of *Fusarium udum* stress-responsive WRKY transcription factors in *Cajanus cajan* under the influence of NaCl stress and *Pseudomonas fluorescens* OKC. Sci Rep 9(1):1–9

Lemanceau P, Blouin M, Muller D, Moënne-Loccoz Y (2017) Let the core microbiota be functional. Trends Plant Sci 22:583–595

Li CH, Yan K, Tang LS, Jia ZJ, Li Y (2014) Change in deep soil microbial communities due to long-term fertilization. Soil Biol Biochem 75:264–272

Liu Y, Zuo S, Xu L, Zou Y, Song W (2012) Study on diversity of endophytic bacterial communities in seeds of hybrid maize and their parental lines. Arch Microbiol 194:1001–1012

Lundberg DS, Lebeis SL, Paredes SH, Yourstone S, Gehring J, Malfatti S, Tremblay J, Engelbrektson A, Kunin V, Del Rio TG, Edgar RC (2012) Defining the core *Arabidopsis thaliana* root microbiome. Nature 488(7409):86–90

Macouzet M (2016) Critical aspects in the conception and production of microbial based plant biostimulants (MBPB). Probiot Intelligent 5:29–38

Malusa E, Vassilev N (2014) A contribution to set a legal framework for biofertilizers. Appl Microbiol Biotechnol 98:6599–6607

Malusá E, Sas-Paszt L, Ciesielska J (2012) Technologies for beneficial microorganisms inocula used as biofertilizers. Sci World J 2012:491206

Markets and Markets (2015) Biofertilizers market by type (nitrogen-fixing, phosphate-solubilizing and potash-mobilizing), microorganisms (*Rhizobium*, *Azotobactor*, *Azospirillum*, cyanobacteria and phosphate solubilizing bacteria), application, crop type and by region-Global Forecast to 2020

Mitter B, Pfaffenbichler N, Sessitsch A (2016) Plant–microbe partnerships in 2020. Microb Biotechnol 9:635–640

NAAS (2013) Biopesticides-quality assurance, Policy Paper No 62, New Delhi

Nithya A, Gothandam KM, Babu S (2014) Alternative ecology of human pathogenic bacteria in fruits and vegetables. Plant Pathol J 13:1–7

Odoh CK (2017) Plant growth promoting rhizobacteria (PGPR): a bioprotectant bioinoculant for sustainable agrobiology. A review. Int JAdv Res Biol Sci 4(5):123–142

Odoh CK, Eze CN, Akpi UK, Unah VU (2019) Plant growth promoting rhizobacteria (PGPR): a novel agent for sustainable food production. Am J Agric Biol Sci 14(35):54

Odoh CK, Sam K, Zabbey N, Eze CN, Nwankwegu AS, Laku C, Dumpe BB (2020) Microbial consortium as biofertilizers for crops growing under the extreme habitats. In: Plant microbiomes for sustainable agriculture. Springer, Cham, pp 381–424

Oldroyd EDG, Murray JD, Poole PS, Downie JA (2011) The rules of engagement in the legume-rhizobial symbiosis. Annu Rev Genet 45:119–144

Parnell JJ, Berka R, Young HA, Sturino JM, Kang Y, Barnhart DM, DiLeo MV (2016) From the lab to the farm: an industrial perspective of plant beneficial microorganisms. Front Plant Sci 7:1110

Patel JS, Sarma BK, Singh HB, Upadhyay RS, Kharwar RN, Ahmed M (2016) *Pseudomonas fluorescens* and *Trichoderma asperellum* enhance expression of Gα subunits of the pea heterotrimeric G-protein during *Erysiphe pisi* infection. Front Plant Sci 6:1206

Patel JS, Kharwar RN, Singh HB, Upadhyay RS, Sarma BK (2017) *Trichoderma asperellum* (T42) and *Pseudomonas fluorescens* (OKC)-enhances resistance of pea against *Erysiphe pisi* through enhanced ROS generation and lignifications. Front Microbiol 8:306

Pérez E, Sulbarán M, Ball MM, Yarzabál LA (2007) Isolation and characterization of mineral phosphate-solubilizing bacteria naturally colonizing a limonitic crust in the southeastern Venezuelan region. Soil Biol Biochem 39:2905–2914

Prabhukarthikeyan R, Saravanakumar D, Raguchander T (2014) Combination of endophytic *Bacillus* and *Beauveria* for the management of Fusarium wilt and fruit borer in tomato. Pest Manag Sci 70:1742–1750

Qureshi N, Annous BA, Ezeji TC, Karcher P, Maddox IS (2005) Biofilm reactors for industrial bioconversion processes: employing potential of enhanced reaction rates. Microb Cell Factories 4:24–26

Rasche F, Cadisch G (2013) The molecular microbial perspective of organic matter turnover and nutrient cycling in tropical agroecosystems–what do we know? Biol Fertil Soils 49:251–262

Rudresh DL, Shivaprakash MK, Prasad RD (2005) Effect of combined application of *Rhizobium*, phosphate solubilizing bacterium and *Trichoderma* spp. on growth, nutrient uptake and yield of chickpea (*Cicer arietinum* L.). Appl Soil Ecol 28:139–146

Samad A, Trognitz F, Compant S, Antonielli L, Sessitsch A (2017) Shared and host specific microbiome diversity and functioning of grapevine and accompanying weed plants. Environ Microbiol 19:1407–1424

Sarma BK, Yadav SK, Singh S, Singh HB (2015) Microbial consortium-mediated plant defense against phytopathogens: readdressing for enhancing efficacy. Soil Biol Biochem 87:25–33

Sekar J, Raj R, Prabavathy VR (2016) Microbial consortial products for sustainable agriculture: commercialization and regulatory issues in India. In: Agriculturally important microorganisms. Springer, Singapore, pp 107–132

Sen T, Barrow CJ, Deshmukh SK (2019) Microbial pigments in the food industry challenges and the way forward. Front Nutr 6:7

Seneviratne G, Zavahir JS, Bandara WM, Weerasekara MLMAW (2007) Fungal–bacterial biofilms: their development for novel biotechnological applications. World J Microbiol Biotechnol 24:739–743

Sessitsch A, Mitter B (2015) 21st century agriculture: integration of plant microbiomes for improved crop production and food security. Microb Biotechnol 8:32–33

Shelat HN, Vyas RV, Jhala YK (2017) Biofertilizers and PGPR for evergreen agriculture. In: Microorganisms in sustainable agriculture, food, and the environment. Apple Academic Press, pp 283–312

Singh JS, Pandey VC, Singh DP (2011) Efficient soil microorganisms: a new dimension for sustainable agriculture and environmental development. Agric Ecosyst Environ 140 (3–4):339–353

Singh A, Jain A, Sarma BK, Upadhyay RS, Singh HB (2013) Rhizosphere microbes facilitate redox homeostasis in *Cicer arietinum* against biotic stress. Ann Appl Biol 163:33–46

Singh M, Dotaniya ML, Mishra A, Dotaniya CK, Regar KL, Lata M (2016a) Role of biofertilizers in conservation agriculture. In: Conservation agriculture. Springer, Singapore, pp 113–134

Singh V, Upadhyay RS, Sarma BK, Singh HB (2016b) *Trichoderma asperellum* spore dose depended modulation of plant growth in vegetable crops. Microbiol Res 193:74–86

Singh BN, Dwivedi P, Sarma BK, Singh GS, Singh HB (2018) *Trichoderma asperellum* T42 reprograms tobacco for enhanced nitrogen utilization efficiency and plant growth when fed with N nutrients. Front Plant Sci 9:163

Singh BN, Dwivedi P, Sarma BK, Singh HB (2019a) *Trichoderma asperellum* T42 induces local defence against *Xanthomonas oryzae* pv. *oryzae* under nitrate and ammonium nutrient in tobacco. RSC Adv 9(68):39793–39810

Singh BN, Dwivedi P, Sarma BK, Singh GS, Singh HB (2019b) A novel function of N-signalling in plants with special reference to *Trichoderma* interaction influencing plant growth, nitrogen use efficiency, and cross talk with plant hormones. 3 Biotech 9:109

Sinha RK, Valani D, Chauhan K, Agarwal S (2010) Embarking on a second green revolution for sustainable agriculture by vermiculture biotechnology using earthworms: reviving the dreams of Sir Charles Darwin. J Agric Biotechnol Sustain Dev 2(7):113–128

Smith SE, Smith FA (2011) Roles of arbuscular mycorrhizas in plant nutrition and growth: new paradigms from cellular to ecosystem scales. Annu Rev Plant Biol 62:227–250

Srivastava S, Patel JS, Singh HB, Sinha A, Sarma BK (2015) *Streptomyces rochei* SM 3 induces stress tolerance in chickpea against *Sclerotinia sclerotiorum* and NaCl. J Phytopathol 163 (7–8):583–592

Stewart EJ (2012) Growing unculturable bacteria. J Bacteriol 194:4151–4160

Suresh K (2012) Biopesticides: a need for food and environmental safety. J Biofertil Biopestic 3 (4):1–3

Timmusk S, Behers L, Muthoni J, Muraya A, Aronsson AC (2017) Perspectives and challenges of microbial application for crop improvement. Front Plant Sci 8:49

Vartoukian SR, Palmer RM, Wade WG (2010) Strategies for culture of 'unculturable' bacteria. FEMS Microbiol Lett 309:1–7

Vassilev N, Vassileva M, Nikolaeva I (2006) Simultaneous P solubilizing and biocontrol activity of microorganisms: potentials and future trends. Appl Microbiol Biotechnol 71:137–144

Villaverde JJ, Sevilla-Morán B, Sandín-España P, López-Goti C, Alonso-Prados JL (2014) Biopesticides in the framework of the European Pesticide Regulation (EC) No. 1107/2009. Pest Manag Sci 70(1):2–5

Vorholt JA (2012) Microbial life in the phyllosphere. Nat Rev Microbiol 10:828–840

Wirthmueller L, Maqbool A, Banfield MJ (2013) On the front line: structural insights into plant–pathogen interactions. Nat Rev Microbiol 11:761–776

Yadav SK, Singh S, Singh HB, Sarma BK (2017) Compatible rhizosphere-competent microbial consortium adds value to the nutritional quality in edible parts of chickpea. J Agric Food Chem 65(30):6122–6130

Yadav SK, Prabha R, Singh V, Bajpai R, Teli B, Rashid MM, Sarma BK, Singh DP (2019) Microbes-mediated nutrient use efficiency in pulse crops. In: Microbial interventions in agriculture and environment. Springer, Singapore, pp 447–460

Yang JW, Kloepper JW, Ryu CM (2009) Rhizosphere bacteria help plants tolerate abiotic stress. Trends Plant Sci 14:1–4

Yeoh YK, Dennis PG, Paungfoo-Lonhienne C, Weber L, Brackin R, Ragan MA, Schmidt S, Hugenholtz P (2017) Evolutionary conservation of a core root microbiome across plant phyla along a tropical soil chronosequence. Nat Commun 8(1):1–9

Zarraonaindia I, Gilbert JA (2015) Understanding grapevine-microbiome interactions: implications for viticulture industry. Microb Cell 2(5):171

Chapter 17
Biofertilizers and Biopesticides: A Whole New Dimension for Ameliorating Soil Fertility and Organic Agriculture Practice

Meenakshi Rajput, V. Vivekanand, and Nidhi Pareek

Abstract In the forthcoming decades, maintaining food security, safety, and quality would impose a major challenge for the rapidly growing tropical countries. The excessive employment of the industrialized production methods has contaminated the food chain and water adversely so far by the continuous release of the harmful chemical residues of fertilizers and pesticides. Furthermore, the chemicals released amends the characteristics of the soil to highly acidic/alkaline that bring about the abatement in the number of beneficial soil microorganisms leading to the reduction in soil fertility and crop yields. Thus, to accomplish the aforementioned goals, it is highly desirable to move toward organic agriculture practices producing food with high quality and standards. The utilization of propitious microorganisms (PGPRs) as biofertilizers and biopesticides serves as better organic and eco-friendly alternative for the enhancement of soil fertility with efficient disease and pest control. Biofertilizers help in retaining the soil's macro and micronutrients, nitrogen fixation, antibiotic production, and phytohormone production and in the degradation of organic matter present in the soil. On the other hand, biopesticides are adeptly aid in pest control as they are comprised of the pathogenic microorganism specific to the pest of interest. Both biofertilizers and biopesticides offer ecologically and economically sustainable organic agriculture strategies with the assurance of an increase in soil biodiversity and the safety of food. The chapter highlights the microorganisms and their role in ameliorating soil fertility with the disease and pest control for sustainable organic agriculture.

Keywords Organic agriculture · Biofertilizers · Biopesticides · Microorganisms · Soil fertility

---

M. Rajput · N. Pareek (✉)
Department of Microbiology, School of Life Sciences, Central University of Rajasthan, Bandarsindri, Kishangarh, Ajmer, Rajasthan, India
e-mail: nidhipareek@curaj.ac.in

V. Vivekanand
Centre for Energy and Environment, Malaviya National Institute of Technology, Jaipur, Rajasthan, India

S. K. Dubey, S. K. Verma (eds.), *Plant, Soil and Microbes in Tropical Ecosystems*, Rhizosphere Biology, https://doi.org/10.1007/978-981-16-3364-5_17

17.1 Introduction

The world population will probably rise up to nine billion by 2050, indicating the urgent need for more food production to feed hungry mouths in the near future (Abbey et al. 2019). However, the practice of sustainable food production is still a major challenging task for the world as chemical-based fertilizers and pesticides are being used for crop production which ultimately imposes deteriorating effects on the environment as well as human health. The employment of these chemical fertilizers and pesticides has been accentuated in Indian agriculture by the commencement of the green revolution. The green revolution was a comprehensive collection of several valuable alternatives to enhance crop production such as high-yielding varieties (genetically engineered through modern breeding techniques), chemical-based fertilizers and pesticides, irrigation techniques (tube well and canal), and nutrient management (inorganic or organic). The appropriate utilization of the aforementioned techniques has amplified the crop yield and aided India to become self-sufficient during the hard times of the post-independence period (Singh et al. 2016). Despite the success of the green revolution in improving crop productivity across the globe, the utilization of chemical fertilizers has downgraded the quality of the soil. The chemical fertilizers increase the soil salinity that further obstructs the accessibility of micronutrients to the crops (Kumar 2018). At the same time, the chemical pesticides are also adding to environmental pollution as they are non-biodegradable and their continuous use making the insects resistant that further compelled the production of stronger pesticides. There are various other detrimental consequences of using chemical fertilizers and pesticides which include soil acidification, weakening of plant roots, high disease occurrence due to the death of healthy microorganisms and insects, and eutrophication of water bodies along with groundwater. This occurs because the chemical fertilizers or pesticides sprayed on the crops are not completely utilized by crops. For instance, the widely used chemical fertilizer, urea, when sprayed on the crops is partially used by crops, and remnants contaminate the water bodies through the runoff water (Kumar 2018). The water contamination through nitrate (urea) may lead to terrible ailments among the infants, viz. methemoglobinemia and hypertension, that in some cases make them handicapped also. Besides, the production of urea is even highly expensive as the production, transportation, and application of around 1 kg urea involve the expenditure of 1 L petroleum products. Therefore, it is clear that urea as fertilizer not only causes ill effects on the environment and mankind but is also not feasible economically (Pathak and Kumar 2016). In this regard, the organic framing serves as the best strategy to ensure food safety along with the replenishment of the soil biodiversity.

Nowadays, organic farming is receiving enormous attention globally from the scientific community as well as the public owing to the increased awareness about the harmful effects caused by the indiscriminate use of chemical fertilizers and pesticides. Interestingly, during the past two decades, the total area of organic farmland has been reported to reach up to 69.8 million hectares, and around 1.4% of total agricultural land is used for organic farming (Willer and Lernoud 2019).

Still, it is inexorable to address the ever-increasing demand for food worldwide without the use of chemical fertilizers and pesticides; thus, the need of the hour is to use these chemical-based products judiciously with the bio-based products (biofertilizers and biopesticides). Rather, organic production can be promoted in the selected niche or crops to satisfy the demand of the domestic export market (Mishra et al. 2013). Organic farming primarily relies on the natural soil microflora comprised of all the beneficial bacterial and fungal species including arbuscular mycorrhiza fungi (AMF). Biofertilizers and biopesticides being the fundamental constituents of organic farming enhance crop production as well as protection. Biofertilizers augment the soil fertility by making it affluent in all the essential micro- and macronutrients by the microbial nitrogen fixation, potassium (K), and phosphate solubilization, the release of phytohormones, and degradation of organic matter. Biopesticides are composed of biocontrol agents to prevent crop loss from diseases, weeds, insects, and nematodes (Abbey et al. 2019). Thus, the holistic twin approach of biopesticides and biofertilizers in organic farming would assist in the augmentation of crop yield throughout the globe with the simultaneous maintenance of soil fertility.

17.2 Need of Bio-Based Fertilizer and Pesticides

In the twenty-first century, one of the major tasks is to fulfil the food requirement of the burgeoning population on the planet with the employment of environmental and economically sound agriculture inputs (Meena et al. 2016). Besides this, the blanket use of agrochemicals is severely declining the population of beneficial microorganisms that further makes the crops more susceptible to biotic and abiotic stresses. The promiscuous utilization of agrochemicals is directly affecting the biogeochemical cycles also to the great extent due to the detrimental effects of agrochemicals on the ecosystem. Moreover, the natural reserves of phosphate (phosphate rocks) are on the verge of complete depletion; on the other side, the high energy-consuming Haber-Bosch process of nitrogen fertilizers production depends on the fossil fuels leading to the depletion of natural non-renewable resources with the aggravation of global warming (Erisman et al. 2013; Cordell and White 2014). Thus, the cost of the chemical-based fertilizers is rising dramatically with the increase in prices of petroleum-based products utilized for their production. The production of agrochemicals requires high energy input, for instance, 1.1 kWh phosphorus (P), 11.2 kWh nitrogen (N), and 1 kWh Potash are required for the production of 1 kg of fertilizer (Saritha and Prasad Tollamadugu 2019). Therefore, after being cognizant about the ruinous effects caused by the agrochemicals and their skyrocketing costs, there is an urgent need to exploit the salubrious interaction between plants, soil microflora, and the environment. There are several plant interacting soil microbes that contribute to the plant growth with the significant enhancement in soil fertility utilized as biofertilizers. At the same time, biopesticides also offer multiple advantages including the targeting of specific pests rather than affecting the whole range of pests

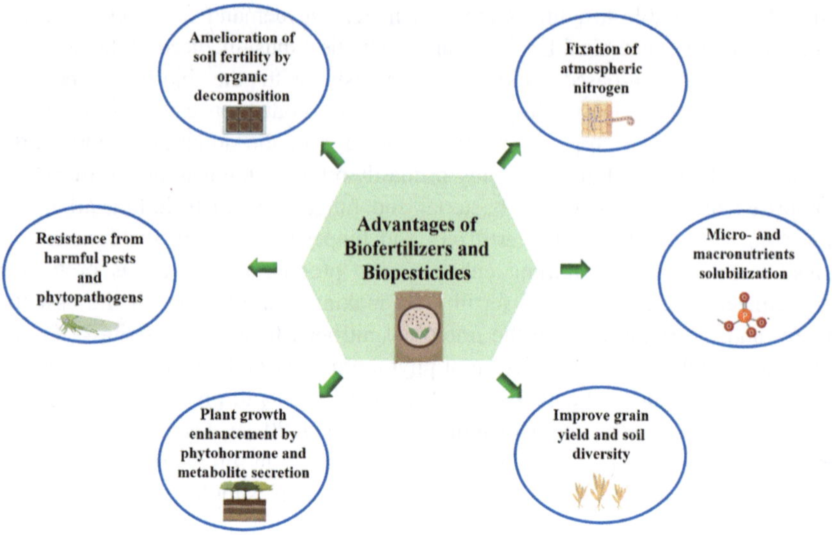

Fig. 17.1 Advantages of biofertilizers and biopesticides in agriculture

together with many birds and animals. The biopesticides are degraded more quickly and required in minimal quantity when compared to chemical-based pesticides, thereby decreasing the exposure. Thus, the biopesticide can be employed as an alternative to the chemically synthesized pesticides in the integrated pest management (IPM) programs, contributing higher crop yield and less harm to the environment (Thakore 2006). The bio-based fertilizers and pesticides formulated by the incorporation of microorganism strains or other natural substances may help in dealing with all the challenges coming in sustainable agriculture practice. The major types of biofertilizers and biopesticides with their mode of action are explained in further sections. Various advantages of biofertilizers and biopesticides are depicted in Fig. 17.1.

17.3 Biofertilizer: A Boon for Sustainable Agriculture Practice

Biofertilizers, generally mentioned as bioinoculants, are the reasonable and eco-friendly microbial preparations that increase the bio-accessibility and bioavailability of plant nutrients. The biofertilizers are prepared from the active or latent strains of microorganisms belonging to the bacterial, fungal, and algal domain. Mostly, bacterial strains are solely employed as bio-inoculants, but in some cases, the combination of bacterial species with fungi or algae has also been used to boost the microbial activity (Suyal et al. 2016). These microorganisms themselves do not serve as the source of nutrition to plants but participate in various rhizospheric interactions to convert the nutrients to plants' utilizable form. These rhizospheric interactions lead to several biochemical processes that involve the fixation of nitrogen (N), solubilization of zinc (Zn) and phosphate, and mobilization of potash, phosphate, and other micronutrients (Suhag 2016; Suyal et al. 2016; Anand et al. 2016; Kamran et al. 2017). Additionally, these microorganisms also assist in the plant growth by secretion of various phytohormones such as auxins, gibberellins, cytokinins, and abscisic acid that directly boost the plant growth (Wong et al. 2015). There are many other roles played by the bacterial species that stimulate the plant growth, viz., secretion of lyases and siderophores, production of antibiotics and low molecular weight metabolites that antagonize other plant pathogens from the colonization on roots, and confer induced systemic resistance (ISR) in plants (Kumar 2018; Gopalakrishnan et al. 2015). Thus, owing to the ability of these microorganisms to promote plant growth together with providing resistance against various stresses, they are generally regarded as plant growth-promoting microorganisms (PGPM). In particular, the fungi and bacteria possessing the potential to alleviate the plant growth are called plant growth-promoting fungi (PGPF) and plant growth-promoting rhizobacteria (PGPR), respectively. The PGPRs have the potentiality to enhance the plant's growth either by direct or indirect mechanisms. The direct secretion of phytohormones and nutrients induces the plant growth directly, whereas the symbiotic association of bacterial species with plants supports the indirect mechanism (Kenneth 2017; Kenneth et al. 2019).

Primarily, the biofertilizers comprised of microorganisms having the potential to fix nitrogen and solubilize phosphate and cellulolytic enzymes secretion. The nitrogen-fixing biofertilizers mainly include *Rhizobium*, *Azolla*, *Azotobacter*, *Cyanobacteria*, and *Azospirillum* having the ability to fix atmospheric nitrogen into the soil in plant utilizable forms. The phosphate-solubilizing biofertilizers such as *Bacillus* and *Pseudomonas* can efficiently solubilize the tricalcium phosphates (TCP) and rock phosphate by secreting various organic acids to make it readily available to the plants (Dotaniya et al. 2013, 2014). The nitrogen-fixing biofertilizers composed of *Rhizobium*, *Azospirillum*, and *Azotobacter* blanket the major portion of the biofertilizers value in the market share. Altogether, the global market value of the biofertilizers was estimated at around USD 1.0 billion in the year 2019, which is expected to evidence a compound annual growth rate (CAGR) of 12.8% between 2020 and 2027 (https://www.grandviewresearch.com/industry-analysis/biofertilizers-industry).

The history of the employment of biofertilizers in agriculture is way too long as the farming community has been continuously using biofertilizers from the generations in rural areas in the form of microbial inoculations of small-scale compost. Still, there is some kind of confusion in the farming community regarding the cost and efficacy of biofertilizers due to the lack of poor handling and storage. Biofertilizers are apparently considered as more expensive than chemical-based fertilizers due to the lack of knowledge about modern technologies that can be utilized to manufacture biofertilizers from available biowastes, short shelf life, suitable carrier material, and instability at high temperatures (Singh et al. 2016). Thus, there is an urgent need to resolve these issues to expand the utilization of biofertilizers in remote areas along with the provision of proper training about the usage and storage of these bio-based products to the farmers.

17.4 Types of Biofertilizers

In natural ecological systems, nutrients such as nitrogen, phosphorus, and sulfur are found in a bound state with the organic molecules which are not utilized directly by plants. Thus, the plants solely rely on the soil microorganism to make these growth-limiting nutrients biologically accessible to them. These soil microorganisms through various metabolic processes convert them into the inorganic forms such as nitrate, ammonium, sulfate, and phosphate and further release them into the soil (Van Der Heijden et al. 2008; Jacoby et al. 2017). Likewise, the biofertilizers composed of these essential soil PGPM can efficiently bring about the nutrient transformations that will enhance the crop productivity with the maintenance of soil diversity. The role and interactions of soil microorganisms with plants in sustainable agriculture practice have been comprehensively reviewed by many researchers worldwide (Meena et al. 2016; Li et al. 2017). At present, biofertilizers as an integral component of organic farming are the center of attraction; thus various types of biofertilizers based on their function and interaction with plants are addressed in the next subsections (Fig. 17.2).

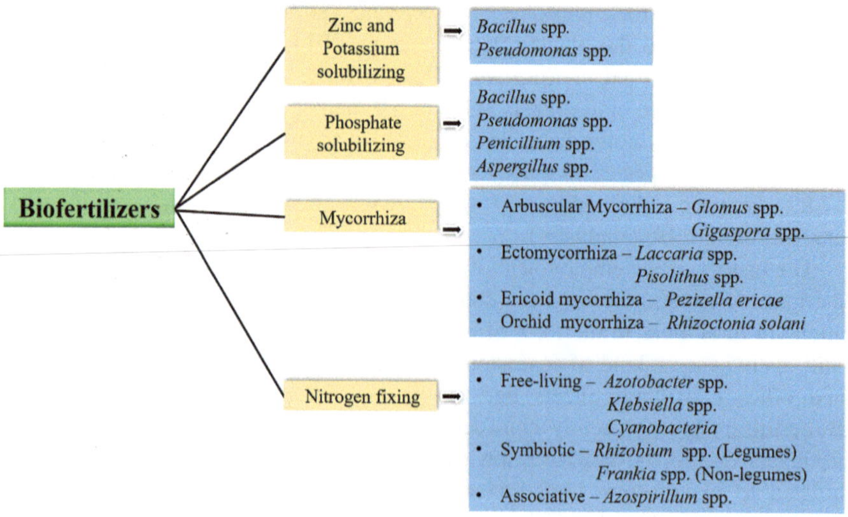

Fig. 17.2 Various types of biofertilizers employed in organic farming

17.4.1 Nitrogen-Fixing Biofertilizers

Nitrogen (N) is one of the main constituents of biomolecules (nucleic acids and proteins) and plays a vital role in the growth and development of all living beings. In plants, it serves as a pivotal element of chlorophyll, alkaloids (colchicine, nicotine, quinine, etc.), plant growth hormones, and glucosinolates. N in the gaseous form makes up approximately 78% of the total Earth's atmosphere, yet cannot be utilized directly by the plants and animals. Thus, it needs to be converted into the relevant organic form (such as ammonium or nitrate) to be utilized in the formation of biomolecules. Several soil microorganisms possess the oxygen-sensitive nitrogenase enzyme for the fixation of atmospheric N into ammonia. This process is generally known as biological N fixation. Mainly, the bacterial species that carry out the process of nitrogen fixation are either free-living (*Azotobacter* and *Azospirillum*) or found in symbiotic association with plants (*Rhizobium* and *Frankia*). *Rhizobium*, *Sinorhizobium*, and *Bradyrhizobium* make symbiotic associations with the leguminous plants and cause root nodule formation. Likewise, *Frankia* forms the root nodule in the non-leguminous actinorhizal plants (Kumar 2018). The cyanobacteria and mycorrhiza have also been reported to participate in the process of nitrogen fixation (Pereira et al. 2009; Püschel et al. 2017).

The N-fixing symbiotic bacteria, *Rhizobium*, is a member of the family *Rhizobiaceae* that can fix nitrogen in legumes at 50–100 kg ha^{-1} and also in some non-leguminous plants such as *Parasponia*. *Rhizobium* gets access in the root system of legumes after germination of seed and colonizes there to form tumor-like growth which is known as root nodules that act as the ammonia manufacturing units. The addition of *Rhizobium* as bio-inoculants in the fields can considerably upsurge the crop yield and benefit several leguminous crops such as lentil, gram, and chickpea; vegetables like sugar beet, pea, and alfalfa; and oilseeds crop including groundnut, soybean, and lentil (Baset Mia and Shamsuddin 2010; Giri and Joshi 2010). Samago et al. (2018) conducted a field experiment on common bean in low-P soil of Ethiopia to examine the effects of *Rhizobium* inoculation and phosphorus application (20 kg P ha^{-1}) on the grain yield, plant growth, and symbiotic performance. The results showed accelerated plant growth and symbiotic performances owing to *Rhizobium* inoculation and high grain yield in the P-fed plants. Similarly, Khan et al. (2018) reported that bio-inoculation of *Rhizobium* strains on three leguminous crops (chickpea, mung bean, and pigeon pea) has positively affected the plant growth, N uptake, nodulation, and leghemoglobin content. Also, the occurrence of galling and reproduction of *Meloidogyne incognita* has been reduced largely in chickpea, mung bean, and pigeon pea through the seed treatment by *Bradyrhizobium japonicum*, *Mesorhizobium ciceri*, and *Rhizobium* sp., respectively. This indicates the dual benefit of *Rhizobium* as biofertilizers by enhancing the crop yield by nitrogen uptake as well as providing protection against biotic stresses. *Azotobacter* belongs to the family *Azotobacteraceae*, which is a heterotrophic, free-living, and aerobic bacteria that colonize on the plant roots and fix around 25 kg N ha^{-1}. The production of antifungal compounds has been observed from

Azotobacter species in the rhizosphere that antagonizes growth of fungal phytopathogen, thereby increasing seeding survival rate (Mishra et al. 2013). Romero-Perdomo et al. (2017) evaluated the influence of *Azotobacter chroococcum* strains AC1 and AC10 on the cotton plant growth, and findings suggested that the co-inoculation of both the strains has reduced the supplementation of N-fertilizers by 50%. The effect of *Azotobacter* sp. and *Azospirillum* sp. on the growth of tomato plants was assessed by Reddy et al. (2018), and results revealed that inoculation of *Azotobacter* sp. and *Azospirillum* sp. with 75% dose of NPK fertilizers displayed the maximum growth in tomato plants. *Azospirillum* (*Rhodospirillaceae)* are heterotrophic and associative bacteria with the potential of 20–40 kg ha^{-1} N-fixing. *Azospirillum* is one of the extensively studied PGPR from the lab to field experiments. It is considered as the safest bacterial species to be utilized as biofertilizer owing to its non-pathogenic behavior. It holds the potential to fix N and solubilize phosphate, phytohormones, and siderophore production (Mehnaz 2015). Sahoo et al. (2014) isolated several strains of *Azospirillum* from the different rhizosphere of rice fields and assessed their effects as biofertilizer. The results revealed that *Azospirillum lipoferum* (As6) has significantly improved the nutrient content, growth, and yield of rice var. *Khandagiri* along with good N-fixing performance, phytohormone production, siderophore secretion, and iron tolerance. Mazhar et al. (2016) evaluated the salinity tolerance and biocontrol potential of the *A. lipoferum* and observed the resistance from *Aspergillus niger* and *Pseudomonas* with considerable salt-stress tolerance in wheat crop. *Azolla* (*Cyanobacteria*) is mostly utilized as green manure or compost. Similar to other N-fixing biofertilizers, it can also assist in the N-fixation as well as phytohormone production for the plant growth promotion. Razavipour et al. (2018) observed that *Azolla filiculoides* compost has notably improved the growth and yield of rice crop for two growing seasons under the water-deficient conditions. The inoculation of 5.0% of total soil has given the highest grain yield which was found to be 13.8% higher than uninoculated crops. Maswada et al. (2020) demonstrated the effect of *A. filiculoides* extract application on maize plants under nitrogen- and water-deficient conditions. The results displayed substantial increase in N uptake, plant growth, grain yield, N-utilization efficiency, and proline accumulation along with notable alleviation in oxidative damage. Additionally, the implementation of urea fertilizer has been decreased by 30% with the application of *A. filiculoides.* Thus, *Azolla* is one of the potential candidates in the development of water saving and low-input agriculture system.

17.4.2 Phosphate Solubilizing Biofertilizers

Phosphorus (P) is the highly essential element for the biosynthesis of phospholipids and nucleic acids and also the most crucial macro-element for the plants after N. It plays important role in the process of photosynthesis and respiration as it is the core component of the "molecular currency," i.e., adenosine triphosphate (ATP). Plants utilize the P in the form of orthophosphates, i.e., $H_2PO_4^-$ and HPO_4^{2-}. P is available in soil in both organic and inorganic forms; out of which, organic form is usually found in decayed organic matter and humus which constitutes a significant reservoir (~30–50%) of P in soil. Most of the P content found in soil is usually fixed, i.e., forms chemical compounds with hydrated oxides or hydroxide of other elements, therefore becoming unavailable for plants. A large part of the fixed P in soil is found due to the application of chemical inorganic phosphate fertilizers which are partly utilized by the plants, and the remaining get immobilized or fixed. There are several microorganisms found in the soil and rhizosphere possessing the ability to solubilize the phosphates of various elements including calcium (Ca), iron (Fe), and aluminum (Al) found in soil. These microorganisms formulate the P mineralizing and solubilizing biofertilizers. They can be aerobic or anaerobic, but in submerged soil, aerobic microbes are more prevalent.

P-fixation and precipitation are highly influenced by soil pH and type. The P-fixation is found to be higher in the acidic or calcareous soil conditions, which can be alleviated by the proper adjustment of soil pH to make phosphorus biologically available to plants (Mahdi et al. 2012). In acidic soils, phosphorus fixation occurs with the hydroxides or oxides of Al and Fe, whereas in alkaline soil conditions, phosphorus fixation occurs by calcium. Phosphate solubilizing biofertilizers secret the organic and inorganic acids, which act on inorganic phosphorus and chelate cations (Ca, Fe, Al) through their acidic hydroxyl and carboxyl groups which further decrease pH in alkaline soil. Phosphate solubilizing biofertilizers secret the organic and inorganic acids which act on inorganic phosphorus and chelate cations (Ca, Fe, Al) through their acidic hydroxyl and carboxyl groups. For the solubilization of mineral phosphates, tri-/di-carboxylic acids have known to be more helpful when compared to monobasic and aromatic acids (Mahdi et al. 2012). The solubilization of organic phosphates in the soil is known as mineralization, which can be achieved by the action of phosphatases derived from soil microorganisms. These phosphatases catalyze the conversion of organic phosphate into inorganic form by utilizing them as a substrate. The most widely used microorganisms as phosphate solubilizing biofertilizers are *Bacillus* spp. (Sharma et al. 2007), *Pseudomonas* spp. (Oteino et al. 2015), *Xanthomonas* spp., *Aspergillus* spp. (Mittal et al. 2008), and *Penicillium* spp. (Reyes et al. 2002; Pradhan and Sukla 2006). Sharma et al. (2007) performed the inoculation of chickpea seeds with *Bacillus megaterium* and *Pseudomonas fluorescens* as phosphate-solubilizing fertilizers with the solubilization efficiency of 128.57 and 200.00, respectively. The findings suggested increased seed germination efficiency, seedling length, and yield; and *P. fluorescens* was found to be more effective, whereas co-inoculation showed more seedling length when compared to single inoculation. The phosphate-solubilizing fertilizer (*Aspergillus niger*) has been reported to enhance the height of the plant, leaf length/width, size of fruit, and number of fruits per plant in okra and bottle guard when utilized together with N-fixing *Azotobacter* sp. (SR-4) (Din et al. 2019). Similarly, the N-fixing *Rhizobium meliloti* and *Klebsiella pneumonia* as phosphate solubilizer as biofertilizers decreased the mortality rate in alfalfa seedlings and increased the root length, shoot height, leaf area, root volume, number of leaves per plant, biomass, and uptake of P in two alfalfa varieties (Li et al. 2013). Oteino et al. (2015) conducted a study on the utilization of endophytic bacteria *Pseudomonas fluorescens* strains as phosphate-solubilizing biofertilizers on pea plants. The results revealed that three strains of *P. fluorescens* L111, L228, and L321 have proficiently solubilized phosphate (400–1300 mg L^{-1}), secreted gluconic acid, and enhanced plant growth. The *P. fluorescens* L321 boosted the plant growth even in the phosphate-limiting conditions, thus considered as the most effective out of all the strains. The studies clear the ability of phosphate-solubilizing biofertilizer in the enhancement of plant growth and soil fertility.

17.4.3 Potassium-Solubilizing Biofertilizers

Potassium (K) is considered as the third most essential nutrient for the plants after N and P. It participates in the opening and closing of stomata, which leads to the regulation of osmotic balance in the plant (Abbey et al. 2019). K-deficient plants possess less developed root system, slow growth, small seeds, and lower product yields (Teotia et al. 2016). In soil, K exists in the various forms which include mineral K, non-exchangeable K, exchangeable K, and ionic K (solution or dissolved form). The K is abundantly present in the soil, yet 1–2% of total K is utilized by the plants because the remaining K cannot be used by the plants as it occurs in silicate mineral form (mica and K feldspar) (Zhang and Kong 2014). The organic acid-producing microorganisms can be utilized as biofertilizers to increase the solubilization of K in soil. The organic acids can readily solubilize K by making a complex with calcium ions or by providing protons (Shanware et al. 2014). *Bacillus* spp. (*B. circulans*, *B. edaphicus*, *B. megaterium*, *B. mucilaginosus*) have been studied extensively for the solubilization of K. Besides, several other bacterial and fungal species have also been reported to have K solubilization ability including *Arthrobacter* sp., *Pseudomonas putida*, *Paenibacillus* sp., and *Aspergillus* spp. (Teotia et al. 2016; Verma et al. 2017). Singh et al. (2010) demonstrated the mobilization of K from the mica waste (MW) by *Bacillus mucilaginous* when inoculated with maize and wheat, whereas the *Rhizobium* spp. and *Azotobacter chroococcum* also displayed K solubilization potential. Likewise, *Bacillus pseudomycoides* isolated from the rhizosphere of tea plants solubilized 33.32 ± 2.40 μg mL^{-1} of K from the broth amended with MW after 7 days incubation, while in soil microcosm, 47.0 ± 7.1 μg kg^{-1} of K was solubilized after 105 days incubation in laboratory conditions (Pramanik et al. 2019). The studies indicate the tremendous potential of these microbial strains to be employed as K-solubilizing biofertilizers.

17.4.4 Zinc-Solubilizing Biofertilizers

Zinc (Zn) is recognized as one of the most important micronutrients for both eukaryotic and prokaryotic organisms as it acts as a cofactor and activator in many enzymes. It participates in protein synthesis, seed development, and growth hormone production (Abbey et al. 2019). The 96–99% of exogenously supplied soluble Zn as fertilizer to the plants convert into the unavailable form and get fixed in the soil. Various parameters of soil such as high pH, organic matter, high $CaCO_3$ content, copper, and phosphate level can fix the soluble Zn into the soil. The solubilization of Zn in the soil can be achieved through the utilization of organic acid-producing microorganisms found in soil. The lowering of soil pH by the release of organic acids such as gluconic acid, glycolic acid, acetic acid, lactic acid, etc. sequester the cations leading to the acidic rhizospheric environment that would help in Zn solubilization. Additionally, the anions can solubilize Zn by its chelation and convert it into a plant usable form, i.e., Zn^{+2} (Kumar 2018). Several microorganisms have proved their potential in Zn solubilization, viz., *Bacillus* sp., *Pseudomonas* sp., *Aspergillus* sp., and *Klebsiella* sp. (Khande et al. 2017; Gontia-Mishra et al. 2017). Four bacterial species (*Pseudomonas aeruginosa*, *Ralstonia pickettii*, *Burkholderia cepacia*, and *Klebsiella pneumoniae*) isolated from the rhizosphere were analyzed for their ability to solubilize the Zn from ZnO and $ZnCO_3$ present in the medium. The results displayed that Zn solubilization promoted the growth in rice seedling and other cereals. Moreover, Zn-solubilizing bacterial species possessed several other plant growth-promoting characteristics also like P and K solubilization, exopolysaccharide production, and 1-aminocyclopropane-1-carboxylic acid (ACC) utilization Gontia-Mishra et al. 2017).

17.4.5 Mycorrhiza Biofertilizers

Mycorrhiza, commonly recognized as fungus root, is the symbiotic association between roots of plant and soil fungal mycelia. In this symbiotic association, the host plant gets benefited with the easy accessibility of growth-limiting nutrients with the help of fine fungal hyphae, and in turn, fungi fulfil its carbon requirements from the plant (Mishra et al. 2013). The AMF possesses a special structure known as arbuscules for the efficient transfer of nutrients from fungus to the root system and vesicles for the storage of P (Dhir 2017). Various types of mycorrhizal associations have been studied so far, namely, ectomycorrhiza, endomycorrhiza (arbuscular mycorrhiza, AMF), ectendomycorrhiza, ericoid mycorrhiza, orchid mycorrhiza, arbutoid mycorrhiza, and monotropoid mycorrhiza. The AMF is highly important as it is prominently found in approximately 85% of terrestrial plant families. The hyphae of AMF reach beyond the nutrition depletion zone in search of the high amount of mineral nutrients for the plant. Thus, AMF benefits the plant by enhancing P content, tolerance to various biotic and abiotic stresses, micronutrients and water uptake, the survival rate of seedling, and resistance against pest and other phytopathogens (Kumar 2018). The effect of four AMF species (*Gigaspora margarita* P18, *Scutellospora heterogama* P29, *Acaulospora longula* P20, and *Funneliformis mossease* P07) isolated from different soils sampled from various fields was investigated on the growth promotion and drought stress tolerance ability of various crops (sorghum, leek, carrot, and red pepper). The AMF conferred the positive effect on the growth and drought-tolerant ability of sorghum and carrot, whereas comparatively lesser growth was observed in red pepper and leek (Kim et al. 2017). Likewise, Oyewole et al. (2017) examined the influence of *Gigaspora gigantea* and *Glomus deserticola* on the growth drought tolerance potential of cowpea. The *G. deserticola* affected the water stress tolerance ability and product yield positively, while the combination of *G. deserticola* and *G. gigantea* has provided resistance against charcoal rot disease of cowpea caused by *Macrophomina phaseolina.* The role of biofertilizers in the amelioration of soil fertility and plant growth promotion has been comprehensively advocated with the implications in various improvements; thus, the contribution of biopesticides as a part of IPM is discussed in further sections.

17.5 Biopesticides

Biopesticides have emerged as a competent alternative for chemically synthesized pesticides. They offer multiple benefits to the crops as compared to chemical pesticides such as environmental safety, target specificity, biodegradability, efficacy, and cost-effectivity (Gupta and Dikshit 2010). Even the continuous use of biopesticides on crops poses no detrimental impacts on the agroecosystems. Biopesticides possess a wide range of microbes and microbes-derived biochemical substances to confer resistance against pests including bacteria, fungi, nematodes, and insects. Biopesticides can be composed of metabolites derived from microorganisms, phytochemicals, or any other microbial by-product that can control pests in an eco-friendly manner through various non-toxic mechanisms. The formulations of microbes containing biopesticides can either be solid or liquid. Solid formulations consist of solid carriers including clay, lignite, talc, etc. and give high crop yield, whereas the liquid formulations are composed of various solvents, namely, water, organic acids, or oil. The liquid formulations have several advantages over solid, as they have longer shelf life, high efficacy and purity, and easy application and handling (Dhir 2017). Broadly, the biopesticides contain the microbial pathogen or natural substances malicious to the target pest, including bioinsecticides, biofungicides, and bioherbicides. They are extensively used in the regions where niche markets, pesticide resistance, and environmental concerns restrict the employment of chemical pesticides. Additionally, biopesticides also serve in the maintenance of beneficial native microbes diversity and insects population owing to the target specificity and non-hazardous implications of biopesticides. Employment of biopesticides in agriculture also aids the farming community to satisfy the demands of enlightened consumers regarding their health and food safety(Abbey et al. 2019).

Global production of biopesticides is approximately 3000 tons per year which is accelerating every year at a rapid pace. In 2014, the US Environment Protection Agency (EPA) has registered over 1320 biopesticides products together with more than 430 active ingredients for biopesticides production (Mehrotra et al. 2017). Asia covers just 5% of total biopesticides sold in the market globally, whereas the US market holds first position in the sale of biopesticides with 200 products. Notwithstanding the environmental safety and low toxicity, the implementation of biopesticides is restricted due to several limitations like short shelf life, high costs, and scarcity. Therefore, it becomes difficult for small and marginal farmers to afford the additional expense of biopesticides. For the growth of the biopesticides market, the pressing priority is to increase research and development along with the ease in procedures for product registration and licensing. Furthermore, regular awareness programs should be organized to make the farmers and growers aware of the leading advantages of biopesticides in agriculture (Mishra et al. 2015).

17.6 Categories of Biopesticides

Biopesticides are broadly categorized into three categories depending on the active biocontrol agent or substance present as microbial pesticides, plant-incorporated protectants (PIPs), and biochemical pesticides. The specific roles of these biopesticides are elucidated in the following subsections.

17.6.1 Microbial Pesticides

The exorbitant use of chemical pesticides in agriculture has led to the development of resistance in many pests leading to the generation of new strains of pests. This phenomenon of resistance development in pest has made the researchers worried, which led to the foundation of biopesticides development (Nawaz et al. 2016). Besides this, the occurrence of acute or chronic poisoning in the developing countries further necessitated the need for bio-alternatives to control the pests. Microbial pesticides are formulated with potent microorganisms (bacteria, viruses, algae, fungi, and protozoans) as active biocontrol ingredients. The microorganisms employed for the construction of microbial pesticides are highly specific to the target pest. Microbial pesticides control the pests by making them diseased through the secretion of specific toxins. Majorly, the toxins secreted by these microorganisms are peptides that are distinct to each other in terms of specificity, toxicity, and chemical structure (Abbey et al. 2019).

The most extensively studied microbial pesticide is the insecticidal bacterium, *Bacillus thuringiensis* (*Bt*). This has been implied for the protection of crops from

Table 17.1 List of various microbial pesticides employed for development of pest-resistant plants

Microbial pesticide	Target pest	Crop improved	References
Bacillus thuringiensis	*Helicoverpa armigera*	Alfalfa	Sharma et al. (2011)
Metarhizium anisopliae *Beauveria bassiana*	*Bemisia tabaci* *Frankliniella occidentalis* *Bactericera cockerelli*	Tomato	Rios-Velasco et al. (2014)
Metarhizium anisopliae *Paecilomyces fumosoroseus*	*Spodoptera exigua*	Chinese cabbage	Han et al. (2014)
Bacillus thuringiensis	*Diabrotica virgifera virgifera*	Maize	Jakka et al. (2016)
Bacillus thuringiensis	*Hyphantria cunea* *Lymantria dispar*	Poplar	Wu et al. (2019)
Metarhizium anisopliae	*Nilaparvata lugens* *Sogatella furcifera*	Rice	Tang et al. (2019)
Metarhizium robertsii *Beauveria bassiana*	*Tetranychus urticae*	Bean	Canassa et al. (2019)
Metarhizium anisopliae	*Frankliniella occidentalis*	Eggplant	Li et al. (2021)

black flies, mosquitoes, and moths (caterpillars/larvae). The enormous amount of research has been conducted on the *Bt*, and it has become the first commercially employed biopesticide across the world. The protein crystals (δ-endotoxin) produced by *Bt* during spore formation are applied to plant foliage. The ingestion of these protein crystals or endotoxin by insects while feeding on plant causes lysis of their gut cells which result in the death of insect (Dhir 2017). *Bacillus subtilis* has also been reported to protect plants against phytopathogens using its antibiosis activity (Romero et al. 2007). Fungi also have the potential to protect the crops against multiple insects and act as a mycoinsecticide agent. Fungi intrude in the insect body by penetrating the cuticle and secret mycotoxins after entering into the hemolymph, thus employed widely to control the insects having piercing mouthparts like whiteflies and aphids. Many fungal species including *Metarhizium anisopliae* (Kern et al. 2010), *Beauveria bassiana* (Jia et al. 2010), and *Paecilomyces fumosoroseus* (Lopez et al. 2014) have efficiently proved their capability to control pests for sustainable agriculture. Several bioinsecticides have been developed using entomopathogenic baculoviruses. Baculoviruses encode many enzymes and proteins that improve its potency to infect and replicate in the host's body. The virus kills the insect by ingestion of virus applied plants that further takeovers the whole metabolic machinery of the insect for its replication and transmission (Hubbard et al. 2014). Baculoviruses hold high specificity toward their hosts and mostly infect insects and a few arthropods. Baculoviruses are categorized into two main genera, namely, *Granulovirus* (GV) and *nucleopolyhedrovirus* (NPV). Interestingly, approximately 13 NPV virus-based microbial insecticides have been registered throughout the world. Various microbial pesticides that helped in the development of pest-resistant plants are enlisted in Table 17.1.

17.6.2 Plant-Incorporated Protectants (PIPs)

PIPs are the substances produced by the genetically engineered plants having toxin encoding genes incorporated in their genome, for instance, the introduction of a gene encoding for *Bt* insecticidal protein or δ-endotoxin into the plant genome. The plant will produce the insecticidal toxin for its protection against various insects. The *Bt* toxin produced by plants gets active in the alkaline environment of the insect's gut. Vaughn et al. (2005) developed corn rootworm-resistant transgenic maize varieties by the introduction of *Cry3Bb1* gene in maize genome. Likewise, *Helicoverpa armigera*- and *Phthorimaea operculella*-resistant transgenic tomato lines were developed through the incorporation of *Cry2Ab* gene via *Agrobacterium*-mediated transformation method (Saker et al. 2011). Siddiqui et al. (2019) conducted a study to develop a double cry gene (*Cry1Ac* + *Cry2Ab*) incorporated cotton plant transgenic lines. The results of the insect assay revealed that these transgenic cotton lines showed 93% mortality rate against armyworm (*Spodoptera litura*).

17.6.3 Biochemical Pesticides

Biochemical pesticides (sometimes called as semiochemical) are composed of naturally occurring substances derived from plants, animals, or insects. This class of biopesticides control pests through non-toxic mechanisms and also obstruct the mating and population growth. For instance, the production of secondary metabolites from plants prevents the consumption of plants by herbivores. Pyrethrin, a secondary metabolite secreted by *Chrysanthemum cinerariaefolium*, acts as a potent insecticidal compound (Silvério et al. 2009). Another most common source of biochemical insecticides is neem (*Azadirachta indica*) oil (Schmutterer 1990). It possesses two organic compounds, namely, salannin and azadirachtin, highly efficacious to kill insects. Azadirachtin has the potential to kill the insect by making it incapable to undergo molting to move in the next life stage. The insect-ingested azadirachtin-treated plants die within a period of 24 h. Liang et al. (2003) demonstrated the insecticidal potential of three commercial neem-based insecticidal preparations, namely, Agroneem, Neemix, and Ecozin, against diamondback moth (*Plutella xylostella* L.). The findings revealed that all three neem-based insecticides exhibited antifeedant effect against the *P. xylostella* and also significantly reduced the size of larvae. The antifeedant and inhibitory effect of neem limonoids (azadirachtin, deacetylnimbin, salannin, 17-hydroxyazadiradione, deacetylgedunin, and gedunin) was assessed against the rice leafroller (*Cnaphalocrocis medinalis*). Azadirachtin has showed the better resistance as compared to other limonoids (Nathan et al. 2005).

17.7 Conclusion

This is the modern era of biotechnology that demands sustainable agriculture practice as the indiscriminate employment of agrochemicals for crop production not only imparting deleterious effects on the environment and human health but also depleting the highly valuable natural non-renewable resources. The depletion of non-renewable resources may lead to a world-food emergency in the next few decades. Therefore, the new vistas of sustainable agriculture need to be explored to develop agriculture-inputs judicious in terms of environment, human health, and cost. Additionally, the burgeoning demand for healthier food across the world has ignited the interest of the farming and research community toward novel organic farming strategies. Thus, the requirement for biofertilizers and biopesticides has also been increased through all these years. These bioproducts serve as commendable alternatives for the agrochemicals with multiple advantages; still meeting the food requirement without agrochemicals is not viable. This is due to some of the demerits of bioproducts such as lack of profiling and narrow target range in biopesticides, selection of appropriate microbial strain for inoculation, high-temperature instability, and shorter life span of biofertilizers due to poor handling techniques. The manufacturing and development of agricultural bioproducts need more attention to drive their journey from the lab to commercial scale. Molecular techniques can aid in the development of biopesticides with a broad spectrum of targets and high activity. For the extension of bioproducts utilization at a wider scale, more research and investment need to be done together with the organization of various seminars and training workshops covering the proper handling, storage, and application strategies of bioproducts for small and marginal farmers.

References

Abbey L, Abbey J, Leke-aladekoba A, Iheshiulo EM, Ijenyo M (2019) Biopesticides and biofertilizers: types, production, benefits, and utilization. In: Simpson BK, Aryee ANA, Toldrá F (eds) Byproducts from agriculture and fisheries: adding value for food, feed, pharma, and fuels. Wiley, pp 479–500. https://doi.org/10.1002/9781119383956.ch20

Anand K, Kumari B, Mallick MA (2016) Phosphate solubilizing microbes: an effective and alternative approach as biofertilizers. Int J Pharm Sci 8:37–40

Baset Mia MA, Shamsuddin ZH (2010) Rhizobium as a crop enhancer and biofertilizer for increased cereal production. African J Biotechnol 9:6001–6009. https://doi.org/10.5897/AJB09.010

Canassa F, Tall S, Moral RA, Lara IAR d, Delalibera I, Meyling NV (2019) Effects of bean seed treatment by the entomopathogenic fungi *Metarhizium robertsii* and *Beauveria bassiana* on plant growth, spider mite populations and behavior of predatory mites. Biol Control 132:199–208. https://doi.org/10.1016/j.biocontrol.2019.02.003

Cordell D, White S (2014) Life's bottleneck: sustaining the world's phosphorus for a food secure future. Annu Rev Environ Resour 39:161–188. https://doi.org/10.1146/annurev-environ-010213-113300

Dhir B (2017) Biofertilizers and biopesticides: eco-friendly biological agents. In: Umar R, Sharma AK, Ahluwalia SS (eds) Advances in environmental biotechnology. Springer, Singapore, pp 167–188. https://doi.org/10.1007/978-981-10-4041-2_10

Din M, Nelofer R, Salman M, Abdullah, Khan FH, Khan A, Ahmad M, Jalil F, Din JU, Khan M (2019) Production of nitrogen fixing *Azotobacter* (SR-4) and phosphorus solubilizing *Aspergillus niger* and their evaluation on *Lagenaria siceraria* and A*belmoschus esculentus*. Biotechnol Rep 22:e00323. https://doi.org/10.1016/j.btre.2019.e00323

Dotaniya ML, Datta SC, Biswas DR, Meena BP (2013) Effect of solution phosphorus concentration on the exudation of oxalate ions by wheat (*Triticum aestivum* L.). Proc Natl Acad Sci India Sect B Biol Sci 83:305–309. https://doi.org/10.1007/s40011-012-0153-7

Dotaniya ML, Datta SC, Biswas DR, Meena HM, Kumar K (2014) Production of oxalic acid as influenced by the application of organic residue and its effect on phosphorus uptake by wheat (T*riticum aestivum* L.) in an inceptisol of North India. Natl Acad Sci Lett 37:401–405. https://doi.org/10.1007/s40009-014-0254-3

Erisman JW, Galloway JN, Seitzinger S, Bleeker A, Dise NB, Petrescu AMR, Leach AM, de Vries W (2013) Consequences of human the global nitrogen cycle. Philos Trans R Soc B Biol Sci 368:1–9. https://doi.org/10.1098/rstb.2013.0116

Giri N, Joshi N (2010) Growth and yield response of chick pea (*Cicer arietinum*) to seed inoculation with *Rhizobium* sp. Nat Sci 8:232–236

Gontia-Mishra I, Sapre S, Tiwari S (2017) Zinc solubilizing bacteria from the rhizosphere of rice as prospective modulator of zinc biofortification in rice. Rhizosphere 3:185–190. https://doi.org/10.1016/j.rhisph.2017.04.013

Gopalakrishnan S, Sathya A, Vijayabharathi R, Varshney RK, Gowda CLL, Krishnamurthy L (2015) Plant growth promoting rhizobia: challenges and opportunities. 3 Biotech 5:355–377. https://doi.org/10.1007/s13205-014-0241-x

Gupta S, Dikshit AK (2010) Biopesticides: an ecofriendly approach for pest control. J Biopest 3:186–188

Han JH, Jin BR, Kim JJ, Lee SY (2014) Virulence of entomopathogenic fungi *Metarhizium anisopliae* and *Paecilomyces fumosoroseus* for the microbial control of *Spodoptera exigua*. Mycobiology 42:385–390. https://doi.org/10.5941/MYCO.2014.42.4.385

Hubbard M, Hynes RK, Erlandson M, Bailey KL (2014) The biochemistry behind biopesticide efficacy. Sustain Chem Process 2:1–8. https://doi.org/10.1186/s40508-014-0018-x

Jacoby R, Peukert M, Succurro A, Koprivova A, Kopriva S (2017) The role of soil microorganisms in plant mineral nutrition—current knowledge and future directions. Front Plant Sci 8:1–19. https://doi.org/10.3389/fpls.2017.01617

Jakka SRK, Shrestha RB, Gassmann AJ (2016) Broad-spectrum resistance to *Bacillus thuringiensis* toxins by western corn rootworm (*Diabrotica virgifera virgifera*). Sci Rep 6:1–9. https://doi.org/10.1038/srep27860

Jia Z, Sun Y, Yuan L, Tian Q, Luo K (2010) The chitinase gene (Bbchit1) from *Beauveria bassiana* enhances resistance to *Cytospora chrysosperma* in *Populus tomentosa* Carr. Biotechnol Lett 32:1325–1332. https://doi.org/10.1007/s10529-010-0297-6

Kamran S, Shahid I, Baig DN, Rizwan M, Malik KA, Mehnaz S (2017) Contribution of zinc solubilizing bacteria in growth promotion and zinc content of wheat. Front Microbiol 8:1–14. https://doi.org/10.3389/fmicb.2017.02593

Kenneth OC (2017) Plant growth promoting rhizobacteria (PGPR): a bioprotectant bioinoculant for sustainable agrobiology. A review. Int J Adv Res Biol Sci 4:123–142. https://doi.org/10.22192/ijarbs

Kenneth OC, Nwadibe EC, Kalu AU, Unah UV (2019) Plant growth promoting rhizobacteria (PGPR): a novel agent for sustainable food production. Am J Agric Biol Sci 14:35–54. https://doi.org/10.3844/ajabssp.2019.35.54

Kern MF, Maraschin SDF, Vom Endt D, Schrank A, Vainstein MH, Pasquali G (2010) Expression of a chitinase gene from *Metarhizium anisopliae* in tobacco plants confers resistance against

Rhizoctonia solani. Appl Biochem Biotechnol 160:1933–1946. https://doi.org/10.1007/s12010-009-8701-1

Khan MR, Mohiddin FA, Ahamad F (2018) Inoculant rhizobia suppressed root-knot disease, and enhanced plant productivity and nutrient uptake of some field-grown food legumes. Acta Agric Scand Sect B Soil Plant Sci 68:166–174. https://doi.org/10.1080/09064710.2017.1374448

Khande R, Sharma SK, Ramesh A, Sharma MP (2017) Zinc solubilizing *Bacillus* strains that modulate growth, yield and zinc biofortification of soybean and wheat. Rhizosphere 4:126–138. https://doi.org/10.1016/j.rhisph.2017.09.002

Kim SJ, Eo JK, Lee EH, Park H, Eom AH (2017) Effects of arbuscular mycorrhizal fungi and soil conditions on crop plant growth. Mycobiology 45:20–24. https://doi.org/10.5941/MYCO.2017.45.1.20

Kumar VV (2018) Biofertilizers and biopesticides in sustainable agriculture. In: Meena V (ed) Role of rhizospheric microbes in soil. Springer, Singapore, pp 377–398

Li J, Xie J, Zeng D, Xia Y, Peng G (2021) Effective control of *Frankliniella occidentalis* by *Metarhizium anisopliae* CQMa421 under field conditions. J Pest Sci. https://doi.org/10.1007/s10340-020-01223-9

Li JF, Zhang SQ, Huo PH, Shi SL, Miao YY (2013) Effect of phosphate solubilizing rhizobium and nitrogen fixing bacteria on growth of alfalfa seedlings under P and N deficient conditions. Pakistan J Bot 45:1557–1562

Li S, Peng M, Liu Z, Shah SS (2017) The role of soil microbes in promoting plant growth. Mol Microbiol Res 7:30–37. https://doi.org/10.5376/mmr.2017.07.0004

Liang GM, Chen W, Liu TX (2003) Effects of three neem-based insecticides on diamondback moth (*lepidoptera*: plutellidae). Crop Prot 22:333–340. https://doi.org/10.1016/S0261-2194(02)00175-8

Lopez DC, Zhu-Salzman K, Ek-Ramos MJ, Sword GA (2014) The entomopathogenic fungal endophytes *Purpureocillium lilacinum* (formerly *Paecilomyces lilacinus*) and *Beauveria bassiana* negatively affect cotton aphid reproduction under both greenhouse and field conditions. PLoS One 9:e103891. https://doi.org/10.1371/journal.pone.0103891

Mahdi SS, Talat MA, Dar MH, Hamid A, Ahmad L (2012) Soil phosphorus fixation chemistry and role of phosphate solubilizing bacteria in enhancing its efficiency for sustainable cropping—a review. J Pure Appl Microbiol 6:1905–1911

Maswada HF, Abd El-Razek UA, El-Sheshtawy ANA, Mazrou YSA (2020) Effect of *Azolla filiculoides* on growth, physiological and yield attributes of maize grown under water and nitrogen deficiencies. J Plant Growth Regul:1–16. https://doi.org/10.1007/s00344-020-10120-5

Mazhar R, Ilyas N, Saeed M, Bibi F, Batool N (2016) Biocontrol and salinity tolerance potential of *Azospirillum lipoferum* and its inoculation effect in wheat crop. Int J Agric Biol 18:494–500. https://doi.org/10.17957/IJAB/15.0115

Meena V, Maurya B, Meena S, Meena R, Kumar A, Verma J, Singh N (2016) Can bacillus species enhance nutrient availability in agricultural soils? In: Islam M, Rahman M, Pandey P, Jha C, Aeron A (eds) Bacilli and agrobiotechnology. Springer, Cham, pp 367–395. https://doi.org/10.1007/978-3-319-44409-3_16

Mehnaz S (2015) *Azospirillum*: a biofertilizer for every crop. In: Arora N (ed) Plant microbes symbiosis: applied facets. Springer, New Delhi, pp 297–314. https://doi.org/10.1007/978-81-322-2068-8_15

Mehrotra S, Kumar S, Zahid M, Garg M (2017) Principles and applications of environmental biotechnology for a sustainable future. Princ Appl Environ Biotechnol Sustain Fut. https://doi.org/10.1007/978-981-10-1866-4

Mishra D, Rajvir S, Mishra U, Kumar S (2013) Role of bio-fertilizer in organic agriculture: a review. Res J Recent Sci 2:39–41

Mishra J, Tewari S, Singh S, Arora NK (2015) Biopesticides: where we stand? In: Arora NK (ed) Plant microbes symbiosis: applied facets. Springer, New Delhi, pp 37–75. https://doi.org/10.1007/978-81-322-2068-8_2

Mittal V, Singh O, Nayyar H, Kaur J, Tewari R (2008) Stimulatory effect of phosphate-solubilizing fungal strains (*Aspergillus awamori* and *Penicillium citrinum*) on the yield of chickpea (*Cicer arietinum* L. cv. GPF2). Soil Biol Biochem 40:718–727. https://doi.org/10.1016/j.soilbio.2007.10.008

Nathan SS, Kalaivani K, Murugan K, Chung PG (2005) Efficacy of neem limonoids on *Cnaphalocrocis medinalis* (Guenée) (Lepidoptera: Pyralidae) the rice leafroller. Crop Prot 24:760–763. https://doi.org/10.1016/j.cropro.2005.01.009

Nawaz M, Mabubu JI, Hua H (2016) Current status and advancement of biopesticides: microbial and botanical pesticides. J Entomol Zool Stud 4:241–246

Oteino N, Lally RD, Kiwanuka S, Lloyd A, Ryan D, Germaine KJ, Dowling DN (2015) Plant growth promotion induced by phosphate solubilizing endophytic *Pseudomonas* isolates. Front Microbiol 6:1–9. https://doi.org/10.3389/fmicb.2015.00745

Oyewole BO, Olawuyi OJ, Odebode AC, Abiala MA (2017) Influence of arbuscular mycorrhiza fungi (AMF) on drought tolerance and charcoal rot disease of cowpea. Biotechnol Rep 14:8–15. https://doi.org/10.1016/j.btre.2017.02.004

Pathak DV, Kumar M (2016) Microbial inoculants as biofertilizers and biopesticides. In: Singh D, Singh H, Prabha R (eds) Microbial inoculants in sustainable agricultural productivity: vol. 1: research perspectives. Springer, New Delhi, pp 197–209. https://doi.org/10.1007/978-81-322-2647-5_11

Pereira I, Ortega R, Barrientos L, Moya M, Reyes G, Kramm V (2009) Development of a biofertilizer based on filamentous nitrogen-fixing cyanobacteria for rice crops in Chile. J Appl Phycol 21:135–144. https://doi.org/10.1007/s10811-008-9342-4

Pradhan N, Sukla LB (2006) Solubilization of inorganic phosphates by fungi isolated from agriculture soil. African J Biotechnol 5:850–854. https://doi.org/10.4314/ajb.v5i10.42884

Pramanik P, Goswami AJ, Ghosh S, Kalita C (2019) An indigenous strain of potassium-solubilizing bacteria *Bacillus pseudomycoides* enhanced potassium uptake in tea plants by increasing potassium availability in the mica waste-treated soil of North-East India. J Appl Microbiol 126:215–222. https://doi.org/10.1111/jam.14130

Püschel D, Janoušková M, Voříšková A, Gryndlerová H, Vosátka M, Jansa J (2017) Arbuscular mycorrhiza stimulates biological nitrogen fixation in two *Medicago* spp. through improved phosphorus acquisition. Front Plant Sci 8:1–12. https://doi.org/10.3389/fpls.2017.00390

Razavipour T, Moghaddam SS, Doaei S, Noorhosseini SA, Damalas CA (2018) Azolla (Azolla filiculoides) compost improves grain yield of rice (*Oryza sativa* L.) under different irrigation regimes. Agric Water Manag 209:1–10. https://doi.org/10.1016/j.agwat.2018.05.020

Reddy S, Singh A, Masih H, Benjamin J, Ojha S, Ramteke P, Singla A (2018) Effect of *Azotobacter* sp and *Azospirillum* sp on vegetative growth of Tomato (*Lycopersicon esculentum*). J Pharmacogn Phytochem 7:2130–2137

Reyes I, Bernier L, Antoun H (2002) Rock phosphate solubilization and colonization of maize rhizosphere by wild and genetically modified strains of *Penicillium rugulosum*. Microb Ecol 44:39–48. https://doi.org/10.1007/s00248-002-1001-8

Rios-Velasco C, Pérez-Corral DA, Salas-Marina MÁ, Berlanga-Reyes DI, Ornelas-Paz JJ, Muñiz CHA, Cambero-Campos J, Jacobo-Cuellar JL (2014) Pathogenicity of the hypocreales fungi *Beauveria bassiana* and *Metarhizium anisopliae* against insect pests of tomato. Southwest Entomol 39:739–750. https://doi.org/10.3958/059.039.0405

Romero D, De Vicente A, Rakotoaly RH, Dufour SE, Veening JW, Arrebola E, Cazorla FM, Kuipers OP, Paquot M, Pérez-García A (2007) The iturin and fengycin families of lipopeptides are key factors in antagonism of *Bacillus subtilis* toward *Podosphaera fusca*. Mol Plant-Microbe Interact 20:430–440. https://doi.org/10.1094/MPMI-20-4-0430

Romero-Perdomo F, Abril J, Camelo M, Moreno-Galván A, Pastrana I, Rojas-Tapias D, Bonilla R (2017) *Azotobacter chroococcum* as a potentially useful bacterial biofertilizer for cotton (Gossypium hirsutum): effect in reducing N fertilization. Rev Argent Microbiol 49:377–383. https://doi.org/10.1016/j.ram.2017.04.006

Sahoo RK, Ansari MW, Pradhan M, Dangar TK, Mohanty S, Tuteja N (2014) Phenotypic and molecular characterization of native *Azospirillum* strains from rice fields to improve crop productivity. Protoplasma 251:943–953. https://doi.org/10.1007/s00709-013-0607-7

Saker MM, Salama HS, Salama M, El-Banna A, Abdel Ghany NM (2011) Production of transgenic tomato plants expressing Cry 2Ab gene for the control of some lepidopterous insects endemic in Egypt. J Genet Eng Biotechnol 9:149–155. https://doi.org/10.1016/j.jgeb.2011.08.001

Samago TY, Anniye EW, Dakora FD (2018) Grain yield of common bean (*Phaseolus vulgaris* L.) varieties is markedly increased by rhizobial inoculation and phosphorus application in Ethiopia. Symbiosis 75:245–255. https://doi.org/10.1007/s13199-017-0529-9

Saritha M, Prasad Tollamadugu NVKV (2019) The status of research and application of biofertilizers and biopesticides: global scenario. Elsevier

Schmutterer H (1990) Properties and potential of natural pesticides from the neem tree, *Azadirachta indica*. Annu Rev Entomol 35:271–297

Shanware AS, Kalkar SA, Trivedi MM (2014) Potassium solubilisers: occurrence, mechanism and their role as competent biofertilizers. Int J of Curr Microbiol Appl Sci 3:622–629

Sharma K, Dak G, Agarwal A, Bhatnagar M, Sharma R (2007) Effect of phosphate solubilizing bacteria on the germination of *Cicer arietinum* seeds and seedling growth. J Herb Med Toxicol 1:61–63

Sharma A, Kumar S, Bhatnagar RK (2011) *Bacillus thuringiensis* protein Cry6B (BGSC ID 4D8) is toxic to larvae of Hypera postica. Curr Microbiol 62:597–605. https://doi.org/10.1007/s00284-010-9749-4

Siddiqui HA, Asif M, Asad S, Naqvi RZ, Ajaz S, Umer N, Anjum N, Rauf I, Sarwar M, Arshad M, Amin I, Saeed M, Mukhtar Z, Bashir A, Mansoor S (2019) Development and evaluation of double gene transgenic cotton lines expressing cry toxins for protection against chewing insect pests. Sci Rep 9:1–7. https://doi.org/10.1038/s41598-019-48188-z

Silvério FO, de Alvarenga ES, Moreno SC, Picanço MC (2009) Synthesis and insecticidal activity of new pyrethroids. Pest Manag Sci 65:900–905. https://doi.org/10.1002/ps.1771

Singh G, Biswas DR, Marwaha TS (2010) Mobilization of potassium from waste mica by plant growth promoting rhizobacteria and its assimilation by maize (*Zea mays*) and wheat (*Triticum aestivum* L.): a hydroponics study under phytotron growth chamber. J Plant Nutr 33:1236–1251. https://doi.org/10.1080/01904161003765760

Singh M, Dotaniya M, Mishra A, Dotaniya C, Regar K, Lata M (2016) Role of biofertilizers in conservation agriculture. In: Bisht J, Meena V, Mishra P, Pattanayak A (eds) Conservation agriculture: an approach to combat climate change in Indian Himalaya. Springer, Singapore, pp 113–134. https://doi.org/10.1007/978-81-322-2647-5_18

Suhag M (2016) Potential of biofertilizers to replace chemical fertilizers. Int Adv Res J Sci Eng Technol 3:163–167. https://doi.org/10.17148/IARJSET.2016.3534

Suyal D, Soni R, Sai S, Goel R (2016) Microbial inoculants as biofertilizer. In: Singh DP, Singh HB, Prabha R (eds) Microbial inoculants in sustainable agricultural productivity: vol. 1: research perspectives. Springer, New Delhi, pp 311–318

Tang J, Liu X, Ding Y, Jiang W, Xie J (2019) Evaluation of *Metarhizium anisopliae* for rice planthopper control and its synergy with selected insecticides. Crop Prot 121:132–138. https://doi.org/10.1016/j.cropro.2019.04.002

Teotia P, Kumar V, Kumar M, Shrivastava N, Varma A (2016) Rhizosphere microbes: potassium solubilization and crop productivity – present and future aspects. In: Meena VS, Maurya BR, Verma JP, Meena RS (eds) Potassium solubilizing microorganisms for sustainable agriculture. Springer, New Delhi, pp 1–331. https://doi.org/10.1007/978-81-322-2776-2_22

Thakore Y (2006) The biopesticide market for global agricultural use. Ind Biotechnol 2:194–208. https://doi.org/10.1089/ind.2006.2.194

Van Der Heijden MGA, Bardgett RD, Van Straalen NM (2008) The unseen majority: soil microbes as drivers of plant diversity and productivity in terrestrial ecosystems. Ecol Lett 11:296–310. https://doi.org/10.1111/j.1461-0248.2007.01139.x

Vaughn T, Cavato T, Brar G, Coombe T, DeGooyer T, Ford S, Groth M, Howe A, Johnson S, Kolacz K, Pilcher C, Purcell J, Romano C, English L, Pershing J (2005) A method of controlling corn rootworm feeding using a *Bacillus thuringiensis* protein expressed in transgenic maize. Crop Sci 45:931–938. https://doi.org/10.2135/cropsci2004.0304

Verma P, Yadav AN, Khannam KS, Saxena AK, Suman A (2017) Potassium-solubilizing microbes: diversity, distribution, and role in plant growth promotion. In: Panpatte D, Jhala Y, Vyas R, Shelat H (eds) Microorganisms for green revolution. Springer, Singapore, pp 125–149. https://doi.org/10.1007/978-981-10-6241-4_7

Willer H, Lernoud J (2019) The world of organic agriculture. Statistics and emerging trends 2019. Reserach Institute of Organic Agriculture FiBL and IFOAM Organics International. https://shop.fibl.org/chde/2020-organic-world-2019.html

Wong W, Tsn S, Ge L, Chen X, Yong J (2015) The importance of phytohormones and microbes in biofertilizers. In: Maheshwari D (ed) Bacterial metabolites in sustainable agroecosystem. Springer, Cham, pp 337–349. https://doi.org/10.1007/978-3-319-24654-3_6

Wu Y, Xu L, Chang L, Ma M, You L, Jiang C, Li S, Zhnag J (2019) *Bacillus thuringiensis* (Bt) cry1C expression from the plastid genome of poplar leads to high mortality of leaf eating caterpillars. Tree Physiol 39:1525–1532. https://doi.org/10.1093/treephys/tpz073

Zhang C, Kong F (2014) Isolation and identification of potassium-solubilizing bacteria from tobacco rhizospheric soil and their effect on tobacco plants. Appl Soil Ecol 82:18–25. https://doi.org/10.1016/j.apsoil.2014.05.002

Index

S. K. Dubey, S. K. Verma (eds.), *Plant, Soil and Microbes in Tropical Ecosystems*, Rhizosphere Biology, https://doi.org/10.1007/978-981-16-3364-5